Advances in Pavement Materials and Civil Engineering

Advances in Pavement Materials and Civil Engineering

Guest Editor

Jiujiang Wu

MDPI

Basel • Beijing • Wuhan • Barcelona • Belgrade • Novi Sad • Cluj • Manchester

Guest Editor
Jiujiang Wu
Shock and Vibration of
Engineering Materials and
Structures Key Laboratory
of Sichuan
Southwest University of
Science and Technology
Mianyang
China

Editorial Office
MDPI AG
Grosspeteranlage 5
4052 Basel, Switzerland

This is a reprint of the Special Issue, published open access by the journal *Coatings* (ISSN 2079-6412), freely accessible at: https://www.mdpi.com/journal/coatings/special_issues/A0VR74950I.

For citation purposes, cite each article independently as indicated on the article page online and as indicated below:

Lastname, A.A.; Lastname, B.B. Article Title. *Journal Name* **Year**, *Volume Number*, Page Range.

ISBN 978-3-7258-8057-7 (Hbk)
ISBN 978-3-7258-8058-4 (PDF)
https://doi.org/10.3390/books978-3-7258-8058-4

Cover image courtesy of Jiujiang Wu

Contents

About the Editor vii

Preface ix

Jiujiang Wu and Wenjie Jiang
Advances in Pavement Materials and Civil Engineering
Reprinted from: *Coatings* **2026**, *16*, 592, https://doi.org/10.3390/coatings16050592 1

Kunpeng Zhao, Yi Leng, Qinxi Dong, Yongling Ding, Huadong Sun, Chunbao Ding, et al.
Effect of Hierarchical ZnO/PAC Nanosheets on the Rheological Performance of SBS-Modified Asphalt
Reprinted from: *Coatings* **2026**, *16*, 520, https://doi.org/10.3390/coatings16050520 7

Yi Zhao and Peixing Yang
Study on Desulfurized Crumb Rubber–Modified Epoxy Asphalt
Reprinted from: *Coatings* **2026**, *16*, 102, https://doi.org/10.3390/coatings16010102 25

Hu Feng, Haisheng Zhao, Dongfang Zhang, Peiyu Zhang, Yindong Ding, Yanping Liu, et al.
Study on the Volatile Organic Compound Emission Characteristics of Crumb Rubber-Modified Asphalt
Reprinted from: *Coatings* **2025**, *15*, 1043, https://doi.org/10.3390/coatings15091043 44

Hao Zhou, Pinhui Zhao, Qiang Xu, Kechao Han, Guixia Zhang, Jipeng Cui, et al.
Study on Infrared Spectrum Analysis of Asphalt Aging Based on Full Spectrum Segmentation
Reprinted from: *Coatings* **2024**, *14*, 1505, https://doi.org/10.3390/coatings14121505 71

Qiqi Zhan, Zuoyang Cheng, Xuejuan Cao, Qing Liu, Ying Yuan, Lihong He and Junfeng Gao
Multiscale Rheological Properties of Pavement Asphalt: A State-of-the-Art Review
Reprinted from: *Coatings* **2026**, *16*, 355, https://doi.org/10.3390/coatings16030355 84

Haijiao Xu, Pinhui Zhao, Wenyu Wang, Zhiyu Zhang, Jingtao Shi, Dongqi Yu, et al.
Effect of Hydrophobic Alkyl Chain Length on the Interfacial Adhesion Performance of Emulsified Asphalt–Aggregate Systems
Reprinted from: *Coatings* **2026**, *16*, 330, https://doi.org/10.3390/coatings16030330 118

Shucheng Tan, Hua Wen, Hua Tang, Wentao Fu, Xiaoyan Guo, Biaotian Bai, et al.
Utilization of Water-Based Drill Cuttings Stabilized by a Novel Composite Stabilizer for Pavement Base Applications
Reprinted from: *Coatings* **2026**, *16*, 406, https://doi.org/10.3390/coatings16040406 141

Kai Yan, Yuanbo Song, Lianfang Sun, Qian Zhang and Tianfeng Yuan
Mechanism of Immersion Crushing on Alkali-Silica Reaction (ASR) in Glass Mortar
Reprinted from: *Coatings* **2025**, *15*, 1398, https://doi.org/10.3390/coatings15121398 163

Hongchang Wang, Haozhe Wu and Congying Liu
Numerical Modeling of the Relationship Between Mechanical Properties and Void Morphology in Porous Asphalt Mixtures
Reprinted from: *Coatings* **2026**, *16*, 214, https://doi.org/10.3390/coatings16020214 174

Shuaishuai Wei, Huan Zhang, Ding Wang, Xuchun Wang and Mengdi Cao
Mesoscale Mechanical Analysis of Concrete Based on a 3D Random Aggregate Model
Reprinted from: *Coatings* **2025**, *15*, 883, https://doi.org/10.3390/coatings15080883 194

He Liu, Wenxi Li, Haonan Zou, Wei Bian, Jingyi Zhang, Ji Zhang and Peng Zhang
Effect of Macroscopic Composition on the Performance of Self-Compacting Concrete
Reprinted from: *Coatings* **2025**, *15*, 161, https://doi.org/10.3390/coatings15020161 . . . . **219**

Haisheng Zhao, Bin Wang, Peiyu Zhang, Yong Liu, Chunhua Su, Mingzhu Xu, et al.
Study on the Dynamic Properties of the Polyurethane Mixture with Open-Graded Gradation
Reprinted from: *Coatings* **2026**, *16*, 153, https://doi.org/10.3390/coatings16020153 . . . . **233**

Jiujiang Wu, Lin Xiao, Jifeng Lian and Lijuan Wang
Settlement Prediction for Cast-in-Place Tubular Piles with Large Diameters Based on the Load Transfer Approach
Reprinted from: *Coatings* **2025**, *15*, 70, https://doi.org/10.3390/coatings15010070 . . . . **257**

Jianlei Zhang, Qiangong Cheng, Haozhen Fan, Mengjie Dai, Yan Li, Jiujiang Wu and Yufeng Wang
A New Method for Evaluating Liquefaction by Energy-Based Pore Water Pressure Models
Reprinted from: *Coatings* **2025**, *15*, 7, https://doi.org/10.3390/coatings15010007 . . . . **274**

Tianjun Xue, Lingxin Liu, Jianlei Zhang, Mengjie Dai, Gengyuan Shi and Xinze Li
Experimental Study on Pipeline–Soil Interaction in Translational Landslide
Reprinted from: *Coatings* **2025**, *15*, 537, https://doi.org/10.3390/coatings15050537 . . . . **296**

About the Editor

Jiujiang Wu

Jiujiang Wu is a Distinguished Professor and PhD/Master's supervisor at Southwest University of Science and Technology, and a postdoctoral researcher at Western University (QS 2026 Global Rank 151). He is recognized as a high-level overseas talent in Sichuan Province and has received awards including the Sichuan Youth Science and Technology Award in Civil and Architectural Engineering, and the Sichuan Environmental Protection Science and Technology Award. He has published over 80 academic papers, including 35 SCI articles as first or corresponding author, holds 22 patents with 17 granted, and leads more than ten national and provincial research projects. Prof. Wu serves as a Guest Editor and Editorial Board Member for multiple SCI, ESCI, and Scopus-indexed journals, including *Coatings*, *Fluid Dynamics & Materials Processing*, and *Scientific Reports*, and is an active member of IGS, ISSMGE, CCES, CSRME, and GSC, contributing widely to research, peer review, and scientific leadership in civil and geotechnical engineering.

Preface

This Reprint, "Advances in Pavement Materials and Civil Engineering," brings together fifteen peer-reviewed articles originally published in the Special Issue of *Coatings*. It addresses recent innovations in pavement materials, sustainable construction practices, multi-scale modeling, and geotechnical engineering, with a focus on enhancing infrastructure durability and performance. The Reprint was compiled to provide researchers, engineers, and practitioners with a cohesive overview of emerging trends, experimental techniques, and numerical approaches in civil engineering. The motivation for assembling this Reprint is to showcase interdisciplinary advances that integrate material science, structural engineering, and environmental sustainability, reflecting both fundamental research and practical applications. We hope this collection serves as a valuable resource for academics, industry professionals, and students seeking insight into the latest developments in high-performance and sustainable infrastructure.

Jiujiang Wu
Guest Editor

coatings

MDPI

Editorial

Advances in Pavement Materials and Civil Engineering

Jiujiang Wu [1,2,*] and Wenjie Jiang [1]

1 Shock and Vibration of Engineering Materials and Structures Key Laboratory of Sichuan Province, Southwest University of Science and Technology, Mianyang 621010, China; 7220240762@mails.swust.edu.cn
2 Department of Civil and Environmental Engineering, Western University, London, ON N6A 5B9, Canada
* Correspondence: wujiujiang1988@swust.edu.cn

1. Introduction and Scope

Pavement materials and civil engineering constitute the backbone of modern infrastructure systems, directly supporting economic development, urbanization, and societal resilience [1–3]. With the continuous expansion of transportation networks and increasing demands for infrastructure durability and serviceability, the performance requirements imposed on engineering materials have become progressively more stringent [4–7].

In recent years, significant advances have been achieved in the development of high-performance pavement materials, including polymer-modified asphalt [8], composite binders [9], and multifunctional cement-based materials [10]. These innovations aim to enhance mechanical properties such as strength, stiffness, and fatigue resistance, while also improving resistance to rutting, cracking, moisture damage, and aging [11]. At the same time, the integration of nanomaterials and advanced additives has enabled the tailoring of material properties at multiple scales, opening new pathways for optimizing pavement performance under extreme service conditions [12].

Parallel to material innovation, sustainability has emerged as a central theme in civil engineering research. The increasing emphasis on circular economy principles has driven the exploration of recycled and waste-derived materials, such as crumb rubber, industrial by-products, and construction waste, for use in pavement and structural applications [13]. These approaches not only reduce environmental burdens but also contribute to resource efficiency and cost-effectiveness. However, the incorporation of such materials introduces additional complexities in terms of durability, environmental impact, and long-term performance, necessitating comprehensive experimental and analytical investigations [14].

Another important trend is the rapid development of advanced modeling and simulation techniques [15]. Traditional empirical design methods are increasingly being complemented or replaced by multi-scale and multi-physics approaches that capture the intrinsic relationships between material composition, microstructure, and macroscopic behavior. Numerical modeling, image-based analysis, and data-driven methods are now widely used to predict performance, optimize material design, and evaluate structural responses under complex loading and environmental conditions [16].

Moreover, civil engineering research is becoming increasingly interdisciplinary, integrating concepts and methods from materials science, geotechnical engineering, environmental science, and computational mechanics [17]. This interdisciplinary perspective is particularly important for addressing complex engineering problems, such as infrastructure resilience under natural hazards, soil–structure interaction, and the environmental impacts of construction materials.

Within this context, the Special Issue entitled "Advances in Pavement Materials and Civil Engineering" in *Coatings* was launched to provide a platform for disseminating recent advances and emerging trends in this field. The collection aims to bring together contributions that address novel pavement materials, sustainable engineering practices, recycling and reuse technologies, advanced modeling and simulation methods, and the influence of environmental factors on material performance and infrastructure durability.

2. Statistics of the Special Issue

A total of fifteen peer-reviewed papers have been published in this collection, including original research articles and one comprehensive review. The proportion of articles in different categories is shown in Figure 1. These contributions cover a wide range of topics, including asphalt modification and aging mechanisms, sustainable material utilization, multi-scale mechanical analysis, and geotechnical engineering applications. Collectively, they provide valuable insights into both fundamental mechanisms and practical engineering solutions, reflecting the current state of the art and future directions in pavement materials and civil engineering.

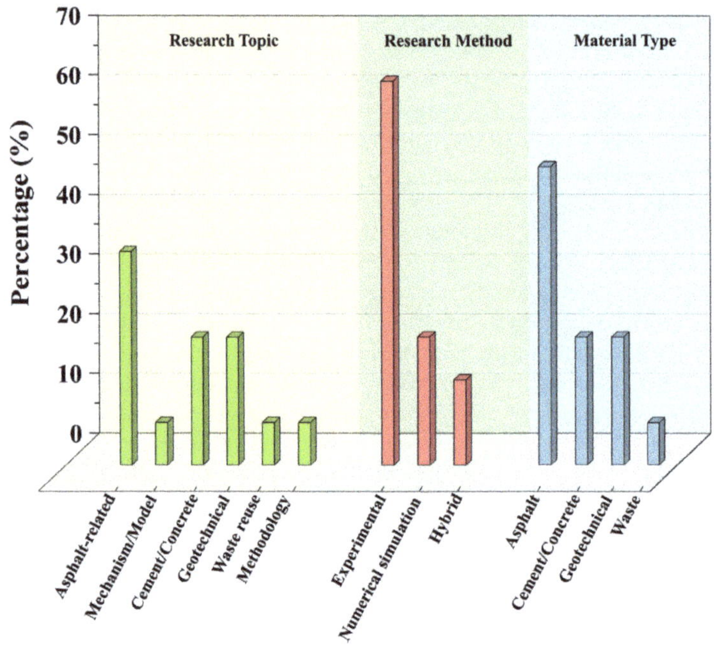

Figure 1. The proportion of articles in different categories.

3. Contributions

3.1. *Advanced Asphalt Materials and Modification Technologies*

A major portion of the Special Issue focuses on the modification and performance enhancement of asphalt materials. Zhao et al. [Contribution 1] constructed a hierarchical ZnO-loaded porous activated carbon (ZnO/PAC) nanostructure and incorporated it into SBS-modified asphalt, achieving a synergistic enhancement in performance. The modified asphalt exhibited superior resistance to both rutting at high temperatures and cracking at low temperatures. Zhao and Yang [Contribution 2] modified epoxy asphalt by introducing desulfurized rubber powder and an organodisulfide plasticizer (WJFL). This approach not

https://doi.org/10.3390/coatings16050592

only improved curing efficiency but also significantly enhanced low-temperature performance while reducing construction costs, thereby effectively addressing the inherent limitations of conventional epoxy asphalt in terms of toughness and low-temperature behavior.

Feng et al. [Contribution 3] systematically investigated the effects of different types of rubber powder, dosages, and additives on volatile organic compound (VOC) emissions from modified asphalt, providing critical insights for environmental assessment and the design of eco-friendly asphalt materials. Zhou et al. [Contribution 4] preprocessed and segmentally analyzed the infrared spectral signals of asphalt to determine an appropriate signal analysis range and developed a corresponding aging model. This enables the estimation of actual pavement service time through infrared spectroscopy, offering a more refined approach for pavement condition evaluation. Zhan et al. [Contribution 5] provided a comprehensive review of the multi-scale rheological properties of asphalt, linking microstructural characteristics to macroscopic performance and highlighting key challenges in multi-scale characterization. Xu et al. [Contribution 6] investigated the effects of emulsifier hydrophobic chain length and aggregate properties on the interfacial adhesion performance between emulsified asphalt and aggregates, providing a theoretical basis for emulsifier structural optimization and interface compatibility design.

These studies collectively demonstrate that asphalt research is evolving toward multi-scale, multi-physics approaches, integrating nanotechnology, environmental considerations, and advanced characterization techniques.

3.2. Sustainable Materials and Recycling Technologies

Sustainability is another core theme across the Special Issue. Tan et al. [Contribution 7] evaluated the feasibility of using water-based drilling cuttings, stabilized with a composite curing agent, as pavement base materials, demonstrating both environmental and engineering viability. Yan et al. [Contribution 8] investigated the effects of different crushing methods on the alkali–silica reaction (ASR) in waste glass aggregate mortar, providing a theoretical basis for the safe utilization of waste glass in cement-based materials and promoting resource recycling in the construction industry.

These contributions highlight the increasing emphasis on circular economy principles, where industrial waste is transformed into valuable construction materials while mitigating environmental risks.

3.3. Mechanical Behavior and Multi-Scale Modeling

Several papers focus on understanding material behavior through advanced experimental and numerical approaches. Wang et al. [Contribution 9] developed a numerical model linking void morphology to mechanical properties in porous asphalt mixtures, emphasizing the importance of internal structure in performance prediction. Wei et al. [Contribution 10] proposed a three-dimensional grid-based simulation method for generating the random distribution of concrete aggregates. The approach effectively avoids particle overlap and significantly improves computational efficiency during the generation process, thereby enhancing both the accuracy and efficiency of aggregate spatial distribution modeling. This provides a reliable tool for investigating the random distribution characteristics of aggregates in concrete. Liu et al. [Contribution 11] proposed a new aggregate spacing coefficient model for self-compacting concrete, which improves the prediction of macroscopic performance based on mix composition. Zhao et al. [Contribution 12] constructed master curves based on the time–temperature superposition principle and multiple rheological models to investigate the dynamic rheological behavior of open-graded and dense-graded polyurethane asphalt mixtures (PUM-OGFC and PUM-AC). The results

https://doi.org/10.3390/coatings16050592

indicate that the open-graded structure combined with a polyurethane binder achieves a favorable balance between elasticity and rutting resistance, demonstrating promising potential for engineering applications.

These studies illustrate the growing role of mesoscale modeling and structural characterization in bridging the gap between material composition and engineering performance.

3.4. Geotechnical Engineering and Infrastructure Performance

This collection also includes important contributions in geotechnical and infrastructure engineering. Wu et al. [Contribution 13] proposed a load-transfer-based settlement prediction model for estimating the settlement of large-diameter cast-in-place pipe piles, which addresses the limitations of conventional settlement prediction approaches and demonstrates good rationality and practical applicability in engineering practice. Zhang et al. [Contribution 14] developed an energy-based pore water pressure model for liquefaction evaluation, offering a novel framework for assessing seismic soil behavior. Xue et al. [Contribution 15] experimentally investigated pipeline–soil interaction under translational landslides, providing insights into failure mechanisms and infrastructure safety.

These works demonstrate the integration of advanced theoretical models with practical engineering applications, addressing critical issues in foundation engineering and hazard mitigation.

4. Concluding Remarks and Outlook

The contributions collected in this Special Issue reflect the multidisciplinary nature of modern pavement materials and civil engineering. Key trends emerging from these studies include:

1. The transition toward multi-scale characterization and modeling, linking microstructure to macroscopic performance;
2. The increasing importance of sustainable materials and recycling technologies;
3. The integration of advanced numerical methods with experimental validation;
4. The expansion of research from traditional pavement materials to broader infrastructure and geotechnical systems.

These developments are consistent with recent progress reported in related journals such as *Construction and Building Materials*, *Journal of Cleaner Production*, and *Materials and Structures*, where sustainability, durability, and data-driven modeling are central research themes.

The Guest Editor sincerely thanks all authors for their valuable contributions and the reviewers for their rigorous evaluations, as well as the editorial team for their professional support throughout the publication process.

With the publication of these papers, this Special Issue is now formally closed. It is expected that the findings presented herein will contribute to future advancements in sustainable infrastructure and high-performance engineering materials.

Conflicts of Interest: The authors declare no conflicts of interest.

List of Contributions

1. Zhao, K.; Leng, Y.; Dong, Q.; Ding, Y.; Sun, H.; Ding, C.; Song, P.; Ni, Y.; Wang, C.; Yin, H. Effect of Hierarchical ZnO/PAC Nanosheets on the Rheological Performance of SBS-Modified Asphalt. *Coatings* **2026**, *16*, 520. https://doi.org/10.3390/coatings16050520.
2. Zhao, Y.; Yang, P. Study on Desulfurized Crumb Rubber–Modified Epoxy Asphalt. *Coatings* **2026**, *16*, 102. https://doi.org/10.3390/coatings16010102.

3. Feng, H.; Zhao, H.; Zhang, D.; Zhang, P.; Ding, Y.; Liu, Y.; Su, C.; Han, Q.; Li, Y. Study on the Volatile Organic Compound Emission Characteristics of Crumb Rubber-Modified Asphalt. *Coatings* **2025**, *15*, 1043. https://doi.org/10.3390/coatings15091043.
4. Zhou, H.; Zhao, P.; Xu, Q.; Han, K.; Zhang, G.; Cui, J.; Meng, W.; Wu, W. Study on Infrared Spectrum Analysis of Asphalt Aging Based on Full Spectrum Segmentation. *Coatings* **2024**, *14*, 1505. https://doi.org/10.3390/coatings14121505.
5. Zhan, Q.; Cheng, Z.; Cao, X.; Liu, Q.; Yuan, Y.; He, L.; Gao, J. Multiscale Rheological Properties of Pavement Asphalt: A State-of-the-Art Review. *Coatings* **2026**, *16*, 355. https://doi.org/10.3390/coatings16030355.
6. Xu, H.; Zhao, P.; Wang, W.; Zhang, Z.; Shi, J.; Yu, D.; Cui, H.; Huang, D.; Su, J. Effect of Hydrophobic Alkyl Chain Length on the Interfacial Adhesion Performance of Emulsified Asphalt–Aggregate Systems. *Coatings* **2026**, *16*, 330. https://doi.org/10.3390/coatings16030330.
7. Tan, S.; Wen, H.; Tang, H.; Fu, W.; Guo, X.; Bai, B.; Wu, J.; Tan, X. Utilization of Water-Based Drill Cuttings Stabilized by a Novel Composite Stabilizer for Pavement Base Applications. *Coatings* **2026**, *16*, 406. https://doi.org/10.3390/coatings16040406.
8. Yan, K.; Song, Y.; Sun, L.; Zhang, Q.; Yuan, T. Mechanism of Immersion Crushing on Alkali–Silica Reaction (ASR) in Glass Mortar. *Coatings* **2025**, *15*, 1398. https://doi.org/10.3390/coatings15121398.
9. Wang, H.; Wu, H.; Liu, C. Numerical Modeling of the Relationship Between Mechanical Properties and Void Morphology in Porous Asphalt Mixtures. *Coatings* **2026**, *16*, 214. https://doi.org/10.3390/coatings16020214.
10. Wei, S.; Zhang, H.; Wang, D.; Wang, X.; Cao, M. Mesoscale Mechanical Analysis of Concrete Based on a 3D Random Aggregate Model. *Coatings* **2025**, *15*, 883. https://doi.org/10.3390/coatings15080883.
11. Liu, H.; Li, W.; Zou, H.; Bian, W.; Zhang, J.; Zhang, J.; Zhang, P. Effect of Macroscopic Composition on the Performance of Self-Compacting Concrete. *Coatings* **2025**, *15*, 161. https://doi.org/10.3390/coatings15020161.
12. Zhao, H.; Wang, B.; Zhang, P.; Liu, Y.; Su, C.; Xu, M.; Zhang, W.; Ma, S. Study on the Dynamic Properties of the Polyurethane Mixture with Open-Graded Gradation. *Coatings* **2026**, *16*, 153. https://doi.org/10.3390/coatings16020153.
13. Wu, J.; Xiao, L.; Lian, J.; Wang, L. Settlement Prediction for Cast-in-Place Tubular Piles with Large Diameters Based on the Load Transfer Approach. *Coatings* **2025**, *15*, 70. https://doi.org/10.3390/coatings15010070.
14. Zhang, J.; Cheng, Q.; Fan, H.; Dai, M.; Li, Y.; Wu, J.; Wang, Y. A New Method for Evaluating Liquefaction by Energy-Based Pore Water Pressure Models. *Coatings* **2025**, *15*, 7. https://doi.org/10.3390/coatings15010007.
15. Xue, T.; Liu, L.; Zhang, J.; Dai, M.; Shi, G.; Li, X. Experimental Study on Pipeline–Soil Interaction in Translational Landslide. *Coatings* **2025**, *15*, 537. https://doi.org/10.3390/coatings15050537.

References

1. Plati, C. Sustainability Factors in Pavement Materials, Design, and Preservation Strategies: A Literature Review. *Constr. Build. Mater.* **2019**, *211*, 539–555. [CrossRef]
2. Tang, Y.; Xiao, J.; Liu, Q.; Xia, B.; Singh, A.; Lv, Z.; Song, W. Natural Gravel-Recycled Aggregate Concrete Applied in Rural Highway Pavement: Material Properties and Life Cycle Assessment. *J. Clean. Prod.* **2022**, *334*, 130219. [CrossRef]
3. Alnedawi, A.; Ullah, S.; Azam, A.; Obaid, I.; Yosri, A.; Mousa, E. Integrated and Holistic Knowledge Map of Resilient Modulus Studies for Pavement Materials: A Scientometric Analysis and Bibliometric Review of Research Frontiers and Prospects. *Transp. Geotech.* **2022**, *33*, 100711. [CrossRef]
4. Djekic, J.; Djukic, A.; Vukmirovic, M.; Djekic, P.; Brankovic, M.D. Thermal Comfort of Pedestrian Spaces and the Influence of Pavement Materials on Warming Up During Summer. *Energy Build.* **2017**, *159*, 474–485. [CrossRef]
5. Georgees, R.N.; Hassan, R.A.; Evans, R.P.; Jegatheesan, P. An Evaluation of Performance-Related Properties for Granular Pavement Materials Using a Polyacrylamide Additive. *Int. J. Pavement Eng.* **2018**, *19*, 153–163. [CrossRef]
6. Yu, C.; Yang, Q. Multiscale Interfacial Interactions in Carbon Nanotube–Asphalt Nanocomposites: Implications for Asphalt Pavement Materials. *ACS Appl. Mater. Interfaces* **2023**, *6*, 7715–7730. [CrossRef]

7. Singh, D.; Chitragar, S.F.; Ashish, P.K. Comparison of Moisture and Fracture Damage Resistance of Hot and Warm Asphalt Mixes Containing Reclaimed Pavement Materials. *Constr. Build. Mater.* **2017**, *157*, 1145–1153. [CrossRef]
8. Yu, J.; Vaidya, M.; Su, G.; Adhikari, S.; Korolev, E.; Shekhovtsova, S. Experimental Study of Soda Lignin Powder as an Asphalt Modifier for a Sustainable Pavement Material. *Constr. Build. Mater.* **2021**, *298*, 123884. [CrossRef]
9. Lu, G.; Liu, P.; Wang, Y.; Faßbender, S.; Wang, D.; Oeser, M. Development of a Sustainable Pervious Pavement Material Using Recycled Ceramic Aggregate and Bio-Based Polyurethane Binder. *J. Clean. Prod.* **2019**, *220*, 1052–1060. [CrossRef]
10. Zhong, R.; Leng, Z.; Poon, C.-S. Research and Application of Pervious Concrete as a Sustainable Pavement Material: A State-of-the-Art Review. *Constr. Build. Mater.* **2018**, *183*, 544–553. [CrossRef]
11. Wang, L.; Gong, H.; Hou, Y.; Shu, X.; Huang, B. Advances in Pavement Materials, Design, Characterisation, and Simulation. *Road Mater. Pavement Des.* **2017**, *18*, 1–11. [CrossRef]
12. Yang, Q.; Yu, C. Multiscale Interfacial Interactions in Graphene-Asphalt Nanocomposites: Implications for Asphalt Pavement Material Applications. *ACS Appl. Nano Mater.* **2023**, *6*, 13048–13062. [CrossRef]
13. Tang, Q.; Zhang, Y.; Gao, Y.; Gu, F. Use of Cement-Chelated Solidified Municipal Solid Waste Incinerator (MSWI) Fly Ash for Pavement Material: Mechanical and Environmental Evaluations. *Can. Geotech. J.* **2017**, *54*, 1553–1566. [CrossRef]
14. Mallick, R.B.; Nivedya, M.; Veeraragavan, R.; Nazarian, S.; Tirado, C. An Alternative Approach to Testing of Moisture Susceptibility of Recycled Pavement Materials. *J. Clean. Prod.* **2019**, *224*, 583–591. [CrossRef]
15. Yao, H.; Liu, J.; Xu, M.; Ji, J.; Dai, Q.; You, Z. Discussion on Molecular Dynamics (MD) Simulations of Asphalt Materials. *Adv. Colloid Interface Sci.* **2022**, *299*, 102565. [CrossRef] [PubMed]
16. Ren, S.; Liu, X.; Lin, P.; Gao, Y.; Erkens, S. Review on the Diffusive and Interfacial Performance of Bituminous Materials: A Molecular Dynamics Perspective. *J. Mol. Liq.* **2022**, *366*, 120363. [CrossRef]
17. Jia, M.; Sha, A.; Jiang, W.; Li, X.; Jiao, W. Developing a Solid–Solid Phase Change Heat Storage Asphalt Pavement Material and Its Application as Functional Filler for Cooling Asphalt Pavement. *Energy Build.* **2023**, *285*, 112935. [CrossRef]

https://doi.org/10.3390/coatings16050592

coatings

MDPI

Article

Effect of Hierarchical ZnO/PAC Nanosheets on the Rheological Performance of SBS-Modified Asphalt

Kunpeng Zhao [1], Yi Leng [2], Qinxi Dong [1], Yongling Ding [1,*], Huadong Sun [1,*], Chunbao Ding [2], Ping Song [2], Yanan Ni [2], Chunyu Wang [2] and Hong Yin [3]

1 Shandong Key Laboratory of Technologies and Systems for Intelligent Construction Equipment, School of Transportation and Civil Engineering, Shandong Jiaotong University, Jinan 250357, China
2 Weihai Municipal Highway Development Center, Weihai 264200, China
3 School of Engineering, Australian National University, Canberra, ACT 2601, Australia
* Correspondence: dingyongling198715@163.com (Y.D.); 204113@sdjtu.edu.cn (H.S.)

Abstract

To improve the rutting resistance and low-temperature cracking performance of polymer-modified asphalt under extreme conditions, hierarchically structured ZnO-loaded porous activated carbon (ZnO/PAC) nanosheets were introduced as a synergistic reinforcing agent for SBS-modified asphalt. The ZnO/PAC hybrids were synthesized via template-assisted carbonization followed by hydrothermal growth, and their effects were evaluated by microscopic characterization and rheological tests, including temperature sweeps, multiple stress creep and recovery (MSCR), and bending beam rheometer (BBR) analyses. ZnO was successfully anchored onto the PAC, forming a three-dimensional flower-like nanostructure. Among the investigated samples, ZPS3 with 3 wt.% ZnO/PAC showed the best overall performance. At 64 °C, the rutting factor increased from 4.2 kPa for the SBS-modified asphalt to 6.8 kPa for ZPS3, representing a ~62% enhancement and indicating markedly improved high-temperature deformation resistance. MSCR results further confirmed the superior rutting resistance of ZPS3, which exhibited the highest recovery and the lowest non-recoverable creep compliance. In addition, BBR results showed that the low-temperature performance grade improved from −12 °C for conventional the SBS-modified asphalt to −18 °C for the ZnO/PAC-modified system. These results demonstrate that ZnO/PAC nanosheets can effectively enhance both the high-temperature rutting resistance and low-temperature cracking resistance of SBS-modified asphalt.

Keywords: ZnO/PAC nanosheets; asphalt binder modification; SBS; rutting resistance; low-temperature cracking; rheological properties

1. Introduction

Asphalt pavements are a critical component of global highway infrastructure, and their performance is largely governed by the properties of the asphalt binder. In particular, resistance to high-temperature rutting and low-temperature cracking directly determines pavement durability and service life [1,2]. To meet the challenges of evolving climatic conditions and increasing traffic loads, extensive efforts have focused on modifying asphalt using materials such as rubber, resins, and polymers. Among these, styrene–butadiene–styrene (SBS) is widely recognized as one of the most effective modifiers for enhancing both high- and low-temperature performance [3,4]. With advances in nanotechnology, nanomaterials, including nano-silica, nano-titanium dioxide, nano-zinc oxide, and nano-clay, have attracted considerable attention due to their high specific surface area and unique interfacial effects, which contribute to improved rutting resistance, aging resistance, and moisture stability [5,6]. More recently, carbon-based nanomaterials such as graphene, graphene oxide (GO), and carbon nanotubes (CNTs) have shown strong potential because of their superior mechanical properties and barrier effects. However, their practical application is often hindered by poor dispersion and limited interfacial compatibility with the asphalt matrix [7,8]. Nano-ZnO exhibits excellent UV-shielding capability and interfacial activity, but its tendency to agglomerate reduces its effectiveness when used alone [9]. To address this limitation, hybridizing nano-ZnO with porous activated carbon (PAC) to form a hierarchical, flower-like ZnO/PAC structure offers a promising strategy. This approach combines the interfacial activity of ZnO with the barrier effect of PAC, thereby improving dispersion and overall modification efficiency. Nevertheless, systematic investigations into the synergistic effects of ZnO/PAC composites on the high- and low-temperature performance of SBS-modified asphalt remain scarce. Therefore, this study aims to synthesize hierarchical flower-like ZnO/PAC composites and comprehensively evaluate their impact on the rheological properties, rutting resistance, and low-temperature cracking resistance of SBS-modified asphalt.

Despite its widespread use as an effective polymer modifier, SBS-modified asphalt still exhibits inherent limitations. The unsaturated butadiene segments in SBS are prone to thermal-oxidative and ultraviolet (UV) degradation during service, leading to reduced elastic recovery, increased brittleness, and diminished durability under harsh environ-mental conditions [10,11]. In addition, the compatibility between SBS and the asphalt matrix remains problematic, as phase separation can occur during high-temperature storage, resulting in uneven performance and reduced modification efficiency [12,13]. Although nano-ZnO and carbon-based materials have both been used in asphalt modification, most previous studies have focused on single-component systems or simple physical blending. Such strategies often suffer from nanoparticle agglomeration, weak interfacial compatibility, and limited synergistic effects, particularly in SBS-modified asphalt systems [14–16]. Moreover, the relationship between the microstructure of hierarchical inorganic/carbon hybrid modifiers and their rheological enhancement mechanisms has not been sufficiently clarified [17–19].

To address this gap, this study proposes a hierarchical flower-like ZnO/PAC hybrid, in which PAC serves as a porous support to anchor ZnO and suppress its aggregation, while also providing barrier and interfacial effects within the asphalt matrix. The objective of this work is to synthesize the ZnO/PAC hybrid, evaluate its influence on the high- and low-temperature rheological properties of SBS-modified asphalt, and reveal the structure–property relationship responsible for the observed performance enhancement. Field durability, storage stability, and scale-up feasibility are important practical issues and will be considered in future work. The ZnO/PAC hybrid will be synthesized via a template-assisted carbonization method combined with hydrothermal growth. Its microstructure, chemical composition, pore characteristics, and thermal stability will be comprehensively characterized using XRD, SEM, EDS, XPS, Raman spectroscopy, BET, and TGA. The dispersion behavior and interfacial interactions of ZnO/PAC within the asphalt matrix will be further analyzed to clarify the modification mechanism. The high-temperature rutting resistance will be evaluated using temperature sweeps and MSCR tests with a dynamic shear rheometer (DSR), while low-temperature cracking resistance will be assessed via bending beam rheometer (BBR) tests. Finally, a structure–performance relationship will

 https://doi.org/10.3390/coatings16050520

be established through multi-scale characterization to reveal the synergistic reinforcement mechanism responsible for enhancing performance across temperature ranges.

2. Materials and Methods

2.1. Materials

1,4-benzenedimethanol, p-toluene sulfonic acid, anhydrous ethanol, N,N-dimethylformamide, zinc acetate ($(CH_3COO)_2Zn$) and N-methyl-2-pyrrolidone were purchased from Shanghai Aladdin Bio-Chem Technology Co., Ltd. (Shanghai, China). The basic properties of base asphalts ($70^{\#}$, SK-70) and SBS polymer modifier are presented in Tables 1 and 2. Coal pitch was acquired from Hebei Fengtaiyuan Energy Technology Co. Ltd (Handan city, Hebei, China). Linear SBS with a block ratio of 30/70 was obtained from Yueyang Petrochemical Co., Ltd (Yueyang city, Hunan, China).

Table 1. Basic properties of SK-70# base asphalt [20].

Index		Results	Standard Values
Penetration (25 °C, 0.1 mm)		64.6	60–80
Softening point (°C)		49.2	≥46
15 °C Ductility (cm)		>100	≥100
60 °C Viscosity (Pa·s)		218	≥180
RTFOT residue	mass loss (%)	0.45	≤±0.8
	Residual penetration ratio (25 °C, %)	75.2	≥61
	Residual ductility (cm)	24.8	≥15

Table 2. Technical indicators of SBS additive.

Sample	S/B	Tensile Strength (MPa)	Constant Tensile Stress (MPa)	Elongation (%)	Permanent Deformation (%)
SBS	30/70	≥18	≥1.9	≥700	≤45

2.2. Synthesis of Modified Coal Pitch

In a typical synthesis, 10 g of coal tar pitch was melted in a 250 mL round-bottom flask at 140 °C. Subsequently, 3.0 g of p-xylylene glycol (crosslinking agent) and 0.5 g of p-toluene sulfonic acid (catalyst) were added. The mixture in the flask was heated to 150 °C under stirring for 3 h, yielding a viscous black product. After cooling to room temperature, the resulting coal tar pitch-based resin was collected for further use.

2.3. Preparation of Pitch Resin-Based Carbon Nanosheets

Micron-sized NaCl templates were synthesized using planetary ball milling at 400 rpm for 30 min. Subsequently, 0.5 g of pitch resin was dissolved in 30 mL of N-methyl-2-pyrrolidone with vigorous stirring, followed by the incorporation of 5 g of ball-milled NaCl templates. Following 1 h of stirring, the mixture was evaporated at 180 °C to eliminate the solvent. The pitch resin/NaCl composite was carbonized at 750 °C for 2 h under argon, with a heating rate of 5 °C per minute. Upon cooling to ambient temperature, the composite was meticulously rinsed with deionized water to remove the template and subsequently dried to obtain pitch resin-based carbon nanosheets (PAC). The schematic design for the synthesis of PAC is illustrated in steps 1 and 2 of Figure 1.

 https://doi.org/10.3390/coatings16050520

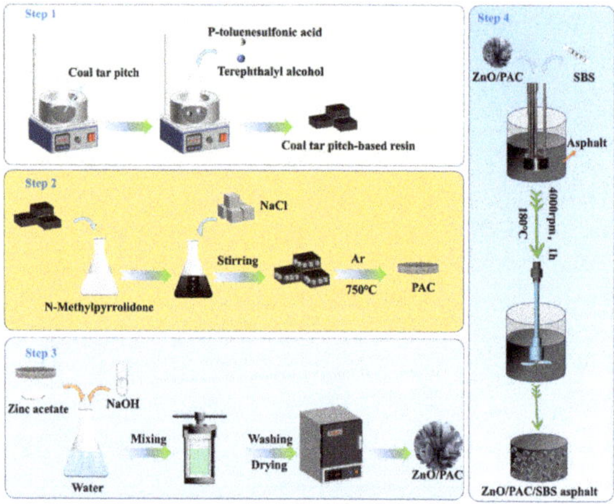

Figure 1. Schematic illustration for pitch resin-based carbon nanosheets.

2.4. Synthesis of ZnO-Loaded PAC (ZnO/PAC)

Initially, 1 g of PAC, infused with 0.88 g of zinc acetate, was dispersed in 20 mL of deionized water while stirring to create a homogenous solution. Subsequently, 20 mL of NaOH solution (2.0 mol/L) was added to the suspension, followed by ultrasonic treatment for 15 min. The mixture was subsequently placed in an autoclave and subjected to hydrothermal treatment at 200 °C for 3 h. The resultant product was sequentially washed with deionized water and ethanol, followed by drying at 70 °C. The dried sample was further calcined in a muffle furnace at 400 °C for 2 h to produce ZnO-loaded PAC (ZnO/PAC). Figure 1 depicts the manufacturing process.

2.5. Preparation of ZnO/PAC and SBS Composite Modified Asphalts

The preparation process of the composite of ZnO/PAC and SBS composite was as follows: firstly, the matrix asphalt was first heated and maintained at 150 ± 5 °C until it melted. The SBS content was fixed at 3 wt.% as a representative dosage commonly used in SBS-modified asphalt, while the ZnO/PAC content (1–3 wt.%) was varied to evaluate its reinforcing effect. Furfural extract oil was heated to 180 °C, followed by the gradual addition of SBS (3% by mass) and ZnO/PAC to obtain the composite modifier mixture. Subsequently, the mixture was subjected to high-speed shearing at 4000 rpm for 2 h at 180 °C using a high-speed shear machine. Finally, the mixtures were further stirred using mechanical agitator for 1 h at 1000 rpm to produce the ZnO/PAC/SBS-composite-modified asphalt. The same processing process as described above was applied to SBS modified asphalt and PAC/SBS-modified asphalt as control groups. The final products were named as ZPS1, ZPS2 and ZPS3, when the dosages of ZnO/PAC is 1%, 2% and 3% by weight of asphalt, respectively.

2.6. Microscopic Analyses

Powder X-ray diffraction (XRD) measurements were employed to examine the phase composition and ascertain the crystalline phase (D8 ADVANCE, Bruker, Germany). The surface chemical characteristics and morphologies of the materials were analyzed utilizing scanning electron microscopy (SEM, Zeiss Sigma-500, Germany). Energy-dispersive X-ray spectroscopy (EDS) was utilized to analyze and quantify the constituents of the samples' surface. High-performance X-ray photoelectron spectroscopy (XPS, Thermo ESCALAB 250, USA) was utilized to analyze the chemical states of the produced materials. The

 https://doi.org/10.3390/coatings16050520

materials' specific surface areas and pore distributions were measured using a specific surface area analyzer (ASAP 2460, Micromeritics, McMurray (Shanghai) Instrument Co., Ltd. China). The Raman spectra were obtained utilizing a laser Raman spectrometer (LabRAM HR Evolution, France). The thermal stabilities of the materials were evaluated using thermogravimetric analysis (TGA) utilizing a synchronous thermal analyzer (STA 449 F3 Jupiter, NETZSCH, Germany). The experiment was performed at temperatures ranging from 0 °C to 800 °C, with a heating rate of 10 °C/min.

2.7. Temperature Sweeping Test

The temperature sweep test was conducted using a dynamic shear rheometer (DSR, Smart Pave 102, Anton Paar) based on AASHTO T315 to evaluate the viscoelastic properties of the modified asphalt binders over a range of temperatures. Specimens with a diameter of 25 mm and a gap thickness of 1 mm were prepared for testing within the linear viscoelastic region. The temperature sweep was performed under controlled-strain mode at a constant frequency of 10 rad/s, covering a temperature range from 46 °C to 82 °C in increments of 6 °C. During the test, the complex shear modulus (G^*) and phase angle (δ) were recorded as functions of temperature to characterize the rheological behavior. Three parallel tests were conducted for each sample.

2.8. MSCR Tests

The anti-deformation potential of asphalt binders containing graphene was also assessed using the multiple stress creep and recovery (MSCR) test at pressures of 0.1 kPa and 3.2 kPa (58 °C), respectively. The procedure was performed ten times during the test, with a stress duration of 1.0 s and a recovery time of 9.0 s. For assessing the recovery (R) and non-recoverable creep compliance (J_{nr}) values based on AASHTO T350, three parallel tests were conducted for each sample.

2.9. BBR Tests

BBR studies were carried out utilizing TE-BBR equipment at three testing temperatures (i.e., 12 °C, 18 °C, and 24 °C) in compliance with AASHTO T313-12 guidelines to examine the low-temperature anti-cracking behavior of the modified binders [21]. For the repeatability test, three duplicate samples were created. Two common indicators of the asphalt binders' capacity to withstand cracking at low temperatures were the creep stiffness and m-value as determined by three-point bending tests.

3. Results and Discussion

3.1. Structural and Morphological Analysis

The XRD patterns of PAC and ZnO/PAC are shown in Figure 2. Two distinctive diffraction peaks for PAC are seen at around 24° and 43°, which correspond to the graphitic carbon crystal planes (002) and (100), respectively. The broad and intense (002) peak at 24° indicates well-ordered graphitic stacking with an interlayer spacing of 0.421 nm calculated using Bragg's law. The relatively weak (100) peak appearing at 43° suggests the coexistence of amorphous carbon characteristics [22,23]. After ZnO incorporation, the ZnO/PAC exhibits distinct crystalline features. The sharpened diffraction peaks confirm the enhanced crystallinity through hydrothermal treatment and subsequent calcination of the ZnO particle precursors. All observed diffraction peaks perfectly match well with the standard hexagonal wurtzite structure of ZnO (JCPDS No. 03-0888), representing high crystallinity.

Raman spectra provide further insight into the graphitization degree of pitch resin-based carbon materials in Figure 2b. Two characteristic peaks are clearly observed: the

https://doi.org/10.3390/coatings16050520

disorder-induced D-band at 1350 cm^{-1}, representing sp^2 hybridized carbon and structural defects, while graphitic G-band at 1580 cm^{-1} reflecting the ordered structure. The intensity ratio (I_D/I_G) serves as a quantitative indicator of the graphitization degree, where lower values correspond to a higher graphitization degree. The calculated I_D/I_G for ZnO/PAC is 0.86, significantly lower than that of PAC (0.92). These results demonstrate that ZnO incorporation promotes graphitization in the pitch-derived carbon matrix [24].

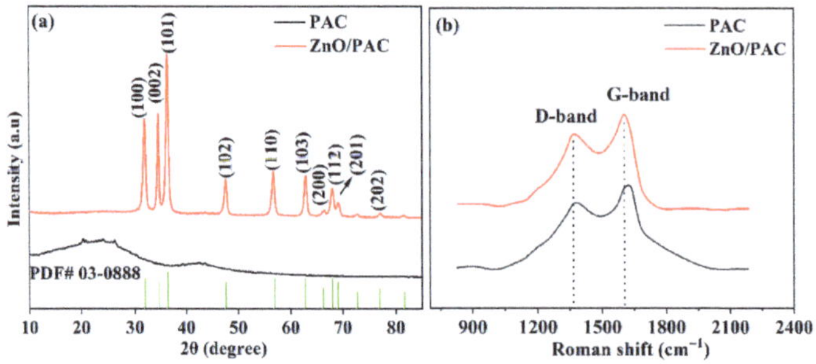

Figure 2. (**a**) Raman spectra and (**b**) XRD patterns of PAC and ZnO/PAC.

The increased graphitization with higher ZnO loading enhances electrical conductivity, while the resulting high thermal conductivity facilitates efficient heat transfer and dissipation under loading. This reduces localized heat accumulation from friction, thereby mitigating asphalt damage [25,26].

Figure 1 shows the PAC manufacturing process. A combination of modified pitch and the NaCl template was carbonized in an Ar environment to create PAC, which was then cleaned and dried. To create a rigid template for the production of carbon frameworks, NaCl particles were first ball-milled from around 500 μm (Figure 3a) to less than 15 μm (Figure 3b). Post-carbonization, the composite exhibited a black coloration, indicating a uniform carbon coating on the NaCl particles. Microscopic analysis revealed a three-dimensional macroporous network structure of the PAC composed of interconnected carbon nanosheets.

Figure 3. SEM image of NaCl particles before (**a**) and after (**b**) ball-milling.

As shown in Figure 4a, PAC exhibits a foam-like skeletal structure with cavities (3–10 μm) formed by the templating effect of NaCl. The high-magnification SEM image (Figure 4b) reveals that the foam-like framework consists of interconnected carbon nanosheets with thicknesses below 100 nm (typically 30–50 nm). PAC possesses a long-range disordered microstructure with enlarged interlayer spacing. The strong π-π interactions among planar coal-tar pitch molecules impart a graphitization tendency to PAC, while the three-dimensional molecular structure of the pitch resin significantly suppresses aromatic stacking during

https://doi.org/10.3390/coatings16050520

carbonization, resulting in higher amorphousness [27]. As a soft carbon material, the large interlayer spacing, amorphous nature, hierarchical porosity, and high specific surface area collectively provide abundant active sites and enhanced mass transport [28]. Furthermore, the multilayered nanosheet architecture facilitates multiple light reflections and refractions between interlayers, making PAC an effective supporting carrier for ZnO decoration and synergistically improving its UV-shielding performance. The micro-lamellar structure of PAC was leveraged to facilitate the in situ growth of ZnO and to inhibit its agglomeration, culminating in a uniformly dispersed nanoscale ZnO/PAC composite.

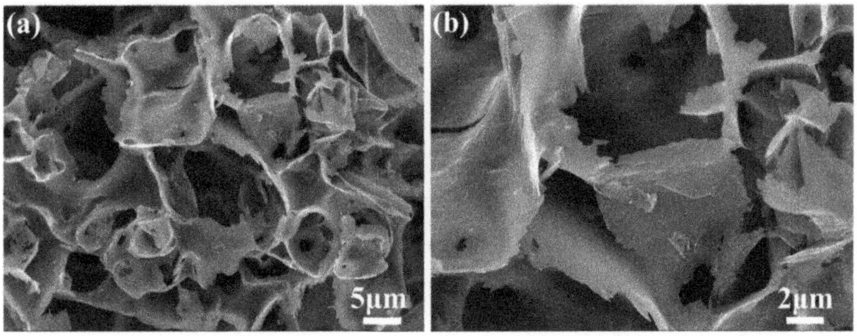

Figure 4. (**a**) SEM images of PAC with different magnification and (**b**) an enlarged view of (**a**).

Figure 5 presents the SEM images of ZnO/PAC with various contents of ZnO loadings. As shown in Figure 5a,b, ZnO are constructed by divergent nanoplates and is homogeneously distributed on the surface of PAC, exhibiting dimensions ranging from 500 nm to 1 μm and a thickness of approximately 30 nm. The synthesized regular hierarchical flower-like ZnO nanostructures assemble by nanosheets and display the uniform sizes, intersecting configurations, and two-dimensional structures perpendicular to the PAC surfaces with irregular edge morphologies. The results indicate that the incorporation of nanocomposite technology facilitates the embedding or supporting of ZnO nanoflakes within carrier matrices, thereby mitigating agglomeration to a certain extent. Superior to its 1D and 2D counterparts, the interwoven-growth 3D flower-like ZnO exhibits a loose and highly porous morphology with a large specific surface area, which offers abundant active sites for chemical reactions and enhances UV-light reflection paths for improved shielding performance [29]. EDS elemental mapping and images (Figure 6) indicate the homogeneous distribution of C, O, and Zn elements, confirming the successful formation of a uniform composite through the in situ growth of ZnO with PAC.

Figure 5. (**a**) SEM images of ZnO/PAC and (**b**) an enlarged view of (**a**).

 https://doi.org/10.3390/coatings16050520

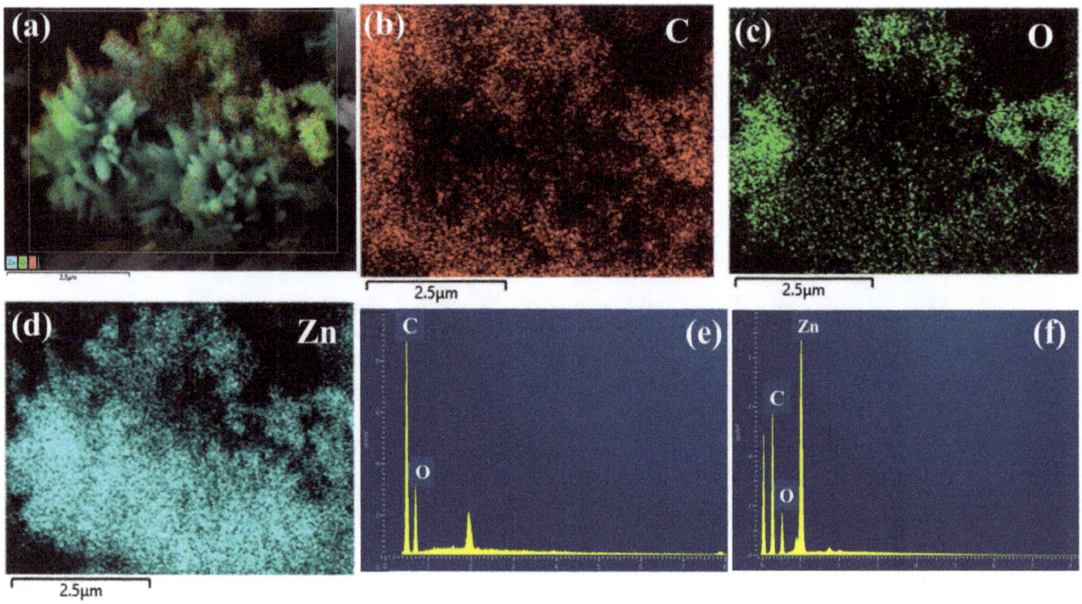

Figure 6. EDS elemental mapping of ZnO/PAC: element layered image (**a**), C (**b**), O (**c**), Zn (**d**); EDS images of (**e**) PAC and (**f**) ZnO/PAC.

The combined SEM and EDS analyses demonstrate that the micron-sized NaCl template facilitates the formation of hollow carbon nanobox structures, within which ZnO nanoflakes uniformly dispersed in an ordered arrangement throughout the carbon matrix. This distinctive multilayered architecture can create either exfoliated or intercalated structures within the asphalt, effectively blocking UV radiation and thermo-oxidative penetration into the asphalt interior [30,31]. The structural configuration not only suppresses the volatilization of light asphalt components but also notably improves the material's resistance to UV-induced and thermo-oxidative aging [32].

XPS was used to further examine the bonding states and surface chemical composition of the as-prepared samples. The survey spectrum (Figure 7a) exhibits dominant peaks corresponding to C, O, and Zn elements of ZnO/PAC. The O 1s spectrum in Figure 7b was deconvoluted into four components at binding energies of 530.5 eV, 531.4 eV, 532.9 eV, and 536.7 eV, respectively. The lower binding energy peaks at 530.5 eV and 531.4 eV are attributed to the lattice oxygen (Zn-O bonds in ZnO), while the higher binding energy peaks at 532.9 eV and 536.7 eV correspond to the surface-adsorbed oxygen species, such as -COOH and -OH groups [33]. The surface functional groups of zinc oxide (ZnO), particularly hydroxyl terminations (OH), significantly influence its optical properties by enhancing absorption and reflectivity in the ultraviolet region [34,35]. The C 1s spectrum in Figure 7c was fitted into four peaks at 284.2 eV, 284.8 eV, 286.9 eV, and 291.2 eV, assigned to C-C, C-O-C, C=O, and O-C=O bonds, respectively [36]. Moreover, the peak observed at 284.2 eV is attributed to the pure graphitic carbon within the PAC matrix [37]. The Zn 2p spectrum in Figure 7d, two spin–orbit splitting peaks were observed at 1044.3 eV (Zn $2p^{1/2}$) and 1021.1 eV (Zn $2p^{3/2}$), confirming the presence of Zn^{2+} in ZnO/PAC [38]. Hitherto, all results unequivocally demonstrate that the as-prepared ZnO/PAC materials are not physical mixtures of separate PAC and ZnO phases, but rather well-integrated

 https://doi.org/10.3390/coatings16050520

composites derived from the calcination of hydrothermally treated precursor mixtures. Evidence confirms the formation of covalent bonds between PAC and ZnO.

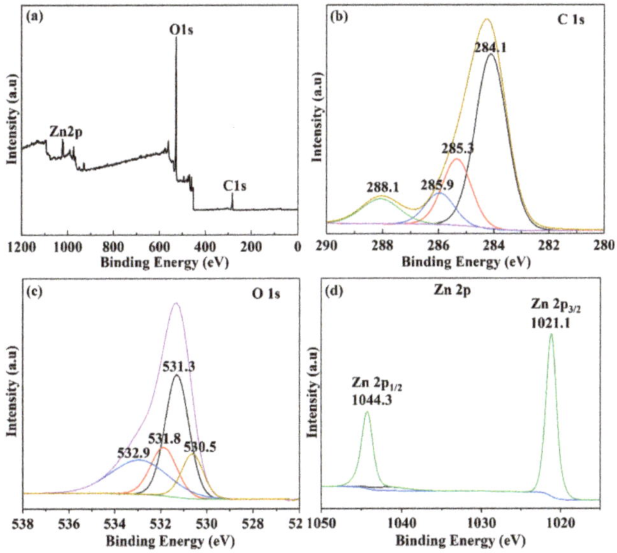

Figure 7. Wide scan (**a**) and high-resolution XPS spectra of as-prepared ZnO/PAC at binding energies corresponding to (**b**) C1s, (**c**) O1s and (**d**) Zn 2p.

3.2. TGA- DTG Analysis

TGA was conducted to analyze the thermal stability of the samples. As observed in Figure 8, all samples exhibited approximately 50% weight loss from room temperature to 650 °C, followed by negligible mass changes up to 800 °C. The weight loss process can be divided into two distinct regions: (1) the first region (100–250 °C) corresponds to the evaporation of adsorbed moisture, showing similar weight loss trends across all samples; (2) the second region (250–650 °C) is associated with the thermal decomposition of PAC in the composites [39]. The DTG analysis reveals a significantly higher maximum weight loss temperature for ZnO/PAC (610 °C) compared to pure PAC (521 °C), demonstrating enhanced thermal stability. This observation aligns with the established correlation between carbon material stability and microstructure [40]. Specifically, the presence of active sites (e.g., intrinsic defects and sp3-hybridized edge sites) in PAC facilitates oxygen interactions, leading to lower thermal stability. These findings corroborate the results of Raman spectroscopy, confirming the lower graphitic crystallinity and higher defect density of PAC. Quantitative analysis revealed a ZnO content of 16.5% in ZnO/PAC. As evidenced in Figure 8, the ZnO/PAC exhibited higher carbon yields compared to PAC. The presence of ZnO significantly increased the pyrolysis temperature of PAC, attributed to its relatively higher graphitization degree, which is consistent with the Raman spectroscopy results.

https://doi.org/10.3390/coatings16050520

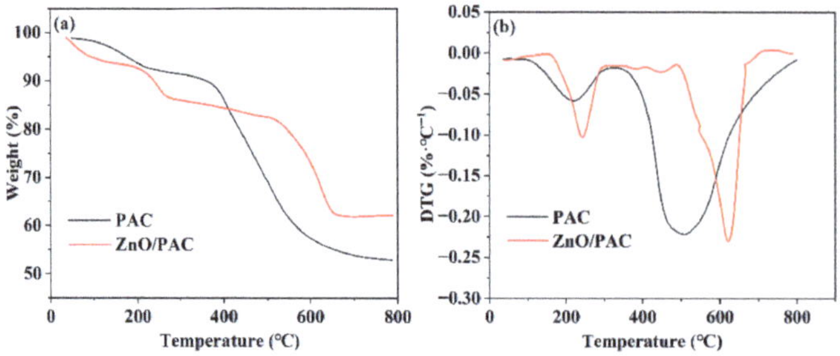

Figure 8. (**a**) TG and (**b**) DTG curves of PAC and ZnO/PAC.

3.3. Pore Structure Analysis

Nitrogen adsorption–desorption analysis was employed to characterize the pore structure and specific surface area of samples. As shown in Figure 9a, the PAC and ZnO/PAC composites exhibited typical Type IV isotherms with three distinct regions: (1) rapid N_2 adsorption capacity at low relative pressures ($P/P_0 < 0.1$), indicating abundant micropores; (2) pronounced adsorption in the 0.4–0.6 P/P_0 range, suggesting capillary condensation in mesopores, with the ZnO/PAC displaying the most significant mesoporous character; and (3) a steep rise at high pressures ($P/P_0 > 0.9$), confirming the presence of macropore (>50 nm). The H_2-type hysteresis loops observed at 0.8–1.0 P/P_0 further verified the well-developed mesoporous structure. BET analysis revealed decreasing specific surface areas with increasing ZnO loading: 1502 $m^2 \cdot g^{-1}$ (PAC) and 942 $m^2 \cdot g^{-1}$ (ZnO/PAC). As summarized in Table 3, both pore volume and average pore diameter followed similar decreasing trends. This textural evolution can be attributed to two primary factors: (1) ZnO crystal growth and particle aggregation progressively blocking existing pore channels, and (2) ZnO infiltration into the interparticle voids of the PAC matrices. These combined effects result in systematic pore filling and a consequent reduction in available surface area and pore volume at higher ZnO concentrations.

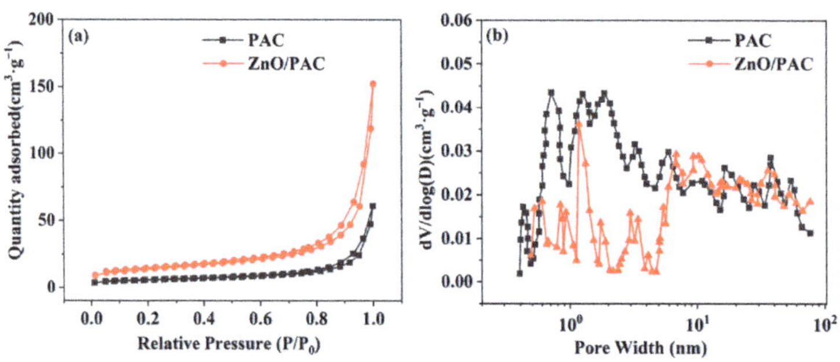

Figure 9. Nitrogen adsorption–desorption isotherms (**a**) and pore width distribution (**b**) of PAC and ZnO/PAC.

Table 3. Pore characteristics of different samples.

Samples	S_{bet} ($m^2 \cdot g^{-1}$)	V_{pore} ($cm^2 \cdot g^{-1}$)	Pore Size (nm)
PAC	1502	0.68	28.9
ZnO/PAC	942	0.46	22.1

The pore size distribution analysis (Figure 9b) confirms a well-developed hierarchical pore structure in synthesized samples, spanning microporous (1.5–2 nm), mesoporous

https://doi.org/10.3390/coatings16050520

(2–50 nm), and macroporous (>50 nm) regimes. The macropores may result from a small amount of nanoscale NaCl particles after ball milling. Previous studies have shown that nanopores within graphite-like carbon nanosheets can regulate the nucleophilic and electrophilic characteristics, as well as the polarity, of graphene nanosheets by altering their molecular electrostatic potential, thereby enhancing the compatibility of complex modified asphalt systems containing components with varying polarity [41]. Nanopores in graphene can impart electrophilic characteristics, enabling the adsorption of nonpolar molecules present in asphalt, whereas the non-porous regions interact with polar molecules through conventional π-π stacking interactions [42].

3.4. *Temperature Sweep*

In Figure 10a, the G* of all asphalt binders decreases progressively with increasing temperature from 52 °C to 82 °C. Among the samples, the SBS-modified asphalt shows the most pronounced decline in G* with increasing temperature, indicating relatively poor resistance to rutting at elevated temperatures. Incorporating PAC into the SBS-modified asphalt produces a modest increase in G*, suggesting that the carbon nanosheets improve the structural rigidity of the binder. In contrast, the ZPS3 composite-modified asphalt consistently exhibits the highest G* values, particularly above 70 °C, where the reduction in modulus is markedly suppressed. This behavior indicates that the introduction of ZnO/PAC modifies the viscoelastic response of the binder by enhancing its resistance to temperature-induced softening. This enhancement can be attributed to the combined structural and dispersion effects of the hierarchical ZnO/PAC hybrid. The three-dimensional-flower-like ZnO nanosheets anchored on the PAC scaffold provide rigid inclusions that restrict the mobility of asphalt molecules and SBS chains. At the same time, the increased interfacial contact area promotes stress transfer within the binder. Consequently, it enhances rutting resistance at high service temperatures and postpones the softening process as well.

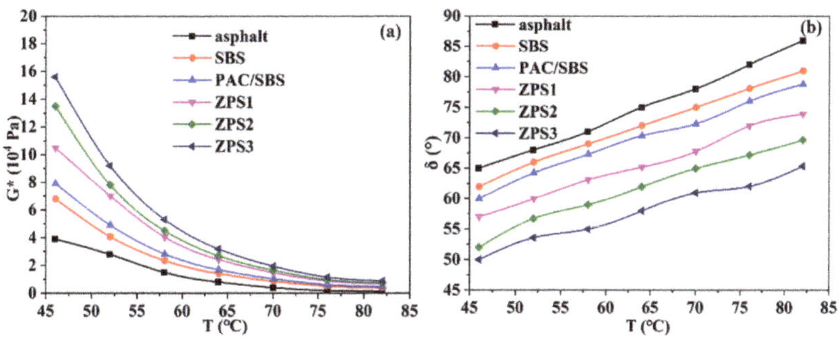

Figure 10. The complex modulus G* (**a**) and phase angle δ (**b**) of the SBS-, PAC/SBS- and ZPS modified asphalt binders.

As shown in Figure 10b, δ increases with temperature for all samples, indicating a gradual transition from predominantly elastic to more viscous behavior. All modified asphalt samples exhibit this increasing trend, suggesting a rapid loss of elastic recovery at elevated temperatures. Compared with SBS, the PAC/SBS system shows a slightly lower δ at equivalent temperatures, implying that the incorporation of PAC enhances elastic recovery. Notably, the ZPS3 composite exhibits the lowest δ values throughout the entire temperature range and the smallest rate of increase, indicating that the binder maintains a predominantly elastic character even at high temperatures. This improved elastic response is mainly attributed to the well-dispersed ZnO/PAC nanofillers, which interact with the SBS polymer network through physical entanglement and possible interfacial

 https://doi.org/10.3390/coatings16050520

bonding. As a result, the three-dimensional polymer network is reinforced, viscous flow is delayed, and resistance to permanent deformation is enhanced. In addition, oxygen-containing functional groups (e.g., -OH and -COOH) on the ZnO/PAC surface can form hydrogen bonds or polar interactions with components in the asphalt and SBS, further anchoring the polymer network to the rigid filler. This effect reduces molecular mobility and delays the transition from elastic to viscous behavior.

3.5. High-Temperature Performance

In Figure 11, the G*/sin δ values for all samples fall as the temperature rises, indicating the transition from an elastic to a viscous condition of the binders. The ZPS3 consistently demonstrates a higher rutting factor (G*/sin δ) throughout the temperature spectrum of 46 °C to 82 °C. For instance, at 64 °C, the G*/sin δ values for SBS, PAC/SBS, and ZPS3 are 4.2 kPa, 5.1 kPa, and 6.8 kPa, respectively. G*/sin δ exhibits a reduced rate of decline at elevated temperatures exceeding 70 °C, thereby indicating enhanced high-temperature rutting resistance relative to both SBS-modified and PAC/SBS-modified binders. This improved performance is mainly due to two complementary mechanisms. The shear-induced in-plane orientation of ZnO/PAC nanosheets establishes a convoluted diffusion pathway that significantly restricts the mobility of polymer chains and light components within the asphalt matrix at elevated temperatures, similar to the barrier enhancement noted in aligned graphene nanosheet composites [19]. Secondly, the multilayered architecture and elevated graphitization level of the ZnO/PAC hybrid create an efficient physical barrier that obstructs heat and oxygen infiltration, thereby maintaining the elastic network structure during high-temperature operation [43,44]. These mechanisms collectively allow the ZPS composite to sustain a high G*/sin δ and remarkable structural stability under thermal stress.

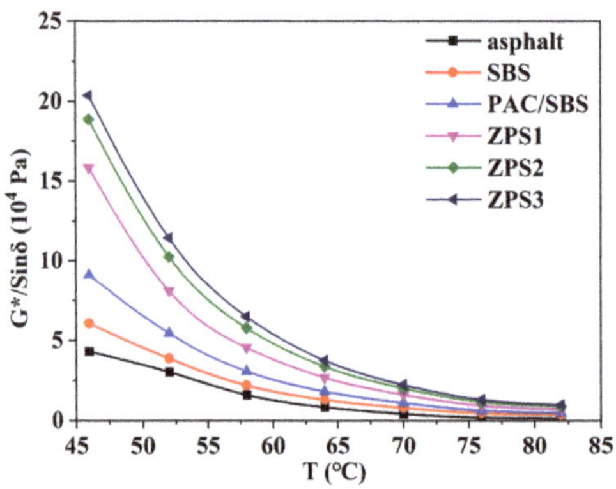

Figure 11. The G*/sin δ of the SBS-, PAC/SBS- and ZPS modified asphalt binders.

Additionally, the MSCR test was used to further examine the high-temperature performance in order to identify significant rutting resistance outside of the linear viscoelasticity range [45]. As illustrated in Figure 12, the ZPS composite-modified asphalt exhibits significantly higher R and lower J_{nr} compared to the SBS-modified and PAC/SBS-modified binders at both stress levels of 0.1 kPa and 3.2 kPa. At 0.1 kPa, the R values for SBS, PAC/SBS, ZPS1, ZPS2, and ZPS3 are 72.6%, 78.5%, 84.9%, 89.5% and 93.3%, respectively, while the corresponding Jnr values are 1.7 kPa^{-1}, 1.1 kPa^{-1}, 0.7 kPa^{-1}, 0.4 kPa^{-1}, and 0.1 kPa^{-1}. At the higher stress level of 3.2 kPa, the R values decrease to 47.7%, 58.5%, 66.8%, 72.7% and 75.1%, and the Jnr values increase to 3.1 kPa^{-1}, 1.5 kPa^{-1}, 0.9 kPa^{-1},

https://doi.org/10.3390/coatings16050520

0.7 kPa^{-1}, and 0.3 kPa^{-1}, respectively. The introduction of PAC and ZnO/PAC markedly alters the high-temperature deformation behavior of the asphalt binder. As the most critical parameter of rutting resistance, the progressive reduction in Jnr indicates that the composite modifiers effectively suppress the irreversible deformation under repeated loading. Among all samples, ZPS3 demonstrates the lowest J_{nr} and highest R, reflecting its superior resistance to permanent deformation and optimal high-temperature performance. In the case of the ZPS modifier, the ZnO/PAC component adsorbs onto the surface of the SBS copolymer, promoting interfacial cross-linking between the SBS phase and the asphalt matrix. The hierarchical ZnO/PAC nanosheets effectively restrict molecular chain mobility and facilitate efficient stress transfer under sustained loading [46]. Moreover, the enhanced interfacial compatibility between the nanofiller and the asphalt matrix, attributed to the oxygen-containing functional groups present on the ZnO/PAC surface, improves the load transfer and reduces energy dissipation during creep–recovery cycles [47,48]. Collectively, these results demonstrate that the ZPS composite-modified binders possess favorable resistance to permanent deformation under high-temperature conditions [49].

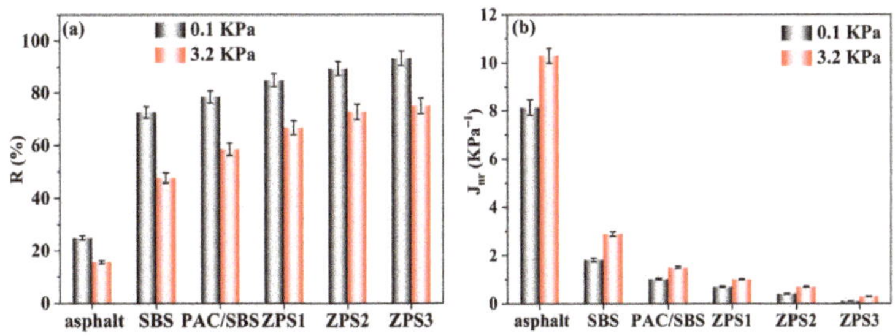

Figure 12. The R (**a**) and Jnr (**b**) of the SBS-, PAC/SBS- and ZPS modified asphalt binders.

3.6. Low-Temperature Performance

The creep stiffness (S) and creep rate (m) obtained from the bending beam rheometer (BBR) test are two effective indicators for evaluating the low-temperature cracking resistance of asphalt binders, with lower stiffness and higher m-values being favorable for enhanced anti-cracking performance. As shown in Figure 12, both the PAC/SBS- and ZPS-modified binders exhibit reduced creep stiffness and increased m-values, indicating improved low-temperature capability. According to ASTM D6648-2008, the specification requires a creep stiffness below 300 MPa and an m-value above 0.3. All asphalt binders demonstrate distinct temperature-dependent variations in S and m, with the ZPS composite-modified asphalt series exhibiting superior low-temperature performance compared to the SBS-modified and PAC/SBS-modified binders. At −18 °C, the S values for base asphalt, SBS, PAC/SBS, ZPS1, ZPS2, and ZPS3 are 345 MPa, 305 MPa, 251 MPa, 240 MPa, 232 MPa and 225 MPa, and the m-values are 0.23, 0.29, 0.31, 0.32, 0.33 and 0.33, respectively. Notably, all ZPS-modified binders satisfy the standard requirements at −18 °C, whereas the SBS-modified binder only meets the criteria at −12 °C. The ZPS series consistently maintains lower S and higher m-values relative to the control binders across all test temperatures, with ZPS3 exhibiting the most favorable combination of low stiffness and high relaxation capability, suggesting superior resistance to low-temperature cracking (Figure 13).

https://doi.org/10.3390/coatings16050520

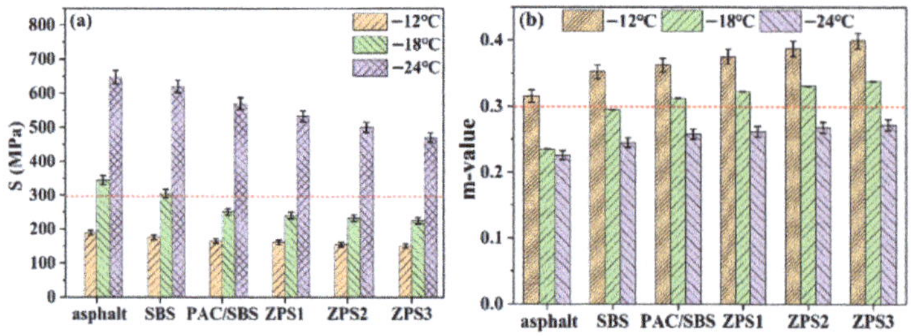

Figure 13. The creep stiffness S (**a**) and m-value (**b**) of the SBS-, PAC/SBS- and ZPS modified asphalt binders.

3.7. Microscopic Analysis of Modified Asphalt

As shown in Figure 14, the SBS-modified asphalt (Figure 6a) exhibits a rough surface with pronounced SBS-rich domains and phase-separated structures, whereas ZPS3 in Figure 6b presents a uniform and compact morphology, in which the flower-like ZnO/PAC structures are homogeneously dispersed throughout the asphalt matrix. This improvement can be mainly attributed to the following factors: PAC, acting as a porous carrier, effectively suppresses the agglomeration of ZnO nanoparticles, while its oxygen-containing surface functional groups form strong interfacial interactions with SBS and the polar components of asphalt, thereby promoting the uniform swelling of SBS and the formation of a continuous network. Meanwhile, the three-dimensional hierarchical structure of ZnO/PAC generates steric hindrance and physical barrier effects within the asphalt, thereby restricting the migration of SBS molecular chains and mitigating phase separation. The optimization of the microstructure provides a structural basis for the synergistic enhancement of the high- and low-temperature performance of ZPS3.

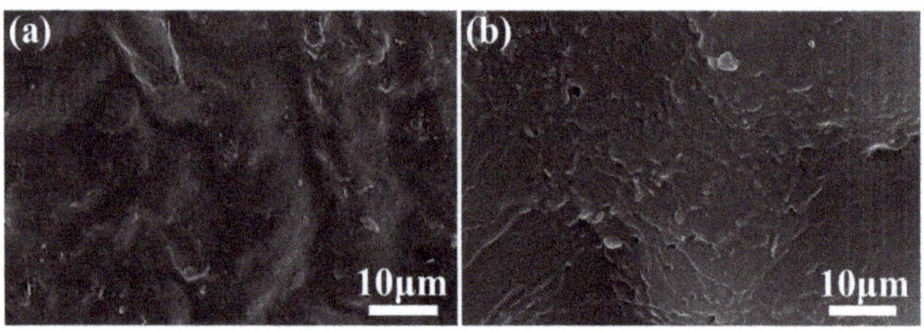

Figure 14. SEM images of SBS-modified asphalt (**a**) and ZPS3 (**b**).

4. Conclusions

(1) The ZnO/PAC hybrid was successfully synthesized via template-assisted carbonization combined with hydrothermal growth. Characterization results confirm that ZnO nanoparticles are uniformly anchored on the porous carbon nanosheets, forming a hierarchical flower-like morphology with a high specific surface area and abundant oxygen-containing functional groups. ZnO incorporation promotes the graphitization of the carbon matrix and enhances thermal stability, as evidenced by Raman spectroscopy and TGA analyses.

(2) Rheological evaluations demonstrate that the ZPS composite modified asphalt exhibits superior high-temperature performance. The ZPS3 formulation shows the highest complex modulus (G*) and lowest phase angle (δ) across the temperature range of 46–82 °C. MSCR test results indicate that the ZPS series achieves significantly higher R

https://doi.org/10.3390/coatings16050520

and lower Jnr compared to SBS-modified and PAC/SBS-modified binders, with ZPS3 exhibiting the most favorable combination, reflecting superior resistance to permanent deformation under repeated loading.

(3) The enhanced high-temperature performance is primarily attributed to the formation of a rigid percolating network by the ZnO/PAC nanosheets within the asphalt matrix. The three-dimensional flower-like morphology and abundant functional groups facilitate strong interfacial interactions with the SBS polymer and asphalt components, restricting molecular chain mobility, promoting efficient stress transfer, and delaying the transition from elastic to viscous behavior.

(4) BBR test results reveal that the ZPS series exhibits superior low-temperature cracking resistance. All ZPS-modified binders satisfy the ASTM D6648-2008 requirements at −18 °C, whereas the SBS-modified binder only meets the criteria at −12 °C. The consistent maintenance of S and higher m-values across all test temperatures indicates enhanced stress relaxation capability, attributed to the homogeneous dispersion of the nanofiller and the preservation of binder flexibility.

(5) The synergistic reinforcement provided by the ZnO/PAC hybrid, combining the barrier effect of the carbon nanosheets and the interfacial reinforcement of the ZnO nanoflowers, effectively mitigates the performance deterioration associated with high-temperature softening and low-temperature embrittlement. This study demonstrates that ZnO/PAC/SBS composite modified asphalt represents a promising solution for high-performance pavement applications subjected to severe thermal and mechanical loading conditions.

This study focuses on binder-scale evaluation of rheological performance and microstructure–property relationships. Further investigations, including chemical interaction analysis (e.g., FTIR), storage stability, standard aging procedures (RTFOT and PAV), and mixture-level performance tests (e.g., wheel tracking and fatigue), are necessary for comprehensive engineering validation and will be addressed in future work. Although the results demonstrate consistent improvements in rheological properties at the binder level, further investigations, including aging evaluation, storage stability analysis, and mixture-level performance tests, are necessary to validate the long-term engineering applicability of the ZnO/PAC-modified asphalt.

Author Contributions: Writing—review and editing, software, Y.L. and K.Z.; Data curation, Formal analysis, Writing—original draft, C.W.; Resources, Data curation, Writing—original draft, K.Z.; Resources, Data curation, Writing—original draft, Q.D.; Project administration, Investigation and Data curation, C.D.; Methodology, Writing—review and editing, Funding acquisition, P.S. and H.Y.; Resources, Data curation, Q.D.; Writing—original draft, Writing—review and editing, Supervision, Project administration, Y.D. and H.S.; Methodology, Investigation, Y.N.; Methodology, Writing—review and editing, C.W. All authors have read and agreed to the published version of the manuscript.

Funding: This research was funded by Shandong Provincial Transportation Science and Technology Program, grant numbers (2025BAI14-02, 2022B101 and 2023B89) and Shandong Province Science and Technology-based Small and Medium-sized Enterprises Innovation Capacity Enhancement Project (Grants No. 2022TSGC2056).

Institutional Review Board Statement: Not applicable.

Informed Consent Statement: Not applicable.

Data Availability Statement: The original contributions presented in this study are included in the article. Further inquiries can be directed to the corresponding authors.

Acknowledgments: Sincere appreciation is extended to all supporting institutions and funding agencies.

https://doi.org/10.3390/coatings16050520

Conflicts of Interest: The authors declare no conflicts of interest.

References

1. Zhao, S.; You, Q.; Ling, J.; Zhang, J.; Chineche, E.B.; Wang, J. Variation of nano-ZnO/SBR modified asphalt features under photo-oxidative aging and optimization of the evaluation indexes based on rheological properties and chemical structure. *Case Stud. Constr. Mater.* **2024**, *21*, e04101. [CrossRef]
2. Zheng, Q.; Wang, H.; Jiang, B.; Yue, S.; Wang, T. Influence mechanism of performance and aging behavior of high-content SBS-modified asphalt. *Buildings* **2025**, *15*, 4430. [CrossRef]
3. Wu, W.; Cavalli, M.C.; Jiang, W.; Kringos, N. Differing perspectives on the use of high-content SBS polymer-modified bitumen. *Constr. Build. Mater.* **2024**, *411*, 134433. [CrossRef]
4. Huang, Y.; Wu, Z.; Liang, L.; Ying, J.; Gui, L.; Shen, P.K.; Tian, Z.Q. Revealing the dependance of mechanical properties of asphalt binder on graphene size via multi-scale methods. *Mater. Des.* **2023**, *236*, 112478. [CrossRef]
5. Shi, B.; Long, K.; Zhang, Z.; Yu, D.; Ai, C.; Yan, C. Investigation of anti-aging performance in nano-sol modified high-SBS content emulsified asphalt and its micro-surfacing mixtures. *Constr. Build. Mater.* **2024**, *453*, 139050. [CrossRef]
6. Rondón-Quintana, H.A.; Zafra-Mejía, C.A.; Urazán-Bonells, C.F. Aging resistance evaluation of an asphalt mixture modified with zinc oxide. *Infrastructures* **2024**, *9*, 81. [CrossRef]
7. Wu, Z.; Huang, Y.; Gui, L.; Liang, L.; Ying, J.; Li, W.; Shen, P.K.; Tian, Z.Q. 3D porous graphene nanosheets as efficient additives for high-performance styrene-butadiene-styrene/crumb rubber blend-modified asphalt. *Mater. Des.* **2023**, *232*, 112157. [CrossRef]
8. Huang, Y.; Wu, Z.; Liang, L.; Ying, J.; Gui, L.; Mo, L.; Shen, P.K.; Tian, Z.Q. Critical role of vacancy defects in graphene nanosheets for enhancing asphalt binder. *Constr. Build. Mater.* **2024**, *438*, 137139. [CrossRef]
9. Xie, W.; Qi, C.; Wang, B.; Li, Q.; Wang, S. Effect of nano-ZnO on the microstructure and application properties of styrene-butadiene-styrene modified asphalt. *Pet. Sci. Technol.* **2025**, *43*, 1659–1676. [CrossRef]
10. Liang, C.H.; Sun, C.J.; Liu, D.; Li, Q.S.; Lin, J.Y.; Wu, X.H.; Zhang, H.L. The role of rejuvenator functional groups in restoring aged SBS modified asphalt: Effects of SBS molecular architecture and block ratio. *Constr. Build. Mater.* **2026**, *516*, 145682. [CrossRef]
11. Zhang, Y.; Wu, C.; Yan, C.; Hu, H.; Liu, Y.; Zhou, Y. Aging inhibition characteristics of boron nitride nanosheets/silicone rubber considering interfacial modification. *Polym. Compos.* **2025**, *46*, 14621–14633. [CrossRef]
12. Chen, J.; Huang, Z.; Wang, H.; Yang, Z.; Zhang, T. Investigating the rheological properties of styrene-butadiene-styrene-based high-viscosity modified asphalt using carbon nanotubes. *Sustainability* **2023**, *15*, 71. [CrossRef]
13. Sun, C.J.; Wang, Z.W.; Li, Q.S.; Liang, C.H.; Guo, S.Y.; Zhu, X.Y.; Zhang, H.L.; Du, P.F. Devulcanized rubber/bio-oil composite rejuvenator for enhanced rheological performance and aging resistance of rejuvenated SBS-modified asphalt. *Constr. Build. Mater.* **2026**, *515*, 145636. [CrossRef]
14. Chen, Y.; Huang, J.; Muhammad, Y.; Zhao, Z.; Fan, S.; Luo, W.; Li, J. Study on the performance and mechanism of carbon nanomaterials incorporated SBS composite modified asphalt. *Constr. Build. Mater.* **2023**, *389*, 131795. [CrossRef]
15. Ullah, Z.; Zhang, Z.; Liu, H.; Ji, J.; Li, Q.; Rukundo, S.P.; Bilal, M. Application of nanomaterials on the performance of SBS modified asphalt binders and mixtures: A comprehensive review. *J. Road Eng.* **2025**, *5*, 323–342. [CrossRef]
16. Luo, Y.; Zhang, H.L.; Chen, Z.H.; Li, Q.S.; Ye, S.X.; Liu, Q.T. Novel multidimensional composite development for aging resistance of SBS-modified asphalt by attaching zinc oxide on expanded vermiculite. *Energy Fuels* **2024**, *38*, 16772–16781. [CrossRef]
17. Xie, Y.; Yu, P.; Zhai, M. Analysis of nano-ZnO-modified asphalt compatibility based on molecular dynamics. *Materials* **2023**, *16*, 4710. [CrossRef] [PubMed]
18. Ding, Y.; Zhang, S.; Luo, X.; Zhou, J. AFM characterization of morphological evolution and adhesion properties of ZnO/C nanosheet-modified asphalt under UV irradiation. *Surf. Topogr. Metrol. Prop.* **2023**, *11*, 015003. [CrossRef]
19. Zhang, J.M.; Zhang, H.L.; Zhu, K.F. Layered double hydroxides intercalated with UV absorbers: Preparation, characterisation, and application in UV-resistant asphalt. *Int. J. Pavement Eng.* **2026**, *27*, 2642958. [CrossRef]
20. Zhao, T.; Ding, Y.; Cai, Q.; Wu, Y.; Jiang, J.; Liu, W.; Huang, J.; Cong, Y.; Yin, X.; Yin, H.; et al. Effects of epoxy-silicone resin on chemistry, morphology and rheology of SBR- modified emulsified asphalt. *Results Eng.* **2026**, *29*, 108990. [CrossRef]
21. *AASHTO T313-12*; Standard Method of Test for Determining the Flexural Creep Stiffness of Asphalt Binder Using the Bending Beam Rheometer (BBR). American Association of State Highway and Transportation Officials: Washington, DC, USA, 2015.
22. Yang, X.; Wang, F.; Li, X.; Zhang, Z.; Wang, C.; Yang, C.; Zhen, Y.; Wang, D.; Fu, F.; Chi, R. Facile and low-cost fabrication of interconnected hierarchically porous carbon for high-performance supercapacitors. *J. Alloys Compd.* **2022**, *921*, 166127. [CrossRef]
23. Shao, J.; Ma, F.; Wu, G.; Dai, C.; Geng, W.; Song, S.; Wan, J. In-situ MgO ($CaCO_3$) templating coupled with KOH activation strategy for high yield preparation of various porous carbons as supercapacitor electrode materials. *Chem. Eng. J.* **2017**, *321*, 301–313. [CrossRef]

 https://doi.org/10.3390/coatings16050520

24. He, H.; He, J.; Yu, H.; Zeng, L.; Luo, D.; Zhang, C. Dual-interfering chemistry for soft-hard carbon translation toward fast and durable sodium storage. *Adv. Energy Mater.* **2023**, *13*, 2300357. [CrossRef]
25. Zhou, X.; Zhang, X.; Xu, S.; Wu, S.P.; Liu, Q.T.; Fan, Z.P. Evaluation of thermo-mechanical properties of graphene/carbon-nanotubes modified asphalt with molecular simulation. *Mol. Simul.* **2017**, *43*, 312–319. [CrossRef]
26. Abo-Shanab, Z.L.; Mohamed, E.A.; Adel, M. Reinforcing bitumen with highly graphitized reduced graphene oxide derived from coal tar pitch via template free sole step. *Sci. Rep.* **2025**, *15*, 41819. [CrossRef]
27. Wang, Y.; Xiao, N.; Wang, Z.; Li, H.; Yu, M.; Tang, Y.; Hao, M.; Liu, C.; Zhou, Y.; Qiu, J. Rational design of high-performance sodium-ion battery anode by molecular engineering of coal tar pitch. *Chem. Eng. J.* **2018**, *342*, 52–60. [CrossRef]
28. Liu, Y.; Lu, Y.X.; Xu, Y.S.; Meng, Q.S.; Gao, J.C.; Sun, Y.G.; Hu, Y.S.; Chang, B.B.; Liu, C.T.; Cao, A.M. Pitch-derived soft carbon as stable anode material for potassium ion batteries. *Adv. Mater.* **2020**, *32*, 2000505. [CrossRef]
29. Zhao, Y.; Xu, Z.; Wang, X.; Lin, T. Superhydrophobic and UV-blocking cotton fabrics prepared by layer-by-layer assembly of organic UV absorber intercalated layered double hydroxides. *Appl. Surf. Sci.* **2013**, *286*, 364–370. [CrossRef]
30. Chen, S.; Liu, Q.; Wu, H.; Yang, C.; Gong, X.; Wu, S.; Li, Y. Simultaneous enhancement of the thermal oxygen and ultraviolet aging resistance of asphalt by incorporating antioxidant intercalated layered double hydroxide. *Constr. Build. Mater.* **2022**, *349*, 128743. [CrossRef]
31. Liu, H.; Zhang, Z.; Tian, Z.; Zhao, Y.; Yang, J.; Zhang, W. UV aging resistance improvement of SBS modified asphalt binder by organic layered double hydroxide and naphthenic oil: Its preparation, properties and mechanism. *Constr. Build. Mater.* **2024**, *449*, 138404. [CrossRef]
32. Xu, S.; Liu, L.; Jia, X.; Tighe, S.; Zhang, C.; Ma, H. Investigation of aging resistance of organic layered double hydroxide/antioxidant composite-modified asphalt. *ACS Sustain. Chem. Eng.* **2023**, *11*, 267–277. [CrossRef]
33. Liu, L.; Kuang, D.; Hou, L.; Luo, H.; Deng, L.; Wang, S. Synthesis and microwave absorption performance of layered hard carbon embedded with ZnO nanoparticles. *J. Alloys Compd.* **2022**, *895*, 162677. [CrossRef]
34. Talouba, N.; Henniche, A.; Liu, L.; Li, J.; Rahouia, N.; Hegazy, M.; Huang, Y. Improving the mechanical properties, UV and hydrothermal aging resistance of PIPD fiber using MXene ($Ti_3C_2(OH)_2$) nanosheets. *Compos. Part B* **2019**, *163*, 260–271. [CrossRef]
35. Berdiyorov, G.R. Optical properties of functionalized $Ti_3C_2T_2$ (T = F, O, OH) MXene: First-principles calculations. *AIP Adv.* **2016**, *6*, 055105. [CrossRef]
36. Yun, S.I.; Kim, S.H.; Kim, D.W.; Kim, Y.A.; Kim, B.H. Facile preparation and capacitive properties of low-cost carbon nanofibers with ZnO derived from lignin and pitch as supercapacitor electrodes. *Carbon* **2019**, *149*, 637–645. [CrossRef]
37. Deng, Q.F.; Liu, L.; Lin, X.Z.; Du, G.H.; Liu, Y.P.; Yuan, Z.Y. Synthesis and CO_2 capture properties of mesoporous carbon nitride materials. *Chem. Eng. J.* **2012**, *203*, 63–70. [CrossRef]
38. Lin, Y.G.; Hsu, Y.K.; Chen, Y.C.; Chen, L.C.; Chen, S.Y.; Chen, K.H. Visible-light-driven photocatalytic carbon-doped porous ZnO nanoarchitectures for solar water-splitting. *Nanoscale* **2012**, *4*, 6515–6519. [CrossRef] [PubMed]
39. Yue, J.; Liu, H.; Zhu, Y.; Xu, Y.; Hu, C.; Cheng, J.; Bai, Y.; Zhao, X. Study on the characterization and thermal conversion behavior of sequential extraction products from medium temperature coal pitch. *Fuel* **2024**, *367*, 131571. [CrossRef]
40. Yolshina, L.A.; Yolshina, V.A.; Pershina, S.V.; Pryakhina, V.I. Study of thermal stability of hierarchical structured carbon composite flakes. *Diam. Relat. Mater.* **2021**, *119*, 108556. [CrossRef]
41. Yan, D.; Xu, X.; Lu, T.; Hu, B.; Chua, D.; Pan, L. Reduced graphene oxide/carbon nanotubes sponge: A new high capacity and long life anode material for sodium-ion batteries. *J. Power Sources* **2016**, *316*, 132–138. [CrossRef]
42. Kolev, S.K.; Aleksandrov, H.A.; Atanasov, V.A.; Popov, V.N.; Milenov, T.I. Interaction of graphene with out-of-plane aromatic hydrocarbons. *J. Phys. Chem. C* **2019**, *123*, 21448–21456. [CrossRef]
43. Xu, Z.H.; Dong, M.; Xu, T. Compatibility, microstructure, and mechanical properties of a dimethyl-sulfoxide-pretreated graphene-modified asphalt binder. *Langmuir* **2024**, *40*, 9809–9818. [CrossRef] [PubMed]
44. Wang, S.; Chen, W.; Li, J.; Song, Z.; Zhang, H.; Zeng, W. Low working temperature of ZnO-MoS_2 nanocomposites for delaying aging with good acetylene gas-sensing properties. *Nanomaterials* **2020**, *10*, 1902. [CrossRef]
45. Zhang, L.; Xing, C.; Gao, F.; Li, T.S.; Tan, Y.Q. Using DSR and MSCR tests to characterize high temperature performance of different rubber modified asphalt. *Constr. Build. Mater.* **2016**, *127*, 466–474. [CrossRef]
46. Chen, W.; Zhu, Z.R.; Fan, S.F.; Su, Y.; Li, Y.B.; Yang, C.; Liu, Q.; Chen, Z.D.; Tang, D.K.; He, X.Y. Preparation of functionalized MXene/carboxylated MWCNTs/SEBS-g-MAH multiscale modified asphalt with superior anti-aging performance. *Constr. Build. Mater.* **2025**, *493*, 143183. [CrossRef]
47. Li, S.W.; Huang, Z.; Fan, J.T.; Zhang, J.P.; Li, Y. Multiscale experimental investigation on the reinforcement mechanism and applications of carbon nanotube-modified asphalt. *J. Mater. Civil. Eng.* **2026**, *38*, 04026099. [CrossRef]

48. Huang, J.X.; Ya, K.Z.; Wan, M.; Shi, X.; Wu, Q.; Chen, Y.W. Enhancing asphalt binder performance and catalytic degradation of exhaust gases with graphene-TiO_2 nanocomposites. *Constr. Build. Mater.* **2025**, *471*, 140711. [CrossRef]
49. Sun, Y.; Lv, B.; Gong, H.; Xu, H.; Wang, W.; Xu, B.; Chen, J. Approach for accurately characterizing nonlinear viscoelastic and nonrecoverable properties of asphalt binders based on MSCR test. *Constr. Build. Mater.* **2023**, *403*, 133098. [CrossRef]

 https://doi.org/10.3390/coatings16050520

coatings

MDPI

Article

Study on Desulfurized Crumb Rubber–Modified Epoxy Asphalt

Yi Zhao and Peixing Yang *

School of Transportation, Southeast University, Nanjing 211189, China; 220233297@seu.edu.cn
* Correspondence: ypxflwc@163.com

Abstract

Epoxy asphalt, as a thermosetting material, has received increasing attention due to its outstanding mechanical properties and durability. However, its insufficient low-temperature resistance, limited toughness, and relatively high material cost still restrict its large-scale application in pavement engineering. To improve its low-temperature performance and reduce construction costs, this study investigates the low-temperature behavior of epoxy asphalt modified with desulfurized crumb rubber. In this study, a functional additive, hereafter referred to as WJFL (a laboratory-designated organic disulfide-based rubber plasticizer), was incorporated during the preparation of the desulfurized rubber–asphalt binder to enhance the curing rate of the modified epoxy asphalt. The addition of WJFL promotes the devulcanization and activation of rubber powder, enhancing the overall performance of the modified epoxy asphalt. When the desulfurized rubber content is 20%, WJFL additive dosage is 2%, and asphalt content is 300% of epoxy resin mass, the modified epoxy asphalt not only meets the specification requirements but also exhibits excellent low-temperature crack resistance and improved economic efficiency. The addition of crumb rubber increased tensile strength by 15.38% and elongation at break by 17.24%. Furthermore, WJFL additive increased tensile strength by 80% and elongation at break by 25% when WJFL content was increased from 0% to 2%. Additionally, optimizing the asphalt-to-epoxy ratio, with asphalt content increased from 100% to 300%, resulted in an 80% increase in tensile strength and a 28.57% improvement in elongation at break. Moreover, desulfurized crumb rubber modification enhanced the low-temperature stiffness modulus, highlighting better performance in cold regions. Relaxation tests conducted at −10 °C, −15 °C, −20 °C, and −25 °C show that the modified epoxy asphalt has significant potential for use in pavement surfacing, particularly in cold climates.

Keywords: desulfurized crumb rubber; epoxy resin; asphalt; additive; low-temperature performance

1. Introduction

With the continuous increase in traffic volume and the frequent occurrence of overloaded vehicles, together with the intensified influence of extreme climate events, asphalt pavement distress—such as cracking, raveling, potholes, and premature failures—has become increasingly severe. These issues highlight the limitations of conventional paving asphalt in meeting the modern requirements for high durability and high performance in road engineering [1–4]. Consequently, modified asphalt has emerged as a major research focus. Common polymer modifiers for asphalt binders include natural rubber (NR), styrene–butadiene rubber (SBR), styrene–butadiene–styrene (SBS) block copolymer, polyethylene (PE), and ethylene–vinyl acetate (EVA). However, these modifiers predominantly interact with asphalt through physical blending rather than chemical reactions, and thus do not fundamentally alter the thermoplastic nature of the binder. As a result, issues such as insufficient high-temperature stability, poor low-temperature toughness, and limited mechanical strength remain [5–7].

Epoxy asphalt, as a new class of thermosetting binder, is formed through the cross-linking reaction between epoxy resin and asphalt in the presence of a curing agent, producing a stable three-dimensional network structure. This significantly enhances strength, stiffness, and high-temperature stability, while providing excellent fatigue resistance and durability [1,8,9]. Recent state-of-the-art reviews and long-term field evaluations have further confirmed the superior functional and in-service performance of epoxy asphalt mixtures on steel bridge decks [10,11]. Nevertheless, due to the intrinsic high stiffness and limited elongation of epoxy networks, conventional epoxy asphalt is prone to brittle fracture under low-temperature or large-deformation conditions. Its high viscosity and narrow workable time window further limit its engineering applicability [12].

To achieve better stiffness–flexibility balance and improve environmental sustainability, the use of waste tire crumb rubber as a modifier has attracted significant attention. On one hand, crumb rubber can enhance the performance of epoxy asphalt and reduce material costs. On the other hand, the continuous growth in global vehicle ownership has resulted in a massive accumulation of waste tires, leading to environmental pollution and land occupation, making their recycling an urgent global issue [13]. Processing waste tires into crumb rubber for use in pavement engineering helps promote resource recycling and alleviates environmental pressure [14–18]. Laboratory and field investigations have consistently shown that waste rubber–modified asphalt mixtures can deliver improved rutting and fatigue resistance compared with conventional mixtures [19,20].

Previous studies have confirmed the beneficial effects of crumb rubber modification in epoxy asphalt. Gong et al. (2022) [21] incorporated waste tire rubber into epoxy asphalt for bridge expansion joint mixtures and found that increased rubber content improved viscosity, deformation resistance, and fatigue performance, although workable time decreased. Similar conclusions were drawn in other epoxy–rubber systems, where the inclusion of tire rubber waste or polymer–rubber composites was shown to improve the toughness and cracking resistance of epoxy asphalt mixtures under service-like loading conditions [22]. Xu et al. (2022)) [23] demonstrated that rubber particles can form effective energy-dissipating zones, thereby enhancing fatigue damage resistance. Li et al. (2019) [7] compared conventional and desulfurized rubber-modified asphalt and reported that desulfurized rubber significantly improves viscosity, storage stability, and high-temperature rutting resistance while maintaining good fatigue performance. Zhang et al. (2022) [24] further revealed that the desulfurization process promotes rubber swelling, refinement, and uniform dispersion within asphalt, resulting in superior modulus–phase angle characteristics across a wide temperature range.

In addition, Xu et al. (2021) [25] showed that the composite modification of desulfurized crumb rubber and SBS enhances the elastic network structure of asphalt and improves both high- and low-temperature performance. Wang et al. (2022) [26] systematically evaluated high-content desulfurized rubber asphalt and confirmed its excellent durability in terms of rutting resistance, crack resistance, fatigue life, and temperature susceptibility. Cong et al. (2023) [27] further demonstrated that incorporating waterborne polyurethane into crumb rubber–modified asphalt can effectively reduce emissions of volatile organic compounds (VOCs) during production, thereby supporting the development of more environmentally friendly asphalt materials. More recently, process-oriented studies have clarified how desulfurization variables and improved rubber–asphalt compatibility govern the rheological response and workability of desulfurized crumb rubber-modified asphalt, especially in high-rubber systems [28–32].

https://doi.org/10.3390/coatings16010102

Despite the above progress, most existing studies have focused on desulfurized rubber-modified base asphalt and conventional asphalt mixtures [33–35], rather than epoxy asphalt systems. Current research on rubber-modified epoxy asphalt predominantly uses untreated crumb rubber and mainly examines the influence of rubber dosage on viscosity, mechanical performance, and fatigue behavior, with relatively limited attention paid to desulfurized crumb rubber. Furthermore, systematic investigations into the coupled effects of desulfurized rubber content, additive dosage, and asphalt proportion on the low-temperature performance of epoxy asphalt binders remain scarce.

Based on these gaps, this study focuses on the low-temperature performance of desulfurized crumb rubber–modified epoxy asphalt. A series of formulations with different rubber contents, additive dosages, and asphalt–epoxy ratios was designed to evaluate their low-temperature mechanical responses, specifically at temperatures of −10 °C, −15 °C, −20 °C, and −25 °C, and identify the optimal composition. The findings deepen the understanding of the toughening mechanism of desulfurized rubber in epoxy asphalt systems and provide scientific guidance for its application in steel bridge deck surfacing and pavement engineering in cold regions.

To address the above research gaps, this study presents several novel aspects. First, desulfurized crumb rubber is introduced into an epoxy asphalt system, extending its application beyond conventional base asphalt and untreated rubber–modified epoxy asphalt. Second, the low-temperature mechanical and rheological behavior of desulfurized crumb rubber–modified epoxy asphalt is systematically investigated, with particular emphasis on relaxation and deformation characteristics under cold conditions. Third, the coupled effects of rubber content, functional additive dosage, and asphalt–epoxy ratio are comprehensively evaluated, rather than considering individual parameters in isolation. Finally, the study adopts an engineering-oriented perspective to identify optimal formulations for steel bridge deck surfacing and pavement applications in cold regions. These contributions provide new insights into the toughening mechanism of desulfurized crumb rubber in epoxy asphalt systems and support the development of durable, low-temperature-resistant epoxy asphalt materials.

2. Materials and Methods

2.1. Materials

The preparation of the desulfurized crumb rubber–modified epoxy asphalt in this study involved the following materials: base asphalt, desulfurized crumb rubber, additive, curing agent, and epoxy resin.

2.1.1. Asphalt

A 70# base asphalt produced in Taizhou was used. Its physical and technical properties are presented in Table 1.

Table 1. Technical properties of 70# base asphalt.

Technical Index	Specification Requirement	Test Result
Penetration/0.1 mm (25 °C, 100 g, 5 s)	60~80	69
Penetration Index	−1.5~1.0	0.2
Softening Point/°C	≥46	47.6
Ductility (15 °C, 5 cm/min/cm)	≥100	165
Dynamic Viscosity (60 °C)/Pa·s	≥180	256

https://doi.org/10.3390/coatings16010102

Table 1. *Cont.*

Technical Index	Specification Requirement	Test Result
Flash Point (Cleveland Open Cup)/°C	≥260	285
Solubility/%	≥99.5	99.8
	Residues after RTFOT	
RTFOT Mass Change/%	≤±0.8	0.26
Retained Penetration (25 °C)/%	≥61	68.7
Retained Ductility (15 °C)/cm	≥15	48

2.1.2. Desulfurized Crumb Rubber

An 80-mesh desulfurized crumb rubber with good modification performance was used. Its physical and chemical properties are listed in Table 2.

Table 2. Properties of desulfurized crumb rubber (80 mesh).

Category	Item	Technical Value (80 mesh)
Physical Properties	Relative Density	1.15
	Moisture Content (%)	0.85
	Metal Content (%)	0.02
	Fiber Content (%)	0.22
	Ash Content (%)	8.4
	Heating Loss (163 °C, 5 h, %)	1.8
Chemical Properties	Acetone Extract (%)	17.6
	Carbon Black Content (%)	29.8
	Rubber Hydrocarbon Content (%)	54.6

2.1.3. Additive

To enhance the curing rate of the modified epoxy asphalt, a functional additive (WJFL) was incorporated during the preparation of the desulfurized rubber–asphalt binder. WJFL can be classified as an organic disulfide-based rubber plasticizer, with a typical chemical structure represented as R–S–S–R′, where R and R′ denote alkyl or phenyl groups.

The incorporation of WJFL introduces disulfide bonds (–S–S–) into the rubber–asphalt system, which can participate in dynamic sulfur exchange reactions under thermal and shear processing. As a result, these disulfide bonds may partially cleave into thioether (–S–) structures and generate a small amount of thiol (–SH) groups, thereby promoting the devulcanization and activation of rubber powder [36].

Consequently, the modified rubber–asphalt binder is able to act not only as a filler in the epoxy system but also as a curing-active component for the epoxy resin. The alkyl groups contribute to the plasticization and flexibility of the rubber phase, while the phenyl groups enhance compatibility with the aromatic components of asphalt, which improves the compatibility between asphalt and epoxy resin and significantly enhances the overall performance of the modified epoxy asphalt.

2.1.4. Epoxy Resin

There are various types of epoxy resins, among which bisphenol A epoxy resin is commonly used as a modifying resin in epoxy asphalt due to its excellent mechanical properties, heat resistance, and cost-effectiveness. It has high tensile strength and flexural strength, which significantly enhance the stiffness, durability, and fatigue resistance of epoxy asphalt. Bisphenol A epoxy resin also has good adhesion, which improves the bonding between the modified asphalt and other materials, ensuring long-term stability under extreme environmental conditions. Moreover, its low cost and availability make it

https://doi.org/10.3390/coatings16010102

an ideal choice for the preparation of epoxy asphalt. Therefore, in this study, bisphenol A epoxy resin is used as the modifying resin in desulfurized crumb rubber-modified epoxy asphalt.

2.2. Experimental Methods

The experimental design followed these steps:

(1) Determine the appropriate crumb rubber content;
(2) Investigate the effect of additive dosage on the performance of the modified epoxy asphalt;
(3) Study the influence of different asphalt–epoxy ratios based on economic considerations;
(4) Simulate practical application conditions to evaluate the combined effects of all parameters.

During modification, the rubber–asphalt binder was stirred for 1 h and sheared for 15 min. The curing agent and epoxy resin dosage remained constant throughout the tests to ensure a fair comparison between different influencing factors. Each test was repeated at least three times to ensure the reliability and reproducibility of the results, and the reported data represent the average values with standard deviations.

2.2.1. Performance Evaluation Methods

Through extensive analysis and research on U.S. epoxy asphalt, domestic scholars have successfully developed a domestically produced version of epoxy asphalt and proposed a set of technical specifications tailored to domestic conditions, as shown in Table 3.

Table 3. Technical specifications for domestic epoxy asphalt.

Technical Index	Specification Requirement	Test Method
Tensile Strength (20 °C, MPa)	≥1.5	ASTM D 638
Elongation at Break (20 °C, %)	≥200	ASTM D 638
Time for Viscosity to Reach 1000 cp (120 °C, min)	≥50	JTJ052-2000

Since desulfurized crumb rubber–modified epoxy asphalt shares similar characteristics with conventional epoxy asphalt, viscosity, tensile strength, and elongation at break were selected as the core performance evaluation indices.

(1) Viscosity

The viscosity of the modified epoxy asphalt directly affects the uniformity of mixture blending. Excessively low viscosity leads to insufficient coating of aggregates, while excessively high viscosity hinders asphalt dispersion during mixing, resulting in poor mixture homogeneity. Maintaining stable viscosity during compaction is essential for achieving desirable pavement performance [37].

In this study, the viscosity–time behavior of the modified epoxy asphalt was measured using a Brookfield rotational viscometer (spindle No. 27).

(2) Tensile Strength and Elongation at Break

The tensile properties of cured epoxy asphalt were evaluated using dumbbell-shaped specimens according to ASTM D638.

Component A (epoxy resin) was heated to 70 ± 5 °C and Component B to 130 ± 5 °C, then mixed thoroughly before casting into rectangular molds. After curing at 120 °C for 4 h, specimens were demolded and cut into standard dumbbell shapes. Tests were conducted on a universal testing machine with a loading rate of 500 ± 50 mm/min. The

https://doi.org/10.3390/coatings16010102

force–displacement response was recorded to calculate tensile strength and elongation at break.

2.2.2. Preparation of Desulfurized Crumb Rubber–Modified Epoxy Asphalt

Based on the preparation processes of both conventional rubberized asphalt and epoxy asphalt, the following steps were used:

1. Heat the base asphalt in an oil bath at 190 °C and add rubber powder at the target proportion; stir for 1 h.
2. Preheat the colloid mill to 190 °C and feed the rubber–asphalt mixture. The mill gap was initially set at its maximum and gradually reduced to 0.2 mm during the 15 min shearing process while maintaining constant temperature.
3. Calculate the mass of Component A and Component B according to predetermined proportions and asphalt dosage. Mix both components under controlled temperature to prepare the desulfurized crumb rubber–modified epoxy asphalt.

2.2.3. Rubber Content

Three rubber contents (10%, 20%, and 25% by mass of base asphalt) were used to prepare rubberized asphalt binders. Each binder was then blended with epoxy resin and curing agent, using a fixed asphalt-to-epoxy mass ratio of 300%. The viscosity, tensile strength, and elongation at break of the cured modified asphalt were tested to evaluate the effect of rubber dosage.

2.2.4. Additive Dosage

After determining the optimal rubber content, different WJFL dosages (0%, 1%, 2%, and 3% of the rubber–asphalt binder) were added during mixing and shearing. The asphalt-to-epoxy ratio remained 300%. Performance tests were conducted to evaluate the influence of additive dosage on the modified epoxy asphalt.

2.2.5. Asphalt Content

To further enhance economic efficiency, the asphalt content in the epoxy system was varied. The rubber–asphalt binder was used at 150%, 300%, 450%, 600%, and 750% of the epoxy resin mass. Brookfield viscosity, tensile strength, and elongation at break were evaluated to determine the optimal asphalt dosage.

2.2.6. Low-Temperature Performance Evaluation

Low-temperature behavior was characterized using a stress relaxation test similar to the Bending Beam Rheometer (BBR) test. Modified epoxy asphalt beams were tested under three-point bending with constant displacement. Since epoxy asphalt has higher stiffness than conventional binders, the displacement was adjusted through multiple trials to match its mechanical range.

The beam dimensions were 10 mm × 20 mm × 160 mm. Tests were conducted at −10 °C, −15 °C, −20 °C, and −25 °C. Specimens were conditioned for 60 min at each target temperature before loading. A constant displacement of 10%–30% of the fracture deformation was applied, and load–time curves were recorded over 240 s.

Relaxation stiffness $S(t)$ and relaxation rate k were calculated using:

$$S(t) = \frac{PL^3}{4bh^3\delta(t)} \quad (1)$$

$$k = \frac{d(\log S)}{d(\log t)} \quad (2)$$

https://doi.org/10.3390/coatings16010102

where $S(t)$ = relaxation stiffness at time t, MPa; P = applied load, N; L = span length, mm; b = specimen width, mm; h = specimen height, mm; $\delta(t)$ = mid-span deflection at time t, mm.

The effects of rubber modification on low-temperature performance were analyzed by comparing relaxation stiffness and relaxation rate among different temperatures and material formulations.

3. Results and Discussion

3.1. Effect of Crumb Rubber Content

3.1.1. Viscosity Analysis

Figure 1 presents the viscosity–time curves of epoxy asphalt modified with three different desulfurized crumb rubber contents (10%, 20%, and 25%). As shown in the figure, all three formulations exhibit similar viscosity-growth patterns. During the initial stage, the viscosity increases slowly. When the curing time reaches approximately 20 min, the viscosity rises to around 1000 mPa·s, after which it enters a rapid growth phase. In the final stage, the viscosity increases sharply and quickly exceeds the measuring range of the rotational viscometer.

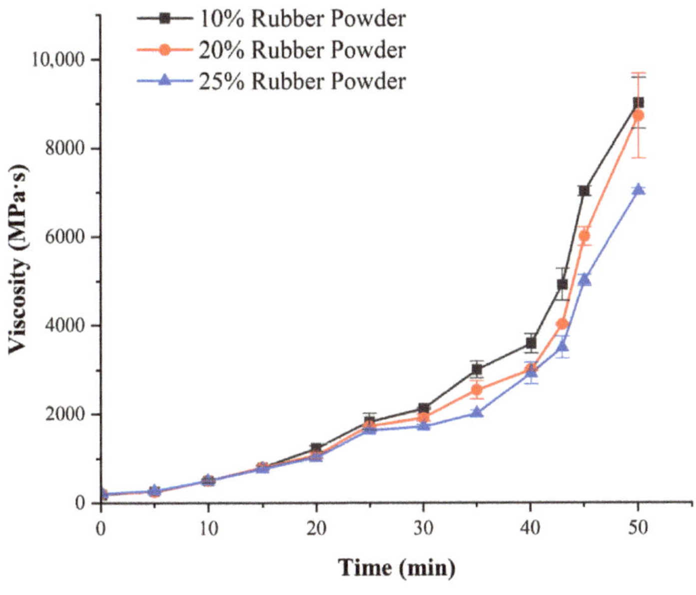

Figure 1. Viscosity–time curves of desulfurized crumb rubber modified epoxy asphalt with different crumb rubber contents.

A comparison of the three curves shows noticeable differences in the initial viscosities. The initial viscosity of the 10% rubber content is 410 mPa·s, followed by 343 mPa·s for the 20% content, and only 230 mPa·s for the 25% content. Moreover, as the rubber dosage increases, the rate of viscosity growth gradually decreases. For example, the time required for the viscosity to reach 8000 mPa·s is 45 min for the 10% formulation, 47 min for the 20% formulation, and 50 min for the 25% formulation.

This phenomenon is attributed to the increasing number of rubber particles within the modified asphalt colloid. At higher dosages, part of the desulfurized crumb rubber is not fully swollen and forms agglomerated particle clusters. The internal particles of

 https://doi.org/10.3390/coatings16010102

these clusters exhibit poor cohesion and insufficient interfacial bonding with asphaltene structures. These free clusters hinder the curing reaction of the epoxy asphalt binder, resulting in reduced initial viscosity and slower viscosity growth.

3.1.2. Tensile Strength and Elongation at Break

Figure 2 shows the tensile test results of epoxy asphalt modified with different crumb rubber contents. The tensile strengths of the 10%, 20%, and 25% formulations are 0.49 MPa, 0.62 MPa, and 0.55 MPa, respectively—much lower than the strength of unmodified epoxy asphalt (approximately 3 MPa). This significant reduction can be explained by two reasons:

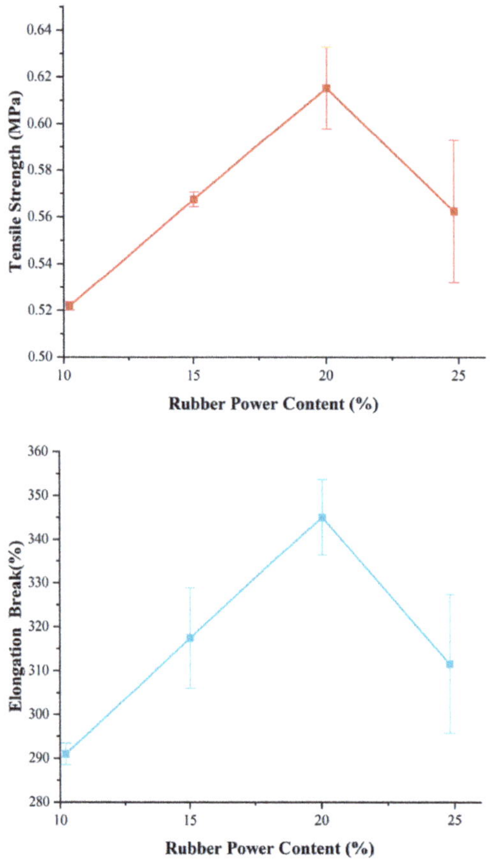

Figure 2. Tensile test results of desulfurized crumb rubber modified epoxy asphalt with different crumb rubber contents.

(1) Incomplete swelling of rubber clusters: Large rubber agglomerates with insufficient development exhibit poor internal cohesion and behave as stress concentration points, reducing the tensile strength of the cured binder.

(2) Intrinsic incompatibility of epoxy resin and asphalt: Epoxy resin and asphalt are inherently poorly compatible due to their large difference in solubility parameters. Introducing rubber particles further worsens phase compatibility, causing additional strength degradation.

https://doi.org/10.3390/coatings16010102

Although the overall strength is low, the variation among the three contents shows a clear trend: the tensile strength first increases and then decreases with increasing rubber dosage, reaching a maximum at the 20% content. This indicates that moderate rubber addition contributes to strength improvement, whereas excessive rubber introduces too many poorly bonded clusters, reducing strength.

Despite the low strength, all three formulations exhibit high elongation at break, exceeding 300%, which demonstrates that desulfurized crumb rubber significantly improves the elastic extensibility of the epoxy asphalt binder.

Based on viscosity, tensile strength, and elongation evaluations, 20% desulfurized crumb rubber was determined to be the optimal dosage, providing a balance between mechanical performance and workability.

3.2. *Effect of Additive (WJFL) on Modified Epoxy Asphalt*

The above results show that although desulfurized crumb rubber improves ductility, the overall strength of the modified epoxy asphalt remains insufficient and cannot meet performance requirements. Simply adjusting the rubber dosage is not enough to achieve desirable properties. Therefore, the additive WJFL was introduced to chemically modify the rubber–asphalt binder, improving its reactivity with epoxy resin and enhancing the performance of the resulting composite binder.

The base formulation contained 20% rubber by mass of asphalt, and different WJFL dosages were introduced during the shearing process. Modified binders were then evaluated to determine the influence of additive dosage.

3.2.1. Viscosity Analysis

Figure 3 shows the viscosity–time curves for formulations with varying WJFL dosages. It is evident that the addition of WJFL accelerates viscosity growth compared with the control sample without WJFL. Higher WJFL dosages produce faster viscosity increases.

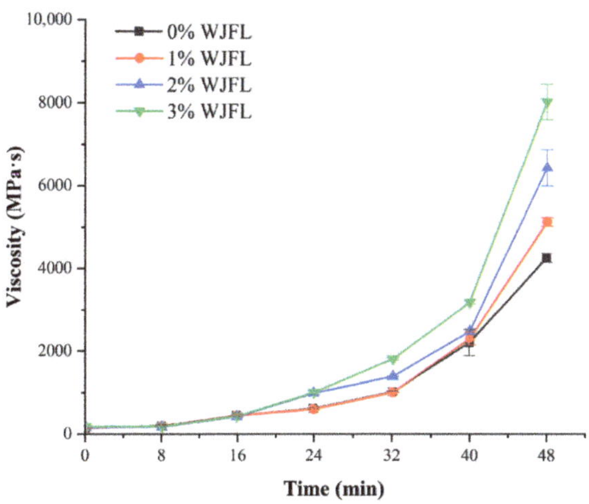

Figure 3. Viscosity–time curves of desulfurized crumb rubber modified epoxy asphalt with different WJFL contents.

WJFL chemically modifies the rubber–asphalt binder, enabling it to serve as a curing component rather than merely as filler in the final epoxy asphalt network. This enhances

 https://doi.org/10.3390/coatings16010102

the compatibility between asphalt and epoxy resin and increases the degree of cross-linking, thereby accelerating curing and causing a faster rise in viscosity.

3.2.2. Tensile Strength and Elongation at Break

Figure 4 shows the tensile properties of epoxy asphalt modified with different WJFL dosages. The tensile strength of the binder without WJFL is approximately 0.99 MPa. Even a small amount of WJFL noticeably improves strength: the tensile strength increases to 1.31 MPa and up to 1.86 MPa, demonstrating a significant enhancement.

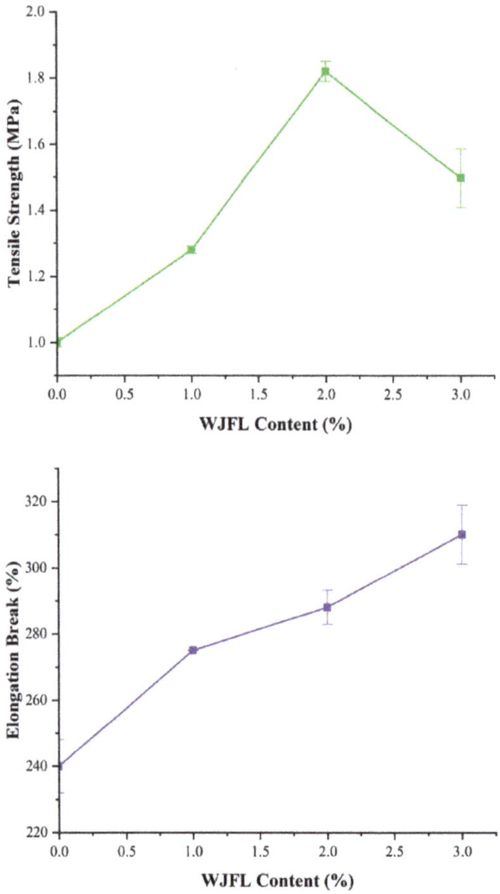

Figure 4. Tensile test results of desulfurized crumb rubber modified epoxy asphalt with different WJFL contents.

This improvement results from the increased reactivity of the modified asphalt binder. The WJFL-modified rubber–asphalt binder participates more effectively in the curing reaction with epoxy resin, forming a stronger cross-linked structure. However, the tensile strength does not increase monotonically with WJFL dosage; instead, it reaches a maximum at 2%, then decreases. This behavior is attributed to incomplete conversion of WJFL during modification. At higher dosages, WJFL may not fully participate in the asphalt modification reactions, and part of the additive can remain in a free or physically dispersed state, which hinders the formation of an effective crosslinked network.

https://doi.org/10.3390/coatings16010102

The elongation at break also increases with higher WJFL content, rising from 241% to 305%. This indicates that WJFL not only enhances strength but also promotes network formation, improving extensibility and overall mechanical performance.

3.3. Effect of Asphalt Content on Modified Epoxy Asphalt

To address the high production and construction costs of conventional epoxy asphalt, this study explored the feasibility of adjusting the asphalt dosage in the desulfurized crumb rubber–modified epoxy asphalt system. The goal is to improve economic efficiency while maintaining acceptable performance. A binder containing 20% desulfurized crumb rubber and 2% WJFL (by mass of rubber–asphalt binder) was prepared. The rubber–asphalt binder was incorporated at 150%, 300%, 450%, 600%, and 750% of the epoxy resin mass. The viscosity and mechanical performance of the resulting epoxy asphalt were evaluated.

3.3.1. Viscosity Analysis

Figure 5 illustrates the viscosity–time curves for epoxy asphalt prepared with different asphalt contents. All binders exhibited an initial viscosity of approximately 300 mPa·s, followed by a gradual increase showing a similar growth trend across formulations.

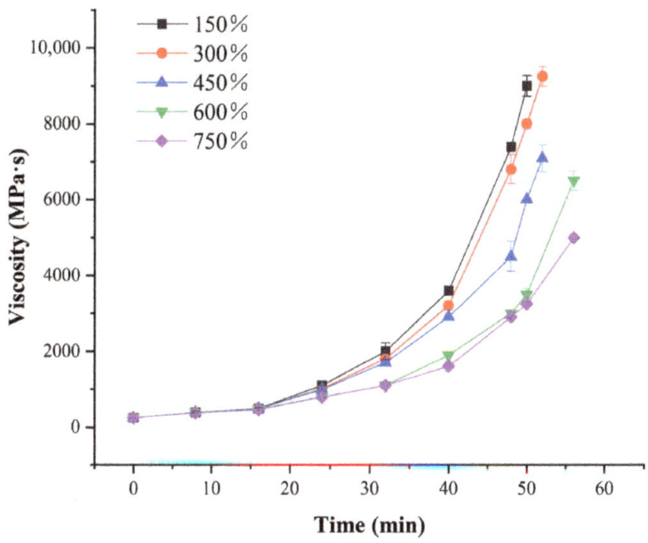

Figure 5. Viscosity–time curves of desulfurized crumb rubber modified epoxy asphalt with different asphalt contents.

A clear relationship was observed between asphalt dosage and viscosity development. When asphalt was added at 150%, 300%, and 450% of the epoxy resin mass, the modified epoxy asphalt reached 1000 mPa·s in roughly 22 min. However, for the 600% and 750% dosages, the time required to reach 1000 mPa·s increased to approximately 32 min, with subsequent viscosity growth occurring much more slowly.

In laboratory preparation, samples containing higher asphalt dosages exhibited significant curing difficulties, including incomplete curing and demolding failure. This is because the fixed amount of curing agent could not fully react with excessive rubber–asphalt binder, resulting in reduced curing efficiency and slower viscosity development.

 https://doi.org/10.3390/coatings16010102

3.3.2. Tensile Strength and Elongation at Break

Figure 6 presents the tensile strength and elongation at break of epoxy asphalt with different asphalt dosages. Unlike the viscosity trend, the tensile strength increased at first and then decreased as the asphalt dosage increased, reaching a peak of 1.86 MPa at the 300% dosage. Beyond this point, the tensile strength dropped sharply, with the 600% and 750% formulations exhibiting values below 0.55 MPa.

This behavior indicates that an optimal asphalt dosage exists for the desulfurized crumb rubber–modified epoxy asphalt. When the asphalt dosage is too low, part of the epoxy resin fails to participate in the curing reaction, resulting in insufficient load-bearing network formation. Conversely, when the asphalt dosage is too high, the fixed curing agent cannot fully cure the rubber–asphalt component. As a result, the final composite binder contains two distinct phases: uncured rubber–asphalt binder and cured epoxy asphalt, leading to severe strength reduction.

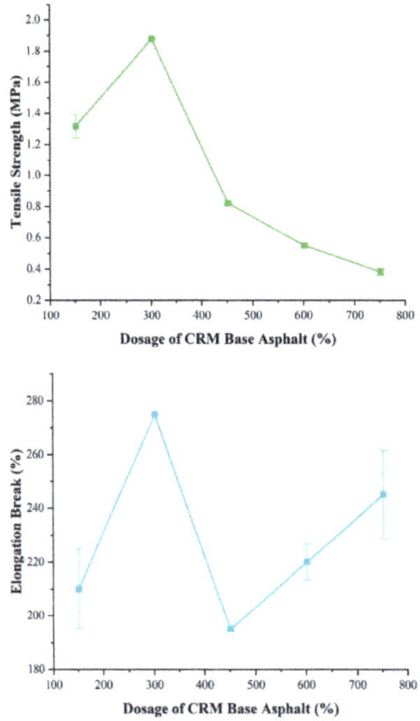

Figure 6. Tensile behavior of desulfurized crumb rubber–modified epoxy asphalt with different asphalt contents.

The elongation at break exhibited a similar non-linear trend, increasing initially and then decreasing. The maximum elongation (272%) was observed for the 300% asphalt dosage. Overall, based on viscosity behavior, tensile strength, and elongation performance, the optimal asphalt dosage is determined to be 300% of the epoxy resin mass.

3.4. Low-Temperature Performance Analysis

The relaxation performance of both unmodified epoxy asphalt and modified crumb rubber epoxy asphalt was tested at four different temperatures using the above-mentioned relaxation tests. The relaxation force-time comparison curves for both are shown in Figures 7 and 8.

https://doi.org/10.3390/coatings16010102

Figures 7 and 8 present the relaxation curves of the desulfurized crumb rubber–modified epoxy asphalt at three different temperatures. Using Equation (1), the relaxation stiffness values at 60 s, 120 s, and 180 s were calculated. Relaxation stiffness (S) reflects the resistance of a binder to imposed deformation under constant displacement. A higher stiffness indicates a more brittle material, which is more susceptible to cracking under low-temperature loading.

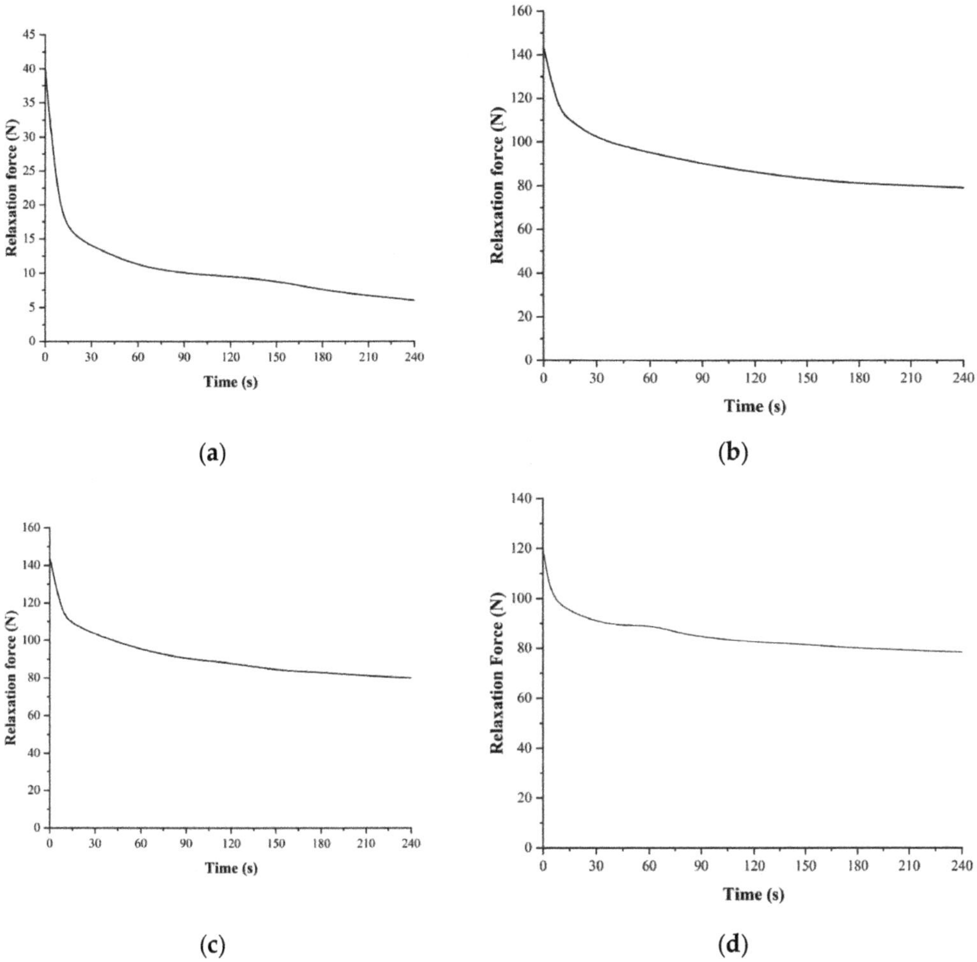

Figure 7. Relaxation curves of epoxy asphalt without crumb rubber at four temperatures. (**a**) Relaxation curve of neat epoxy asphalt at −10 °C. (**b**) Relaxation curve of neat epoxy asphalt at −15 °C. (**c**) Relaxation curve of neat epoxy asphalt at −20 °C. (**d**) Relaxation curve of neat epoxy asphalt at −25 °C.

As shown in Figure 9, the relaxation stiffness of the modified epoxy asphalt is significantly lower than that of the unmodified epoxy asphalt at all test temperatures. This demonstrates that the internal stress generated within the modified binder during low-temperature deformation is substantially reduced, resulting in superior low-temperature performance.

Based on Equation (2), the relaxation rates k at 30 s, 60 s, and 120 s were calculated. The relaxation rate represents the rate of stiffness variation with time during the displacement-

https://doi.org/10.3390/coatings16010102

controlled loading process. By combining the relaxation curves shown in Figures 7 and 8 with the analysis presented in Figure 10, it can be observed that, when larger time control points are selected, the relaxation rates of the crumb rubber–modified epoxy asphalt are consistently lower than those of the epoxy asphalt without crumb rubber. This behavior indicates that the modified epoxy asphalt undergoes most of its deformation relaxation within a relatively short period and subsequently enters a stable deformation stage, leading to a reduced rate of stiffness change over time. Consequently, the relaxation rates obtained at 60 s and 120 s are not suitable indicators for characterizing stiffness evolution.

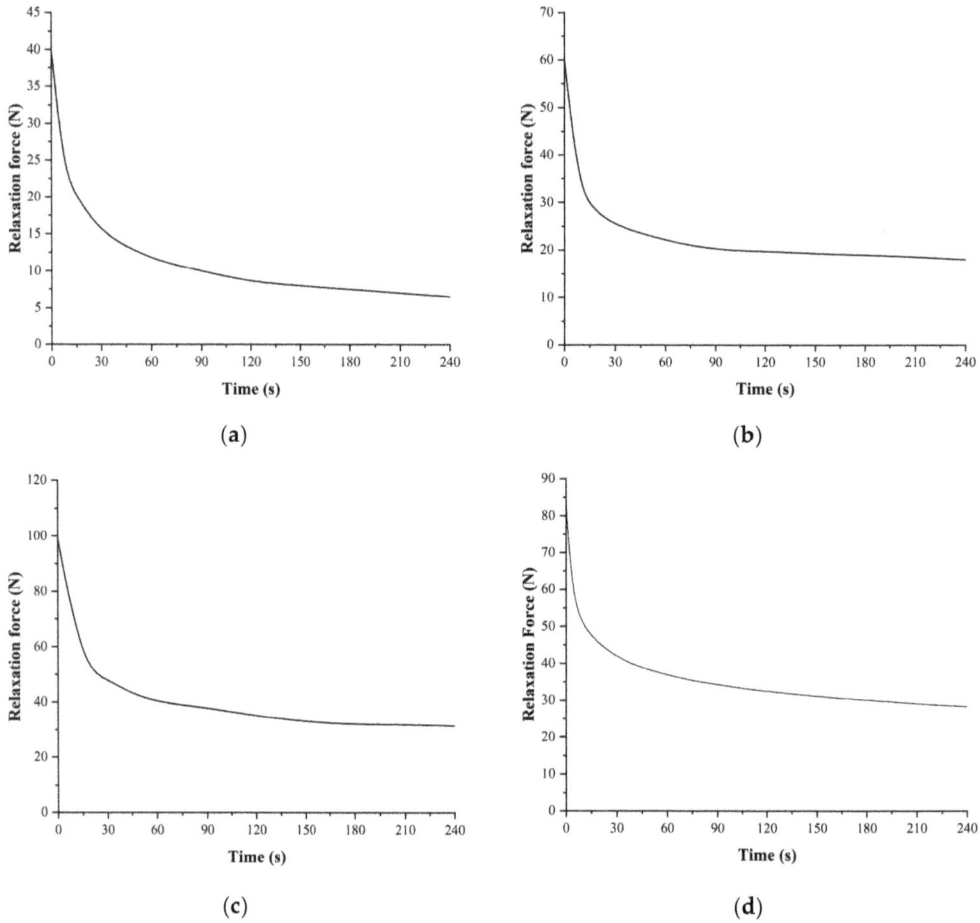

Figure 8. Relaxation curves of desulfurized crumb rubber–modified epoxy asphalt at four temperatures. (**a**) Relaxation curve of desulfurized crumb rubber–modified epoxy asphalt at −10 °C. (**b**) Relaxation curve of desulfurized crumb rubber–modified epoxy asphalt at −15 °C. (**c**) Relaxation curve of desulfurized crumb rubber–modified epoxy asphalt at −20 °C. (**d**) Relaxation curve of desulfurized crumb rubber–modified epoxy asphalt at −25 °C.

Therefore, 30 s was selected as the control point for evaluating the relaxation rate in this study. The results show that, at extremely low temperatures, the relaxation rate k of the crumb rubber–modified epoxy asphalt is relatively high. This suggests that, when temperature reduction induces pavement contraction, the asphalt mixture responds in

 https://doi.org/10.3390/coatings16010102

a manner similar to a material with reduced stiffness, resulting in lower tensile stress development and, consequently, a reduced susceptibility to low-temperature cracking.

Furthermore, the experimental results indicate that the incorporation of crumb rubber significantly decreases the low-temperature stiffness modulus of the modified epoxy asphalt, accompanied by a corresponding reduction in the k value. These findings demonstrate that the addition of crumb rubber effectively enhances the low-temperature creep and deformation performance of crumb rubber–modified epoxy asphalt.

The improvement can be explained by several mechanisms. During the swelling process, the desulfurized crumb rubber particles absorb light components (such as aromatics and saturates) from the base asphalt, leading to an increased proportion of colloidal structures and reduced brittleness at low temperatures. In addition, the fine rubber particles promote the formation of micro-scale "crazing zones" under tensile stress at low temperature. These zones help dissipate external energy and delay the initiation and propagation of cracks.

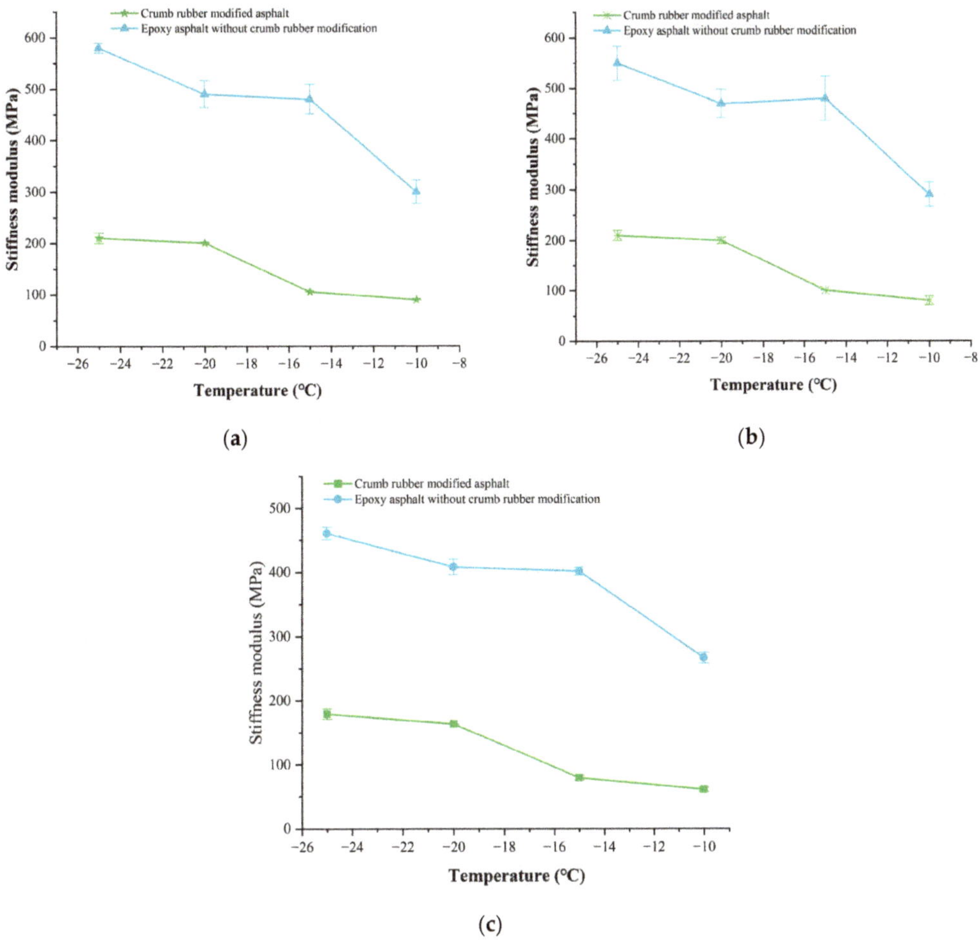

Figure 9. Comparison of stiffness modulus between the two materials. (**a**) Comparison of the stiffness modulus of the two materials at 60 s. (**b**) Comparison of the stiffness modulus of the two materials at 60 s. (**c**) Comparison of the stiffness modulus of the two materials at 60 s.

 https://doi.org/10.3390/coatings16010102

Furthermore, the rubber hydrocarbons present in the desulfurized crumb rubber contribute additional elasticity and flexibility, significantly enhancing the binder's crack-resistance capability at low temperatures. Overall, the incorporation of desulfurized crumb rubber markedly improves the inherent low-temperature limitations of conventional epoxy asphalt.

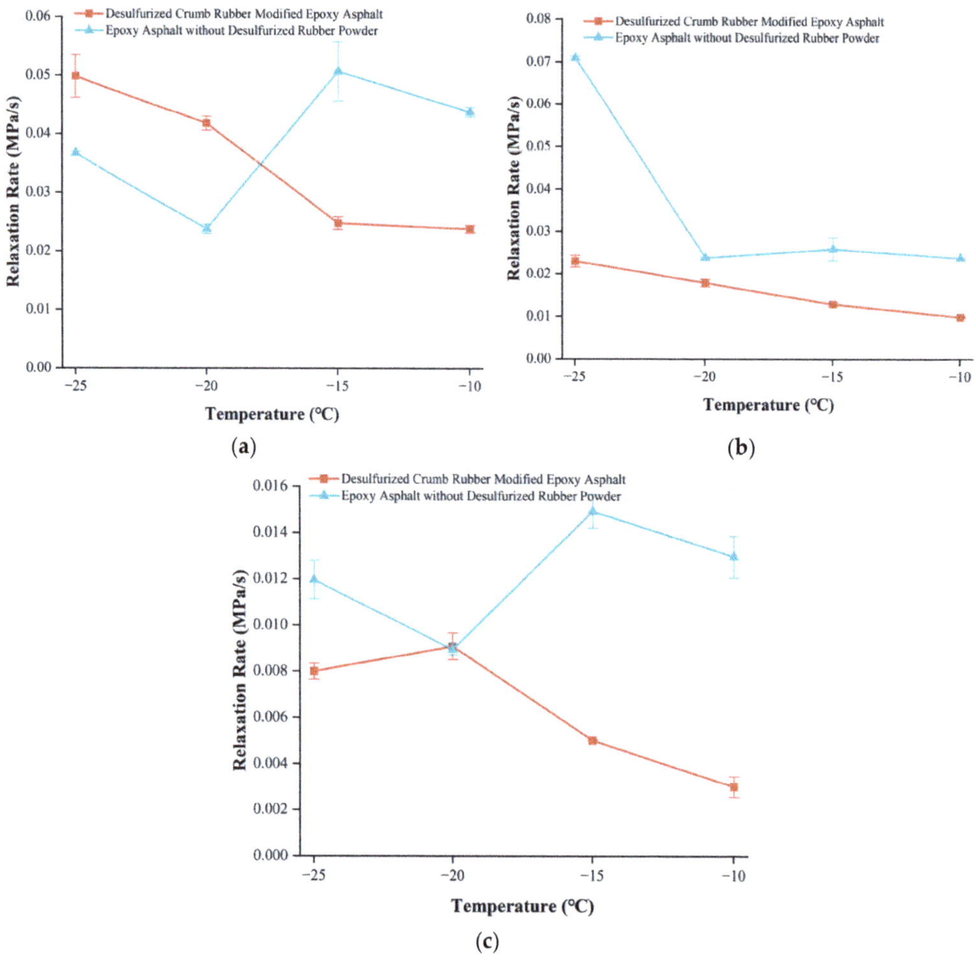

Figure 10. Comparison of the relaxation rates of the two materials. (**a**) Relaxation rate of the two materials at 30 s. (**b**) Relaxation rate of the two materials at 60 s. (**c**) Relaxation rate of the two materials at 120 s.

4. Conclusions

This study investigated the influence of desulfurized crumb rubber content, additive dosage, and asphalt content on the performance of modified epoxy asphalt. The conclusions are summarized as follows:

(1) Without additional modification, incorporating desulfurized crumb rubber reduces the tensile strength of epoxy asphalt but significantly improves its ductility. An optimal dosage of approximately 20% was identified, balancing strength and extensibility.

https://doi.org/10.3390/coatings16010102

(2) The additive WJFL effectively accelerates viscosity growth and significantly enhances the tensile strength of the modified epoxy asphalt. The optimum dosage was determined to be 2% of the rubber–asphalt binder, at which the degree of cross-linking and mechanical performance are maximized.

(3) Adjusting the asphalt dosage showed that excessive rubber–asphalt content leads to incomplete curing and a sharp reduction in tensile strength, along with brittle failure at low temperatures. The optimal dosage was determined to be 300% of the epoxy resin mass, providing the best balance between performance and economic feasibility.

(4) Low-temperature relaxation tests revealed that desulfurized crumb rubber considerably enhances the low-temperature crack-resistance of epoxy asphalt. Mechanistic analysis based on viscoelastic behavior confirmed that rubber modification reduces stress concentration and improves deformation recovery under low-temperature loading. This effectively addresses the inherent brittleness of conventional epoxy asphalt in cold environments.

Author Contributions: Y.Z.: Conceptualization, methodology, visualization, writing—original draft preparation; P.Y.: writing—review and editing. All authors have read and agreed to the published version of the manuscript.

Funding: This research received no external funding.

Institutional Review Board Statement: Not applicable.

Informed Consent Statement: Not applicable.

Data Availability Statement: The original contributions presented in this study are included in the article. Further inquiries can be directed to the corresponding author.

Conflicts of Interest: The authors declare that they have no known competing financial interests or personal relationships that could have appeared to influence the work reported in this paper.

References

1. Huang, W. Integrated Design Procedure for Epoxy Asphalt Concrete-Based Wearing Surface on Long-Span Orthotropic Steel Deck Bridges. *J. Mater. Civ. Eng.* **2016**, *28*, 04015189.
2. Xue, Y.; Qian, Z.; Yao, Z. Development and Performance Evaluation of Epoxy Asphalt Concrete Modified with Mineral Fiber. *Constr. Build. Mater.* **2016**, *102*, 378–383. [CrossRef]
3. Qian, Z.; Liu, Y.; Liu, C.; Zheng, D. Design and Skid Resistance Evaluation of Skeleton-Dense Epoxy Asphalt Mixture for Steel Bridge Deck Pavement. *Constr. Build. Mater.* **2016**, *114*, 851–863. [CrossRef]
4. Min, W.; Lu, P.; Liu, S.; Wang, H. A Review of Crack Sealing Technologies for Asphalt Pavement: Materials, Failure Mechanisms, and Detection Methods. *Coatings* **2025**, *15*, 836. [CrossRef]
5. Polacco, G.; Filippi, S.; Merusi, F.; Stastna, J. A Review of the Fundamentals of Polymer-Modified Asphalts: Asphalt/Polymer Interactions and Principles of Compatibility. *Adv. Colloid Interface Sci.* **2015**, *224*, 72–112.
6. Bilema, M.; Karim, M.R.; Mashaan, N.S. A Review of Rubberised Asphalt for Flexible Pavement Applications. *Sustainability* **2023**, *15*, 14481. [CrossRef]
7. Li, H.; Dong, B.; Zhao, D.; Guo, P.; Zhang, J. Physical, Rheological and Stability Properties of Desulfurized Rubber Asphalt and Crumb Rubber Asphalt. *Arab. J. Sci. Eng.* **2019**, *44*, 5043–5056. [CrossRef]
8. Xu, X.; Li, Y.; Huang, W.; Chen, D.; Zhang, C.; Shi, W. Fatigue Design of Steel Bridge Deck Asphalt Pavement Based on Nonlinear Damage Accumulation Theory. *Appl. Sci.* **2021**, *11*, 5668. [CrossRef]
9. Yang, J.; Yi, X.; Chen, H.; Wong, Y.D.; Fan, W.; Huang, W. Homogeneity Enhancement of Mixtures Containing Epoxy Polymer and 100% Reclaimed Asphalt Pavement. *Polymers* **2023**, *15*, 4261. [CrossRef]
10. Jamshidi, A.; White, G.; Kurumisawa, K. Functional and Field Performance of Epoxy Asphalt Technology—State-of-the-Art. *Road Mater. Pavement Des.* **2023**, *24*, 881–918.

11. Han, Y.; Zhang, Z.; Tian, J.; Ni, F.; Gu, X. Long-Term In Situ Performance Evaluation of Epoxy Asphalt Concrete for Long-Span Steel Bridge Deck Pavement. *Coatings* **2023**, *13*, 545. [CrossRef]
12. Min, Z.; Wang, Q.; Zhang, K.; Shen, L.; Lin, G.; Huang, W. Investigation on the Properties of Epoxy Asphalt Mixture Containing Crumb Rubber for Bridge Expansion Joint. *Constr. Build. Mater.* **2022**, *331*, 127344. [CrossRef]
13. Duan, K.; Wang, C.; Liu, J.; Song, L.; Chen, Q.; Chen, L. Research progress and performance evaluation of crumb-rubber-modified asphalts and their mixtures. *Constr. Build. Mater.* **2022**, *361*, 129687. [CrossRef]
14. Liu, B.; Zhang, K.; Fan, X.; Tu, C. Performance Evaluation of Desulfurized Rubber Powder and Styrene–Butadiene–Styrene Composite-Modified Asphalt. *Coatings* **2025**, *15*, 607. [CrossRef]
15. Zhao, Z.; Wu, S.; Xie, J.; Yang, C.; Yang, X.; Wang, F.; Liu, Q. Utilization of High Contents Desulfurized Crumb Rubber in Developing an Asphalt Rubber Pellets Modified Asphalt. *Constr. Build. Mater.* **2023**, *402*, 133043. [CrossRef]
16. Zhao, Y.; Chen, M.; Wu, S.; Jiang, Q.; Xu, H.; Zhao, Z.; Lv, Y. Effects of Waterborne Polyurethane on Storage Stability, Rheological Properties and VOCs Emission of Crumb Rubber Modified Asphalt. *J. Clean. Prod.* **2022**, *340*, 130682. [CrossRef]
17. Min, Z.; Li, F.; Zhang, X.; Wang, L.; Zhai, R. The Preparation Process, Service Performances and Interaction Mechanisms of Crumb Rubber Modified Asphalt (CRMA) by Wet Process: A Comprehensive Review. *Constr. Build. Mater.* **2022**, *354*, 129168. [CrossRef]
18. Gong, Y.; Wang, X.; Jia, D. Waste Tire Crushing Technology and Its Application Progress. *Rubber Ind.* **2021**, *68*, 66–72.
19. Xia, C.; Lv, S.; Cabrera, M.B.; Wang, X.; Zhang, C.; You, L. Unified Characterizing Fatigue Performance of Rubberized Asphalt Mixtures Subjected to Different Loading Modes. *J. Clean. Prod.* **2021**, *279*, 123740. [CrossRef]
20. Ban, I.; Barišić, I.; Cuculić, M.; Zvonarić, M. Performance Evaluation of Waste Rubber-Modified Asphalt Mixtures: A Comparative Study of Asphalt Concrete and Stone Mastic Asphalt Gradings. *Infrastructures* **2025**, *10*, 107. [CrossRef]
21. Gong, J.; Jing, F.; Zhao, R.; Li, C.; Cai, J.; Wang, Q.; Xie, H. Waste Cooking Oil-Modified Epoxy Asphalt Rubber Binders with Improved Compatibility and Extended Allowable Construction Time. *Molecules* **2022**, *27*, 7061. [CrossRef]
22. Hasan, S.S.; Abd Al-Ameer, R.H.; Hassani, H.A. Laboratory Assessment of Epoxy Asphalt Mixture Incorporating Tire Rubber Waste. *Wasit J. Eng. Sci.* **2021**, *9*, 1–10. [CrossRef]
23. Xu, G.; Kong, P.; Yu, Y.; Yang, J.; Zhu, M.; Chen, X. Rheological Properties of Rubber Modified Asphalt as a Function of Waste Tire Rubber Reclaiming Degree. *J. Clean. Prod.* **2022**, *332*, 130113. [CrossRef]
24. Zhang, H.; Zhang, Y.; Chen, J.; Liu, W.; Wang, W. Effect of Desulfurization Process Variables on the Properties of Crumb Rubber Modified Asphalt. *Polymers* **2022**, *14*, 1365. [CrossRef]
25. Xu, G.; Yu, Y.; Yang, J.; Wang, T.; Kong, P.; Chen, X. Rheological and Aging Properties of Composite Modified Bitumen by Styrene–Butadiene–Styrene and Desulfurized Crumb Rubber. *Polymers* **2021**, *13*, 3037. [CrossRef] [PubMed]
26. Wang, S.; Huang, W.; Liu, X.; Lin, P. Influence of High Content Crumb Rubber and Different Preparation Methods on Properties of Asphalt Under Different Aging Conditions: Chemical Properties, Rheological Properties and Fatigue Performance. *Constr. Build. Mater.* **2022**, *327*, 126937. [CrossRef]
27. Cong, P.; Liu, C.; Han, Z.; Zhao, Y. A Comprehensive Review on Polyurethane Modified Asphalt: Mechanism, Characterization and Prospect. *J. Road Eng.* **2023**, *3*, 16–34. [CrossRef]
28. Zhang, S.; Yang, Y.; Guo, R.; Yan, Y.; Huan, H.; Wan, B. Study on the Low-Temperature Pre-Desulfurization of Crumb Rubber-Modified Asphalt. *Polymers* **2023**, *15*, 2273. [CrossRef] [PubMed]
29. Xie, S.; Cheng, Z.; Zhou, Y.; Cao, Y.; Wang, T.; Zhang, Z.; Dai, Y.; Zhang, W. Performance of Asphalt Mixtures Modified with Desulfurized Rubber and Rock Asphalt Composites. *Buildings* **2024**, *14*, 3026. [CrossRef]
30. Abualia, A.; Akentuna, M.; Mohammad, L.N.; Cooper, S.B., III; Cooper, S.B., Jr. Improving Asphalt Binder Durability Using Sustainable Materials: A Rheological and Chemical Analysis of Polymer-, Rubber-, and Epoxy-Modified Asphalt Binders. *Sustainability* **2024**, *16*, 5379. [CrossRef]
31. Wang, Z.; Li, H.; Jia, M.; Du, Q. Emission Risk and Inhibition Technology of Asphalt Fume from Crumb Rubber Modified Asphalt. *Sustainability* **2024**, *16*, 8840. [CrossRef]
32. Lee, S.-Y.; Le, T.H.M. Laboratory and Full-Scale Testbed Study in the Feasibility of Styrene-Butadiene-Styrene Asphalt Pavement Having Epoxy Resin and Crumb Rubber Powder. *Buildings* **2023**, *13*, 652. [CrossRef]
33. Zhao, P.; Liang, H.; Wu, W.; Yang, Y.; Liu, Y.; Li, C.; Meng, W.; Zhang, R.; Song, X.; Wang, C.; et al. Rheological properties of high-asphalt-content emulsified asphalt. *Constr. Build. Mater.* **2024**, *419*, 135511. [CrossRef]
34. Zhao, P.; Zhou, X.; Yang, Y.; Wang, C.; Wang, C.; Liang, H.; Li, C.; Ren, H. Molecular dynamics simulation study of the effect of molecular structure of warm mix additive on lubricating properties. *Constr. Build. Mater.* **2023**, *407*, 133529. [CrossRef]
35. Zhao, P.; Song, X.; Dong, M.; Sun, H.; Wu, W.; Zhang, R.; Sun, M.; Zhao, X. Preparation and characterization of CQDs/SBS composites and its application performance as asphalt modifier. *Constr. Build. Mater.* **2022**, *320*, 126312. [CrossRef]

https://doi.org/10.3390/coatings16010102

36. Yang, P.; Min, Z.; Chen, W.; Yang, S.; Huang, W.; Tao, R.; Guo, K.; Chen, H. Crosslinking sites of sulfur and asphalt molecules: A DFT and macroscopic experimental study. *Constr. Build. Mater.* **2025**, *477*, 141364. [CrossRef]
37. Wu, W.; Wang, C.; Zhao, P.; Xiu, L.; Fan, L.; Bi, F.; Song, X.; Zhou, X. The Fingerprint Identification of Asphalt Aging Based on 1H-NMR and Chemometrics Analysis. *Materials* **2022**, *15*, 19. [CrossRef] [PubMed]

 https://doi.org/10.3390/coatings16010102

coatings

MDPI

Article

Study on the Volatile Organic Compound Emission Characteristics of Crumb Rubber-Modified Asphalt

Hu Feng [1], Haisheng Zhao [2,3,*], Dongfang Zhang [1], Peiyu Zhang [2], Yindong Ding [4], Yanping Liu [1], Chunhua Su [2], Qingjun Han [4] and Yiran Li [2]

1 Shandong Ludong High-Tech Materials Technology Co., Ltd., Dongying 250098, China; 18654669160@126.com (H.F.); 13954608572@163.com (D.Z.); 13385463377@163.com (Y.L.)
2 Key Laboratory of Highway Maintain Technology Ministry of Communication, Jinan 250102, China; zhangpeiyu@sdjtky.cn (P.Z.); suchunhua@sdjtky.cn (C.S.); liyiran@sdjtky.cn (Y.L.)
3 School of Highway, Chang'an University, Xi'an 710064, China
4 Shandong Ludong Transportation Construction Group Co., Ltd., Dongying 250098, China; 18954013811@163.com (Y.D.); 529931476@163.com (Q.H.)
* Correspondence: zhaohaisheng@sdjtky.cn; Tel.: +86-150-5319-0628

Abstract: Crumb rubber used in asphalt modification can generally improve the road performance of asphalt mixture pavement while offering substantial environmental and economic benefits. This study investigates the volatile organic compound emissions from crumb rubber-modified asphalt binders via gas chromatography–mass spectrometry, focusing on the effects of crumb rubber types (e.g., activated crumb rubber, non-activated crumb rubber), contents, and additives (warm-mix agents, deodorants, styrene–butadiene–styrene (SBS)). The analysis encompasses total volatile organic compound emissions, compositional variations, secondary organic aerosol and ozone formation potentials, and carcinogenic risks. Results indicate that non-activated crumb rubber increases volatile organic compound emissions initially, peaking at a 15% content (3.99 times higher than base asphalt), dominated by trichloroethylene. The surfactant-based warm-mix additive significantly reduces emissions by 73%, whereas deodorants exhibited limited efficacy. At equivalent contents, activated crumb rubber-modified asphalt emits more volatile organic compounds than non-activated crumb rubber-modified asphalt and leads to a higher ozone formation potential. Activated crumb rubber/SBS-modified asphalt blends reduce emissions by 69%–81% due to synergistic effects. In contrast, non-activated crumb rubber/SBS blends increase emissions, likely due to phase separation. All samples contain carcinogens, primarily trichloroethylene (20%–79%) and benzene (0.1%–9%). These findings underscore the critical importance of crumb rubber activation status and SBS addition in controlling volatile organic compound diffusion. The activated crumb rubber/SBS combination achieves a synergistic reduction exceeding the sum of individual effects ("1 + 1 > 2"). These findings provide valuable insights for designing eco-friendly asphalt.

Keywords: VOCs; crumb rubber modification; emission characteristic; SBS

1. Introduction

The increased demand for vehicles generates growing volumes of end-of-life tires [1]. The waste rubber tires are difficult to dispose of owing to their chemical inertness and excellent resistance to thermal, mechanical, and biological degradation [2], resulting in black pollution. Accumulated end-of-life tires pose serious risks, including environmental pollution, land occupation, auto-ignition hazards, and resource wastage [3]. Numerous studies demonstrate that end-of-life tires can be processed into crumb rubber (CR) powder and incorporated into asphalt binder at optimal quantities. This enhances overall performance through improved elasticity, recoverability, and reduced glass transition temperature [4–14]. Studies internationally confirm that CR modified asphalt binder (CRMA) outperforms conventional asphalt by (1) enhancing resistance to high-temperature rutting, low-temperature cracking, fatigue, and aging; (2) improving pavement skid resistance and noise reduction; and (3) reducing distresses, delaying maintenance procedures, and extending service life [5,15–28]. Utilizing CR in asphalt provides an environmentally sustainable recycling solution for end-of-life tires, mitigating black pollution [29]. Evolving CRMA technology addresses environmental issues from end-of-life tire accumulation through resource-efficient recycling [6,27,30–34]. These demonstrated outstanding environmental and socio-economic benefits promote the sustainable utilization of end-of-life rubber resources, offering comprehensive economic and environmental advantages [11,19,35–37].

However, CRMA technology faces significant challenges limiting its widespread utilization and potential applications [14,27,38–43]. During production, mixing, transportation, and paving, CRMA generates substantial odorous and hazardous gases—primarily volatile organic compounds (VOCs) and volatile sulfur compounds—resulting from partial swelling and degradation of CR particles at high temperature [44–54]. CRMA's higher viscosity [55] requires elevated construction temperature to achieve workability, significantly increasing odorous and hazardous gases [45,46,56,57]. These emitted toxic odor gases pose greater environmental and occupational health risks than base asphalt [58–61]. Studies report that the CRMA mixture produces offensive odors and carcinogens [43,46,62–65] during production and paving. Growing demand for cleaner asphalt paving and environmental concerns have intensified focus on controlling CMRA's toxic odor emissions [60,66,67]. Significant research focuses on environmental hazards from CR-modified asphalt [68–70].

Pretreatment, warm-mix additive, and deodorant are internationally recognized as the most effective solutions to inhibit the total released VOC emissions [22,46,53,56,71–77]. Specific pretreatment methods include microwave activation [78–81], surface active treatment and chemical modification [82–84] of CR, plasma treatment [85], microbial desulfurization [86], supercritical carbon dioxide (CO_2) treatment [87], pre-swelling, pre-degradation, pre-oxidation, graft activation, polymer coating, solution immersion, and radiation [88–92]. These methods enhance CR–asphalt compatibility by modifying physical properties (e.g., altering cross-link network structures) and chemical structures (e.g., activating surfaces of CR particles). CR pretreatment significantly reduces polycyclic aromatic hydrocarbons (PAHs) and sulfur-containing compounds, thereby minimizing odor during CRMA production [80,81]. Asphalt exhibits temperature-sensitive properties. Warm-mix additives improve fluidity and modify viscosity characteristics, reducing VOC generation through lower construction temperature [46,74,77,93–102]. Chemical reactants may also neutralize toxic gases emitted from CRMA [103,104]. Deodorants reduce fume emissions through (a) high internal specific surface area, (b) porous structure, and (c) strong intermolecular forces on pore walls [8,76,105–108]. These properties enable physical absorption of lightweight fractions and fumes via lattice energy and van der Waals forces [27,38,53,62,71,80,109–114]. Alternatively, deodorants chemically capture emitted fumes by reacting with small molecules and volatile components in CRMA [20]. Researchers have incorporated absorption materials—activated carbon, organic mon–montmorillonite, nano-calcium carbonate, and metal–organic frameworks (MOFs) materials, and zeolite—into CRMA to surpass volatilization of lightweight fractions (particularly hydrocarbons and benzene) [38,95,111–113,115–124].

Previous studies have revealed that various factors influence the emission behavior of CRMA. Refs. [125–129] identified that base asphalt binder composition critically determines CRMA emission behavior. Substantial research [125,129] demonstrated that temperature dramatically affects the composition and concentration of VOC emissions from CRMA. Furthermore, other factors significantly influence CRMA emission behavior and must be considered during fume assessment, including CR particle size [125], CR types and activated method [130,131], CR content, warm-mix additive, preparation process [125], deodorant type, and content [28,38,95,111–113,115–124]. However, previous studies typically focus on one or a few factors influencing CRMA emission behavior. Comparing results and conducting in-depth exploration across studies is challenging due to variations in base asphalt binders, test temperatures, CR sources, contents, or additives. Therefore, this study employs a consistent preparation process to fabricate CRMA with various compositions. It aims to investigate the influence of CR types, CR content, warm-mix additive type, deodorant content, and SBS polymer on CRMA emission release behavior, providing insights into the VOCs release mechanism. Using an indoor fume collection system, GC–MS was employed to identify and quantify compositions within the fume emissions. A hazardous evaluation of the CRMA asphalt fumes was also performed. The findings enhance understanding of the multifaceted impacts of CRMA asphalt fumes and propose optimal, targeted measures to mitigate VOC emissions.

2. Materials and Methods

2.1. Raw Materials

The crumb rubber particles were recycled and activated through the combined action of heat, chemical agents, and controlled mechanical shearing using a twin-screw extruder. The activated process breaks the S–S and C–S bonds to the greatest possible extent while preserving the C–C main chain, ensuring the high strength of activated CR materials. Two types of rubber particles were used to modify the base asphalt. The first type is non-activated crumb rubber (NCR) particles produced through mechanical grinding of end-of-life tires at ambient temperature. The second type is activated reclaimed rubber (ACR) particles produced using twin-screw extruder at high temperatures (200–220 °C). The characteristics and particle size distributions of ACR and NCR are presented in Tables 1 and 2, respectively.

Table 1. The characteristics of activated and non-activated CR particles.

NO.	Test Item	Test Result	
		ACR	NCR
1	Relative density	1.18	1.17
2	Ash content/%	5.40	6
3	Acetone extract content/%	13.40	11
4	Rubber hydrocarbon content/%	53	51
5	Carbon black content/%	30.60	30

Table 2. The screening results of the activated and non-activated CR particles.

Mesh	30	40	50	60	70	80	>80
ACR (%)	3.1	47.6	16.1	5.7	12.4	5.1	8.7
NCR (%)	10.3	42.2	15.9	5.4	11.5	4.3	8.5

The base asphalt binder used was 70#, supplied by Qingdao Luhuitong Industrial Co., Ltd., Qingdao, China. Two warm-mix additives were used: (1) liquid warm-mix additive

(W1) (recommended content of 0.5%) from Beijing Road Animal Husbandry Technology Co., Ltd., Beijing, China; and (2) liquid warm-mix additive (W2) (recommended content of 0.5%) from Meidewei Shiwek Company. The deodorant additive was developed in solid form by Qingdao University of Science and Technology. The CR particles, warm-mix additives, and deodorant are shown in Figure 1.

(a) (b) (c) (d) (e)

Figure 1. The additives used in this paper. (**a**) Activated CR particles; (**b**) non-activated CR particles; (**c**) liquid warm-mix additive; (**d**) solid warm-mix additive; (**e**) deodorizer.

2.2. Preparation of CR Modified Asphalt

This study investigated the influence of multiple factors, including CR type, CR content, warm-mix additive type, deodorant content, and SBS modifier. The experimental design (Table 3) comprised 18 sample groups, with BA and SBS-modified asphalt samples serving as controls. Additive content is expressed as a percentage relative to the mass of the asphalt binder.

Table 3. The descriptions of all tested groups.

NO.	Label	Description
1	BA	70# base asphalt
2	SBS	70# base asphalt + 4 wt.% SBS
3	SA10	70# base asphalt + 2 wt.% SBS + 10 wt.% ACR
4	SA15	70# base asphalt + 2 wt.% SBS + 15 wt.% ACR
5	SA20	70# base asphalt + 2 wt.% SBS + 20 wt.% ACR
6	SN15	70# base asphalt + 2 wt.% SBS + 15 wt.% NCR
7	SN10	70# base asphalt + 2 wt.% SBS + 10 wt.% NCR
8	N10	70# base asphalt + 10 wt.% NCR
9	N15	70# base asphalt + 15 wt.% NCR
10	N20	70# base asphalt + 20 wt.% NCR
11	N15W1	70# base asphalt + 15 wt.% NCR + 0.5 wt.% W1
12	N15W2	70# base asphalt + 15 wt.% NCR + 0.5 wt.% W2
13	N15D1	70# base asphalt + 15 wt.% NCR + 1 wt.% deodorizer
14	N15D2	70# base asphalt + 15 wt.% NCR + 2 wt.% deodorizer
15	N15D3	70# base asphalt + 15 wt.% NCR + 3 wt.% deodorizer
16	A10	70# base asphalt + 10 wt.% ACR
17	A15	70# base asphalt + 15 wt.% ACR
18	A20	70# base asphalt + 20 wt.% ACR

Samples were prepared according to the following procedure:

(a) Base asphalt heating

Base asphalt was heated in an oven until flowable, then transferred to a reaction vessel maintained at 175 °C with continuous stirring (500 rpm).

(b) SBS modifier addition

Control Group: At 175 °C, 4 wt% SBS modifier was added and mixed at 500 rpm for 60 min while maintaining temperature. The blend was then processed through a colloid mill five times, followed by additional mixing (500 rpm, 120 min).

Test Group: At 175 °C, 2% wt SBS modifier was added and stirred at 500 rpm for 60 min.

(c) Crumb rubber addition

Following SBS modification, the set temperature was increased to 190 °C. Pre-weighed CR was added incrementally over 60 min with continuous stirring (500 rpm), while maintaining the temperature at 190 $\pm$ 5 °C.

(d) Final processing and additive incorporation

After CR incorporation, blends were processed through a colloid mill for five cycles, then mixed at 500 rpm and 190 °C for 120 min to ensure complete dispersion and the desired microstructure. Other additives were incorporated during this final mixing stage.

2.3. Generation and Collection Process for VOC Emissions

A custom-designed gas generation and collection apparatus (Figure 2) was employed to simulate high-temperature VOC release behavior from asphalt binder under laboratory conditions while preventing fugitive emissions. The collecting apparatus is composed of (a) a temperature-controlled fume generation platform and (b) a VOC collection/absorption module using a ZR–3922 environmental air particulate matter sampler with a mercury suction intake, which captures both direct and indirect asphalt VOC emissions. A Teflon tube connects the fume generation platform to the collection module. During heating/stirring, fresh air was continuously pumped into the flask to (1) maintain constant internal pressure and (2) simulate the VOC–air reaction occurring during actual mixing/construction.

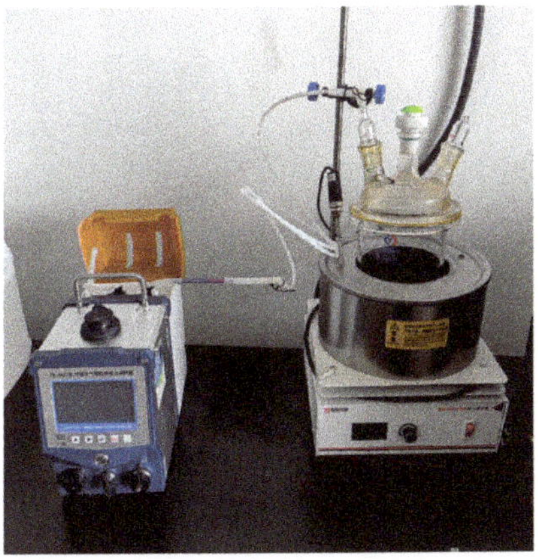

Figure 2. The VOC collection apparatus.

The VOC emission generation and collection procedure of all samples, including N15W1 and N15W2, comprised the following:

(a) Asphalt samples were melted to a fluid state. Next, 600 g±5 g was transferred into a glass flask sealed with a three-neck glass lid.

(b) The flask was placed in a temperature-controlled oil bath (180 °C) with a magnetic stirrer. The asphalt binder was stirred at 500 rpm to ensure uniform heating and facilitate VOC fume generation.

In this study, all samples, including those with and without the warm-mix additive, were tested under identical temperature conditions. This approach was taken to control for the temperature variable and to isolate the effect of the additive itself on viscosity and VOC reduction.

(c) The VOC collection system (0.2 L/min) was connected, using pre-aged Tenax TA tubes (6 mm× 150 mm). After temperature stabilization, absorption was processed for 30 min, collecting 6 L of VOC-laden air.

2.4. VOC Emission Composition Analysis

Collected VOC emissions were analyzed using thermal desorption gas chromatography–mass spectrometry (GC–MS; Trace GC Ultra DSQII, Thermo Fisher Scientific, Waltham, MA, USA), which identifies compound types and relative proportions. VOC samples underwent thermal desorption using Tenax TA tubes, with separation on an HP–5 MS ultra inert capillary column (30 m × 0.25 mm × 0.25 μm) and detection via quadrupole mass spectrometer (PerkinElmer, Waltham, MA, USA). Thermal desorption conditions were as follows: initial adsorbent tube temperature: room temperature; initial focusing trap temperature: room temperature; dry purge flow rate: 40 mL/min; dry purge time: 2 min; desorption temperature of adsorbent tube: 300 °C; desorption time of adsorbent sampling tube: 3 min; desorption flow rate: 40 mL/min; focusing trap temperature: −10 °C; desorption temperature of focusing trap: 300 °C; trap desorption time: 5 min; and transfer line temperature: 150 °C.

Compounds were identified by matching mass spectra against the NIST database (match threshold >85%), with quantification based on characteristic ion peak areas. Concentrations were quantified against liquid standards (2000 mg/L) using external calibration. Calibration employed 34 compounds per HJ 644-2013 standard, predominantly benzene derivatives and chlorinated organics.

3. Results

3.1. Total Concentration

VOCs, the primary asphalt fume emissions, are defined as organic compounds with >70 Pa vapor pressure at 25 °C and a boiling point < 260 °C. VOCs comprise complex mixtures classified by functional groups and molecular weight: oxygenated hydrocarbons (OXHs), halogenated alkenes (HKE), halogenated alkanes (HKA), aromatic hydrocarbons (ARHs), aromatic hydrocarbon derivatives (ARHD), and sulfides (SCs) [22,23,38,76,125,132–135]. Using 34 standard solutions, compounds in fume samples were categorized into four groups: HKE, HKA, ARHs, and ARHD. Analysis of all 18 samples detected only 21 compounds. All detected compounds fell into these four groups. Group concentrations per sample are shown in Figure 3 and Table 4. Total VOC concentrations (sum of all groups) appear above each bar.

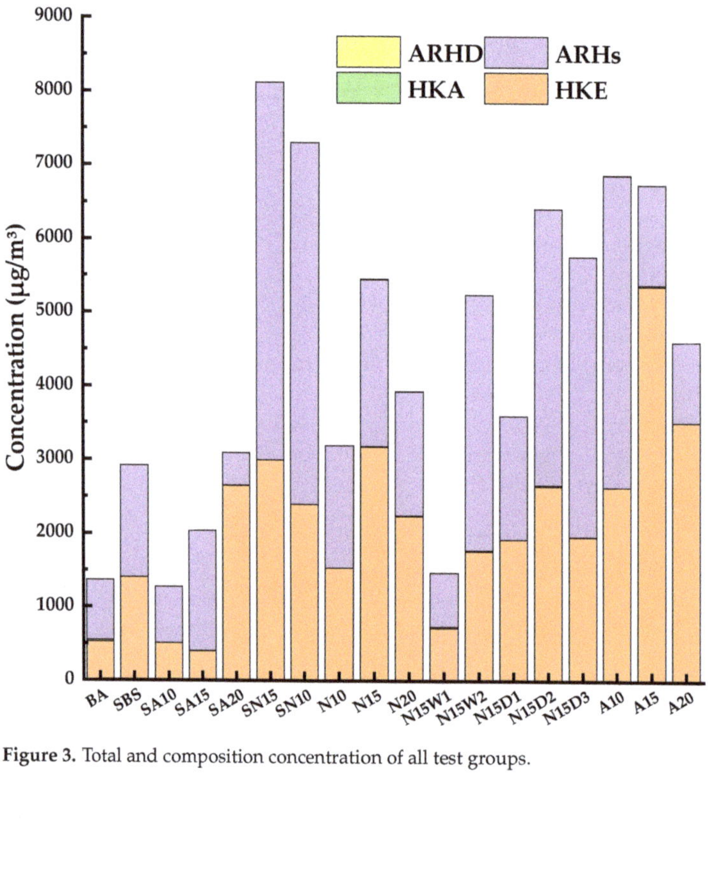

Figure 3. Total and composition concentration of all test groups.

Table 4. The total and composition concentration of all test groups (μg/m^3).

Group	Total Concentration (mg/m^3)	HKE (mg/m^3)	HKA (μg/m^3)	ARHs (mg/m^3)	ARHD (μg/m^3)
BA	1.4	0.5	20.2	0.8	0.0
SBS	2.9	1.4	0.0	1.5	0.0
SA10	1.3	0.5	3.0	0.8	0.0
SA15	2.0	0.4	2.5	1.6	0.0
SA20	3.1	2.7	7.2	0.4	0.0
SN15	8.1	3.0	1.1	5.1	0.0
SN10	7.3	2.4	1.9	4.9	0.0
N10	3.2	1.5	0.0	1.7	0.0
N15	5.5	3.2	7.2	2.3	7.2
N20	3.9	2.2	6.9	1.7	3.9
N15W1	1.5	0.7	16.4	0.7	0.0
N15W2	5.2	1.8	11.8	3.5	0.0
N15D1	3.6	1.9	0.0	1.7	0.0
N15D2	6.4	2.6	15.8	3.7	0.0
N15D3	5.8	2.0	2.9	3.8	0.0
A10	6.9	2.6	0.0	4.2	2.6
A15	6.7	5.4	13.9	1.4	0.0
A20	4.6	3.5	0.0	1.1	0.0

3.2. Composition Concentration Variation

Figure 4 and Table 5 show, for each sample group, the number of VOC compounds categorized by concentration proportion: >10%, 1%–10%, and <1%. It also displays the cumulative concentration proportion from compounds exceeding 10%.

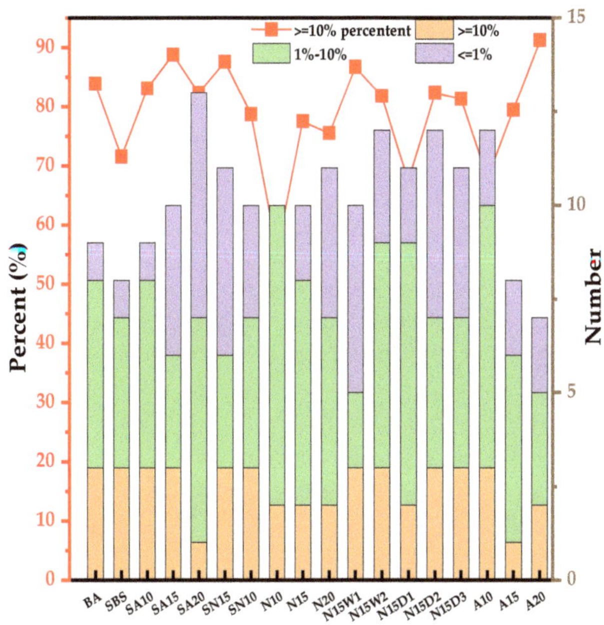

Figure 4. Total and proportion of compounds in VOC emissions of all test groups.

Table 5. Total and proportion of compounds in VOC emissions of all test groups.

Group	>10% Percent (%)	>10% Number	1%–10% Number	<1% Number
BA	83.9	3	5	1
SBS	71.6	3	4	1
SA10	83.1	3	5	1
SA15	88.8	3	3	4
SA20	82.3	1	6	7
SN15	87.6	3	3	5
SN10	78.7	3	4	4
N10	55.6	2	8	0
N15	77.5	2	6	2
N20	75.5	2	5	4
N15W1	86.7	3	2	5
N15W2	81.8	3	6	3
N15D1	67.5	2	7	2
N15D2	82.3	3	4	5
N15D3	81.3	3	4	5
A10	67.9	3	7	2
A15	79.4	1	5	2
A20	91.2	2	3	2

3.3. Potential of Secondary Organic Aerosols (SOA) and Ozone (O_3) Formation

VOC emissions harm humans and the environment through photochemical reactions that generate ozone (O_3) and secondary organic aerosols (SOA) [136]. Two indices—ozone (O_3) formation potential (OFP) and secondary organic aerosol formation potential (SOAFP)—identified key VOC species contributing to O_3/SOA formation and qualified environmental burden across all samples. OFP was calculated using the maximum incremental reactivity (MIR) coefficient (Equation (1)) [137–144].

$$OFP_i = VOCs_i \times MIR_i \quad (1)$$

where OFP_i is the ozone formation potential of certain compound i in VOCs ($\mu g/m^3$); VOC_{si} is the corresponding compound concentration ($\mu g/m^3$); and MIR_i is the maximum incremental reactivity coefficient of certain substance i in VOCs from William P.L. Carter [140].

The fractional aerosol coefficient (FAC) method estimates SOA formation using reported FAC values (the ratio of the aerosol yield from a specific compound to its initial concentration [138,139]) and the measured VOC concentrations [145–147]. The potential of SOA formation (or SOA concentration, $\mu g/m^3$) of a species i can be calculated by the following equation: [148].

$$SOAFP_i = VOCs_i \times FAC_i \quad (2)$$

where $SOAFP_i$ is the SOA formation potential of certain compound i in VOCs ($\mu g/m^3$); $VOCs_i$ is the compound concentration ($\mu g/m^3$), and FAC_i is the aerosol formation coefficient of certain compound i in VOCs from Derwent et al. [138,148–150].

To assess the environmental burden of compounds in VOC emissions, Table 6 summarizes the contribution ratios of VOC species to SOA and O_3 formation across all sample groups. "N" in Table 6 denotes a negligible contribution to SOA/O_3 formation. Consistent with U.S. Environmental Protection Agency (EPA) stationary sources AP-42 and California Air Resources Board (CARB) VOC exemptions, the HKE compounds exhibit negligible SOA formation potential. The ozone (O_3) and SOA formation potential of the VOC emissions are calculated using Equations (1) and (2), based on the contribution ratios for all tested compounds (Table 6) and their concentrations (Figure 3 and Table 4). All calculated results are presented in Figure 5.

Table 6. The contribution ratios of detected substances in VOC emissions.

Serial No.	Compound	Contribution Value		Serial No.	Compound	Contribution Value	
		SOA/%	O_3			SOA	O_3
1	1,1-dichloroethylene	N	1.7	11	tetrachloroethene	N	3.1×10^{-2}
2	chloropropene	N	12.22	12	chlorobenzene	N	0.32
3	methylene chloride	N	1.87×10^{-2}	13	ethylbenzene	5.4	3.04
4	cis-1,2-dichloroethene	N	1.79	14	p,m-xylene	3.15	9.75
5	chloroform	N	0.03	15	styrene	5.7	1.73
6	1,2-dichloroethane	N	0.21	16	o-toluene	5	7.64
7	benzene	2.6	0.72	17	4-ethyltoluene	2.5	4.44
8	carbon tetrachloride	N	0	18	1,3,5-triethyltoluene	2.9	11.76
9	toluene	5.4	4	19	1,2,4-triethyltoluene	2	8.87
10	1,1,2-trichloroethane	N	8.6×10^{-2}	20	1,2,4-trichlorobenzene	N	N

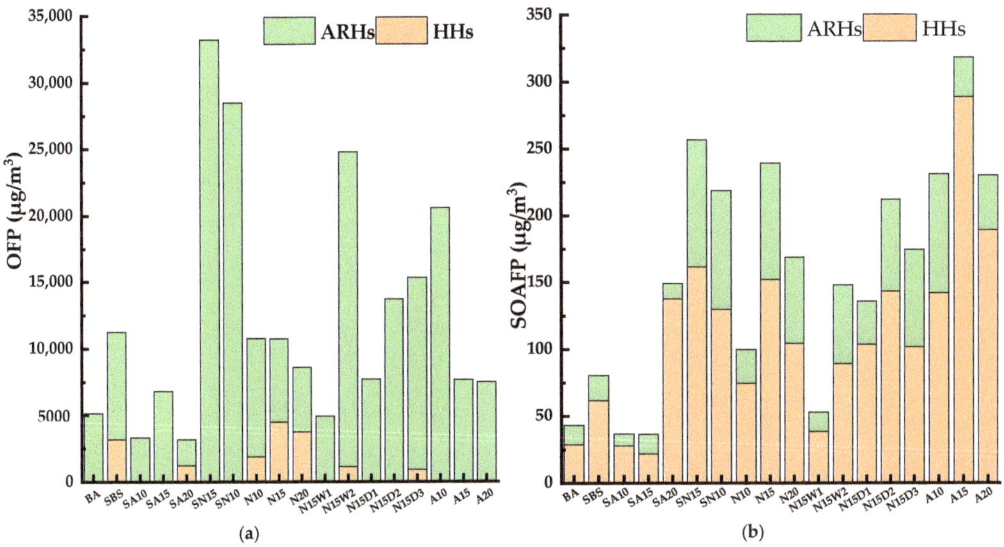

Figure 5. Formation potential of ozone (O_3) (**a**) and SOA (**b**) by VOC emissions of all test groups.

3.4. Human Carcinogen Risk Analysis

According to the International Agency for Research on Cancer (IARC) carcinogen list of the World Health Organization, substances classified as group 1 carcinogens are confirmed anthropogenically determined human carcinogens with unequivocal evidence of carcinogenicity from extensive scientific investigation. Group 2A carcinogens are probable human carcinogens, with limited human evidence but sufficient evidence in experimental animals. Group 2B carcinogens are possible human carcinogens with inadequate human evidence and insufficient animal data. Carcinogen risk indices for detected compounds in VOCs with potential human carcinogenic risk are presented in Table 7. Figure 6 presents human carcinogenic and non-carcinogenic risk analysis results of VOC emissions across all sample groups.

Table 7. Human carcinogen risk analysis results.

No.	Name	Carcinogen Risk
1	Benzene	IARC 1
2	1,2-dichloroethane	IARC 1
3	Trichloroethylene	IARC 1
4	Tetrachloroethylene	IARC 2A
5	Methylene chloride	IARC 2A
6	Styrene	IARC 2A
7	Carbon tetrachloride	IARC 2B
8	1,1–Dichloroethylene	IARC 2B
9	Chloroform	IARC 2B
10	Ethylbenzene	IARC 2B

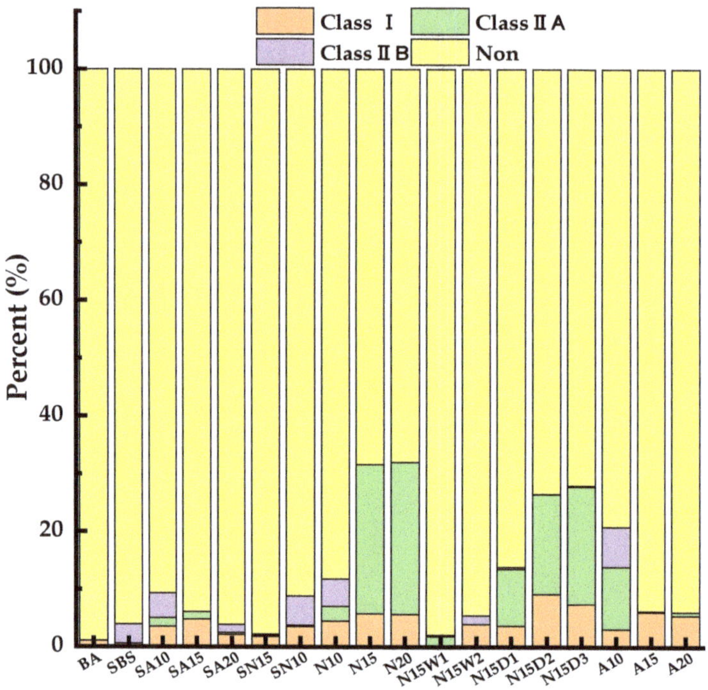

Figure 6. Human carcinogen risk analysis of VOC emissions of all test groups.

4. Discussion

4.1. Total Concentration Analysis

Figure 3 shows that not all four compound groups appear in every sample; some contain only two or three groups. ARHD and HKA concentrations fell below the 20 μg/m^3 detection threshold and thus are not displayed. Consequently, HKA and ARHD did not constitute primary sample components. Since HKA and HKE are halogenated hydrocarbons (HHs) and ARHD derives from ARHs, compounds were regrouped as HHs and ARHs. VOC results are therefore categorized as HHs and ARHs, with their proportions showing a consistent trend across samples. Asphalt fume emissions constitute a complex system influenced by multiple factors.

Total VOC concentrations in Figure 3 represent the sum of four compound groups. SBS-modified asphalt emitted 2.9 mg/m^3 total VOC emissions, 2.1 times higher than the

BA sample (BA, 1.4 mg/m^3). This result could be explained by the fact that the resin and asphaltene in SBS-modified asphalt binder began to volatilize at a higher temperature (180 °C) [151], while its spatial network structure only constrains the evaporation of certain compounds ≈ 180 g/mol. Previous work [23] confirms higher VOC emissions from SBS-modified versus base asphalt mixture. However, other studies [102,134,152,153] reported that SBS could mitigate the VOC emissions and exhibit a fume suppression effect; the analyzed VOC profiles in these prior studies differed substantially from those in this paper.

Non-activated CR-modified sample groups (N10, N15, and N20) showed total VOC emissions 2.3 times, 4.0 times, and 2.9 times higher than BA, respectively. This aligns with the existing literature [28,125,129,154–156]. Prolonged high-temperature exposure causes non-activated CR particles to swell and degrade, resulting in significant VOCs. At 10%–15% CR content, increased non-activated CR particle loading required greater absorption of base asphalt's lightweight fractions for swelling. Consequently, the molecular chain (main chain/crosslink bond) approached maximum tensile strength, increasing susceptibility to disruption [157]. Elevated temperature ruptured certain main chain and crosslink bonds within non-activated CR particles, releasing additional VOCs [44]. Above 15% non-activated CR content, excessive non-activated CR particles continued to absorb more lightweight fractions, inhibiting VOCs evaporation [80,130,158]. Depleted lightweight fractions prevent full non-activated CR particle swelling and degradation, also suspending volatilization of small/medium asphalt components. Similar non-activated CR-dependent emission behavior at higher temperatures is documented [154,159].

N15W1 and N15W2 samples contained 0.5% liquid (W1) and solid (W2) warm-mix additives, respectively, added to the N15 group. At identical temperatures, the N15W1 sample emitted significantly fewer VOCs than N15 and N15W2, while remaining slightly above BA sample levels. N15W2 sample emissions were marginally lower than the N15 sample but 3.58 times that of N15W1. It can be calculated that the doping of W1 warm-mix additive reduced the total VOC emissions by 73%. The W1 warm-mix additive is a surfactant-based additive containing both lipophilic and hydrophilic groups. When blended with asphalt, surfactant gel groups envelop smaller asphalt molecules via lipophilic interactions. This encapsulation inhibits the emission of smaller molecular or lightweight fractions in the asphalt binder [151]. Simultaneously, the nitrogenous organic compounds (Lewis bases) in the W1 warm-mix additive react with asphalt sulfides (Lewis acids), further prompting the decline of VOC emissions. The W2 warm-mix additive's stable structure limits reactivity with asphalt. Consequently, the N15W2 sample showed minimal VOC emission reduction versus the N15 sample. In this paper, samples containing warm-mix additives were tested with the same temperature as those without additives. Consequently, the isolated effect of the warm-mix additives was analyzed. Future studies should investigate the combined influence of temperature and warm-mix additives.

Solid deodorant was added to the N15 sample at 0.1% (N15D1), 0.2% (N15D2), and 0.3% (N15D3). The N15D1 sample showed 34% lower VOC emissions than the N15 sample at 0.1% deodorant loading. However, the VOC emissions increased by 18% and 6% at 0.2% and 0.3% deodorant loading. Without a micropore/layered structure, the deodorant's VOC absorption ability would not be enhanced regardless of content.

Activated CR particle was incorporated at 10% (A10), 15% (A15), and 20% (A20). The A10 sample showed peak VOC emissions, with the A15 and A20 samples decreasing by 2% and 33%, respectively, versus the A10 sample. Higher activated CR content absorbs more lightweight fractions in the asphalt binder, increasing asphalt viscosity. The increased asphalt viscosity restricts molecular mobility, limiting VOC emissions. Activated CR samples emitted 2.15 times (A10), 1.24 times (A15), and 1.17 times (A20) more VOC emissions than non-activated CR counterparts at equivalent CR content conditions. This result can be explained by the fact that the activation procedure used in this paper breaks certain chemical bonds of the CR particle; the fume would more easily escape, which contributes to the increased VOC emissions. This outcome was inconsistent with the findings in reference [130], in which CR pre-treated with microwave activation, waste cooked oil (WCO) swelling, microwave/WCO combined, HCl solution soaking, and NaOH solution soaking reduced VOC emissions in the corresponding modified asphalt. Microbial pretreatment (7–14 days) for CR helped in reducing VOC emissions of the corresponding CR-modified asphalt [131]. The outcome suggested that solution immersion removes VOC precursor substances from CR particles but minimally weakens the chemical bond of CR particles. The chemical bond, like the S–S bond, prefers to rupture at high temperature due to lower dissociation energies [160].

Adding 2% SBS modifier to non-activated CR-modified asphalt increased VOC emissions by 128% (SN10) and 49% (SN15), respectively. SN15 and SN10 samples showed the highest and the second-highest VOC emissions overall. The SBS modifier absorbs the lightweight fractions in the asphalt binder. Leaving insufficient lightweight fractions for non-activated CR particles swelling, this prolongs the degradation of untreated CR particles under thermal conditions and increases VOCs release. The CR particles undergo simultaneous swelling/degradation during prolonged thermal conditions [161]. Degradation weakens chemical bonds and ruptures the net structures of the CR particles, releasing additional fumes.

With activated CR, SBS addition reduced VOC emissions by 81% (SA10), 69% (SA15), and 33% (SA20), respectively. Gel permeation chromatography (GPC) analysis [151] showed VOC emission molecular weight ranging from 0 to 1000 across binders. According to the analysis results, most VOC emissions (83%) comprised 100–299 g/mol compounds, with 15% at 300 to 399 g/mol. Thus, low molecular weight species dominated the VOC profiles. Consequently, the additive or the structure of modified asphalt that immobilizes lightweight asphalt fractions effectively suppresses VOC emissions. Activated CR's broken chemical bonds enhance reactivity with SBS modifiers, forming a stable, easily spatial network structure that traps saturates/aromatics as less volatile compounds [162]. The improved net-shaped structure encapsulates lightweight fractions or smaller molecular substances, preventing leakage. It further prompts small molecule conversion to macromolecular, enhancing the cohesive force of molecules and increasing the viscosity of the combined modified asphalt [153]. This mechanism inhibits VOCs volatilization, aligning with [151], where absorption of lightweight components reduced 200–400 g/mol volatiles.

4.2. Composition Concentration Variation Analysis

As evident from Figure 4, the number of VOC compounds emitted from all tested samples ranged from 7 to 13. In samples SA10, SA15, and SA20, the number of emitted VOC compounds increased with the proportion of activated CR. A similar trend occurred in groups of SN10, SN15, and N10, N15, and N20. An increased proportion of CR particles enhanced interaction with the asphalt binder, consequently prompting the release of VOCs compounds.

Compositional analysis revealed that only one to three kinds of compounds per sample exceeded 10% concentration. However, these dominant compounds collectively accounted for 56% to 91% of total VOC emissions. In three samples (N10, N15D1, and A10), dominant compounds (>10% concentration) accounted for less than 70% of total VOC emissions. This indicates that higher-concentration compounds contribute over half of VOC emissions in all samples, while other species are trace substances. Targeting these dominant compounds is therefore the most effective strategy for reducing total VOC emissions.

Six VOC compounds consistently exceeded 10% concentration: trichloroethylene, toluene, p,m-xylene, styrene, o-xylene, and 4-ethyltoluene. Trichloroethylene was detected in all samples, with concentrations ranging from 20% to 79%, consistent with reference [133]. Therefore, based on convergent evidence for its definitive identification, trichloroethylene—which exhibited the highest concentration in most samples—is the primary target for mitigation, with the exception of samples SA15 and N15W2. However, numerous previous studies have not detected the trichloroethylene in VOCs of tested samples. It is widely accepted that trichloroethylene should not be present in asphalt binder, SBS, or CR. This consensus contrasts sharply with the findings of this study. It is hypothesized that the trichloroethylene may originate from the base asphalt, which was contaminated with trichloroethylene. Further studies should focus on the original base asphalt resource to eliminate the influence of unusual composition. Toluene was detected in 13 groups (excluding SA20, N10, N15, A15, and A20), with concentrations ranging from 12% to 55%, making it the second mitigation priority. Styrene and o-xylene were detected in five samples, 4-ethyltoluene in three groups, and p-xylene/m-xylene solely in sample A20. Trichloroethylene and toluene were the most prevalent compounds across all tested samples.

4.3. Potential of Secondary Organic Aerosols (SOA) and Ozone (O_3) Formation Analysis

According to the analysis in Section 4.2, although trichloroethylene is a compound detected at relatively high concentrations, the previous literature suggests that its presence may result from contamination of the base asphalt. In order to more accurately assess the potential risks of VOCs in asphalt, trichloroethylene will be excluded from the discussion in Sections 4.3 and 4.4. Figure 5a shows that ARH compounds were the main contributors to OFP. Aromatic hydrocarbons (ARHs) contribute to OFP by participating in tropospheric photochemical reactions with nitrogen oxides (NO_x = NO + NO_2) and radicals. This process involves sunlight-driven chain reactions, where ARHs act as key precursors. The OFP contribution of VOC emissions did not correlate with their total concentration of distribution across all samples. Sample N15W1 exhibited lower OFP values, indicating that the W1 warm-mix additive inhibits ozone formation. Samples SN10 and SN15 contributed most to OFP, whereas groups N10, N15, and N20 contributed less. This difference arises because the SBS modifier prompts emissions of toluene and o-xylene, compounds exhibiting high concentrations and high ozone formation potential. The W2 warm-mix additive had an adverse effect on ozone formation inhibition. OFP results for samples N15D1, N15D2, and N15D3 indicate that deodorant content did not significantly influence ozone formation inhibition. Activated CR-modified asphalt sample (A10) showed higher OFP values than non-activated samples, while samples of A15 and A20 showed smaller OFP values. Samples SA10, SA15, and SA20 exhibited lower OFP values than the corresponding activated CR-modified asphalt samples (A10, A15, and A20). This demonstrates that the combined CR and SBS modification enhanced ozone formation inhibition, thereby reducing the environmental impact of CR-modified asphalt application.

Since not all compounds contribute to SOA formation, the SOAFP distribution differed from both OFP and total VOCs concentration distribution across samples. HH compounds contributed most significantly to SOA formation. Samples N10, N15, and N20 exhibited higher SOAFP values than the BA and SBS samples. Similarly, activated CR-modified samples (A10, A15, and A20) exhibited higher SOAFP, indicating that CR particles comparably promote SOA formation in the asphalt binder. The W1 warm-mix additive significantly inhibited SOA formation, reducing its harmful environmental effects. Conversely, the W2 warm-mix additive and deodorant showed a moderate inhibitory effect on SOA formation. CR/SBS-modified asphalt samples (SA20, SN10, and SN15) exhibited elevated SOAFP values due to high trichloroethylene concentration. Trichloroethylene, with an aerosol formation coefficient (AFC) of 5%, emerged as the primary SOA formation contributor, heightening the risk of haze pollution (e.g., PM2.5). Therefore, a VOC species' contribution to SOA depends not only on its AFC value but also on its concentration. Consequently, trichloroethylene is the primary target for SOA pollution control; reducing its emissions would effectively mitigate SOA pollution.

4.4. Human Carcinogen Risk Analysis

According to the carcinogen risk indices of compounds in VOCs detected in all samples, Table 5 identifies three class 1 carcinogens in VOC emissions: benzene and 1,2-dichloroethane. As previously noted, benzene and trichloroethylene were detected in all samples. Trichloroethylene concentration varied significantly across samples, ranging from 0.4 mg/m^3 (SA15) to 5.3 mg/m^3 (A15). The percentage of concentration of trichloroethylene in total VOC emissions exhibited an enlarged range, accounting for 20% (SA15) to 79% (A15). Only the SA15 sample contained <30% trichloroethylene, while five samples exceeded 50%. Thus, trichloroethylene represents the highest-priority carcinogen due to its dominant concentration and risk. Therefore, it is crucial to verify the origin of trichloroethylene. Consequently, trichloroethylene will be excluded from the discussion in this section to enable a more precise analysis. Benzene (also a class 1 carcinogenic substance) was detected in all samples, with concentrations ranging from 1.9 $\mu g/m^3$ (N15W1) to 0.6 mg/m^3 (N15D1). Benzene concentrations were consistently lower than trichloroethylene's concentration, constituting 0.1% to 9% of total VOC emissions. Given its lower but non-negligible concentration, benzene reduction measures are warranted. 1,2-Dichloroethane appeared in six samples (2.5–11.8 $\mu g/m^3$), contributing <0.3% of total VOC emissions—insignificant for carcinogen risk.

Tetrachloroethane, methylene chloride, and styrene are class 2A carcinogens (probable human carcinogens). Methylene chloride appeared only in the N15 sample (1.1 $\mu g/m^3$) and the N15W1 sample (13 $\mu g/m^3$). Tetrachloroethane occurred in six samples (0.8–11.7 $\mu g/m^3$), exceeding 2 $\mu g/m^3$ only in the N15W1 sample. Dichloromethane and tetrachloroethane (class 2A carcinogens) showed lower concentrations and limited occurrence compared to class 1 carcinogens. Thus, tetrachloroethane and methylene chloride (class 2A carcinogens) pose minimal carcinogen risk. Styrene was the predominant class 2A carcinogen, detected in 13 samples at 12.8 $\mu g/m^3$–1.1 mg/m^3 (0.2%–26% of total VOC emissions).

Four class 2B carcinogens were identified: carbon tetrachloride, 1,1-dichloroethene, chloroform, and ethylbenzene, showing variable occurrence and concentrations. Carbon tetrachloride appeared solely in N15W1 (3.3 $\mu g/m^3$). 1,1-Dichloroethene occurred in the SA20 sample (4.6 $\mu g/m^3$) and the N15W2 sample (21.9 $\mu g/m^3$). Chloroform appeared in six samples (0.8 $\mu g/m^3$), contributing <1% of total VOC emissions. Ethylbenzene occurred in ten samples (7–474 $\mu g/m^3$, 0.1%–7%), making it the most significant class 2B carcinogen.

HHs constituted a significantly larger proportion of Table 5 compounds than ARHs. Thus, HH contributed most to the human carcinogen risk for all VOC emissions. Mitigation strategies should target HHs volatilization and concentration to reduce human carcinogen risk in CR-modified asphalt.

According to previous research, not all VOC compounds pose a human health risk. Non-carcinogenic risks encompass VOCs' health effects excluding cancer and genetic mutations [133,163]. Compounds of VOC emissions should be classified via the globally harmonized system of classification and labeling of chemicals (GHS) for health hazards, acute toxicity, or environmental hazards. Figure 6 shows varying proportions of non-carcinogenic risk substances across samples, with all samples exceeding 70% (72%–98%). Therefore, non-carcinogens dominate VOC emissions in most samples.

4.5. VOC Emission Mechanism Analysis

VOC emissions in non-activated CR-modified asphalt arise from physical volatility and chemical decomposition. Non-activated CR particles contain networks of stable S–S bonds and C–S bonds, resulting in a smooth surface and inertia characteristic that exhibits poor compatibility with asphalt binder [130,154]. Non-activated CR particles (crosslink density >1.5×10^4 mol/cm^3) [154] illustrate poor interfacial interaction with asphalt binder and limited swelling (<120%), absorbing minimal lightweight asphalt fractions. Non-activated CR modification relies on physical filling and limited swelling, exhibiting poor absorption of lightweight asphalt fractions. Most lightweight fractions remain suspended in the asphalt binder and volatilize at elevated temperatures. However, unevenly distributed non-activated CR particles partially inhibit the volatilization of lightweight fractions through the physical barrier [76]. The volatilized substances in processing additives (vulcanization accelerators, antioxidants, plasticizers) remain trapped within non-activated CR's dense crosslink net structure due to limited swelling rate (<120%), inhibiting volatilization [154]. Elevated temperatures and shear degrade swollen non-activated CR particles, breaking part of S–S/C–S bonds and releasing low molecular weight additive decomposition products that increase VOC emissions [134,154,155].

Activated CR modification employs absorption inhibition and structural stability to control VOC emissions. Activation disrupts CR particles' crosslink networks and dense structure, therefore increasing specific surface area (30%–50%) while reducing crosslink density (<0.3×10^{-4} mol/cm^3) [154]. This enhances asphalt compatibility and swelling capacity (from <120% to 200%–300%) [130,154,155]. Activated CR modification involves multi-level swelling and interface enhancement, where VOC adsorption occurs when the pore diameter of porous materials exceeds molecular size [164]. Activated CR's spongy-like micropore structure and swelling ability absorb lightweight asphalt components (saturated and aromatic components) via van der Waals force and capillary effect, preventing high-temperature volatilization and reducing VOC emissions [131,155]. Excessive activation (too many crosslink bonds were disrupted) or high-temperature cleavage of unstable bonds (e.g., single sulfur bond) overwhelms the absorption ability of lightweight asphalt fractions, increasing VOC emissions through accelerated molecule mobility at high temperature [154,155].

Activated CR particles exhibit significantly enhanced swelling, absorbing substantial lightweight asphalt fractions during the early modification stage. Over-enhanced compatibility between activated CR particles and asphalt binder causes (1) excessive swelling loosens activated CR structure, weakening van der Waals force/capillary absorption forces/lightweight asphalt fraction absorption ability. Absorbed lightweight fractions by activated CR particles volatilize under increased high-temperature molecular motion [154]. (2) Active surface groups (–SH, C=C) on the surface of activated CR particles generated during the activation effectively bond with polar asphalt components (e.g., resin), improving activated CR/asphalt compatibility, but compatibility remains limited. Small gaps, existing in the interface of activated CR particles and asphalt binder, became the evaporating tunnel for VOC substances [76]. The activated CR particles aggregated due to limited compatibility, which displaces the lightweight asphalt fraction within the activated CR particles aggregation area to free phases [61,154], thus, high-temperature VOC volatilization was enhanced [76,130].

SBS modifier forms a continuous 3D elastic network: styrene hard segments in SBS create physical crosslinking while butadiene soft segments establish an elastic continuous phase. This network encapsulates lightweight asphalt fractions (saturated and aromatic components) and volatile substances evaporated from the swelling of CR particles [76,134]. SBS reduces total VOC emissions by 47%. The core mechanism of VOC decline was that the network formed by the SBS modifier significantly reduces the diffusion rate of lightweight fractions and inhibits high-temperature volatilization [134].

The emission suppression of non-activated CR/SBS combined modified asphalt binder relies on limited physical absorption. The match between non-polar rubber and polar segments of SBS is low; therefore, non-activated CR particles and SBS modifier exhibit poor compatibility and cannot form a collaborative network. The modification mechanism of non-activated CR/SBS combined modified asphalt binder is an independent network and a limited hybrid. The inert surface of non-activated CR particles is difficult to combine with the chain segment of the SBS modifier; the SBS modifier constructs a network alone, and non-activated CR particles would be embedded into the network in the form of an isolated island. The network of SBS cannot effectively package the non-activated CR particles; the effect of the SBS network is influenced by the reunion of non-activated CR particles, and the synergistic effect of the SBS network and non-activated CR particles is difficult to exhibit. The gap that existed in the network due to the interface separation phenomenon in the combined modification system became a volatilizing tunnel of VOCs and cannot hinder the VOCs dispersion [76].

The introduction of the SBS modifier increases swelling in non-activated CR particles but inadvertently promotes VOC emissions. In the combined modification system, the SBS acts as an interface compatibilizer: its butadiene segments bond to the rubber chain segment in non-activated CR particles, increasing swelling from <120% to 150%–180%. Enhanced swelling disrupts the dense crosslinking structure in non-activated CR particles, releasing part of the encapsulated volatile substances (vulcanization accelerators, antioxidants, plasticizers) [154]. Improved compatibility of asphalt/non-activated CR particles enables more uniform dispersion of non-activated CR particles, partially disrupting agglomeration-induced physical barriers. Lightweight asphalt fraction volatilizes through the phase gap between asphalt and non-activated CR particles [61,76]. At 2% concentration, SBS forms a discontinuous elastic network that inadequately encapsulates volatile substances. The compatibility effect of SBS outweighs the inhibition effect; therefore, SBS may turn out to be the evaporating tunnel for VOCs, increasing the total VOC emissions concentration. Low SBS content thus reduces the inhibition effect on VOC emissions in non-activated CR/SBS combined modified asphalt binder.

Activated CR/SBS systems inhibit VOC emissions through three mechanisms: physical absorption, chemical control, and network collaboration. Enhanced physical absorption/chemical control of activated CR particles was previously described. The effect of SBS addition will be discussed. Activated CR particles exhibit a higher swelling rate. The high temperatures, excessive swelling rate, and broken micropore structure of non-activated CR particles cause the non-activated CR structure to become loose, leading to the desorption and volatilization of absorbed lightweight asphalt fractions. The SBS spatial network restricts excessive swelling of activated CR particles through steric hindrance. SBS chain segments interpenetrate pores in activated CR particles, forming mechanical constraints and interpenetrating structures between SBS and activated CR particles that maintain swelling at 180%–220% and prevent the activated CR structure from loosening [76]. SBS acts as an interface compatibilizer, enhancing asphalt-activated CR particle compatibility. Its butadiene segments (non-polar) bind surface-activated groups (–SH, C=C, –OH, –COOH) on activated CR particles [165], immobilizing absorbed VOC precursors at the interface of CR–SBS and reducing desorption probability [61,76]. Its styrene segments in SBS (highly polar) bond polar asphalt fractions (asphaltenes, resin) via hydrogen bonding/dipole interactions, strengthening the interface bond. SBS butadiene segments (non-polar) combined with a rubber chain (polyisoprene, butadiene) [61,76] to form a dense, three-phase continuous structure by the interpenetration of the swelling phase of the activated CR particles and the SBS spatial network [61,154]. This structure prompts uniform activated CR particle dispersion and minimized phase gaps that serve as VOC migration pathways. VOC diffusion coefficient emissions decreased by 40%–60% due to restricted volatilization through phase interfaces in the dense three-phase structure [61,76]. The network physically blocks VOC emissions, particularly small-molecule alkanes and aromatics. Uniform activated CR particle dispersion prevents localized high-temperature areas that concentrate lightweight fractions evaporation [130,154]. Activated CR particles exhibit enhanced micropore thermal stability, and the swelling–shrinking process of activated CR particles is more gentle. The interpenetration networks formed by SBS distribute thermal stress, maintaining structural integrity. Thus, this network cannot be easily destroyed, preventing sudden VOC release at high temperatures. Consequently, activated CR/SBS systems maintain stable, low-concentration VOC emissions [61,154]. Table 8 summarizes VOC emission inhibition mechanisms across all CR-modified asphalt samples.

Table 8. The inhibition mechanism of different CR-modified asphalt.

System Type	Structural Characteristics	VOC Emission Mechanism	Key Difference Symbols
Non-activated CR-modified asphalt	Dense crosslink network (intact S–S/C–S bonds)	1. Physical volatility: Light fractions escape through gaps (★★→↑↑)	★★: Floating light fractions ╳╳: Volatilized VOCs ↑↑: Emission direction
	Limited swelling (<120%)	2. Chemical decomposition: Additive release (╳╳)	
	Gaps exist in the asphalt matrix	3. High volatility (gaps act as emission channels)	
Activated CR-modified asphalt	Porous sponge-like structure (broken crosslinks)	1. Adsorption: Porous structure traps light fractions (▒▒)	▒▒: Adsorbed fractions ‖ ‖: Capillary retention ⊔: Structural collapse
	High swelling (200%–300%)	2. Capillary retention: Restricts light fraction escape (‖ ‖)	
	Capillary adsorption effect	3. Over-activation: Structural collapse leads to desorption (⊔→↑↑)	
Non-activated CR/SBS combined modified asphalt	CR agglomerates into "islands"	1. Gaps act as VOC tunnels (///→╳╳)	///: Gap channels ╳╳: VOC emissions
	Fragmented SBS network	2. SBS-enhanced emission (additive release)	
	Obvious phase separation with gaps	3. Low SBS (2%) causes network discontinuity, failing to contain emissions	
Activated CR/SBS combined modified asphalt	Activated CR and SBS form an interpenetrating continuous network	1. Synergistic suppression: Physical adsorption + chemical bonding (-SH/SBS binding) + network encapsulation	
	Triple suppression mechanism (adsorption, bonding, encapsulation)	2. VOCs diffusion reduced by 40%–60%	
	Good high-temperature stability	3. Uniform heat distribution and stable structure at high temperatures	

5. Conclusions

This study analyzed VOC emission characteristics from CR-modified asphalt binder with varying CR types, CR contents, and additives using GC–MS. How CR-modified asphalt binder parameters influence total VOC emissions, composition concentration variation, potential of secondary organic aerosols (SOA) and ozone (O_3) formation, and human carcinogen risk analysis was evaluated. The VOC emission mechanisms across CR modification types were analyzed, with conclusions summarized below.

(1). Increasing non-activated CR content initially elevates total VOC emissions. Peak VOC emissions occur at 15% non-activated CR content (3.99 times that of base asphalt). At 20% CR content, VOC emissions decrease due to the lightweight asphalt fraction. Higher non-activated CR contents increase compound diversity and shift the composition of VOC emissions toward the HHs group, dominated by trichloroethylene (20%–79%). These also elevate SOA formation potential while showing lower OFP values than SBS-modified asphalt.

(2). Surfactant warm-mix additive (W1) reduces the total VOC emission concentration by 73% in 15% non-activated CR content asphalt through lightweight fractions encapsulation and sulfide reaction. Warm-mix additive (W2) shows a negligible reduction. W1 significantly suppresses OFP, while W2 marginally reduces SOAFP, demonstrating differential efficacy. W1 minimizes carcinogenic trichloroethylene and benzene.

(3). Additionally, 0.1% deodorant reduces VOC emissions by 34% in 15% non-activated CR-modified asphalt, but higher doses (0.2%–0.3%) increase emissions by 6%–18% due to the non-porous structure. Deodorants show limited VOC emission inhibition efficacy and negligible OFP/SOAFP impact.

(4). Activated CR-modified asphalt shows peak VOC emissions at 10% CR content (A10); A15 and A20 show 2% and 33% reductions versus A10. However, activated CR-modified asphalt VOC emissions exceed non-activated counterparts at equal contents. Trichloroethylene dominates, with diversity increasing with CR content. OFP increases due to ARHs, while SOAFP remains high (HHs-driven).

(5). Non-activated CR/SBS compound blends increase VOC emissions by 128% (10% CR content) and 49% (15% CR content) due to phase gaps and additive release. OFP peaks are due to toluene/o–xylene. Activated CR/SBS compound blends reduce VOC emissions by 81% (10% CR content) and 69% (15% CR content) through triple synergy: adsorption, SBS–CR bonding, and network encapsulation. OFP decreases, while SOAFP correlates with trichloroethylene.

(6). Class 1 carcinogen benzene (0.1%–9%) was ubiquitous. HHs contribute >70% of the carcinogenic risk. Non-carcinogenic risks exceed 70% in all samples. Activated CR/SBS combinations reduce carcinogen proportions.

(7). Non-activated CR particles feature dense crosslinks and low swelling (<120%), releasing VOCs through bond breakage and poor lightweight asphalt fraction absorption. Activated CR exhibits a porous structure and high swelling (200%–300%), absorbing VOCs but risking structure collapse at high temperatures, which prompts VOCs. SBS addition creates interpenetrating networks with activated CR, reducing VOCs by 40%–60%. Non-activated CR/SBS blends suffer phase separation, creating VOCs diffusion pathways.

Non-activated CR's dense crosslink structure and poor compatibility create barriers that inhibit inner compound release, lowering total VOC emissions. SBS improves compatibility with asphalt and prompts non-activated CR swelling (150%–180%), disrupting physical barriers and creating an incomplete network. Then, the diffusion tunnels and releasing efficiency of VOCs will be enhanced, and the volatilizing composition in non-activated CR particles and lightweight asphalt fractions will evaporate by the phase gap between non-activated CR and asphalt. SBS/non-activated CR interactions decompose part of the stable fractions (heavy aromatic hydrocarbons) into small molecular VOC compounds, increasing VOC emissions. Thus, SBS dose and CR activation status jointly influence VOC emissions, with low SBS dose in a non-activated CR system weakening VOC emission inhibitions.

Activated CR particles rely only on porous structures to absorb part of the VOC precursors. Without physical network restraint, excessive swelling loosens the structure of activated CR particles, increasing VOC emissions. In an activated CR/SBS system, porous structure, network isolation, optimized compatibility, inhibition of excessive swelling, and reduced temperature sensitivity collectively inhibit VOC volatilization. SBS forms a higher effective collaborative inhibiting mechanism with activated CR particles, yielding lower VOC emissions than the non-activated counterparts at equal CR content. This highlights CR activation's critical role in VOC emission inhibition. Activation maximizes SBS network inhibition and CR absorption, demonstrating a synergistic ("1 + 1 > 2") emission reduction effect.

While this study provides a comprehensive profile of VOC emissions, certain limitations should be acknowledged. Firstly, although GC–MS is the standard technique for such analyses, the quantitative accuracy of MS can be influenced by factors such as ionization efficiency and matrix effects. Although we employed internal standards to calibrate and improve quantitative reliability, these inherent limitations of the method may affect the absolute concentrations reported. Secondly, the unexpected presence of compounds like trichloroethylene, while strongly indicated by convergent evidence (including characteristic isotope patterns), could be further confirmed in future work using high-resolution mass spectrometry to achieve unambiguous identification. These points highlight avenues for more precise methodological development in subsequent studies.

Author Contributions: Conceptualization, H.Z. and P.Z.; methodology, H.Z. and P.Z.; software, P.Z. and Y.L. (Yiran Li); validation, H.Z. and P.Z.; formal analysis, H.Z., P.Z., C.S., and Y.L. (Yiran Li); investigation, H.F. and D.Z.; resources, H.F., Y.D., Y.L. (Yanping Liu), and Q.H.; data curation, H.Z. and P.Z.; writing—original draft preparation, H.Z.; writing—review and editing, H.Z.; visualization, C.S. and Y.L. (Yiran Li); supervision, H.F. and D.Z.; project administration, H.F. and Y.D.; funding acquisition, Y.L. (Yanping Liu), and Q.H. All authors have read and agreed to the published version of the manuscript.

Funding: This research received no external funding.

Institutional Review Board Statement: Not applicable.

Informed Consent Statement: Not applicable.

Data Availability Statement: The data presented in this study are available on request from the corresponding author.

Acknowledgments: We thank Jingtao Dong for their assistance with experiments and valuable discussion.

Conflicts of Interest: Authors Hu Feng, Dongfang Zhang, and Yanping Liu were employed by the company Shandong Ludong High-Tech Materials Technology Co., Ltd.; authors Yindong Ding and Qingjun Han were employed by the company Shandong Ludong Transportation Construction Group Co., Ltd. The remaining authors declare that the research was conducted in the absence of any commercial or financial relationships that could be construed as a potential conflict of interest.

Abbreviations

The following abbreviations are used in this manuscript:

CR	crumb rubber
CRMA	crumb rubber-modified asphalt binders
GC–MS	gas chromatography–mass spectrometry
ACR	activated crumb rubber
NCR	non-activated crumb rubber
SOA	secondary organic aerosol

References

1. Mitchell, C.J.; Jayakaran, A.D. Mitigating tire wear particles and tire additive chemicals in stormwater with permeable pavements. *Sci. Total Environ.* **2024**, *908*, 168236. [CrossRef] [PubMed]
2. Huang, R.; Ren, Q.; Zhang, J.; He, L.; Su, S.; Wang, Y.; Jiang, L.; Xu, J.; Hu, S.; Xiang, J. Adjusting effects of pyrolytic volatiles interaction in char to upgrade oil by swelling waste nylon-tire. *Waste Manag.* **2023**, *169*, 374–381. [CrossRef] [PubMed]
3. Zerin, N.H.; Rasul, M.G.; Jahirul, M.I.; Sayem, A.S.M. End-of-life tyre conversion to energy: A review on pyrolysis and activated carbon production processes and their challenges. *Sci. Total Environ.* **2023**, *905*, 166981. [CrossRef] [PubMed]
4. Qiu, Y.; Gao, Y.; Zhang, X.; Wei, Y.; Cao, J.; Wang, X.; Wang, S. Conventional properties, rheological properties, and storage stability of crumb rubber modified asphalt with WCO and ABS. *Constr. Build. Mater.* **2023**, *392*, 131987. [CrossRef]
5. Xu, G.; Yao, Y.; Ma, T.; Hao, S.; Ni, B. Experimental study and molecular simulation on regeneration feasibility of high-content waste tire crumb rubber modified asphalt. *Constr. Build. Mater.* **2023**, *369*, 130570. [CrossRef]
6. Zhao, Z.; Wu, S.; Xie, J.; Yang, C.; Yang, X.; Wang, F.; Liu, Q. Utilization of high contents desulfurized crumb rubber in developing an asphalt rubber pellets modified asphalt. *Constr. Build. Mater.* **2023**, *402*, 133043. [CrossRef]
7. Ma, Y.; Wang, S.; Zhang, M.; Jiang, X.; Polaczyk, P.; Huang, B. Weather aging effects on modified asphalt with rubber-polyethylene composites. *Sci. Total Environ.* **2023**, *865*, 161089. [CrossRef]
8. Zhao, S.; Zhang, H.; Feng, Y.; Li, J.; Gao, M.; Wu, G. Study on rheological and micro-structural properties of different modified asphalt by using organic and inorganic modifier. *Case Stud. Constr. Mater.* **2023**, *18*, e02072. [CrossRef]
9. Xie, J.; Zhao, X.; Lv, S.; Zhang, Y.; He, W.; Yu, F. Research on performance and mechanism of terminal blend/grafting activated crumb rubber composite modified asphalt. *Constr. Build. Mater.* **2023**, *394*, 132225. [CrossRef]
10. Zhu, Y.; Xu, G.; Ma, T.; Fan, J.; Li, S. Performances of rubber asphalt with middle/high content of waste tire crumb rubber. *Constr. Build. Mater.* **2022**, *335*, 127488. [CrossRef]
11. Duan, K.; Wang, C.; Liu, J.; Song, L.; Chen, Q.; Chen, Y. Research progress and performance evaluation of crumb-rubber-modified asphalts and their mixtures. *Constr. Build. Mater.* **2022**, *361*, 129687. [CrossRef]
12. Yang, S.; Zhu, H.; Tan, Q.; Yang, X.; Chen, Y.; Lei, L. Application of TOR as a secondary modifier for the preparation of crumb rubber modified asphalt with excellent storage stability. *Constr. Build. Mater.* **2024**, *428*, 135863. [CrossRef]
13. Jin, T.; Sun, L.; Liu, L.; Li, M.; Li, J.; Yuan, J.; Yang, R.; Cheng, H. Development of dynamic shear-rheology-based method to improve evaluation of swelling degree of recycled crumb rubber in asphalt rubber. *J. Clean. Prod.* **2024**, *449*, 141851. [CrossRef]
14. Zhao, Y.; Chen, M.; Wu, S.; Jiang, Q.; Xu, H.; Zhao, Z.; Lv, Y. Effects of waterborne polyurethane on storage stability, rheological properties, and VOCs emission of crumb rubber modified asphalt. *J. Clean. Prod.* **2022**, *340*, 130682. [CrossRef]
15. Zou, Y.; Wu, S.; Chen, A.; Liu, Q.; Amirkhanian, S.; Xu, S.; Yang, C.; Wan, P.; Xu, H.; Lu, Z. Comprehensive properties assessment of asphalt binder under aqueous solutions with different pH values and its gradient damage behaviors. *Constr. Build. Mater.* **2024**, *414*, 134938. [CrossRef]
16. Xu, L.; Ni, H.; Tian, Y.; Sun, D.; Chen, Z.; Jin, H. Multi-scale analysis of damping characteristics of dry mixed rubberized porous asphalt mixtures for tire-pavement noise reduction. *J. Clean. Prod.* **2023**, *425*, 138969. [CrossRef]
17. Saberi, K.F.; Fakhri, M.; Azami, A. Evaluation of warm mix asphalt mixtures containing reclaimed asphalt pavement and crumb rubber. *J. Clean. Prod.* **2017**, *165*, 1125–1132. [CrossRef]
18. Zhou, Y.; Xu, G.; Leng, Z.; Kong, P.; Wang, H.; Yang, J.; Qiu, W.; Xu, Z.; Chen, X. Investigation on storage stability and rheological properties of environment-friendly devulcanized rubber modified asphalt. *Phys. Chem. Earth Parts A/B/C* **2023**, *129*, 103332. [CrossRef]
19. Li, H.; Cui, C.; Temitope, A.A.; Feng, Z.; Zhao, G.; Guo, P. Effect of SBS and crumb rubber on asphalt modification: A review of the properties and practical application. *J. Traffic Transp. Eng. Engl. Ed.* **2022**, *9*, 836–863. [CrossRef]
20. Li, L.; Wang, P.; Zhou, T.; Cao, L.; Ding, X.; Zhang, B.; Wei, L. Performance evaluation and multiscale investigation of deodorant on mechanism of rubber modified asphalt. *J. Clean. Prod.* **2023**, *425*, 138954. [CrossRef]

21. Zhao, Z.; Wu, S.; Xie, J.; Yang, C.; Wang, F.; Li, N.; Liu, Q.; Amirkhanian, S. Effect of direct addition of asphalt rubber pellets on mixing, performance and VOCs of asphalt mixtures. *Constr. Build. Mater.* **2024**, *411*, 134494. [CrossRef]
22. Lv, Y.; Wu, S.; Li, N.; Liu, Q.; Yang, C.; Zou, Y.; Amirkhanian, S. Flue gas suppression and environmental evaluation of deodorizer-modified rubber asphalt based on radar method. *Constr. Build. Mater.* **2024**, *411*, 134526. [CrossRef]
23. Zhang, J.; Chen, M.; Zhou, X.; Cao, Z.; Wu, S.; Zhao, Y. Performance and VOCs emission inhibition of high-content waste rubber powder modified asphalt and its mixture. *Int. J. Pavement Eng.* **2023**, *24*, 2282031. [CrossRef]
24. Gong, F.; Guo, S.; Chen, S.; You, Z.; Liu, Y.; Dai, Q. Strength and durability of dry-processed stone matrix asphalt containing cement pre-coated scrap tire rubber particles. *Constr. Build. Mater.* **2019**, *214*, 475–483. [CrossRef]
25. Li, J.; Yu, J.; Wu, S.; Xie, J. The Mechanical Resistance of Asphalt Mixture with Steel Slag to Deformation and Skid Degradation Based on Laboratory Accelerated Heavy Loading Test. *Materials* **2022**, *15*, 911. [CrossRef]
26. Hu, J.; Ma, T.; Zhu, Y.; Huang, X.; Xu, J.; Chen, L. High-viscosity modified asphalt mixtures for double-layer porous asphalt pavement: Design optimization and evaluation metrics. *Constr. Build. Mater.* **2021**, *271*, 121893. [CrossRef]
27. Zhang, J.; Chen, M.; Wu, S.; Zhou, X.; Zhao, G.; Zhao, Y.; Cheng, M. Evaluation of VOCs inhibited effects and rheological properties of asphalt with high-content waste rubber powder. *Constr. Build. Mater.* **2021**, *300*, 124320. [CrossRef]
28. Chang, X.; Long, Y.; Wang, C.; Xiao, Y. Chemical fingerprinting of volatile organic compounds from asphalt binder for quan-titative detection. *Construction and Building Materials* 2023. [CrossRef]
29. Xu, H.; Zou, Y.; Airey, G.; Wang, H.; Zhang, H.; Wu, S.; Chen, A. Wetting of bio-rejuvenator nanodroplets on bitumen: A molecular dynamics investigation. *J. Clean. Prod.* **2024**, *444*, 141140. [CrossRef]
30. Kocak, S.; Kutay, M.E. Fatigue performance assessment of recycled tire rubber modified asphalt mixtures using viscoelastic continuum damage analysis and AASHTOWare pavement ME design. *Constr. Build. Mater.* **2020**, *248*, 118658. [CrossRef]
31. Duan, H.; Zhu, C.; Li, Y.; Zhang, H.; Zhang, S.; Xiao, F.; Amirkhanian, S. Effect of crumb rubber percentages and bitumen sources on high-temperature rheological properties of less smell crumb rubber modified bitumen. *Constr. Build. Mater.* **2021**, *277*, 122248. [CrossRef]
32. Zhao, Z.; Wu, S.; Xie, J.; Yang, C.; Yang, X.; Chen, S.; Liu, Q. Recycle of waste tire rubber powder in a novel asphalt rubber pellets for asphalt performance enhancement. *Constr. Build. Mater.* **2023**, *399*, 132572. [CrossRef]
33. Li, F.; Zhang, X.; Wang, L.; Zhai, R. The preparation process, service performances and interaction mechanisms of crumb rubber modified asphalt (CRMA) by wet process: A comprehensive review. *Constr. Build. Mater.* **2022**, *354*, 129168. [CrossRef]
34. Luan, Y.; Ma, Y.; Ma, T.; Wang, C.; Xia, F. Research on the effects of asphalt performance on rutting resistance and its correlation with rutting performance indicators. *Constr. Build. Mater.* **2023**, *400*, 132773. [CrossRef]
35. Adnan, A.M.; Wang, J. Enhancement of rheological performance and compatibility of tire rubber-modified asphalt binder with the addition of graphene. *Constr. Build. Mater.* **2023**, *402*, 133038. [CrossRef]
36. Xu, X.; Leng, Z.; Lan, J.; Wang, W.; Yu, J.; Bai, Y.; Sreeram, A.; Hu, J. Sustainable Practice in Pavement Engineering through Value-Added Collective Recycling of Waste Plastic and Waste Tyre Rubber. *Engineering* **2021**, *7*, 857–867. [CrossRef]
37. Yang, C.; Wu, S.; Cui, P.; Amirkhanian, S.; Zhao, Z.; Wang, F.; Zhang, L.; Wei, M.; Zhou, X.; Xie, J. Performance characterization and enhancement mechanism of recycled asphalt mixtures involving high RAP content and steel slag. *J. Clean. Prod.* **2022**, *336*, 130484. [CrossRef]
38. Lv, Y.; Wu, S.; Li, N.; Cui, P.; Wang, H.; Amirkhanian, S.; Zhao, Z. Performance and VOCs emission inhibition of environmentally friendly rubber modified asphalt with UiO-66 MOFs. *J. Clean. Prod.* **2023**, *385*, 135633. [CrossRef]
39. Xue, Y.; Ge, D.; Lv, S.; Duan, D.; Deng, Y. Evaluation of asphalt modified with bio-oil and high rubber content: Low temperature and short mixing time production condition. *Constr. Build. Mater.* **2023**, *408*, 133656. [CrossRef]
40. Gamboa, C.J.O.; Ruiz, P.A.C.; Kaloush, K.E.; Linares, J.P.L. Life cycle assessment including traffic noise: Conventional vs. rubberized asphalt. *Int. J. Life Cycle Assess.* **2021**, *26*, 2375–2390. [CrossRef]
41. Wang, T.; Xiao, F.; Zhu, X.; Huang, B.; Wang, J.; Amirkhanian, S. Energy consumption and environmental impact of rubberized asphalt pavement. *J. Clean. Prod.* **2018**, *180*, 139–158. [CrossRef]
42. Nanjegowda, V.H.; Biligiri, K.P. Recyclability of rubber in asphalt roadway systems: A review of applied research and advancement in technology. *Resour. Conserv. Recycl.* **2020**, *155*, 104655. [CrossRef]
43. Chang, X.; Xiao, Y.; Long, Y.; Wang, F.; You, Z. Temperature dependency of VOCs release characteristics of asphalt materials under varying test conditions. *J. Traffic Transp. Eng. Engl. Ed.* **2022**, *9*, 280–292. [CrossRef]
44. Li, J.; Xiao, X.; Chen, Z.; Xiao, F.; Amirkhanian, S.N. Internal de-crosslinking of scrap tire crumb rubber to improve compatibility of rubberized asphalt. *Sustain. Mater. Technol.* **2022**, *32*, e00417. [CrossRef]
45. Tang, N.; Zhang, Z.; Dong, R.; Zhu, H.; Huang, W. Emission behavior of crumb rubber modified asphalt in the production process. *J. Clean. Prod.* **2022**, *340*, 130850. [CrossRef]

46. Yang, X.; You, Z.; Perram, D.; Hand, D.; Ahmed, Z.; Wei, W.; Luo, S. Emission analysis of recycled tire rubber modified asphalt in hot and warm mix conditions. *J. Hazard. Mater.* **2019**, *365*, 942–951. [CrossRef]
47. Xiao, Z.; Pramanik, A.; Basak, A.K.; Prakash, C.; Shankar, S. Material recovery and recycling of waste tyres—A review. *Clean. Mater.* **2022**, *5*, 100115. [CrossRef]
48. Yang, Q.; Lin, J.; Wang, X.; Wang, D.; Xie, N.; Shi, X. A review of polymer-modified asphalt binder: Modification mechanisms and mechanical properties. *Clean. Mater.* **2024**, *12*, 100255. [CrossRef]
49. Li, J.; Chen, Z.; Xiao, F.; Amirkhanian, S.N. Surface activation of scrap tire crumb rubber to improve compatibility of rubberized asphalt. *Resour. Conserv. Recycl.* **2021**, *169*, 105518. [CrossRef]
50. Li, J.; Jiang, H.; Han, F.; Lin, Z.; Zhao, Z.; Jin, X.; Liu, Y. Investigation of the effects of chemical modification and oxidative aging on the properties and compatibility of rubber asphalt based on thermodynamic principles. *J. Clean. Prod.* **2023**, *428*, 139070. [CrossRef]
51. Kong, P.; Xu, G.; Fu, L.; Feng, H.; Chen, X. Chemical structure of rubber powder on the compatibility of rubber powder asphalt. *Constr. Build. Mater.* **2023**, *392*, 131769. [CrossRef]
52. Gągol, M.; Boczkaj, G.; Haponiuk, J.; Formela, K. Investigation of volatile low molecular weight compounds formed during continuous reclaiming of ground tire rubber. *Polym. Degrad. Stab.* **2015**, *119*, 113–120. [CrossRef]
53. Cao, L.; Yang, C.; Li, A.; Wang, P.; Zhang, Y.; Dong, Z. Flue gas composition of waste rubber modified asphalt (WRMA) and effect of deodorants on hazardous constituents and WRMA. *J. Hazard. Mater.* **2021**, *403*, 123814. [CrossRef] [PubMed]
54. Zhou, T.; Yu, F.; Li, L.; Dong, Z.; Fini, E.H. Swelling-degradation dynamic evolution behaviors of bio-modified rubberized asphalt under thermal conditions. *J. Clean. Prod.* **2023**, *426*, 139061. [CrossRef]
55. Guo, F.; Zhang, J.; Pei, J.; Zhou, B.; Falchetto, A.C.; Hu, Z. Investigating the interaction behavior between asphalt binder and rubber in rubber asphalt by molecular dynamics simulation. *Constr. Build. Mater.* **2020**, *252*, 118956. [CrossRef]
56. Zhou, B.; Gong, G.; Liu, Y.; Guo, M.; Wang, C. Technical and environmental performance assessment of VOCs inhibited asphalt binders and mixtures. *Transp. Res. Part D Transp. Environ.* **2023**, *124*, 103931. [CrossRef]
57. Li, H.; Zhang, Y.; Zhou, L.; Feng, Z.; Sun, J.; Hao, G.; Li, Z. Effect of rubber powder desulfurization on the compatibility between rubber and bitumen: Combination and insight of experiments and molecular dynamics simulation. *Constr. Build. Mater.* **2024**, *414*, 134966. [CrossRef]
58. Formela, K. Strategies for compatibilization of polymer/waste tire rubber systems prepared via melt-blending. *Adv. Ind. Eng. Polym. Res.* **2024**, *7*, 466–481. [CrossRef]
59. Zou, F.; Leng, Z.; Li, D.; Zhang, S.; Yaphary, Y.L.; Lu, G. Leaching potential of metals and PAHs of asphalt rubber paving materials. *Transp. Res. Part D Transp. Environ.* **2023**, *125*, 103984. [CrossRef]
60. Chlebnikovas, A.; Marčiulaitienė, E.; Šernas, O.; Škulteckė, J.; Januševičius, T. Research on air pollutants and odour emissions from paving hot-mix asphalt with end-of-life tyre rubber. *Environ. Int.* **2023**, *181*, 108281. [CrossRef]
61. Xie, J.; Ding, Z.; Luo, H.; Zhao, X.; Li, S.; Ma, Y. Interaction of composite fume suppression and odor elimination agents with crumb rubber modified asphalt: Inhibition behavior of volatile organic compounds (VOCs) and inorganic fume. *Sci. Total Environ.* **2024**, *935*, 173459. [CrossRef] [PubMed]
62. Chang, X.; Wan, L.; Long, Y.; Xiao, Y.; Xue, Y. Optimal zeolite structure design for VOC emission reduction in asphalt materials. *Constr. Build. Mater.* **2023**, *366*, 130227. [CrossRef]
63. Boczkaj, G.; Przyjazny, A.; Kaminski, M. Characteristics of volatile organic compounds emission profiles from hot road bitumens. *Chemosphere* **2014**, *107*, 23–30. [CrossRef] [PubMed]
64. Celebi, U.B.; Vardar, N. Investigation of VOC emissions from indoor and outdoor painting processes in shipyards. *Atmos. Environ.* **2008**, *42*, 5685–5695. [CrossRef]
65. Wang, F.; Li, N.; Hoff, I.; Wu, S.; Li, J.; Maria Barbieri, D.; Zhang, L. Characteristics of VOCs generated during production and construction of an asphalt pavement. *Transp. Res. Part D Transp. Environ.* **2020**, *87*, 102517. [CrossRef]
66. Xing, B.; Feng, Y.; Sun, S.; Qian, C.; Fang, C.; Lv, X.; Song, A.; Lyu, Y. Investigations on the rheological and swelling-degradation behavior of crumb rubber within the bituminous matrix. *Constr. Build. Mater.* **2023**, *367*, 130262. [CrossRef]
67. Wang, Z.; Wu, S.; Yang, C.; Xie, J.; Xiao, Y.; Zhao, Z.; Wang, F.; Zhang, L. Quantitative Assessment of Road Performance of Recycled Asphalt Mixtures Incorporated with Steel Slag. *Materials* **2022**, *15*, 5005. [CrossRef]
68. Yan, Y.; Guo, R.; Liu, Z.; Yang, Y.; Wang, X.-Y. Property improvement of thermosetting natural rubber asphalt binder by mineral oil. *J. Mater. Res. Technol.* **2023**, *24*, 8807–8825. [CrossRef]
69. Zhao, X.; Li, F.; Zhang, X.; Cao, J.; Wang, X. Rheological properties and viscosity reduction mechanism of aromatic/naphthenic oil pre-swelling crumb rubber modified asphalt. *Constr. Build. Mater.* **2023**, *398*, 132545. [CrossRef]
70. Zhang, X.; Wang, X.; Wan, C.; Yang, B.; Tang, Z.; Li, W. Performance evaluation of asphalt binder and mixture modified by pre-treated crumb rubber. *Constr. Build. Mater.* **2023**, *362*, 129777. [CrossRef]

71. Wang, G.; Yang, X.; Yang, D.; Rong, H.; Meng, Y.; Liu, G. Effect of heating history on the emission of volatile organic compounds from asphalt materials. *Sci. Total Environ.* **2023**, *900*, 165692. [CrossRef] [PubMed]
72. Meng, Y.; Fang, G.; Hu, Y.; Qin, Y.; Xu, R.; Yang, F.; Lei, J.; Zhang, C. Study on the effect of different aldehyde modifiers on the fume suppression effect, mechanism and road performance of SBS modified asphalt. *Sci. Total Environ.* **2024**, *912*, 169162. [CrossRef] [PubMed]
73. Zhang, Y.; Qi, J.; Sun, Y.; Zhu, Z.; Wang, C.; Sun, X.; Li, J. Anchoring nanosized MOFs at the interface of porous millimeter beads and their enhanced adsorption mechanism for VOCs. *J. Clean. Prod.* **2022**, *353*, 131631. [CrossRef]
74. Xiu, M.; Wang, X.; Morawska, L.; Pass, D.; Beecroft, A.; Mueller, J.F.; Thai, P. Emissions of particulate matters, volatile organic compounds and polycyclic aromatic hydrocarbons from warm and hot asphalt mixes. *J. Clean. Prod.* **2020**, *275*, 123094. [CrossRef]
75. Chen, Q.; Wang, C.; Wen, P.; Sun, X.; Guo, T. Performance evaluation of tourmaline modified asphalt mixture based on grey target decision method. *Constr. Build. Mater.* **2019**, *205*, 137–147. [CrossRef]
76. Liu, G.; Fang, S.; Wang, Y.; Liu, J.; Liang, Y.; Cao, T.; Liu, Q. Emission of Volatile Organic Compounds in Crumb Rubber Modified Bitumen and Its Inhibition by Using Montmorillonite Nanoclay. *Polymers* **2023**, *15*, 1513. [CrossRef]
77. Zeng, S.; Mao, S.; Xu, S.; He, Y.; Yu, J. Investigation on DOPO as reactive fumes suppressant to reduce the fumes emission of asphalt. *J. Hazard. Mater.* **2024**, *463*, 132878. [CrossRef]
78. Yao, Z.; Yang, R.; Shi, J.; Zhang, C.; Zhang, Y. Separation mechanism of asphalt rubber and improvement with microwave irradiation. *Constr. Build. Mater.* **2022**, *342*, 127862. [CrossRef]
79. Hosseinnezhad, S.; Kabir, S.F.; Oldham, D.; Mousavi, M.; Fini, E.H. Surface functionalization of rubber particles to reduce phase separation in rubberized asphalt for sustainable construction. *J. Clean. Prod.* **2019**, *225*, 82–89. [CrossRef]
80. Yang, X.; Shen, A.; Li, B.; Wu, H.; Lyu, Z.; Wang, H.; Lyu, Z. Effect of microwave-activated crumb rubber on reaction mechanism, rheological properties, thermal stability, and released volatiles of asphalt binder. *J. Clean. Prod.* **2020**, *248*, 119230. [CrossRef]
81. Liang, M.; Ren, S.; Fan, W.; Wang, H.; Cui, W.; Zhao, P. Characterization of fume composition and rheological properties of asphalt with crumb rubber activated by microwave and TOR. *Constr. Build. Mater.* **2017**, *154*, 310–322. [CrossRef]
82. Phiri, M.M.; Phiri, M.J.; Formela, K.; Wang, S.; Hlangothi, S.P. Grafting and reactive extrusion technologies for compatibilization of ground tyre rubber composites: Compounding, properties, and applications. *J. Clean. Prod.* **2022**, *369*, 133084. [CrossRef]
83. Szerb, E.I.; Nicotera, I.; Teltayev, B.; Vaiana, R.; Rossi, C.O. Highly stable surfactant-crumb rubber-modified bitumen: NMR and rheological investigation. *Road Mater. Pavement Des.* **2017**, *19*, 1192–1202. [CrossRef]
84. Wang, Z.; Xu, X.; Wang, X.; Huo, J.; Guo, H.; Yang, B. Performance of modified asphalt of rubber powder through tetraethyl orthosilicate (TEOS). *Constr. Build. Mater.* **2021**, *267*, 121032. [CrossRef]
85. Li, J.; Xiao, F.; Amirkhanian, S.N. Rheological and chemical characterization of plasma-treated rubberized asphalt using customized extraction method. *Fuel* **2020**, *264*, 116819. [CrossRef]
86. Saputra, R.; Walvekar, R.; Khalid, M.; Mubarak, N.M.; Sillanpää, M. Current progress in waste tire rubber devulcanization. *Chemosphere* **2021**, *265*, 129033. [CrossRef]
87. Wang, Z.; Pan, C.; Hu, Y.; Zeng, D.; Huang, M.; Jiang, Y. High-quality ground tire rubber production from scrap tires by using supercritical carbon dioxide jet pulverization assisted with diphenyl disulfide. *Powder Technol.* **2022**, *398*, 117061. [CrossRef]
88. Badughaish, A.; Li, J.; Amirkhanian, S.; Xiao, F. Adhesion and segregation characteristics of crumb rubberized binders based on solution-soaked methods. *J. Clean. Prod.* **2022**, *354*, 131762. [CrossRef]
89. Dong, R.; Zhao, M.; Tang, N. Characterization of crumb tire rubber lightly pyrolyzed in waste cooking oil and the properties of its modified bitumen. *Constr. Build. Mater.* **2019**, *195*, 10–18. [CrossRef]
90. Li, J.; Yao, S.; Xiao, F.; Amirkhanian, S.N. Surface modification of ground tire rubber particles by cold plasma to improve compatibility in rubberised asphalt. *Int. J. Pavement Eng.* **2020**, *23*, 651–662. [CrossRef]
91. Xie, J.; Yang, Y.; Lv, S.; Zhang, Y.; Zhu, X.; Zheng, C. Investigation on Rheological Properties and Storage Stability of Modified Asphalt Based on the Grafting Activation of Crumb Rubber. *Polymers* **2019**, *11*, 1563. [CrossRef] [PubMed]
92. Xu, M.; Liu, J.; Li, W.; Duan, W. Novel Method to Prepare Activated Crumb Rubber Used for Synthesis of Activated Crumb Rubber Modified Asphalt. *J. Mater. Civ. Eng.* **2015**, *27*, 04014173. [CrossRef]
93. Liu, H.; Zhang, Z.; Xie, J.; Gui, Z.; Li, N.; Xu, Y. Analysis of OMMT strengthened UV aging-resistance of Sasobit/SBS modified asphalt: Its preparation, characterization and mechanism. *J. Clean. Prod.* **2021**, *315*, 128139. [CrossRef]
94. Xiao, F.; Hou, X.; Amirkhanian, S.; Kim, K.W. Superpave evaluation of higher RAP contents using WMA technologies. *Constr. Build. Mater.* **2016**, *112*, 1080–1087. [CrossRef]
95. Lv, Y.; Wu, S.; Li, N.; Wang, H.; Cui, P.; Xu, H.; Zhao, Y.; Yang, C.; Zhou, X.; Amirkhanian, S. Environmental hazard reduction and anti-aging enhancement of steel slag powder-asphalt mastic based on TiO_2@UiO-67 composites. *J. Clean. Prod.* **2023**, *419*, 138331. [CrossRef]

96. Tao, Z.; Shen, S.; Yu, H.; Sun, Y.; Zou, Y. Rejuvenating aged asphalt using surfactant-foaming warm recycling technology. *Constr. Build. Mater.* **2023**, *384*, 131297. [CrossRef]
97. Sesay, T.; You, Q.; Chuan, J.; Qiao, H.; Zhang, H.; Tian, S. High and intermediate temperature performance of warm asphalt rubber containing conventional warm mix additives and novel chemical surfactant. *Constr. Build. Mater.* **2023**, *394*, 132214. [CrossRef]
98. Amirkhani, M.J.; Fakhri, M.; Amirkhani, A. Evaluating the use of different fillers and Kaowax additive in warm mix asphalt mixtures. *Case Stud. Constr. Mater.* **2023**, *19*, e02489. [CrossRef]
99. Zaremotekhases, F.; Sadek, H.; Hassan, M.; Berryman, C. Impact of warm-mix asphalt technologies and high reclaimed asphalt pavement content on the performance of alternative asphalt mixtures. *Constr. Build. Mater.* **2022**, *319*, 126035. [CrossRef]
100. Jiang, Q.; Li, N.; Yang, F.; Ren, Y.; Wu, S.; Wang, F.; Xie, J. Rheology and volatile organic compounds characteristics of warm-mix flame retardant asphalt. *Constr. Build. Mater.* **2021**, *298*, 123691. [CrossRef]
101. Wang, S.; Yang, C.; Zhao, J.; Li, C.; Fan, X. Rapid and Direct Assessment of Asphalt Volatile Organic Compound Emission Based on Carbon Fiber Ionization Mass Spectrometry. *ACS Omega* **2023**, *8*, 12968–12979. [CrossRef]
102. Wang, M.; Li, P.; Nian, T.; Mao, Y. An overview of studies on the hazards, component analysis and suppression of fumes in asphalt and asphalt mixtures. *Constr. Build. Mater.* **2021**, *289*, 123185. [CrossRef]
103. Guo, P.; Feng, Y.; Wei, W.; He, L.; Tang, B. Adhesion of warm-mix recycled asphalt aggregate mixtures based on surface free energy theory. *J. Mater. Civ. Eng.* **2019**, *31*, 04019209. [CrossRef]
104. Tang, N.; Yang, K.-k.; Alrefaei, Y.; Dai, J.-G.; Wu, L.-M.; Wang, Q. Reduce VOCs and PM emissions of warm-mix asphalt using geopolymer additives. *Constr. Build. Mater.* **2020**, *244*, 118338. [CrossRef]
105. Zhao, Y.; Wang, M.; Liang, N.; Wang, Y.; Li, Z.; Li, X. Polyphosphazene functionalized magnesium hydroxide core-shell architecture fire retardant: Toward a fire-safe multifunctional styrene-butadiene-styrene asphalt. *Fuel* **2024**, *370*, 131822. [CrossRef]
106. Xu, G.; Chen, X.; Zhu, S.; Kong, L.; Huang, X.; Zhao, J.; Ma, T. Evaluation of Asphalt with Different Combinations of Fire Retardants. *Materials* **2019**, *12*, 1283. [CrossRef]
107. Wang, Y.; Huang, G.; Zhu, X. Emission reduction effect of eco-friendly asphalt modified with isocyanate-nanoclay composites. *J. Clean. Prod.* **2024**, *458*, 142530. [CrossRef]
108. Sheng, Y.; Wu, Y.; Yan, Y.; Jia, H.; Qiao, Y.; Underwood, B.S.; Niu, D.; Kim, Y.R. Development of environmentally friendly flame retardant to achieve low flammability for asphalt binder used in tunnel pavements. *J. Clean. Prod.* **2020**, *257*, 120487. [CrossRef]
109. Qi, L.; Tang, X.; Wang, Z.; Peng, X. Pore characterization of different types of coal from coal and gas outburst disaster sites using low temperature nitrogen adsorption approach. *Int. J. Min. Sci. Technol.* **2017**, *27*, 371–377. [CrossRef]
110. Cao, Z.; Gong, G.; Liu, Y.; Cao, X.; Tang, B.; Wang, C. Effects of different inhibitors on emission characteristics, environmental impact and health risk of volatile organic compounds from asphalt binder. *J. Clean. Prod.* **2023**, *428*, 139427. [CrossRef]
111. Lv, Y.; Wu, S.; Xu, H.; Li, N.; Liu, Q.; Wang, H.; Zou, Y.; Jiang, Q.; Amirkhanian, S. Simulation of Inhibition and Targeted Adsorption of Rubberized Asphalt VOCs by UiO-66 Based on Molecular Dynamics. *J. Clean. Prod.* **2024**, *447*, 141541. [CrossRef]
112. Liu, J.; Hao, P.; Dou, Z.; Wang, J.; Ma, L. Rheological, healing and microstructural properties of unmodified and crumb rubber modified asphalt incorporated with graphene/carbon black composite. *Constr. Build. Mater.* **2021**, *305*, 124512. [CrossRef]
113. Wang, H.; Liu, Q.; Wu, S.; Lv, Y.; Wan, P.; Gong, X.; Liu, G. Study of synergistic effect of diatomite and modified attapulgite on reducing asphalt volatile organic compounds emission. *Constr. Build. Mater.* **2023**, *400*, 132827. [CrossRef]
114. Rubio, M.C.; Martínez, G.; Baena, L.; Moreno, F. Warm mix asphalt: An overview. *J. Clean. Prod.* **2012**, *24*, 76–84. [CrossRef]
115. Tan, H.; Shan, G. Computational screening and functional tuning of chemically stable metal organic frameworks for I2/CH3I capture in humid environments. *iScience* **2024**, *27*, 109096. [CrossRef]
116. Liang, J.; Tan, H.; Liu, J.; Qi, H.; Li, X.; Wu, L.; Xue, X.; Shan, G. Combined experimental and theoretical studies on iodine capture of Zr-based metal-organic frameworks: Effect of N-functionalization and adsorption mechanism. *Mater. Today Sustain.* **2023**, *24*, 100574. [CrossRef]
117. Vo, T.K.; Le, V.N.; Quang, D.T.; Song, M.; Kim, D.; Kim, J. Rapid defect engineering of UiO-67 (Zr) via microwave-assisted continuous-flow synthesis: Effects of modulator species and concentration on the toluene adsorption. *Microporous Mesoporous Mater.* **2020**, *306*, 110405. [CrossRef]
118. Wu, R.; Xiao, Y.; Zhang, P.; Lin, J.; Cheng, G.; Chen, Z.; Yu, R. Asphalt VOCs reduction of zeolite synthesized from solid wastes of red mud and steel slag. *J. Clean. Prod.* **2022**, *345*, 131078. [CrossRef]
119. Li, J.; Zhang, X.; Wei, H.; Zhu, H.; Tang, B. Evaluation of volatile organic compounds emissions and rheological properties of Buton rock asphalt modified asphalt. *Constr. Build. Mater.* **2023**, *408*, 133575. [CrossRef]
120. Wang, S.; Tan, L.; Xu, T. Synergistic effects of developed composite flame retardant on VOCs constituents of heated asphalt and carbonized layer compositions. *J. Clean. Prod.* **2022**, *367*, 133107. [CrossRef]

121. Bi, F.; Ma, S.; Gao, B.; Liu, B.; Huang, Y.; Qiao, R.; Zhang, X. Boosting toluene deep oxidation by tuning metal-support interaction in MOF-derived Pd@ZrO_2 catalysts: The role of interfacial interaction between Pd and ZrO_2. *Fuel* **2024**, *357*, 129833. [CrossRef]
122. Zhang, X.; Gao, B.; Rao, R.; Bi, F.; Li, C.; Yue, K.; Wang, Y.; Xu, J.; Feng, X.; Yang, Y. Defects materials of Institut Lavoisier-125(Ti) materials enhanced photocatalytic activity for toluene and chlorobenzene mixtures degradation: Mechanism study. *J. Colloid Interface Sci.* **2024**, *660*, 423–439. [CrossRef] [PubMed]
123. Chen, W.; Zhao, H.; Xue, Y.; Chang, X. Adsorption Effect and Adsorption Mechanism of High Content Zeolite Ceramsite on Asphalt VOCs. *Materials* **2022**, *15*, 6100. [CrossRef]
124. Li, L.; Wu, S.; Liu, G.; Cao, T.; Amirkhanian, S. Effect of organo-montmorillonite nanoclay on VOCs inhibition of bitumen. *Constr. Build. Mater.* **2017**, *146*, 429–435. [CrossRef]
125. Li, L.; Zhou, T.; Cao, L.; Zhou, J.; Liu, Z.; Dong, Z. Characterization of emissions from rubber modified asphalt and their impact on environmental burden: Insights into composition variability and hazard assessment. *J. Hazard. Mater.* **2024**, *477*, 135336. [CrossRef]
126. Shi, H.; Xu, T.; Zhou, P.; Jiang, R. Combustion properties of saturates, aromatics, resins, and asphaltenes in asphalt binder. *Constr. Build. Mater.* **2017**, *136*, 515–523. [CrossRef]
127. Xia, W.; Xu, T.; Wang, H. Thermal behaviors and harmful volatile constituents released from asphalt components at high temperature. *J. Hazard. Mater.* **2019**, *373*, 741–752. [CrossRef]
128. Xu, T.; Shi, H.; Wang, H.; Huang, X. Dynamic evolution of emitted volatiles from thermal decomposed bituminous materials. *Constr. Build. Mater.* **2014**, *64*, 47–53. [CrossRef]
129. Kitto, A.M.; Pirbazari, M.; Badriyha, B.N.; Ravindran, V.; Tyner, R.; Synolakis, C.E. Emissions of Volatile and Semi-Volatile Organic Compounds and Particulate Matter from Hot Asphalts. *Environ. Technol.* **1997**, *18*, 121–138. [CrossRef]
130. Zhang, J.; Chen, M.; Leng, B.; Wu, S.; Chen, D.; Zhao, Z. Investigation on storage stability, H2S emission and rheological properties of modified asphalt with different pretreated waste rubber powder. *J. Clean. Prod.* **2024**, *456*, 142469. [CrossRef]
131. Xie, J.; Zhao, X.; Liu, Y.; Ge, D.; Wang, S.; Ding, Z.; Lv, S. Microbial treatment of waste crumb rubber: Reducing energy consumption and harmful emissions during asphalt production process. *J. Clean. Prod.* **2024**, *464*, 142778. [CrossRef]
132. Liu, J.; Qi, C.; Liu, Q.; Wang, H.; Wang, C.; Zhang, Y.; Wu, S.; Li, N. Study on the effects of organic-inorganic composite inhibitor on fume emissions of rubber modified asphalt. *Constr. Build. Mater.* **2024**, *455*, 139206. [CrossRef]
133. Lv, Y.; Wu, S.; Li, N.; Cui, P.; Liu, Q.; Amirkhanian, S. Screening and Environmental Risk Assessment of VOCs Preferential Pollutants in Net Flavor Rubber-Modified Asphalt Based on the Information Entropy Method. *Constr. Build. Mater.* **2024**, *420*, 135616. [CrossRef]
134. Zhang, R.; Tang, N.; Zhu, H.; Xi, Y.; Cheng, H.; Liu, J.; Li, R. Compositional analysis and quantitative evaluation of organic emissions from asphalt materials: Improvements and refinements. *J. Clean. Prod.* **2024**, *467*, 142936. [CrossRef]
135. Wu, W.; Zhao, B.; Wang, S.; Hao, J. Ozone and secondary organic aerosol formation potential from anthropogenic volatile organic compounds emissions in China. *J. Environ. Sci.* **2017**, *53*, 224–237. [CrossRef]
136. Zhao, X.; You, F. Cascading polymer macro-debris upcycling and microparticle removal as an effective life cycle plastic pollution mitigation strategy. *Environ. Sci. Technol.* **2023**, *57*, 6506–6519. [CrossRef]
137. Li, N.; Jiang, Q.; Wang, F.; Xie, J.; Li, Y.; Li, J.; Wu, S. Emission behavior, environmental impact and priority-controlled pollutants assessment of volatile organic compounds (VOCs) during asphalt pavement construction based on laboratory experiment. *J. Hazard. Mater.* **2020**, *398*, 122904. [CrossRef]
138. Grosjean, D.; Seinfeld, J.H. Parameterization of the formation potential of secondary organic aerosols. *Atmos. Environ.* **1989**, *23*, 1733–1747. [CrossRef]
139. Grosjean, D. In situ organic aerosol formation during a smog episode: Estimated production and chemical functionality. *Atmos. Environ. Part A Gen. Top.* **1992**, *26*, 953–963. [CrossRef]
140. Carter, W.P. *Updated Maximum Incremental Reactivity Scale and Hydrocarbon Bin Reactivities for Regulatory Applications*; California Air Resources Board: Sacramento, CA, USA, 2009; Volume 339.
141. Carter, W.P.L. Development of the SAPRC-07 chemical mechanism. *Atmos. Environ.* **2010**, *44*, 5324–5335. [CrossRef]
142. Gao, Z.; Hu, G.; Wang, H.; Zhu, B. Characterization and assessment of volatile organic compounds (VOCs) emissions from the typical food manufactures in Jiangsu province, China. *Atmos. Pollut. Res.* **2019**, *10*, 571–579. [CrossRef]
143. Yan, Y.; Peng, L.; Li, R.; Li, Y.; Li, L.; Bai, H. Concentration, ozone formation potential and source analysis of volatile organic compounds (VOCs) in a thermal power station centralized area: A study in Shuozhou, China. *Environ. Pollut.* **2017**, *223*, 295–304. [CrossRef]
144. Wang, Q.; Li, S.; Dong, M.; Li, W.; Gao, X.; Ye, R.; Zhang, D. VOCs emission characteristics and priority control analysis based on VOCs emission inventories and ozone formation potentials in Zhoushan. *Atmos. Environ.* **2018**, *182*, 234–241. [CrossRef]

145. Huang, C.; Wang, H.L.; Li, L.; Wang, Q.; Lu, Q.; de Gouw, J.A.; Zhou, M.; Jing, S.A.; Lu, J.; Chen, C.H. VOC species and emission inventory from vehicles and their SOA formation potentials estimation in Shanghai, China. *Atmos. Chem. Phys.* **2015**, *15*, 11081–11096. [CrossRef]
146. Na, K.; Song, C.; Cockeriii, D. Formation of secondary organic aerosol from the reaction of styrene with ozone in the presence and absence of ammonia and water. *Atmos. Environ.* **2006**, *40*, 1889–1900. [CrossRef]
147. Sato, K.; Nakao, S.; Clark, C.H.; Qi, L.; Cocker Iii, D.R. Secondary organic aerosol formation from the photooxidation of isoprene, 1,3-butadiene, and 2,3-dimethyl-1,3-butadiene under high NO_x conditions. *Atmos. Chem. Phys.* **2011**, *11*, 7301–7317. [CrossRef]
148. Zhang, Z.; Wang, H.; Chen, D.; Li, Q.; Thai, P.; Gong, D.; Li, Y.; Zhang, C.; Gu, Y.; Zhou, L.; et al. Emission characteristics of volatile organic compounds and their secondary organic aerosol formation potentials from a petroleum refinery in Pearl River Delta, China. *Sci. Total Environ.* **2017**, *584–585*, 1162–1174. [CrossRef]
149. Zheng, H.; Kong, S.; Chen, N.; Niu, Z.; Zhang, Y.; Jiang, S.; Yan, Y.; Qi, S. Source apportionment of volatile organic compounds: Implications to reactivity, ozone formation, and secondary organic aerosol potential. *Atmos. Res.* **2021**, *249*, 105344. [CrossRef]
150. Derwent, R.G.; Jenkin, M.E.; Utembe, S.R.; Shallcross, D.E.; Murrells, T.P.; Passant, N.R. Secondary organic aerosol formation from a large number of reactive man-made organic compounds. *Sci. Total Environ.* **2010**, *408*, 3374–3381. [CrossRef]
151. Li, H.; Jia, M.; Zhang, X.; Wang, Z.; Liu, Y.; Yang, J.; Yang, B.; Sun, Y.; Wang, H.; Ma, H. Laboratory Investigation on Fumes Generated by Different Modified Asphalt Binders. *Transp. Res. Part D-Transp. Environ.* **2023**, *121*, 103828. [CrossRef]
152. Wang, M.; Wang, C.; Huang, S.; Yuan, H. Study on asphalt volatile organic compounds emission reduction: A state-of-the-art review. *J. Clean. Prod.* **2021**, *318*, 128596. [CrossRef]
153. Cui, P.Q.; Wu, S.P.; Xiao, Y.; Zhang, H.H. Experimental study on the reduction of fumes emissions in asphalt by different additives. *Mater. Res. Innov.* **2015**, *19*, S1-158–S151-161. [CrossRef]
154. Xie, J.; Ding, Z.; Luo, H.; Li, S.; He, W.; Zhao, X.; Ma, Y. Study on high-temperature emission behaviors of crumb rubber modified asphalt: Experimental analysis and molecular simulation approach. *J. Clean. Prod.* **2024**, *478*, 143977. [CrossRef]
155. Zhang, J.; Chen, M.; Wan, L.; Wu, S.; Zhao, Y.; Chen, D.; Fan, Y. Effect of recycled agricultural waste on fume emission and physicochemical properties of crumb rubber modified asphalt. *Constr. Build. Mater.* **2024**, *438*, 137093. [CrossRef]
156. Wang, Z.; Li, H.; Jia, M.; Du, Q. Emission Risk and Inhibition Technology of Asphalt Fume from Crumb Rubber Modified Asphalt. *Sustainability* **2024**, *16*, 8840. [CrossRef]
157. Betancur, M.; Martínez, J.D.; Murillo, R. Production of activated carbon by waste tire thermochemical degradation with CO_2. *J. Hazard. Mater.* **2009**, *168*, 882–887. [CrossRef] [PubMed]
158. Li, H.; Feng, Z.; Liu, H.; Ahmed, A.T.; Zhang, M.; Zhao, G.; Guo, P.; Sheng, Y. Performance and inorganic fume emission reduction of desulfurized rubber powder/styrene–butadiene–styrene composite modified asphalt and its mixture. *J. Clean. Prod.* **2022**, *364*, 132690. [CrossRef]
159. Cheng, Z.; Li, M.; Li, J.; Lin, F.; Ma, W.; Yan, B.; Chen, G. Transformation of nitrogen, sulfur and chlorine during waste tire pyrolysis. *J. Anal. Appl. Pyrolysis* **2021**, *153*, 104987. [CrossRef]
160. Tozzi, K.A.; Canto, L.B.; Scuracchio, C.H. Reclaiming of Vulcanized Rubber Foam Waste from the Shoe Industry Through Solid-state Shear Extrusion and Compounding with SBR. *Macromol. Symp.* **2020**, *394*, 2000094. [CrossRef]
161. Zhou, T.; Li, L.; Liu, R.; Yu, F.; Dong, Z. Thermal effects on rheological characteristics and its evolution mechanism of bio-modified rubberized asphalt for sustainable road system: A focus on rubber particle size. *J. Clean. Prod.* **2024**, *452*, 142168. [CrossRef]
162. Borinelli, J.B.; Portillo-Estrada, M.; Costa, J.O.; Pajares, A.; Blom, J.; Hernando, D.; Vuye, C. Emission reduction agents: A solution to inhibit the emission of harmful volatile organic compounds from crumb rubber modified bitumen. *Constr. Build. Mater.* **2024**, *411*, 134455. [CrossRef]
163. Urban, P.L. Quantitative mass spectrometry: An overview. *Philos. Trans. R. Soc. A Math. Phys. Eng. Sci.* **2016**, *374*, 20150382. [CrossRef] [PubMed]
164. Chi, K.H.; Chang, S.H.; Huang, C.H.; Huang, H.C.; Chang, M.B. Partitioning and removal of dioxin-like congeners in flue gases treated with activated carbon adsorption. *Chemosphere* **2006**, *64*, 1489–1498. [CrossRef]
165. Li, S.; Liu, Q.; Wang, H.; Wang, J.; He, L.; Wu, S. Effect of kaolin and sepiolite on fume emissions of rubber modified asphalt. *Constr. Build. Mater.* **2024**, *416*, 135276. [CrossRef]

coatings

MDPI

Article

Study on Infrared Spectrum Analysis of Asphalt Aging Based on Full Spectrum Segmentation

Hao Zhou [1,*], Pinhui Zhao [1], Qiang Xu [1], Kechao Han [2], Guixia Zhang [3], Jipeng Cui [4], Weikun Meng [5] and Wenxin Wu [5]

1 School of Transportation Engineering, Shandong Jianzhu University, Jinan 250101, China; 13024@sdjzu.edu.cn (Q.X.)
2 Jinan Engineering Consulting Institute, Jinan 250000, China; hankechao@163.com
3 Shandong Institute of Highway Technician, Jinan 250104, China
4 Shandong Provincial Transportation Planning and Design Institute Group Co., Ltd., Jinan 250101, China; jipengcui@163.com
5 Shandong High-Speed Transportation Construction Group Co., Ltd., Jinan 250000, China; 2165690@163.com (W.W.)
* Correspondence: zhouhao@sdjzu.edu.cn

Abstract: Infrared spectroscopy is an effective method for studying asphalt aging. To improve the effectiveness of infrared spectroscopy in asphalt aging, this paper first performs preprocessing of the infrared spectral signals and divides the full spectrum into six segments. By quantitative analysis of the relation of noise and signals of different segments, a correlation model between noise and errors was established and a reasonable signal analysis range was determined. A stable method for describing the aging degree, which is based on the area change values of the infrared spectral curve over a wide wavelength range, was determined, and the aging model was established. Based on the relationship between PAV aging and natural aging, a model has been established to describe the relationship between the spectral area increase and aging in the asphalt pavement aging process. This model can be used to determine the actual service time of the asphalt pavement through infrared spectral analysis.

Keywords: asphalt aging; infrared spectrum analysis; signal preprocessing

1. Introduction

The asphalt pavement, with superior performance and driving comfort, accounts for a large proportion in high-grade highways and urban roads around the world [1,2]. However, asphalt pavements generally suffer from early diseases such as cracking, rutting, and raveling, as well as short service life [3]. Apart from design and construction reasons, the main factors affecting the performance [4] and durability of asphalt pavements under normal construction conditions are aging and water damage, especially the aging problem, which is the most common and difficult to prevent. Asphalt aging refers to the physical and chemical changes that occur on the asphalt's chemical composition and microstructure due to long-term exposure to air during storage, transportation, construction, and road use processes, under the influence of environmental factors such as heat, oxygen, sunlight (UV), and water. In the early stages of asphalt aging, stiffness and elasticity increase, and viscosity decreases [5–7]. While in further aging, asphalt modulus increases, leading to the formation of numerous cracks in the asphalt pavement [8], which deteriorates rapidly and eventually fails to meet the requirements of traffic loads. In conclusion, asphalt aging significantly affects the performance of pavement. Therefore, studying the aging process of asphalt is of great significance [9].

Currently, a large amount of research has been carried out on asphalt aging, and many experimental methods and indicators have been proposed to evaluate the aging degree of asphalt. So far, the most widely studied asphalt aging degree indicators are still the changes in physical property indicators. Many scholars have studied asphalt aging by conducting aging experiments on asphalt and then analyzing the changes in asphalt physical properties such as penetration, softening point, ductility, viscosity, four component contents, creep stiffness, and dynamic viscoelasticity [10,11]. After years of uninterrupted experimental exploration, pavement researchers have gained a profound understanding of the changes in physical properties and performance degradation of asphalt after aging. It is generally believed that the aging performance changes of different asphalts follow a similar pattern, namely, with the increase of aging time, the penetration of asphalt gradually decreases, the penetration index PI increases, the softening point increases, the ductility decreases, while the viscosity, complex shear modulus, and creep stiffness gradually increase. This indicates that aging enhances the elasticity of asphalt, weakens temperature sensitivity, impairs fatigue cracking resistance, and shortens the service life of the pavement.

Although the aging performance of asphalt changes consistently, the differences in anti-aging performance and durability among different asphalts are significant, which is the external manifestation of the asphalt's chemical composition, molecular structure, and transformation. Therefore, studying the influence of the asphalt's chemical composition and molecular structure on its aging performance is of great significance for revealing the aging laws and mechanisms of asphalt [12,13]. With the advancement of chemical analysis techniques, research on asphalt aging has gradually shifted from changes in macroscopic indicators to understanding at the microscopic level. Asphalt belongs to organic molecular materials, and many researchers use chemical microscopic characterization methods such as infrared spectroscopy, gel chromatography, elemental analysis, thermogravimetry, mass spectrometry, etc., to study the changes in the microscopic properties of asphalt before and after aging and explore the aging mechanism of asphalt. Studies have shown that the aging of asphalt is mainly due to the reaction of active groups in asphalt with oxygen in the air, and the oxidation products are mainly carbonyl and sulfoxide functional groups, with the possibility of generating carboxyl functional groups during deep aging; in addition, aging also causes changes in the composition and structure of asphalt [14], such as the volatilization of light components, the transformation of aromatic components into resins, and the transformation of resins into asphaltene (A (aromatic fraction) $\rightarrow$ R (resin) $\rightarrow$ As (asphaltene)); at the same time, due to the association and condensation between molecules, the average molecular weight of aged asphalt increases.

However, there are some problems in current research, whether based on physical performance indicators or microstructure characterization, such as long testing times, the need for expensive precision instruments, and stringent test conditions. Especially in the research of the long-term aging performance of asphalt, there are still significant limitations [15,16]. Furthermore, predicting and evaluating the aging performance changes of asphalt during long-term use is one of the key technologies for studying the service life of asphalt pavements [17]. The structural performance of asphalt pavements is directly related to the aging degree of the asphalt in order to clarify the actual situation of the pavement, reverse research should be conducted through the indicators of asphalt aging to achieve optimal control of the timing for preventive maintenance of asphalt pavements. However, there is currently limited research on the identification of asphalt aging levels and the prediction of macro performance indicators, which hinders the development of the road construction and maintenance industry. Therefore, it is crucial to study a fast and convenient method to predict the service time and performance indicators of asphalt.

The infrared spectroscopy technique is widely used in asphalt material research for its outstanding advantages, such as easy sample preparation, being label free, fast, non-destructive, and having low analysis cost [18,19]. Infrared spectroscopy can reflect the fundamental frequency information of the vibration and rotation of material molecules, which is the functional group information. By analyzing the intensity of the absorption peaks, it is possible to qualitatively and quantitatively identify the functional groups present in asphalt materials. The infrared spectra of asphalt samples generally exhibit

12 distinct characteristic absorption peaks, corresponding to the following wavenumbers: 723 cm^{-1}, 812 cm^{-1}, 872 cm^{-1}, 969 cm^{-1}, 1032 cm^{-1}, 1166 cm^{-1}, 1380 cm^{-1}, 1460 cm^{-1}, 1600 cm^{-1}, 1700 cm^{-1}, 2850 cm^{-1}, 2920 cm^{-1}. Among them, the strong absorption peaks near 2920 cm^{-1} and 2850 cm^{-1} are attributed to the stretching vibration of the methylene C-H bond; the very weak absorption peak near 1700 cm^{-1} corresponds to the stretching vibration of the C=O bond; at 1600 cm^{-1}, and in the critical region below 1300 cm^{-1}, the absorption peak at 1032 cm^{-1} is caused by the stretching vibration of the S=O bond; the weak absorption peak near 969 cm^{-1} is due to the stretching vibration of C-H; the absorption peaks appearing between 900–700 cm^{-1} are attributed to the out-of-plane rocking vibration absorption peaks of =C-H on the benzene ring. Ren et al. [16,19] identified asphalt samples from different oil sources by using Fourier transform infrared spectroscopy and chemometrics. Petersen [20] found that the content of sulfoxide functional groups reaches a maximum value during asphalt aging. Hda B et al. [21] discovered that the peak positions in the 2919–2856 cm^{-1} range shift to lower wavenumbers at lower temperatures, based on Fourier transform infrared spectroscopic analysis of aging experiments at 20 h and 60 h. In contrast, the peak positions in the 1456–1465 cm^{-1} range shift to higher wavenumbers. Ge DD et al. [22] found that the IC=O bond and IS=O bond are important indicators to evaluate the extent of asphalt aging at different temperatures. Xu Meng et al. [23] also found that the IC=O bond has the greatest impact on asphalt aging behavior. Despite many researchers using infrared spectroscopy to study asphalt aging behavior.

However, when using infrared spectroscopy to test complex compounds such as asphalt, there are often large differences in absorption peak areas and serious interference from noise, which leads to significant impacts on the accuracy of infrared analysis by the selection of test spectrum segments and algorithms. Therefore, many scholars have adopted correlation algorithms to process the spectra and have achieved good results. Zhihui Li, et al. [15], after comparing various algorithms, including Monte Carlo uninformative variable elimination (MC-UVE), moving window partial least squares (mwPLS) with support vector machine (SVM) classifiers, determined 42 discriminant wavenumbers and examined their stability, providing a feasible method for the rapid identification of asphalt quality on road construction sites. B. Hofko, et al. [24], based on a large amount of data, conducted a statistical study on the reproducibility of Fourier transform infrared (FTIR) spectroscopy analysis, using the coefficient of variation (CV) as the benchmark parameter, and calculated the carbonyl index and sulfoxide index using baseline correction, tangent integration, and baseline integration methods. Kang Wang, et al. [25], combined partial least squares regression (PLS), background correction, and log-scaled unit variance (LSUV) to achieve detection across the entire spectrum range, demonstrating robustness in determining the SBS content in modified asphalt in the mid-infrared region. Chuanqi Yan, et al. [26] used PLS for quantitative analysis of REOB content in asphalt binders and compared the accuracy with traditional PIB peak analysis, obtaining an accurate prediction model. Sun XT, et al. [27], combined attenuated total reflection Fourier transform infrared spectroscopy (ATR-FTIR) with spectral representation algorithms to quickly determine spectral characteristics.

Although researchers have provided many calibration models for asphalt infrared spectroscopy, the establishment of these models often requires a large amount of sample data, and they have disadvantages such as high workload, high maintenance costs, and poor stability and cannot be related to the macroscopic performance indicators of asphalt. To address these issues, this study proposes a new method combining spectral segmentation and signal preprocessing, which solves the problem of poor stability of infrared measurement technology. Specifically, this method divides the full spectrum into six spectral segments for quantitative analysis of noise and signals, establishes the correspondence between noise and errors, and achieves detection of the entire spectral range. This method has the advantages of accuracy, speed, and directness.

2. Materials and Methods

2.1. Material

The matrix asphalt was selected from China, ZH AH-70 asphalt, referring to the Chinese standard JTG E20-2011 [28]. The matrix asphalt was short-term aged using the thin film oven test (TFOT), followed by long-term aging for 5 h, 10 h, 15 h, and 20 h to obtain six asphalt samples with different aging levels. According to JTG E20-2011, the physical performance indicators such as penetration, softening point, ductility, rotational viscosity, and dynamic viscosity of the 6 asphalt samples were tested, and the results are shown in Table 1.

Table 1. Physical performance indicators of asphalt samples.

Test Items	Test Results	Technical Requirements
Penetration (25 °C, 100 g, 5 s)/0.1 mm	64.4	60–80
Softening point (R&B)/°C	48.7	≥46
Ductility (15 °C, 5 cm/min)/cm	>100	≥100
Flash point/°C	267	≥260
Solubility/%	99.8	≥99.5
Relative density (25 °C/25 °C)	1.030	As measured

2.2. ATR-FTIR Analysis

Fourier transform infrared spectroscopy (FTIR), also known as mid-infrared spectroscopy, has a wavelength range of 4000 cm^{-1}~400 cm^{-1}. It is based on the vibrational and rotational fundamental frequency information of molecules, specifically functional group information, combined with the strength of absorption peaks to identify compounds [29]. Infrared spectroscopy includes four acquisition modes: transmission, attenuated total reflectance (ATR), diffuse reflectance, and specular reflectance (Microscope). ATR is generally considered the best FTIR test mode for characterizing asphalt, as it provides faster sample preparation and excellent sample-to-sample reproducibility. Therefore, this paper used ATR-FTIR (Thermo fisher IS5, Thermo Fisher Scientific, USA) to measure the infrared spectrum of asphalt samples. The scan number is 128, the resolution is 4 cm^{-1}, and the range is 4000–400 cm^{-1}. The initial infrared files obtained from the test were pre-processed using Thermo Scientific™ OMNIC Specta™ software, including baseline correction to eliminate the effects of baseline differences, and the spectrum files were saved in CSV format for future use.

2.3. Signal Preprocessing

The original signal of the infrared spectrum is influenced by electrical noise, sample background, and stray light, all of which contain certain errors. Therefore, in order to improve the accuracy and stability of the model, noise in the spectrum should be eliminated and the signal-to-noise ratio of the spectrum should be improved before establishing a regression model for the infrared spectrum. This process is called spectral preprocessing. Currently, widely used preprocessing methods in near-infrared spectroscopic analysis techniques include mean centering, standardization, derivatives, smoothing, standard normal variate transformation, and multivariate scatter correction.

(1) Mean centering (MC)

This method is commonly used in the process of establishing quantitative and qualitative models. It can increase the differences between sample infrared spectra, thereby improving the stability and predictive ability of the models. It is one of the most commonly used preprocessing methods for infrared spectra. The process of mean centering transformation of sample infrared spectra is shown in Equation (1) below:

$$x_u = x_{ik} - \overline{x}_k \tag{1}$$

The $\overline{x}_k$—mean value of n samples.
x_u—mean-centered feature values.
x_{ik}—original feature values.

(2) Standardization

The standardized variation of the spectrum is based on the mean-centered infrared spectrum, and the specific calculation process is shown in Equation (2):

$$x\prime_{ik} = \frac{x_{ik} - \overline{x}_k}{S_k}; S_k = \sqrt{\frac{1}{n-1}\sum_{i=1}^{n}(x_{ik} - \overline{x}_k)^2} \quad (2)$$

x'_{ik}—standardized feature value.
x_{ik}—original feature value.
S_k—standard deviation.

(3) Normalization Processing

The purpose of normalization is to make data points fill the data space. The commonly used method for normalization is interval normalization. This method involves subtracting each element in the original data set by the minimum value in the column and then dividing by the range of that column, as shown in Equation (3):

$$x'_{ik} = \frac{x_{ik} - \min(xk)}{\max(x_k) - \min(x_k)} \quad (3)$$

x'_{ik}—standardized feature value.
x_{ik}—original feature value.
$\min(xk)$—minimum value in the original data.
$\max(x_k)$—maximum value in the original data.
This method can express variables with different dimensions and ranges as data that are all within the range of 0 to 1. However, this method is very sensitive to outliers, and if outliers exist, all processed data will be nearly equal.

(4) Derivative method

Spectra may undergo shifts, tilts, and other positional changes due to noise interference, which can lead to spectral overlap and affect subsequent modeling. The derivative preprocessing method can effectively address this issue by eliminating background interference, correcting drift in baselines, separating overlapping spectra, enhancing spectral resolution, and sensitivity. However, at the same time, the derivative method can introduce noise, reduce the signal-to-noise ratio of spectra, and ultimately lower the predictive ability and regression effect of models. There are two common derivative methods: first-order derivative (1st derivative) and second-order (2nd derivative) derivative, as in Equations (4) and (5):

1st derivative

$$X_{k,1st} = \frac{x_{k+i} - x_{k-i}}{i} \quad (4)$$

2nd derivative

$$X_{k,2nd} = \frac{x_{k+i} - 2x_k + x_{k-i}}{i^2} \quad (5)$$

x_{k+i}—a certain data point in the original spectral data.
$X_{k,1st}$—derivative value.
$X_{k,2nd}$—second derivative value.

(5) Multiple scattering correction

The purpose of multiple scattering correction is the same as standard normal variable transformation, mainly to eliminate the scattering effects caused by the uneven distribution of particles and the sizes of particles, often used to process solid diffuse reflection and transmittance (reflection) near-infrared spectra of pasty materials, with the specific process as shown in Equations (6)–(8):

$$\text{Average spectrum}: \ \overline{A} = \sum_{i=1}^{n} A_i \Big/ n \tag{6}$$

$$\text{Linear regression}: \ A_i = m_i\overline{A} + b_i \tag{7}$$

$$\text{MSC calibration}: \ A_{i(MSC)} = \frac{A_i - b_i}{m_i} \tag{8}$$

A_i—original spectral data.
n—the number of samples.

(6) Smoothing Filter

The basic idea of this method is to measure spectral data multiple times to obtain the average distribution error in order to reduce the impact of errors on useful information in the spectrum. The relationship between the two is as in Equation (9):

$$y*_k = \frac{1}{2m+1}\sum_{j=-m}^{m} y_{k+j} \tag{9}$$

y_k—original signal value.
$y*_k$—filtered value.

In the formula, k represents the index of the original data point, $2m + 1$ indicates the size of the window (filter bandwidth), and m is the parameter for adjusting the filter bandwidth. As the window moves, a series of filtered signal values are obtained. Filtering reduces noise in the data, but it is important to carefully select the filter bandwidth in practical applications to avoid distorting the data structure.

After testing, five times of smoothing filtering can effectively filter out the spectral noise while preserving the original signal information, achieving the purpose of noise reduction. The specific process of five times smoothing filtering is as Equations (10)–(14):

$$\overline{Y_{-2}} = \frac{1}{70}(69Y_{-2} + 4Y_{-1} - 6Y_0 + 4Y_1 - Y_2) \tag{10}$$

$$\overline{Y_{-1}} = \frac{1}{30}(2Y_{-2} + 27Y_{-1} + 12Y_0 - 8Y_1 + 2Y_2) \tag{11}$$

$$\overline{Y_0} = \frac{1}{35}(-3Y_{-2} + 12Y_{-1} + 17Y_0 + 12Y_1 - 3Y_2) \tag{12}$$

$$\overline{Y_1} = \frac{1}{35}(2Y_{-2} - 8Y_{-1} + 12Y_0 + 27Y_1 + 2Y_2) \tag{13}$$

$$\overline{Y_2} = \frac{1}{70}(-Y_{-2} + 4Y_{-1} - 6Y_0 + 4Y_1 + 69Y_2) \tag{14}$$

The $\overline{Y_i}$ is improvement value of Y_i.

2.4. Asphalt Infrared Spectral Analysis

During the aging process of asphalt, the absorbance peak strength of certain functional groups may increase. Therefore, based on the Lambert–Beer law, quantitative analysis of the FTIR spectra of asphalt binders can be performed [30]. Currently, there is no unified standard for the quantitative analysis of FTIR data of asphalt binders. Many studies use the peak area of absorbance peaks for analysis. To establish a stable and systematic FTIR spectroscopic quantitative analysis method, this study plans to select different spectral areas as reference benchmarks. As shown in Figure 1, the area enclosed by O_1, O_2, O_3, and O_4 is used as the reference point to calculate the difference in the infrared spectra between aged asphalt and original asphalt in the 900–1020 cm^{-1} range. By systematically comparing the influence of different reference benchmarks on the quantitative analysis of the infrared spectra of asphalt at different aging levels, a method for quantitative analysis of asphalt infrared spectra will be established. Through the quantitative analysis of infrared spectra of aged asphalt during the aging process, the aging law of asphalt and the prediction of changes in performance indicators of asphalt will be studied.

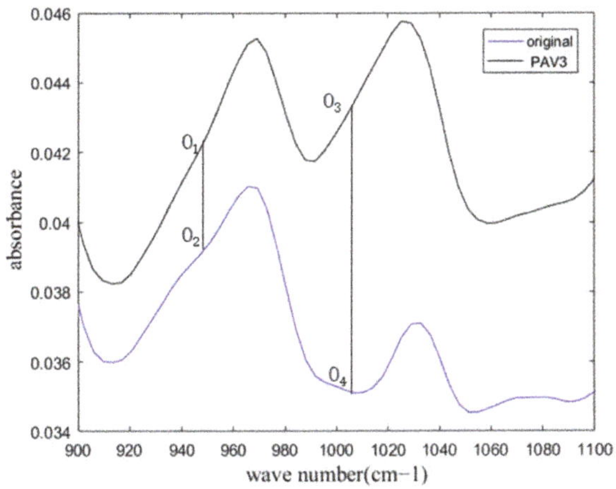

Figure 1. Infrared spectroscopy projection area.

Establishment of an asphalt performance prediction model based on infrared spectroscopy.

Based on the analysis above, a linear fitting is performed between the area of the infrared spectrum curve of asphalt not passing the aging degree and the macro performance indicators in order to obtain the correlation between the two and use it for predicting the performance of asphalt. The model is as Equation (15):

$$A = \mathrm{a} \times \mathrm{T} + \mathrm{b} \tag{15}$$

A is the area of the infrared absorption peak, with the unit of cm^{-1}%, and T is the aging time, with the unit of h.

And evaluate the model's fitting accuracy by using the coefficient of determination R^2, where the calculation method of R^2 is as Equation (16):

$$R^2 = 1 - \frac{\sum_{i=1}^{n} (y_i - \hat{y}_i)^2}{\sum_{i=1}^{n} (y_i - \hat{y}_m)^2} \tag{16}$$

3. Results

3.1. Signal Noise Anslysis

As shown in Figure 2a, the blue curve represents the original infrared spectrum values before asphalt aging, and the red curve represents the difference in spectral transmittance before and after denoising. It can be concluded that the bands with drastic changes in the original curve waveform have relatively high noise errors, while the bands with gentle changes in the original curve have relatively small errors. In order to quantitatively analyze the infrared spectrum, the full spectrum is divided into six spectral segments and calculated separately to investigate the relationship between noise errors in different spectral segments and the wavenumber curvature of the curve, as shown in Figure 2b. It can be observed that noise errors have a good correlation with spectral wavenumber curvature. The larger the curvature, the larger the noise error, and a relationship between the two is established, as shown in Equation (17).

$$E_n = 0.1521S_c + 0.0047 \quad (17)$$

E_n—error of noise.

S_c—spectral wavenumber curvature.

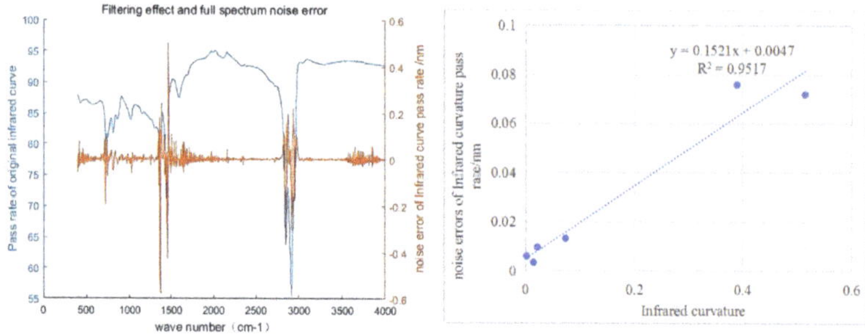

(a) (b)

Figure 2. Infrared spectrum noise error analysis. (**a**) Filtering effect and full spectrum noise error; (**b**) relationship between noise errors and spectral wavenumber curvature.

By the infrared spectrum noise error figure, it could be concluded that the error increases as the spectral curvature increases. It should be noted in the subsequent signal analysis that errors are larger at locations where the spectral curvature changes dramatically, which can be avoided or dealt with by performing in-depth preprocessing.

The quantitative analysis results of the signals and noise for the six segments of the full spectrum range of asphalt spectrum are shown in Table 2, with the signal in the 400–650 waveband serving as the baseline value. The signal of each segment compared to the 400–650 waveband indicates its signal is a multiple of the first segment. It can be concluded that the signal-to-noise ratio is higher in the 650–1800, 2750–3000 wavebands, which can be used as key analysis bands; while the signal-to-noise ratio is lower in the 3000–4000 waveband, special attention should be paid to the impact of noise.

Table 2. Study on the noise laws of infrared spectrum.

Band	Noise	Signal
400–650	1	1
650–1300	1.3838	3.1956
1300–1800	7.7475	17.2311
1800–2750	0.3939	0.7244
2750–3000	7.3333	22.7556
3000–4000	0.5657	0.3600

3.2. *Analysis of Spectral Changes During Asphalt Aging Process*

3.2.1. Comparison of Spectra Before and After Asphalt Aging

FTIR tests were conducted on asphalt binders with different aging periods to research the changes in infrared spectra of asphalt during the aging process, as shown in Figure 3. In Figure 3a, the original asphalt, short-term aged asphalt, and four long-term aged asphalt samples are presented. It can be observed from the figure that the overall shape of the asphalt spectrum remains consistent before and after aging, with an increase in absorbance.

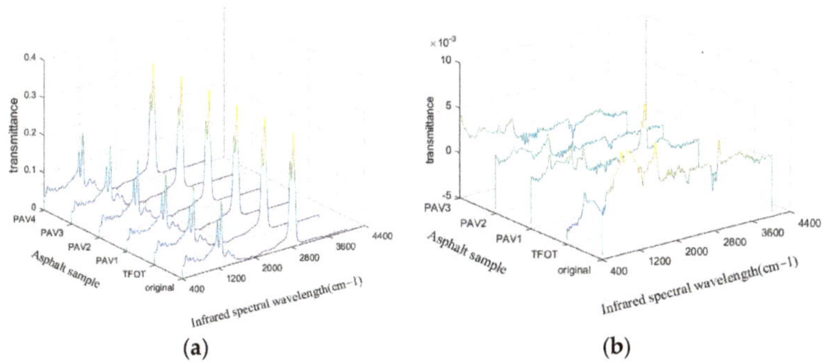

Figure 3. Comparison of infrared spectra between original asphalt and aged asphalt. (**a**) original bitumen and aged bitumen Infrared spectra; (**b**) difference spectra between aged bitumen and original bitumen.

Figure 3b shows the results of the infrared spectroscopy absorbance values of short-term aging and four different aging periods of asphalt minus the absorbance values of the original asphalt. It can be concluded from the figure that, from different spectral regions, the difference curves differ significantly from the shape of the asphalt spectrum itself, with less prominent peak protrusion and the peak region not corresponding to the peak region of the original spectrum curve. In terms of the aging process, the growth trend of the difference curves for the four different aging periods of long-term aging is relatively stable.

In order to accurately compare the spectral difference curve and the original value curve, the PAV-4-original difference curve is compared with the original asphalt curve. As shown in Figure 4, the blue curve represents the original spectral curve of asphalt, and the red curve represents the difference curve. It can be concluded that the difference curve has 4 peaks, with two major peaks (800–1300, 1600–1750) on both sides of the original curve peak at 1400, and the position of the peak in the original curve is exactly the valley in the difference curve. However, the overall difference in the 600–1750 wavenumber range is relatively high (about 3/4 of the whole curve) when viewed in full spectrum, which can achieve a more stable area calculation value. The local band difference curve is negative, indicating that the value of the difference spectral transmittance may decrease locally during the aging process.

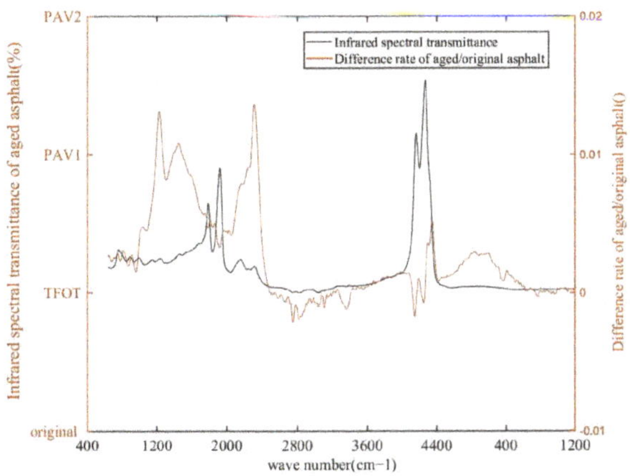

Figure 4. Comparison of asphalt spectrum difference–original spectrum curve.

3.2.2. Study on the Different Sectional Variation Laws of Infrared Curves

The asphalt spectrum varies greatly in different spectral regions. In order to accurately analyze the changes in the asphalt spectrum, this paper conducts a segmented study of the spectrum, and the results are shown in Figure 5, comparing the spectra of aged asphalt and original asphalt. The spectra of asphalt aged in the short term cannot be analyzed together with asphalt aged in the long term and require separate research. The regularity of asphalt aged in different age periods in the long term is very good, with a stable increase in the difference values, and the results of the subdivision comparison indicate that the curves in the 600–1600 wavebands have large values and stable growth rates, showing the best regularity. It indicates that in the long-term aging process, the infrared curve of asphalt varies regularly, and the analysis of infrared spectra of aged asphalt can well characterize the aging process of asphalt.

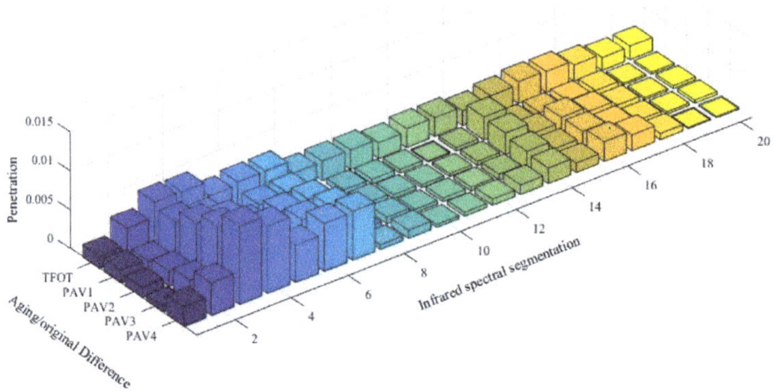

Figure 5. Spectral difference curve segmented comparison.

3.3. *Study on Variation Law of the Infrared Curve with Aging Time in the Aging Process*

Increase the area of the red infrared spectrum curve and focus on the aging process of asphalt with aging durations of 5 h, 10 h, 15 h, and 20 h.

The specific fitting accuracy is as Table 3.

Table 3. Accuracy of fitting of asphalt infrared curve aging degree with time segments.

Segment	R^2 Value	Segment	R^2 Value
1	0.1038	11	0.3832
2	0.9434	12	0.9778
3	0.9956	13	0.9035
4	0.9978	14	0.9764
5	0.9663	15	0.4722
6	0.9826	16	0.8505
7	0.9949	17	0.8878
8	0.8974	18	0.2834
9	0.9592	19	0.3210
10	0.9904	20	0.1171

Segments 3, 4, 7, and 10 have fitting accuracies of above 0.99. This suggests that the infrared changes in asphalt during the aging process are very stable, and these four sections can be used to reflect the aging time of the asphalt.

The relationship between the increase in spectral area and aging time is represented by Equations (18) and (19) in the 3rd and 4th paragraphs.

$$A = 16.89T - 10.14 \tag{18}$$

$$A = 19.13T - 9.527 \quad (19)$$

The aging time can also be estimated by increasing the area value, specifically as in Equations (20) and (21):

$$T = 0.05920A + 0.003335 \quad (20)$$

$$T = 0.05225A + 0.002766 \quad (21)$$

Referring to Geng Jiuguang's [31] research on the relationship between PAV aging and natural aging, a general correspondence between the increase in spectral area and aging time can be established during the aging process of asphalt, as in Equations (22) and (23).

$$T_m = 0.2344A + 0.01321 \quad (22)$$

$$T_m = 0.2069A + 0.01095 \quad (23)$$

Among Equations (18)–(23), A stands for area, with the unit of: T stands for time, with the unit of month.

4. Conclusions

1. The noise error is not significantly related to the spectral wavenumber but shows a good correlation with the curvature of the spectral wavenumber. A relationship model between noise error and spectral wavenumber curvature has been established.
2. Through quantitative analysis using signal-to-noise ratio (SNR), it was determined that in the 1350–1480 cm^{-1} and 2850–2950 cm^{-1} regions, the infrared spectrum noise is higher, and the area increase during aging is small, making these regions unsuitable for area analysis of asphalt aging. In contrast, in the 900–1300 cm^{-1} and 1550–1800 cm^{-1} regions, the noise is lower, and the area increase during aging is larger, making them suitable for area analysis of asphalt aging.
3. Based on the analysis of the area change pattern of the infrared spectrum curves in selected key segments, a reliable model for asphalt aging has been established.
4. Based on the relationship between PAV aging and natural aging, a model has been established to describe the relationship between the spectral area increase and aging in the asphalt pavement aging process. This model can be used to determine the actual service time of the asphalt pavement through infrared spectral analysis.

Author Contributions: Conceptualization, H.Z.; methodology, P.Z.; software, H.Z. and K.H.; validation, Q.X. and J.C.; formal analysis, Q.X.; investigation, W.M.; resources, W.W.; data curation, J.C.; writing—original draft preparation, K.H. and G.Z.; writing—review and editing, H.Z. and P.Z.; visualization, G.Z.; supervision, Q.X.; project administration, P.Z.; funding acquisition, H.Z. All authors have read and agreed to the published version of the manuscript.

Funding: This research received no external funding.

Institutional Review Board Statement: Not applicable.

Informed Consent Statement: Not applicable.

Data Availability Statement: Data are contained within the article.

Conflicts of Interest: Author Jipeng Cui was employed by the company Shandong Provincial Transportation Planning and Design Institute Group Co., Ltd. Authors Weikun Meng and Wenxin Wu were employed by the company Shandong High-Speed Transportation Construction Group Co., Ltd. The remaining authors declare that the research was conducted in the absence of any commercial or financial relationships that could be construed as a potential conflict of interest.

References

1. Wang, K.; Yuan, Y.; Han, S.; Yang, H. Application of attenuated total reflectance Fourier transform infrared (ATR-FTIR) and principal component analysis (PCA) for quick identifying of the bitumen produced by different manufacturers. *Road Mater. Pavement Des.* **2017**, *19*, 1940–1949. [CrossRef]
2. Wang, J.; Yuan, J.; Kim, K.W.; Xiao, F. Chemical, thermal and rheological characteristics of composite polymerized asphalts. *Fuel* **2018**, *227*, 289–299. [CrossRef]
3. Guo, M.; Ren, X.; Jiao, Y.; Liang, M. Summary of aging and anti-aging research on asphalt and asphalt mixture. *J. China Highw. Eng.* **2022**, *35*, 19. [CrossRef]
4. Editorial Department of China Journal of Highways. Academic Research Review of Chinese Pavement Engineering. *China J. Highw.* **2020**, *33*, 1–66. [CrossRef]
5. Wang PE, Y.; Wen, Y.; Zhao, K.C.; Chong, D.; Wong, A.S.T. Evolution and locational variation of asphalt binder aging in long-life hot-mix asphalt pavements. *Constr. Build. Mater.* **2014**, *68*, 172–182. [CrossRef]
6. Petersen, J.; Harnsberger, P. Asphalt Aging: Dual Oxidation Mechanism and Its Interrelationships with Asphalt Composition and Oxidative Age Hardening. *Transp. Res. Rec. J. Transp. Res. Board* **1998**, *1638*, 47–55. [CrossRef]
7. Harnsberger, R.E. Robertson, Factors affecting the kineticsand mechanisms of asphalt oxidation and the relative effects of oxidationproducts on age hardening, Prepr. *Pap. Am. Chem. Soc. Div. Fuel Chem.* **1996**, *41*, 1232–1244.
8. Petersen, J.C. *A Review of the Fundamentals of Asphalt Oxidation Chemical, Physicochemical, Physical Property, and Durability Relationships*; Transportation Research Board: Washington, DC, USA, 2009; 78p.
9. Wang, Y. Study on Indicators Evaluating Asphalt Aging and the Aging Degree. *Subgrade Eng.* **2014**, *6*, 63–68.
10. Zhang, H.; Chen, Z.; Li, L.; Zhu, C. Evaluation of aging behaviors of asphalt with different thermochromic powders. *Constr. Build. Mater.* **2017**, *155*, 1198–1205. [CrossRef]
11. Mokhtari, A.; Lee, H.D.; Williams, R.C.; Guymon, C.A.; Scholte, J.P.; Schram, S. A Novel Approach to Evaluate Fracture Surfaces of Aged and Rejuvenator-Restored Asphalt Using Cryo-Sem and Image Analysis Techniques. *Constr. Build. Mater.* **2017**, *133*, 301–313. [CrossRef]
12. Li, Z.; Fa, C.; Zhao, H.; Zhang, Y.; Chen, H.; Xie, H. Investigation on evolution of bitumen composition and micro-structure during aging. *Constr. Build. Mater.* **2020**, *244*, 118322. [CrossRef]
13. Li, Z.; Zeng, J.; Li, Y.; Zhao, Z.; Cong, P.; Wu, Y. Effect of bitumen composition on micro-structure and rheological properties of styrene-butadiene-styrene modified asphalt before and after aging. *Mater. Struct.* **2022**, *55*, 165. [CrossRef]
14. Fernández-Gómez, W.D.; Quintana, H.A.R.; Reyes-Lizcano, F. A review of asphalt and asphalt mixture aging. *Ing. E Investig.* **2013**, *33*, 5–12. [CrossRef]
15. Zhihui, L.; Rui, Z.; Yonghua, Z.; Qian, C.; Weijun, Q. Discriminating wavenumbers selection of ATR-FTIR spectra for identifying graded asphalt. *Constr. Build. Mater.* **2019**, *214*, 565–573. [CrossRef]
16. Ren, R.B.; Fan, W.M.; Zhao, P.H.; Zhou, H.; Meng, W.K.; Ji, P. Crude Oil Source Identification of Asphalt via ATR-FTIR Approach Combined with Multivariate Statistical Analysis. *Adv. Mater. Sci. Eng.* **2020**, *2020*, 2025072. [CrossRef]
17. Wang, F.; Xiao, Y.; Cui, P.; Lin, J.; Li, M.; Chen, Z. Correlation of asphalt performance indicators and aging Degrees: A review. *Constr. Build. Mater.* **2020**, *250*, 118824. [CrossRef]
18. Fahrenfort, J.; Visser, W.M. On the determination of optical constants in the infrared by attenuated total reflection. *Spectrochim. Acta* **1962**, *18*, 1103–1116. [CrossRef]
19. Ren, R.B.; Han, K.C.; Zhao, P.H.; Shi, J.T.; Zhao, L.; Gao, D.X.; Zhang, Z.Y.; Yang, Z.Q. Identification of asphalt fingerprints based on ATR-FTIR spectroscopy and principal component-linear discriminant analysis. *Constr. Build. Mater.* **2019**, *198*, 662–668. [CrossRef]
20. Petersen, J.C. Asphalt oxidation—AH overview including a new modelfor oxidation proposing that physicochemical factors dominate the oxidation kinetics. *Fuel Sci. Technol. Int.* **2007**, *11*, 57–87. [CrossRef]
21. Ding, H.; Hesp, S.A. Variable-temperature Fourier-transform infrared spectroscopy study of asphalt binders from the SHRP Materials Reference Library. *Fuel* **2021**, *298*, 120819. [CrossRef]
22. Ge, D.D.; Chen, S.Y.; You, Z.P.; Yang, X.; Yao, H.; Ye, M.X.; Yap, Y.K. Correlation of DSR Results and FTIR's Carbonyl and Sulfoxide Indexes: Effect of Aging Temperature on Asphalt Rheology. *J. Mater. Civ. Eng.* **2019**, *31*, 04019115. [CrossRef]
23. Xu, M.; Zhang, Y.; Zhao, P.; Liu, C. Study on aging behavior and prediction of SBS modified asphalt with various contents based on PCA and PLS analysis. *Constr. Build. Mater.* **2020**, *265*, 120732. [CrossRef]
24. Hofko, B.; Porot, L.; Falchetto Cannone, A.; Poulikakos, L.; Huber, L.; Lu, X.; Mollenhauer, K.; Grothe, H. FTIR spectral analysis of bituminous binders: Reproducibility and impact of ageing temperature. *Mater. Struct.* **2018**, *51*, 45. [CrossRef]
25. Wang, K.; Yuan, Y.; Han, S.; Yang, Y. Application of FTIR spectroscopy with solvent-cast film and PLS regression for the quantification of SBS content in modified asphalt. *Int. J. Pavement Eng.* **2017**, *20*, 1336–1341. [CrossRef]
26. Yan, C.; Huang, W.; Xu, J.; Yang, H.; Zhang, Y.; Bahia, H.U. Quantification of re-refined engine oil bottoms (REOB) in asphalt binder using ATR-FTIR spectroscopy associated with partial least squares (PLS) regression. *Road Mater. Pavement Des.* **2020**, *23*, 958–972. [CrossRef]
27. Sun, X.T.; Yuan, H.F.; Song, C.F.; Deng, X.X.; Lv, G.L.; Li, X.Y.; Hu, A.Q. Rapid and simultaneous determination of physical and chemical properties of asphalt by ATR-FTIR spectroscopy combined with a novel calibration-free method. *Constr. Build. Mater.* **2020**, *230*, 116950. [CrossRef]

28. *JTG E20-2011*; Standard Test Methods of Bitumen and Bituminous Mixtures for Highway Engineering. Ministry of Transport of the People's Republic of China: Beijing, China, 2011.
29. Zhao, P.H.; Wu, W.X.; Song, X.Q.; Sun, H.M.; Ren, R.B.; Liu, F.K. Synthesis of imidazoline biosurfactants and application as an asphalt emulsifier. *J. Dispers. Sci. Technol.* **2023**, *44*, 600–607. [CrossRef]
30. Huang, G.; He, J.X.; Zhang, X.; Feng, M.M.; Tan, Y.; Lv, C.C.; Huang, H.; Jin, Z. Applications of Lambert-Beer law in the preparation and performance evaluation of graphene modified asphalt. *Constr. Build. Mater.* **2021**, *273*, 121582. [CrossRef]
31. Geng, J. Research on Aging Mechanism and Regeneration Technology of Asphalt. Ph.D. Thesis, Chang'an University, Xi'an, China, 2009.

coatings

MDPI

Review

Multiscale Rheological Properties of Pavement Asphalt: A State-of-the-Art Review

Qiqi Zhan [1], Zuoyang Cheng [1], Xuejuan Cao [2,3,*], Qing Liu [1], Ying Yuan [1], Lihong He [2] and Junfeng Gao [3,*]

[1] School of Civil Engineering, Chongqing Jiaotong University, Chongqing 400074, China; 13883967470@163.com (Q.Z.); chengzuoy@163.com (Z.C.); cdxpslhwcbl@163.com (Q.L.); yuanying618@foxmail.com (Y.Y.)

[2] School of Materials Science and Engineering, Chongqing Jiaotong University, Chongqing 400074, China; sunnyhlh@cqjtu.edu.cn

[3] National and Local Joint Engineering Laboratory of Traffic Civil Engineering Materials, Chongqing Jiaotong University, Chongqing 400074, China

* Correspondence: caoxj@cqjtu.edu.cn (X.C.); jfgao@cqjtu.edu.cn (J.G.); Tel.:+86-153-1092-1209 (X.C.)

Highlights

What are the main findings?

- A full review of the chemicals in asphalt, ideas about its structure, and how it flows when subjected to different forces.

What are the implications of the main findings?

- A revelation of the rheology of colloidal structures and supramolecular models.
- A thorough and detailed evaluation of large-amplitude oscillatory shear testing.

Abstract

Asphalt rheological properties are fundamental to pavement performance, yet their accurate assessment requires multi-scale characterization due to asphalt's inherent complexity. This article reviews the connections between asphalt rheology across chemical, microstructural, and macro-mechanical scales, employing a methodological analysis of supramolecular and colloidal models for micro-scale behavior and dynamic shear rheometry for macro-scale properties. Current research confirms asphalt as a complex multiphase continuum, where micro-scale rheology is explained by intermolecular interactions and colloidal structures, while macro-scale analysis successfully characterizes linear viscoelasticity through established empirical and mechanical models. However, the study identifies critical gaps: nonlinear viscoelastic characterization under large-amplitude oscillatory shear (LAOS) remains underdeveloped, and fundamental issues like directly probing molecular interactions and the origin of microstructures like the "bee structure" are unresolved. The primary conclusion is that a comprehensive understanding of asphalt rheology hinges on future research that integrates experimental and simulation data across these scales to bridge the gaps between chemical composition, microstructure, and macroscopic performance.

Keywords: asphalt binder; colloid; rheological properties; multiscale; supramolecular

1. Introduction

Asphalt rheology primarily focuses on studying the properties of asphalt, such as elasticity, viscosity, viscoelasticity, and deformation, and their responses to variations in temperature, time, and loading conditions [1], which is closely connected to the asphalt pavement performance. The asphalt's rheological properties are influenced by various factors, including atmospheric conditions, temperature, sunlight, moisture, and loading. Understanding these properties is crucial for improving asphalt pavement performance and preventing pavement deterioration.

Asphalt compounds can be categorized into different components based on specific properties. Researchers often study the rheological mechanisms of asphalt by examining the rheological properties of these individual components. Extensive research indicates that at room temperature, polar components primarily contribute to elasticity, while non-polar components primarily contribute to viscosity [2]. The 12-molecule model, commonly used in numerical simulations of asphalt, is based on the four asphalt components and the eight average molecular models proposed by the Strategic Highway Research Program (SHRP). This model enables a more in-depth investigation of asphalt rheological mechanisms, including intermolecular interactions and molecular aggregation behavior. However, the asphalt system is complex, consisting of hundreds of thousands of compounds [3]. Each component remains intricate, making it difficult to create precise molecular models for numerical simulations. As a result, simulation results often show only trends similar to those of experimental results. Scattering experiments have confirmed the presence of asphalt micelles, indicating that the macroscopic properties of asphalt materials depend not only on the physicochemical properties of individual molecules but also on molecular aggregation behavior and aggregation structures.

The action of wheel loading on asphalt pavements can be understood as low-frequency vertical oscillations and localized impacts. Asphalt, which serves as a binder in pavements, exhibits linear or nonlinear viscoelastic behavior under environmental and loading conditions [4]. As a result, researchers often use rheological parameters obtained through dynamic mechanical analysis to assess the performance of asphalt pavements. Viscoelasticity is the most prominent rheological property of asphalt. Currently, numerous well-established empirical algebraic and mechanical element models describe the linear viscoelasticity of asphalt over a wide frequency or temperature range. However, these models have drawbacks, including a lack of universality across different types of asphalt, an inability to describe viscoelasticity under extreme conditions, and excessive parameterization and inflexibility. Creep tests under steady shear loads are the most commonly used method to study the nonlinear viscoelasticity of asphalt. However, these testing methods cannot fully reflect the actual stress conditions experienced by asphalt pavements. In addition, methods for predicting the fatigue life of asphalt often underestimate it by neglecting factors such as nonlinear viscoelasticity and thixotropy. This leads to overly conservative predictions.

The research methods for the rheological properties of polymer materials can be divided into structural and phenomenological approaches. The structural approach uses thermodynamics and statistical mechanics to construct microscale models that describe the internal structure of materials, enabling the study of microscale rheological properties. Phenomenological approaches, on the other hand, do not consider the microstructure. Instead, they employ mathematical methods from continuum mechanics to study the macroscopic rheological properties. In the realm of asphalt materials, ongoing research is predominantly centered on macroscopic rheology [5]. Indeed, establishing a connection between rheological mechanisms and macroscopic rheological properties is essential for a better understanding of asphalt's constitutive behavior.

In recent years, research in asphalt rheology has progressively advanced towards more complex nonlinear behaviours and deeper multiscale correlations [6,7]. In the realm

 https://doi.org/10.3390/coatings16030355

of nonlinear rheology (LAOS), investigations have transcended the linear viscoelastic regime to systematically characterise asphalt's intricate responses under high-amplitude shear. This includes quantifying nonlinearity through higher-order harmonic analysis and Lissajous curve decomposition, revealing pronounced higher-order harmonic intensity within the nonlinear domain for specific modified asphalts.

Regarding the time–temperature superposition principle (TTSP) in modified bitumen, research focus has shifted towards identifying its failure boundaries [8,9]. For instance, certain polymer modifiers or high wax content may disrupt temperature-frequency equivalence, leading to failure in constructing the master curve. This is often assessed by examining the continuity of data within black plots. However, most current studies remain focused on observing macroscopic rheological responses through alterations in microstructure, whilst research systems for reverse-predicting or quantifying specific microstructural features based on rheological parameters remain underdeveloped.

Compared to recent reviews focusing on specific modification techniques or fatigue damage mechanisms, this work provides a multiscale correlative framework centred more squarely on the rheological mechanisms themselves. In light of this, this work thoroughly reviews significant advancements in the field and methodically contrasts research findings on asphalt rheological mechanisms by conducting a systematic search of the Elsevier library using the keyword "asphalt rheology" from the 1950s to 2024, with each section encompassing approximately 50 publications. It not only systematically reviews the microscopic systems—from chemical composition, intermolecular interactions, and colloidal structures to honeycomb structures—alongside linear and nonlinear macroscopic rheology, but also places particular emphasis on elucidating the specific, mechanistic connections between these scales.

The core contributions lie in: (1) explicitly proposing a physical picture for the transformation of microscopic parameters (such as interaction strength and colloid index) into macroscopic rheological parameters (such as $|G^*|$ and δ); (2) emphasising the importance of establishing structure-property relationships in the nonlinear regime, alongside the dynamic transition of colloidal structure from 'sol' to "gel" to 'aged gel' and its direct influence on rheological properties; (3) It systematically identifies the current research gap between linear viscoelastic testing and microstructural prediction, alongside the limitations of existing rheological models in terms of universality, thereby providing a clear direction for future research.

The structure–performance relationship of asphalt materials is depicted in Figure 1. Figure 1 outlines the multiscale structure-property relationship framework for asphalt discussed herein. From left to right, it clearly illustrates the logical chain extending from the molecular scale (chemical composition, intermolecular interactions) to the mesoscale (microscopic structures such as micelles and honeycomb structures), then to the macroscale (rheological parameters), ultimately influencing pavement performance at the engineering scale. The figure particularly emphasises the intervention of 'dynamic regulation' (such as ageing and modification) at each scale, alongside the pivotal role of rheological parameters as the core bridge linking microscopic mechanisms to macroscopic resistance against external stimuli (wheel loads, environmental factors).

https://doi.org/10.3390/coatings16030355

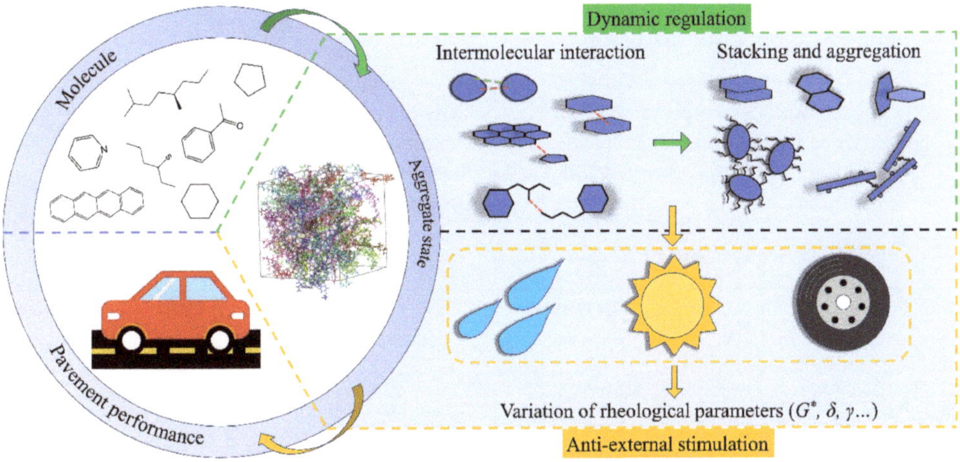

Figure 1. Structure–performance relationship based on rheological properties of asphalt.

2. Chemical Composition

2.1. Element Composition and Functional Groups

Chemically, asphalt is a complex mixture primarily composed of carbon (80%~88%) and hydrogen (8%~12%). It also contains sulfur (0%~9%), oxygen (0%~2%), nitrogen (0%~2%), as well as trace amounts of metals such as nickel, iron, vanadium, and manganese [10–13]. Asphalt primarily contains aliphatic compounds, aromatic compounds, or a combination of both [3,14]. Other elements present in lower amounts combine with these compounds to form heteroatomic compounds with various functional groups. Sulfur exists in the form of sulfides or dibenzothiophenes [15]. Alkaline nitrogen exists in the form of pyridine, while non-alkaline nitrogen exists in the form of pyrrole [2]. When asphalt undergoes oxidation, it introduces oxygen elements and forms new heteroatomic compounds. Benzyl groups are prone to oxidation to ketones, and further oxidation can lead to the formation of carboxyl groups and even anhydrides [10]. Oxygen atoms also exist in the form of phenols, quinones, and sulfoxides. Metal elements are often present in the form of porphyrins [16]. Figure 2 illustrates common compounds and characteristic functional groups found in asphalt. Due to the complex composition of asphalt, its compounds are often categorized into different fractions based on their properties, including molecular size, ionic nature (acidic, basic, or amphoteric), and polarity.

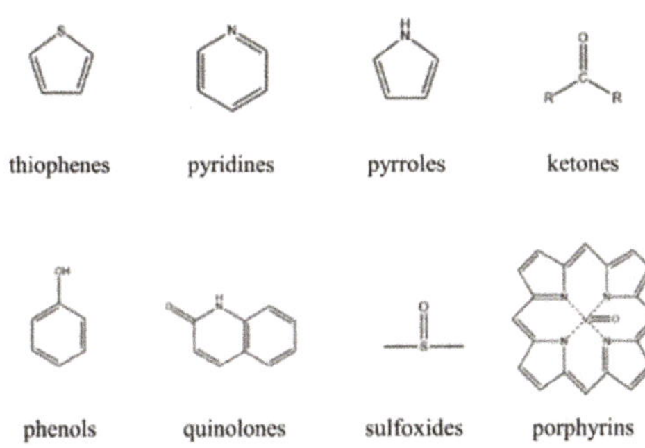

Figure 2. Typical functional groups in asphalt.

https://doi.org/10.3390/coatings16030355

2.2. *SARA Fractions (Saturates, Aromatics, Resins and Asphaltenes)*

2.2.1. Basic Properties of SARA Fractions

Corbett's SARA separation method is the most commonly used technique for separating compounds in asphalt based on their polarity differences. This method uses solvent precipitation and chromatographic column techniques to separate asphalt into asphaltenes, resins, aromatics, and saturates [17]. The four fractions mentioned above are referred to as SARA fractions. The properties of the SARA fractions are presented in Table 1 [10,11,18–20].

Table 1. Properties of SARA fractions.

	Saturates	Aromatics	Resins	Asphaltenes
State (room temperature)	white translucent liquid	yellow or red liquid	black semisolid	black powder
Density/(g·cm^{-3})	0.90	1.00	1.07	1.15
Weight percent/%	3.0~19.1	22.4~46.6	30.0~50.0	4.0~22.9
Molar weight/(g·mol^{-1})	600~780	500~800	830~1100	800~3500
Solubility parameter/(MPa$^{-0.5}$)	15.0~17.0	17.0~18.5	18.5~20.0	17.6~21.7
Carbon-hydrogen ratio	2.00	1.5	1.38~1.69	1.15
Carbon percentage/%	78.0~84.0	80.0~87.0	67.0~88.0	80.0~88.6
Hydrogen percentage/%	8.0~12.0	9.0~13.0	9.0~12.0	7.1~10.0
Nitrogen percentage/%	<0.1	0~0.4	0.2~1.7	0.3~0.4
Oxygen percentage/%	<0.1	0.2	0.3~2.0	0.3~5.0
Sulfur percentage/%	<0.1	0~4.0	0.4~5.0	3.0~10

NOTE: Due to the complexity of the chemical composition of asphalt, the physical and chemical properties of the SARA fractions of different asphalts may be outside the range in the table.

The aromatics and saturates exhibit similar Fourier-transform infrared (FTIR) spectra, indicating that they have similar covalent bonds or functional groups. The differences in peak intensities in their spectra suggest variations in the content of these similar bonds or functional groups. Figure 3a demonstrates that aging leads to a notable rise in the carbonyl peak (at wavenumber 1700 cm^{-1}) and sulfone peak (at wavenumber 1030 cm^{-1}) within the aromatics. This suggests that oxidative aging affects the aromatic fraction more than the saturate fraction.

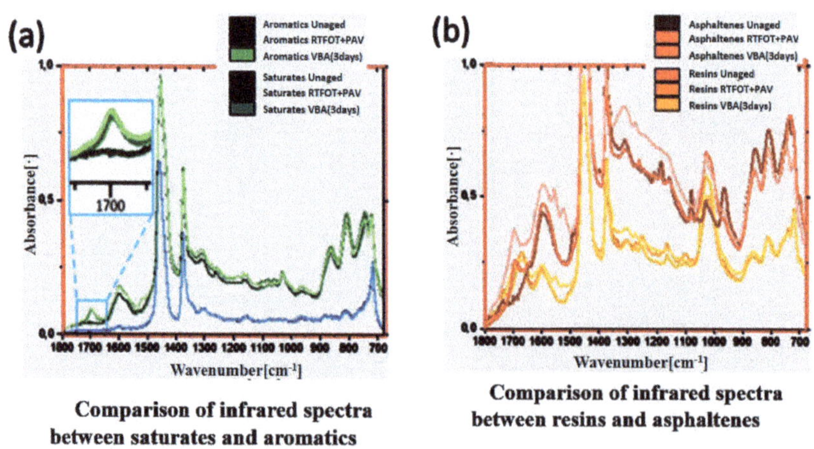

Figure 3. (**a**) Comparison of infrared spectra between saturates and aromatics [21]. (**b**) Comparison of infrared spectra between resins and asphaltenes. NOTE: In Figure (**a**): RTFOT—Rolling Thin Film Oven Test; PAV—Pressure Aging Vessel, VBA—Vienna Binder Aging [21].

https://doi.org/10.3390/coatings16030355

The resins and asphaltenes are the polar fractions. Resins contain a significant number of ketone functional groups (Figure 3b). Asphaltenes molecules primarily consist of island-type structures (typically consisting of 4 to 10 aromatic rings). Sabbah et al. [22] utilized two-step laser desorption-ionization mass spectrometry (L2MS) to characterize asphaltenes. Their findings revealed that under high-power laser irradiation, archipelago-type compounds undergo single-molecule decomposition.

2.2.2. Correlation Between SARA Fractions and Rheological Properties

Robertson et al. [23] conducted a study on the correlation between chemical composition and rheological properties of asphalt. They found that polar molecules play a significant role in the elasticity of asphalt materials, whereas nonpolar molecules primarily affect viscosity. Dealy [24] studied how the ratio of asphaltenes to maltene affects the rheological properties of asphalt. The results showed that as the asphaltenes content increased, the asphalt viscosity also increased significantly. Conversely, an increase in maltene led to a significant decrease in viscosity. In the study by Sultana and Bhasin, a similar pattern was observed. They found that the modulus of asphalt increased with an increase in the content of polar components, particularly influenced by the presence of asphaltenes [25]. Eberhardsteiner et al. [26] proposed a micromechanical model to predict the viscoelastic behavior of asphalt based on the rheological properties of SARA fractions. The model's predictions showed good agreement with the experimental data.

However, methodological variations and any inconsistencies should be carefully considered when comparing findings from various investigations of the relationship between chemical composition and rheological characteristics. For example, SARA separation results for bitumens from various sources show intrinsic heterogeneity, despite numerous studies confirming the main influence of polar components (especially asphaltenes) on elastic modulus. Additionally, the definition and quantification of components are direct.

This might lead to inconsistent reported relationships between rheological parameters and the "colloidal index (IC)" or "asphaltic index (IA)" across investigations using various separation procedures. Some of the research listed in Table 5 investigated rheological parameters only in the linear viscoelastic (LVE) domain, suggesting that these correlations might not apply directly to the high-strain, nonlinear conditions frequently encountered in real paving. This restriction underscores how insufficient it is to construct universal "structure-property correlations" based solely on LVE parameters. Non-linear viscoelastic (NLVE) characterization should be used.

Among the four fractions, saturates have the weakest polarity, while asphaltenes have the strongest polarity, with aromatics and resins falling in between. Asphalt oxidation aging increases the amount of polar components in asphalt. The content of saturates remains consistent before and after asphalt aging, while aromatics and resins decrease, and asphaltenes content increases. This is because, as asphalt ages, the number of functional groups, such as sulfoxides, in aromatic components and resins increases. This causes aromatic components with higher polarity or molecular weight to be categorized as asphaltenes [11,26,27]. As a result, aged asphalt typically has lower flowability compared to the original asphalt.

2.2.3. Waxes

Wax, as a special component, has similarities in chemical composition and properties with the saturates. It typically makes up less than 5% of asphalt used for roads but significantly affects its rheological properties [28]. Wax in asphalt can be defined in various ways, and this paper distinguishes it into macrocrystalline wax and microcrystalline wax.

https://doi.org/10.3390/coatings16030355

Macrocrystalline wax (typically containing 24 to 40 carbon atoms) is composed of n-alkanes with few or no side chains [29,30]. It forms a uniform, ordered, periodic crystalline structure, which gives it strong shear resistance but poor plasticity. The main type of wax found in asphalt is microcrystalline wax (typically with more than 40 carbon atoms). It is composed of isoalkanes and cycloalkanes, which make it difficult for the wax to form a consistent, organized crystalline structure. This lack of uniformity leads to improved plasticity [31]. When the temperature falls below the wax's melting point, the wax will crystallize, grow, and interconnect to form a network-like structure. This structure surrounds and adsorbs other components of the asphalt, making it less fluid and more deformable, thereby diminishing its low-temperature performance [32–34]. However, in the high-temperature environment of asphalt production and mixing for pavement laying, wax behaves as a low-viscosity Newtonian fluid, reducing asphalt viscosity. Additionally, wax exhibits good compatibility with asphalt [35]. Therefore, despite the various adverse effects of wax on asphalt, in recent years, research has also focused on using wax as an additive to improve asphalt rheological properties and facilitate the development of warm-mix asphalt technology. The petroleum waxes mentioned above have large, flat plate-like crystals that can negatively impact the adhesion between asphalt and aggregates. On the other hand, waxes used as additives are commonly referred to as "commercial waxes", including synthetic waxes, mineral waxes, modified montan wax (similar to petroleum wax but with a higher molecular weight [35]), and functionalized waxes (containing polar functional groups such as amide groups, and so on [36]).

Asphalt is a chemically continuous system without distinct boundaries between its components [37]. Any separation method (for example, Ion Exchange Chromatography (IEC), Size-Exclusion Chromatography (SEC), Thin-Layer Chromatography (TLC), etc.) [2] results in components that remain heterogeneous, with different components separated under specific conditions. When conditions change, the separation results also change accordingly.

2.3. Molecular Simulation Models

2.3.1. All-Atom Simulation

Molecular Dynamics (MD) simulation is an effective tool for studying the relationship between molecular structure and function. In MD simulations, all atoms follow the laws of Newtonian motion. Interactions between atoms are constrained using force field functions. Kinetic and mechanical properties of materials are obtained using statistical mechanics and by calculating atomic trajectories and relative positions through simulations [38]. The establishment of molecular models is a prerequisite for conducting MD simulations. SHRP has proposed average models for eight different types of asphalt (AAA-1, AAB-1, AAC-1, AAD-1, AAF-1, AAG-1, AAK-1, and AAM-1) based on average molecular weight, elemental analysis, and nuclear magnetic resonance data [39]. Li et al. [40] further refined the average molecular models of AAA-1, AAK-1, and AAM-1 based on asphalt's SARA fractions and 12 typical molecular models. The improved AAA-1 model shows a density close to that of AAA-1 asphalt, differing by 0.06 g/cm^3. This provides a more reliable model for establishing the connection between asphalt chemical composition and rheological properties. You et al. [41] employed a 12-component all-atom model to predict the rheological properties of asphalt at the microscale. The results demonstrated a good agreement between the simulated viscosity and laboratory data.

Additionally, as the temperature increased, the simulation results closely approximated the experimental data. Kang et al. [42] used full-atom MD simulations to investigate the glass transition of asphalt. They concluded that asphalt undergoes a transition from viscoelasticity to a Newtonian fluid state at 400 K. Given the intricate nature of asphalt

 https://doi.org/10.3390/coatings16030355

systems, it is presently unfeasible to construct asphalt models that precisely replicate real conditions. Furthermore, cumulative errors may arise during simulations. Therefore, it is essential to validate simulation results with experimental data.

2.3.2. Coarse-Grained Simulation

The coarse-grained model simplifies atom interactions, providing a lower resolution than the full-atom model. However, it enables more efficient long-term mesoscale simulations. Standard mesoscale simulation methods for modeling asphalt molecule behavior include coarse-grained molecular dynamics (CGMD), Monte Carlo (MC) methods, and dissipative particle dynamics (DPD) [43]. Li et al. [44] constructed a coarse-grained asphalt model suitable for CGMD and explored the aggregation behavior of asphalt molecules employing the Martini force field. The simulation results showed significant changes in diffusion coefficients across different temperatures, spanning several orders of magnitude, consistent with the viscoelastic behavior of asphalt. Aguilera-Mercado et al. [45] employed the (MC) method to investigate the aggregation behavior of asphaltenes and resins in crude oil. Compared to the methods mentioned above, DPD is capable of simulating large, complex colloidal systems and is better suited for asphalt-related research. This method maps multiple atoms into a single bead based on the structure of the all-atomic model. It calculates interaction parameters between beads using Flory–Huggins parameters and maintains the rigidity of the aromatic system by setting average bond lengths and spring constants. Guan et al. [38] employed the DPD method to explain the self-assembly behavior of polycyclic aromatic hydrocarbon systems within heavy petroleum. Tang et al. [46] utilized the DPD method to investigate the compatibility between SBS modifiers and asphalt. The simulation results provided molecular-level insights into the modification mechanism, demonstrating good agreement with rheological experiments.

Given the intricate nature of asphalt systems, it is presently unfeasible to construct asphalt models that precisely replicate real conditions. Furthermore, cumulative errors may arise during simulations. Therefore, it is essential to validate simulation results with experimental data [47,48].

3. Microstructure

3.1. Intermolecular Interactions

Intermolecular interactions in asphalt cause molecules to bond together, forming a network structure. The strength of these bonds affects the viscosity and hardness of asphalt, making it a crucial factor in its rheological properties [14,49]. Typical intermolecular interactions in asphalt systems are summarized in Table 2.

Table 2. Intermolecular interactions in the asphalt system.

Interactions	**Existence**
dipole–dipole interaction	Between molecules with uneven electron distribution
dispersion force	Between all molecules
hydrogen bond	O, S, N and other atoms with large electronegativity often appear in asphaltenes, so N-H...O, O-H...N, S-H...O, O-H...S and other forms of hydrogen bonds often appear in asphaltenes [50].
π-π interaction	π-π stacking often occurs between aromatic compounds in asphalt, which is a non-covalent bond interaction as important as hydrogen bonds. The number of benzene rings, arrangement and side chain length of aromatic compounds in asphalt have a certain influence on π-π interaction [50].
metal coordination	S, N, O heteroatoms with lone pair electrons in asphaltene molecules, which provides the possibility of coordination between asphaltene molecules and porphyrin nickel and porphyrin vanadium containing empty valence orbitals [50].

 https://doi.org/10.3390/coatings16030355

The intermolecular interactions mentioned above not only dictate the form of asphalt and its fractions (especially asphaltenes) but also determine the molecular behavior and microstructure within asphalt. The intermolecular interactions in asphalt restrain certain molecules, giving asphalt its elastic characteristics. When relatively free molecules overcome these constraints and begin to move, asphalt exhibits its viscous characteristics. Functional groups within asphalt compounds also contribute to the intermolecular network binding. Asphaltenes, containing nearly all functional groups present in asphalt, thus have the strongest binding capability [51–53]. Furthermore, temperature influences the establishment of a new equilibrium in intermolecular interactions. At low temperatures, asphalt molecules with lower energy tend to form tighter network connections. Conversely, higher temperatures provide asphalt molecules with greater kinetic energy, allowing them to break free from these constraints and move more freely, resulting in a less compact network structure [2].

3.2. Supramolecular Structures

While characterizing the exact strength of intermolecular forces in asphalt systems remains difficult, it is clear that the properties of asphalt materials are influenced not only by individual molecule properties but also by the aggregation states formed through intermolecular interactions, i.e., the supramolecular structures consisting of covalent bonds and intermolecular interactions. Molecular interactions are considered the key determinants of the behavior of supramolecular structures. These interactions are highly sensitive and fragile when exposed to external factors such as temperature and loading [54]. Cho et al. [55] established a supramolecular assembly model for asphalt based on a solvent-solute system using the host-guest system from supramolecular chemistry. In this model, host molecules act as the solvent and guest molecules act as the solute. Together, they form the preorganized supramolecular body (PSB). PSBs can explain some fundamental rheological properties of asphalt. External actions, such as stirring or shearing, can disrupt PSBs, leading to deformation of the asphalt. PSBs with dipole–dipole interactions contribute to the viscosity of asphalt, while those with π-π interactions enhance the ductility of asphalt. The mutual adhesion between PSBs allows for self-assembly, resulting in the formation of supramolecular structures. The PSBs will assemble into two types of structures: one is an irreversible structure with chain-like or linear morphology containing strong covalent bonds, while the other is a reversible structure with certain sliding ability.

The asphalt supramolecular model primarily explains asphalt's fundamental rheological properties through intermolecular interactions. However, due to current limitations in characterizing intermolecular interactions, this model remains largely conceptual. When explaining the rheological properties of asphalt under complex conditions, researchers often turn to another higher-scale model, the colloidal structure. This model is also based on a solvent-solute system.

3.3. Early Stage Colloidal Structure

In 1923, based on the characteristics of the Tyndall effect, Brownian motion, and the inability of asphaltene solutions to pass through a semipermeable membrane, Nellensteyn proposed the colloid structure of asphalt [56]. Mack conducted further research into the colloidal structure of asphalt. He proposed that asphaltene constituted the dispersed phase, while the resin-oil mixture served as the dispersed medium [57]. In 1940, Pfeiffer introduced a new colloidal structure model for asphalt, building upon previous findings. This novel structure is centered on asphaltene, with soluble components adsorbed on its surface or within its interior, forming micelles. In micelles, the maltenes with the high-

 https://doi.org/10.3390/coatings16030355

est molecular weight and strongest aromaticity encapsulate the asphaltenes, while these maltenes are surrounded by other maltenes with lower molecular weight and aromaticity until reaching the inter-micellar phase [58]. Based on the content of each fraction in asphalt and its rheological properties, asphalt can be classified into the following structures: (1) Sol Structure: When there is a sufficient amount of oil and resins, asphaltene clusters are completely dispersed and can move freely within a viscosity permissible range. The rheological behavior resembles that of a Newtonian fluid. (2) Gel Structure: When the amount of oil and resin is low, and the asphaltene content is high, the concentration of asphaltene clusters increases, forming an irregular network structure. The rheological behavior is that of a non-Newtonian fluid. (3) Sol–Gel Structure: Most pavement asphalts belong to the sol–gel structure, which is an intermediate state between sol and gel structures [59]. In the sol–gel structures, the asphaltene content is moderate, and there is an appropriate amount of resin to maintain colloid stability. The rheological behavior exhibits viscoelasticity.

The interaction and aggregation of asphaltene micelles are important factors determining the colloid type and influencing the rheological properties. Gaestel et al. [60] introduced the colloid index IC to describe the degree of colloidization, which is expressed as shown in Equation (1).

$$I_C = \frac{x_{asph} + x_{floc}}{x_{surf}} \tag{1}$$

In Equation (1), xi represents the mass of various fractions in asphalt, *asph* represents asphaltene, floc represents flocculant, referring to components in the maltenes that cause asphaltenes flocculation, and surf represents surfactant, referring to components that generate surface activity and promote asphaltenes dispersion. For SARA fractions, Equation (1) can be rewritten as:

$$I_C = \frac{x_{Asphaltenes} + x_{Saturates}}{x_{Resins} + x_{Aromatics}} \tag{2}$$

As I_C increases, colloidal stability decreases. The I_C of road asphalt is typically between 0.5 and 2.7. When $I_C > 1.2$, asphalt exhibits typical gel-like structural characteristics, with high stiffness and an elastic network. When $I_C < 0.7$, it exhibits typical sol structural characteristics [11,61]. In general, aging will gradually transform the sol–gel structure of asphalt into a gel structure, leading to an increase in I_C. Li et al. [62] found that although some asphalts had similar I_C, they displayed different levels of I_C reduction after aging. This indicates that even though these asphalts have similar colloid structures, their resistance to aging may differ.

In addition to the colloid index I_C, the asphaltene index IA (Equation (3)) can also be used to describe colloidal stability [63].

$$I_A = \frac{x_{Asphaltenes} + x_{Resins}}{x_{Saturates} + x_{Aromatics}} \tag{3}$$

The methods mentioned above all incorporate asphaltenes as a critical factor, since micelles are dispersed in the maltenes, with asphaltenes at their cores. The flocculation ratio-dilution method (Equation (4)) is also employed to assess colloidal stability. This method offers a simpler experimental procedure, as it eliminates the need to separate asphalt fractions [64].

$$P = \frac{P_0}{1 - P_a} \tag{4}$$

where Pa represents the degree of anti-flocculation, which refers to the difficulty of asphalt being dissolved and dispersed in soluble solvents; a smaller value indicates that asphalt is more difficult to form a stable colloidal structure. P_0 denotes the ability of the solvent to

 https://doi.org/10.3390/coatings16030355

disperse the asphaltenes, i.e., the solvent's capacity to dissolve asphaltene. P represents a comprehensive indicator of the colloidal state of asphalt, with higher values indicating a more stable colloidal structure that is less prone to flocculation [64].

3.4. Modern Colloidal Structures

3.4.1. Asphaltene Micelles

With advances in experimental methods, more precise descriptions of colloid structures have emerged. X-ray, small-angle scattering, and neutron small-angle scattering studies have shown that asphaltenes will form micelles in organic solvents, crude oil, and asphalt [65–71]. Yen et al. [72] proposed the Yen model, suggesting that asphaltene molecules aggregate into micelles driven by π-π interactions or weak covalent bonds. In the Yen model, multiple aromatic layers stack parallel to each other and are surrounded by extending aliphatic chains. Asphaltene micelles can further aggregate into supermicelles, clusters, and flocs. Wang et al. [73] used transmission electron microscopy (TEM) to observe the microstructure of asphaltene. Asphaltene exhibited two forms under TEM, namely, needle-like and plate-like structures. At higher resolutions (10 nm), both forms displayed a highly ordered layered structure, indicating the presence of a crystalline structure in asphaltene, consistent with the Yen model.

Mullins revised the Yen model in 2010 based on previous research findings. The modified Yen model suggests that asphaltene molecules form two levels of aggregation: nanoclusters and clusters. The asphaltenes are in a dispersed state in a low-concentration asphaltene solution. As the solution concentration increases, dimers, trimers, and other oligomers begin to appear. When the concentration of the asphaltene solution reaches a certain level, the additional asphalt molecules aggregate to form new aggregates. The newly formed aggregates remain relatively constant in size and number. At this stage, they are referred to as nanoaggregates. As the concentration of the asphaltene solution increases, the nanoaggregates come closer together to form clusters (<30 nm) and even flocculate (>300 nm) [31,74,75]. The shape, size, and stability of aggregates are significantly influenced by temperature, solution concentration, and the source of asphaltenes, which in turn affect the colloidal structure of asphalt.

3.4.2. Correlation Between Colloidal Structures and Rheological Properties

The proposal and modification of the Yen model provide strong evidence for explaining the rheological properties of asphalt and offer a clearer description of the transition process between colloidal structures.

Lesueur et al. [76] studied the relationship between asphalt's linear viscoelastic properties and its internal structure. The study found that the Newtonian fluid behavior of asphalt at high temperature is caused by the Brownian motion of colloidal particles, which can be fully described by zero shear viscosity and temperature dependence. When the asphalt is at a lower temperature, longer loading time, or higher loading frequency, the viscosity of the maltenes increases and hinders the diffusion of the asphaltene micelles, so that the asphalt exhibits elasticity. This transition from Newtonian fluid behavior to viscoelasticity is called α relaxation [11]. The elastic behavior of asphalt becomes more significant as the temperature decreases, eventually exhibiting glassy elasticity. This process is known as β relaxation. Although the main reason for this phenomenon is the solute undergoing a glass transition, the interaction between micelles also plays an important role, as it is inversely related to the relaxation rate and asphaltenes content [27].

Aging leads to the transformation of non-polar fractions into polar groups, thereby altering the original colloidal structure. The change in colloidal structure is reversible with

https://doi.org/10.3390/coatings16030355

the aid of additives; that is, the aged asphalt can reconstruct its colloidal structure with the help of additives (rejuvenators [77], warm-mix agents [78], etc.), bringing its rheological properties closer to those of the original asphalt. Filippelli et al. [78] combined the inverse Laplace transform with nuclear magnetic resonance relaxation times to calculate P(T2), which represents the probability density function (a function that describes the structure of substances). By comparing the relaxation times of original asphalt, aged asphalt, and aged asphalt with warm mix additives, it was observed that the additives effectively extended the relaxation times of various fractions in the asphalt. This indicates that additives reconstructed the colloidal structure of aged asphalt, making it more fluid, but they could not restore the light fraction lost during the aging process.

Colloidal structure models explain why asphalt exhibits rheological properties and provide a theoretical foundation for the relationship between its chemical composition and rheology. As illustrated, the transition from sol-type to sol–gel-type, and ultimately to gel-type structures, is accompanied by a shift in micelles from a dispersed state to a continuous network. This structural evolution directly leads to systematic changes in macroscopic rheological parameters: the complex modulus and viscosity increase significantly, while the phase angle decreases (i.e., enhanced material elasticity). This theory links the increase in colloidal index, chemical composition changes due to ageing, and the resulting mechanical property hardening, providing a clear physical picture for understanding processes such as ageing and modification.

3.5. Bee Structure

3.5.1. Morphology of Asphalt Under Atomic Force Microscopy

Asphalt is a heterogeneous multiphase material with a complex microstructure [79]. Understanding the origin, chemical composition, and mechanical properties of asphalt's microstructure helps further elucidate its rheological characteristics. Currently, various optical and electron microscopy techniques, such as scanning electron microscopy, environmental scanning electron microscopy, fluorescence microscopy, and atomic force microscopy (AFM), are commonly used to investigate the microstructure of asphalt materials. In contrast to other observation techniques mentioned above, AFM not only provides the ability to capture the microstructure of samples at sub-nanometer resolution but also allows for the quantitative assessment of mechanical properties, including modulus and deformation [80], which is more conducive to the construction of the correlation between the microstructure and rheological properties of asphalt.

In 1996, Loeber et al. [81] were the first to utilize AFM to observe the "bee structure" of gel-type asphalt. Masson and co-workers [82] used a combination of AFM and phase-contrast microscopy to investigate the structure and behavior of asphalt at low temperatures, categorizing AFM-imaged asphalt into three phases: Catanaphase, Periphase, and Paraphase.

In addition to the typical bee structure, AFM also observed other structures. Across different asphalts and sample preparation methods, dendrite structures, flower-like domains, flake-like domains, and other microscopic morphologies can be observed (Table 3).

Table 3. AFM imaging of asphalt under different conditions and different sample preparation methods.

Morphology	Asphalts	Sample Preparation Methods
Bee structure [81,83]	A gel-type binder (unknown crude source)	Heat-cast
Dendrite structure [84]	Sasobit modified asphalt	Solution-cast
Flower-like domain [85]	Binder B1 from a high-sulphur Middle East crude source	Solution-cast
Flake-like domain [61]	SHRP binder AAM	Heat-cast

 https://doi.org/10.3390/coatings16030355

However, there is no consensus on the origin of asphalt's bee structure, which is mainly divided into three viewpoints: asphaltenes [86,87], trace metals in asphalt [82], and crystalline waxes [88–90].

These contradictory theories about the structure's origin highlight the shortcomings of the characterization methods used today. It is challenging to perform non-destructive in situ chemical characterization of specific microdomains, as atomic force microscopy (AFM) primarily provides surface topography information. As a result, based on sample parameters (e.g., high wax or high asphaltene concentration) and observed topographical correlations, various study teams may propose different dominant mechanisms. This discrepancy also raises the possibility that honeycomb patterns are a collective surface expression of phase separation and aggregation behaviors across several ingredients rather than the result of a single chemical component. To settle this dispute, AFM must be more widely integrated with supplementary methods that can provide chemical information, such as AFM-IR, to directly establish it.

3.5.2. Correlation Between Bee Structure and Rheological Properties

As outlined in Section 3.5.1, AFM has revealed diverse microscopic morphologies on bitumen surfaces, including honeycomb, dendritic, and floral structures. However, owing to the "honeycomb structure's" most pronounced characteristics and the overwhelming focus of existing research on its correlation with macroscopic rheological properties, this section will primarily review findings on honeycomb structures. The rheological significance of other morphologies remains to be elucidated by future studies.

Many research teams have employed AFM to study the mechanical behavior of different phases of the bee structure [59,91–94]. This imaging can be described as follows: the Catanaphase, composed of asphaltenes, is surrounded by the Periphase, consisting of resins and aromatics, and dispersed within the Paraphase, composed of saturates. This is consistent with modern theories of asphalt colloid structure, further supporting the validity of using colloid structure models to explain asphalt rheological properties. In the study by Dourado et al. [92], it was observed that the overall elastic modulus of the honeycomb structure is lower than that of the phase gaps, and the elasticity recovery of the bee structure is highly dependent on the colloidal structure of asphalt. Zhang et al. [95] obtained different derived asphalts by blending the SARA fraction content of asphalt, and studied the micro-morphology and micro-rheological properties of the samples by AFM. The results show that the number of bee structures in asphalt is related to the asphaltenes content. The storage modulus and loss modulus increase with the increase in asphaltenes and resins content and decrease with the increase in saturates and aromatics content, which is consistent with the results obtained from the dynamic shear rheometer (DSR) test.

The effect of aging on the rheological properties of asphalt is also reflected in AFM imaging. Asphalt tends to form more small bee structures with lower peaks after aging. These small bee structures tend to aggregate and may even form cracks [62]. Xing et al. [96] obtained the distribution of functional groups before and after aging of asphalt at the nanoscale using AFM-based infrared spectroscopy (AFM-IR; chemical analysis can be performed at the resolution of AFM). Ganter et al. [97] found that, compared with unaged asphalt, the AFM images of short- and long-term-aged asphalt showed more bee structures, and the moduli of each phase changed significantly.

3.6. Summary

In conclusion, the colloidal network structure is determined by the strength of intermolecular contacts, and its properties directly translate into quantifiable rheological parameters.

https://doi.org/10.3390/coatings16030355

From colloidal structure to interactions: Bituminous nano-aggregate formation and stability are determined by the strength of intermolecular forces in bitumen, such as hydrogen bonding and π-π stacking. By affecting colloidal indices (such as IC), the total of these forces ultimately decides whether bitumen displays a sol, sol–gel, or gel structure. Age-related increases in polar functional groups, such as carbonyl groups, for example, improve intermolecular polar contacts and encourage the transformation of resins and aromatics into bituminous components. IC rises as a result, and the structure becomes more gel-like.

The type and strength of the colloidal structure directly govern macroscopic rheological responses. A reinforced gel network implies greater structural rigidity and stronger constraints on molecular motion. Interactions and aggregation behaviours at the microscopic level are "materialised" through the mesoscopic scale of the colloidal structure, ultimately captured by dynamic shear rheometers (DSR) as specific parameters such as $|G^*|$ and δ.

Among these: Complex modulus $|G^*|$: Characterizes the material's overall resistance to deformation. Stronger colloidal networks provide a more robust framework to resist shear, directly leading to higher $|G|$ values.

Phase angle δ: Reflects the ratio of viscous (dissipative) to elastic (storage) response. Within gel structures, deformation induces reversible elastic deformation of the network rather than irreversible viscous flow; hence, δ diminishes. The commonly observed increase in $|G^*|$ and decrease in δ following aging precisely manifests the macroscopic manifestation of microstructural gelation.

Relaxation behavior: α-relaxation correlates with the glass transition and corresponds to large-scale cooperative motion of colloidal network units. A denser gel network significantly prolongs its relaxation time. β-relaxation relates to more localized molecular motion (e.g., movement of small-molecule components) and is less directly influenced by the network structure.

Viscosity: As a measure of flow resistance, viscosity increases sharply with enhanced interaction strength and network connectivity within the colloidal network. This explains the marked viscosity increase observed in high-asphalt-content or aged bitumens.

4. Rheological Properties of Asphalt Under Dynamic Shear Loading

Asphalt exhibits a range of rheological properties, including viscoelasticity, shear thinning, and thixotropy [98]. Various microscale studies can help elucidate the rheological mechanisms of asphalt and provide a theoretical basis and direction for its modification. However, these methods often have strict requirements for instruments and the environment, and cannot directly reflect asphalt performance.

Therefore, a series of research methods for asphalt rheological properties based on dynamic mechanics and quasi-static mechanics has also been developed. Among them, the dynamic mechanical analysis method measures the material's mechanical response by applying a periodically varying stress or strain. When the vehicle passes through a specific point on the asphalt pavement, the point can be regarded as bearing a load with a minor frequency. At the same time, due to the unevenness of the asphalt pavement, wheel loads induce vertical vibration and localized impacts on the pavement. The complex dynamic mechanics between the vehicle and pavement can be simulated by superposing multiple simple steady shear loads. Hence, it is more practical to study the rheological properties of asphalt by the dynamic mechanics analysis method.

4.1. Linear Viscoelasticity

Over five years, SHRP conducted a series of studies and established a comprehensive set of methods for grading and evaluating asphalt performance. One of SHRP's key recom-

 https://doi.org/10.3390/coatings16030355

mendations was the utilization of DSR testing to test the dynamic viscoelastic properties of asphalt binder under dynamic shear loading [99]. The main output parameters of the DSR test are dynamic modulus and phase angle. The dynamic modulus characterizes the ability of viscoelastic materials to resist deformation, that is, the elastic part of asphalt. The phase angle characterizes the time lag in the response, representing the viscous part of asphalt [100].

In the 1950s, Van der Poel developed a nomogram for asphalt based on penetration and softening-point test results, and determined the stiffness modulus of asphalt as a function of loading process, frequency, and temperature [101]. However, a nomogram cannot effectively characterize the viscoelasticity of modified asphalt [102]. At the same time, with the development of modern computer technology, to better describe the dynamic viscoelastic mechanical characteristics of asphalt materials, researchers established functions of asphalt viscoelasticity as a function of frequency, temperature, and strain. These viscoelastic mechanical models are mainly divided into empirical algebraic models and mechanical element models.

Viscoelasticity is the most prominent rheological behavior of asphalt. It is generally believed that the strain produced by asphalt under vehicle loading is linear viscoelastic, but under high temperature and heavy loading, the asphalt strain far exceeds its linear viscoelastic region [103,104]. Therefore, research on the viscoelasticity of asphalt is primarily divided into two categories: linear viscoelasticity (LVE) and nonlinear viscoelasticity (NLVE). The methods of dividing linear and nonlinear viscoelastic regions are mainly done in two ways: (1) Linear and nonlinear viscoelastic regions in asphalt are defined by the transition point in the stress or strain sweep curves from the plateau to the descending region [105]. (2) Linear and nonlinear viscoelastic regions are divided according to the decrease in dynamic modulus to 95% of the initial modulus [3].

4.1.1. Time–Temperature Superposition Principle

Throughout the entire lifecycle of asphalt materials, they experience a wide range of temperature and frequency loads. However, existing testing equipment and methods are challenging to replicate such broad testing conditions [4]. The time–temperature superposition principle (TTSP) states that the mechanical behavior of viscoelastic materials under long-term (or low-frequency) loading can be approximated by short-term (or high-frequency) tests at higher temperatures (or loading frequencies) [11]. In general, the TTSP is also applicable to asphalt materials [106,107]. Based on the TTSP, data measured at different temperatures (within the same testing frequency range) can be shifted appropriately relative to a reference temperature to form a continuous, smooth curve, known as the master curve. The master curve is a function that characterizes a material's viscoelasticity and depends on factors such as temperature, frequency, and strain. It serves as a tool for studying the viscoelastic properties of asphalt materials over a wide temperature and frequency range.

The master curve constructed from the TTSP must use the temperature-shift function to calculate the shift factor α(T). As shown in Equation (5), the shift factor is multiplied by the frequency to obtain the reduced frequency ξ of the master curve.

$$\xi = f(a(T)) \tag{5}$$

The shift factor plays an important role in constructing the master curve. At present, the methods for calculating the shift factor mainly include: numerical methods (no functional form shift factor), the WLF equation, log-linear, quadratic polynomial, and Viscosity Temperature Susceptibility (VTS, mainly for unaged asphalt). The commonly used calculation methods are listed in Table 4.

 https://doi.org/10.3390/coatings16030355

Table 4. Calculation equations of the shift factor.

Methods	Equations	Parameter
WLF equation [108]	$loga(T) = \frac{-C_1(T-T_0)}{C_2+T-T_0}$, $T > T_g$ (glass transition temperature)	$a(T)$: shift factor T: temperature (°C) T_0: reference temperature (°C) C_1, C_2: constant
Arrhenius equation [109]	$loga(T) = \frac{\triangle E_a}{2.303R}\left(\frac{1}{T} - \frac{1}{T_0}\right)$, $T < T_g$	$\triangle E_a$: Apparent activation energy R: Universal gas constant, (8.315 K·mol^{-1}) T: temperature (K) T_0: reference temperature (K)
Log-linear [110]	$loga\left(\frac{T}{T_0}\right) = \beta(T - T_0)$	β: slope
VTS [111]	$loga(log\eta) = A + VTS[\log(T_R)]$	η: viscosity (*cPoise*) A: regression intercept VTS: VTS equation regression slope T_R: temperature (°R)
Quadratic polynomial [4]	$log\xi_r = log\xi + C_1(T_0 - T) + C_2(T_0 - T)^2$	ξ_r: educed frequency (Hz) T: temperature (°C) T_0: reduced frequency (°C) ξ: frequency (Hz)

When selecting displacement factor equations, their applicable conditions must be considered. The WLF equation is an empirical formula that is generally effective for many bitumens over a temperature range approximately Tg + 100 °C above the glass transition temperature Tg. However, it is unsuitable for regions where temperatures are close to Tg or significantly exceed its upper applicability limit. For the latter, the Arrhenius equation may prove more appropriate. The VTS method is explicitly intended for use with unaged bitumen. Different asphalts, particularly modified or high-wax varieties, may exhibit temperature dependencies deviating from these classical equations. This can cause failure or errors in master curve construction, manifesting in black plots where data points fail to overlap into a single curve. Consequently, no displacement factor model is universally applicable; its suitability must be rigorously validated by superimposing experimental data.

The methods for constructing the master curve based on the TTSP allow data comparison across different test conditions (temperature or frequency). To reduce errors caused by the calculation and use of the shift factor, Wicket diagrams [112,113], the least-squares method [114], Kramers–Kronig relations [107,115], Bueche–Rouse theory [116], and other methods are used to optimize the calculation of the shift factor. However, the TTSP is not applicable to all asphalt materials, such as asphalt with high wax content, high asphaltenes content, and some modified asphalt [11]. When constructing the master curve, it is necessary to select an appropriate model based on the material's viscoelastic mechanical characteristics. The viscoelastic properties of asphalt materials are typically categorized into two types: viscoelastic liquids and viscoelastic solids. Under dynamic loading with sufficiently low frequencies, the complex modulus of liquid materials approaches 0°, and the phase angle approaches 90°, while for solid materials, the complex modulus approaches a constant value, and the phase angle approaches 0° [102]. Generally, asphalt binder is regarded as a viscoelastic liquid, and a few polymer-modified asphalt binders are regarded as viscoelastic solids [4]. In addition, the master curve needs to be established in the LVE region, where the macromolecules in the asphalt material do not undergo structural rearrangement with temperature changes [117], to ensure the effectiveness of the TTSP.

The black diagram is a rheological relationship diagram based on phase angle and complex modulus. It does not need to use the TTSP to shift the test data and can reflect

https://doi.org/10.3390/coatings16030355

the original data characteristics. Therefore, the validity of the TTSP is often judged by the shape of the black space diagram (Figure 4).

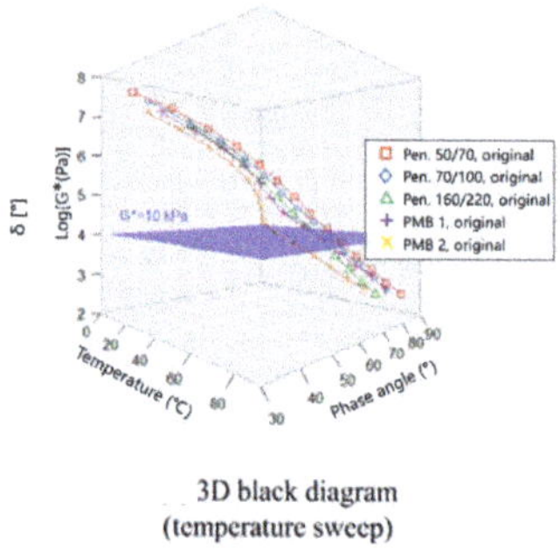

Figure 4. Black diagram [118].

From a microscopic perspective, the efficacy of TTSP stems from the fact that molecular motion rates (relaxation times) at different temperatures follow the same principles of activation energy. In other words, elevated temperatures impart greater kinetic energy to molecular segments or colloidal aggregates, enabling them to overcome energy barriers and rearrange more rapidly. This effectively accelerates the relaxation process on the temporal scale. Consequently, shifting high-frequency data at elevated temperatures towards lower frequencies via a horizontal displacement factor (e.g., the WLF equation [26]) can simulate the response observed under prolonged loading at lower temperatures.

However, when asphalt's microstructure undergoes fundamental changes within the test temperature range (e.g., wax crystallisation/melting or colloidal structure phase transitions), the activation energy for molecular motion becomes non-constant. Consequently, TTSP becomes ineffective, as evidenced by Black's diagram, in which data from different temperatures do not overlay into a single curve. This further underscores that microstructural stability is a prerequisite for the validity of macroscopic rheological models.

4.1.2. Empirical Algebraic Models

The empirical algebraic model usually uses a concise mathematical formula to fit the master curve. The commonly used empirical algebraic models are summarized in Table 5.

When fitting the linear viscoelastic behavior of asphalt using empirical algebraic models, the selection of an appropriate model is essential based on the test data. In most instances, these models provide a good fit within specific conditions. However, it is critical to recognize that these models are not universally applicable and their validity is constrained by temperature, frequency range, and binder type. For instance, some models may not effectively characterize the LVE properties of modified asphalt at low temperatures or of asphalt with high wax content. The widely used Christensen-Anderson (CA) model and its derivatives are typically calibrated in the intermediate temperature range relevant to pavement performance and may not accurately capture behavior near the glass transition or at very high temperatures. Furthermore, these empirical algebraic models employ mathematical analytical formulas that may not inherently satisfy the fundamental Kramers–Kronig relations between the storage and loss moduli, limiting their physical basis and extrapolation reliability [4].

To describe the linear viscoelasticity of bitumen, researchers have proposed various empirical algebraic models. Among these, the Christensen-Anderson (CA) model and its variants are widely applied due to their formal simplicity and minimal parameter requirements. This model characterises the variation of the complex modulus with equivalent frequency using two asymptotes (glass and rubber moduli) and a shape parameter. However, the CA model exhibits reduced fitting accuracy at extremely low temperatures near the glass transition and at very high frequencies. The sigmoidal model, by introducing additional parameters, offers greater flexibility in fitting and is particularly suitable for modified bitumen. A comprehensive list of models is provided in Appendix A.1.

4.1.3. Mechanical Element Models

The mechanical element model investigates the dynamic viscoelastic behavior by elucidating the constitutive relationship of asphalt materials. Researchers typically employ spring elements, which adhere to Hooke's law, to represent the material's elasticity, and dashpot elements equipped with Newtonian fluid to represent its viscosity. These two types of mechanical elements are combined using various connection modes (series, parallel, or both) to depict the material's mechanical behavior, and the material's complex mechanical characteristics are described by adding or removing elements. The classical mechanical element models employed to characterize the viscoelastic properties of asphalt include the Kelvin, Maxwell, Kelvin–Voigt, Burgers, modified Burgers, generalized Kelvin, and generalized Maxwell models, among others [119–121]. While these classical models offer intuitive theories with well-defined physical concepts, they occasionally do not align closely with experimental findings [122–125].

In the definition of a Newtonian dashpot, for viscoelastic materials, it is assumed that they adhere to the constitutive relationship given by Equation (6). Building upon this assumption, a fractional derivative constitutive equation is introduced.

$$\sigma(t) \sim \frac{d^{\alpha}\varepsilon(t)}{dt^{\alpha}}, \ (0 < \alpha < 1) \tag{6}$$

where $\sigma(t)$ denotes shear stress, $\varepsilon(t)$ denotes shear strain, t denotes time, and α denotes differential order.

Mechanical element models originate from the material's constitutive relations. Whilst classical models such as Burgers offer physical intuitiveness, they struggle to accurately describe asphalt behaviour across its broad temperature and frequency domains. Fractional-order derivative models (e.g., Huet and 2S2P1D models) achieve precise characterisation of broadband viscoelastic behaviour with fewer parameters by incorporating differential operators of non-integer orders. For instance, the 2S2P1D model successfully simulates the transition from the glassy to the flow state through a combination of two springs, two parabolic elements, and a damping pot. Key parameters of these models—such as the fractional order—are closely related to the colloidal structure of asphalt, though their physical interpretation requires further elaboration. The complete model equations and parameters are detailed in Appendix A.2.

Although these mechanical element models can describe the viscoelastic properties of asphalt materials, they lack flexibility for simulating complex materials such as asphalt due to the large number of parameters [104]. In addition, these models cannot accurately describe the viscoelasticity under extreme conditions (i.e., the fitting degree of the two ends of the master curve is poor), and they also lack versatility in the face of different asphalt materials [117]

4.1.4. Correlation Between Microstructures and Rheological Properties

The rheological properties of polymers are susceptible to molecular weight distribution [126]. Themeli et al. [127] used the δ-method to determine the molecular weight distribution of asphalt based on the phase angle. The method's validity was verified by gel permeation chromatography (GPC) results. Furthermore, they introduced the concept of aging molecular distribution displacement as a parameter to assess the aging progression

https://doi.org/10.3390/coatings16030355

of asphalt. Pipintakos et al. [128] used a variety of statistical methods to demonstrate a significant relationship between the chemical composition of asphalt and its rheological properties. Merusi et al. [129] used the multiphase filler model to explain the stress yield behavior of asphalt at low shear rates. Obviously, the rheological parameters are the bridge between the microscopic mechanism and the macroscopic performance.

Table 5 summarizes some studies on the relationship between microstructure and asphalt rheological properties, organized into two sections. (a) focuses on chemical composition and colloidal structure parameters, while (b) summarises studies correlating microstructural features with rheological properties.

Interpretation of Table 5a: This table demonstrates that the colloid index or bitumen index generally exhibits a positive correlation with the complex modulus. Interpretation of Table 5b: This table is primarily based on atomic force microscopy (AFM) observations. Studies by Soenen et al. and Li et al. both found that the dimensions, quantity, and distribution of the 'honeycomb-like structures' are closely correlated with variations in the asphalt's storage modulus and loss modulus.

Table 5. (**a**) Correlation between microstructures and rheological properties. (**b**) Correlation between microstructures and rheological properties.

Reference	Factors	Methods	Findings
		(**a**)	
Wang et al. [130]	Colloid index (I_c)	Characterizing various types of asphalt and their fractions using DSR to establish the relationship between colloid index and rheological parameters.	The complex modulus of asphalt is significantly correlated with the colloid index, increasing with the increase in colloid index.
Siroma et al. [131]	Molecular weight	Based on the phase angle master curve of asphalt, the molecular weight distribution of asphalt was calculated by δ-method.	The molecular weight distribution (MWD) calculated by δ-method has the same trend as the GPC test results. The molecular aggregation index (MAI) was proposed to quantify the aggregation rate of asphaltene.
Krolkral et al. [132]	Molecular weight	Based on the phase angle master curve of asphalt, the MWD of asphalt is calculated by δ-method. The molecular weight distribution is decomposed into four Gaussian functions, and the asphalt structure is inversely calculated by viscoelastic master curve.	During the aging process of asphalt, the proportion of low molecular weight components gradually decreased, while that of high molecular weight components increased.
Paliukaite et al. [133]	Chemical composition	The strength of functional groups in asphalt was determined semi-quantitatively using Fourier transform infrared spectroscopy, and the metal content in asphalt was indirectly assessed using an organic element analyzer. The impact of chemical composition on asphalt was evaluated by comparing the changes in these parameters and the master curve before and after aging.	Carbon, oxygen, and sulfur are the primary elements that significantly impact asphalt performance. The presence of heavy metals in asphalt results in reduced asphaltenes content and heightened sensitivity of sulfur to oxygen within the structure.
Weigel et al. [134]	Chemical composition	Through a variety of statistical methods to explore whether there is a relationship between the chemical composition of asphalt and its rheological parameters.	There is a significant relationship between asphaltenes content and phase angle and complex modulus, but the accuracy of the established prediction model decreases above 50 °C.

 https://doi.org/10.3390/coatings16030355

Table 5. *Cont.*

Reference	Factors	Methods	Findings
		(b)	
Yu et al. [135]	Chemical composition, Bee structure	Derived asphalt was prepared by blending four groups of components of two kinds of original asphalt. Based on the AFM and DSC test results of original asphalt, SARA fractions and derived asphalt, as well as their viscoelastic properties, the relationship between chemical composition, microstructure and rheological properties was established.	Increasing the asphaltenes content in the derived asphalt can approximately simulate asphalt aging. With the increase in asphaltenes content, the micelles in the asphalt gradually increase, and the transition from sol structure to gel structure is gradually realized. The higher the asphaltenes content, the higher the glass transition temperature of asphalt, and the smaller the phase angle.
Oldham et al. [136]	Bee structure	By comparing the morphology and rheological parameter changes in asphalt before and after aging, as well as after the addition of rejuvenators using AFM and DSR, a connection between asphalt's morphology and rheological properties was established.	The length of bee structure is significantly related to the hardness and viscosity of asphalt, and the addition of rejuvenator will significantly affect the length of bee structure.
Soenen et al. [137]	Interactions between polycyclic aromatic structures	The size of aromatic structure was semi-quantitatively analyzed by spectral techniques, and the relationship between the interaction force between aromatic structures and the rheological properties of asphalt was established by combining the master curve.	Under higher temperature or longer loading times, the viscoelasticity of asphalt is primarily influenced by its larger conjugated aromatic structure. Conversely, at lower temperatures or shorter loading times, the viscoelasticity of asphalt is notably associated with its smaller aromatic structure.
Li et al. [138]	Asphaltene particle size	Establish a continuous relaxation spectrum using the storage modulus and loss modulus from the master curve. Then, calculate the equivalent asphalt particle size based on the longest relaxation time in the relaxation spectrum.	The equivalent asphaltene particle size a_{eq} was proposed. The longest relaxation time and equivalent asphaltene particle size of asphalt increase with the increase in aging degree.

NOTE: The rheological parameters in these works are derived from the LVE region of asphalt, and only the complex modulus at this time has a clear and precise physical meaning.

4.2. Nonlinear Viscoelasticity

4.2.1. Nonlinear Viscoelastic Characterization Methods

Under high strain or stress, asphalt is more likely to exhibit NLVE behavior [139]. Masad and Somadevan [140] developed a finite element model of the microstructure of hot-mix asphalt to investigate the impact of local strain on the strain distribution within aggregates and the binder. The findings indicate that the average binder strain is approximately 8 times greater than the overall mixture strain. The linear viscoelastic strain of certain asphalt binders is 1% [141], yet the strain amplitude in the linear amplitude scanning (LAS) test can reach 30%. This significant deviation from linear behavior in the strain amplitude during testing can potentially impact the assessment of asphalt fatigue performance due to NLVE. The simplified viscoelastic continuum damage theory (S-VECD) is commonly used to study the fatigue performance of asphalt. However, NLVE effects are rarely accounted for, thereby diminishing the accuracy of asphalt fatigue life prediction [142].

Currently, NLVE of asphalt is commonly assessed through multiple stress creep recovery (MSCR) [143], repeated stress sweep (RSS) [139], and incremental stress sweep (ISS) tests [142], with analysis using models such as the Schaper model [144], Delgadillo model [104], Narayan model [98], and others. The mechanical behavior of asphalt damage is similar to the nonlinear viscoelastic behavior, as reflected in the increase in the phase angle and the decrease in the dynamic modulus. To improve the prediction accuracy of

 https://doi.org/10.3390/coatings16030355

asphalt fatigue life, Safaei et al. [142] used the LAS test to calculate the nonlinear dynamic shear modulus based on the S-VECD theory. An et al. [139] determined the NLVE strain threshold of asphalt binder using an ISS test and separated it from the damage process.

In addition to the steady-shear load-based test methods mentioned earlier, the large-amplitude oscillatory shear (LAOS) test is also used to assess nonlinear viscoelasticity [145]. This approach is not only more versatile but also more representative of the actual stresses experienced by road asphalt materials [146,147].

In a stress-controlled LAOS test, the strain-time curve of asphalt can be divided into three stages: the unstable zone, the stable zone, and the damage zone. In strain-controlled mode, the stress-time curve can be divided into two stages: the stable zone and damage zone [103]. With an increase in LAOS loading cycles, the microstructure of the asphalt material changes, leading to damage [148,149]. Therefore, it is necessary to select undamaged data for analysis of LAOS test results.

The input strain and input strain rate under the strain-controlled mode are shown as formula (7):

$$\gamma(t) = \gamma_0 \sin \omega t, \dot{\gamma}(t) = \omega \gamma_0 \cos \omega t \tag{7}$$

where γ(t) denotes the dimensionless input shear strain, $\dot{\gamma}$(t) represents the input shear strain rate (in units of s^{-1}), and γ0 denotes the strain amplitude, and ω is the angular frequency (in units of rad/s).

The corresponding output stress can be expressed by Equation (8):

$$\sigma(t) = \sigma_0 \sin(\omega t + \delta) \tag{8}$$

where σ(t) denotes the output stress, σ0 denotes the stress amplitude, and δ denotes the phase angle.

4.2.2. Nonlinear Viscoelastic Evaluation Methods in the LAOS Test

Unlike LVE test methods, in the nonlinear viscoelastic region of stress–strain response during LAOS testing, the stress–strain response is no longer a sine waveform, and the complex modulus loses its precise physical meaning at this point. Therefore, it is necessary to develop alternative methods to qualitatively and quantitatively analyze the NLVE of asphalt under LAOS testing. The commonly used evaluation methods are shown in Table 6.

Table 6. Evaluation methods of NLVE under the LAOS test.

Methods	Equations or Definitions	Parameters	Uses and Legends
Relative intensity of the higher order harmonic [103]	Strain-controlled mode: $I_{n/1} = \frac{I_n}{I_1} = \frac{\tau_n}{\tau_1} = \frac{\sqrt{(\tau_n \cos \delta_n)^2 + (\tau_n \sin \delta_n)^2}}{(\tau_1 \cos \delta_1)^2 + (\tau_1 \sin \delta_1)^2}$, n = 1,3,5... Stress-controlled mode: $I_{n/1} = \frac{I_n}{I_1} = \frac{\gamma_n}{\gamma_1} = \frac{\sqrt{(y_n \cos \delta_n)^2 + (y_n \sin \delta_n)^2}}{(y_1 \cos \delta_1)^2 + (y_i \sin \delta_1)^2}$, n = 1,3,5...	$I_{n/1}$: n-order harmonic relative intensity ratio τ_n: n-order harmonic stress γ_n: n-order harmonic strain δ_n: n-order harmonic phase angle	Evaluate the degree of NLVE of asphalt. Legend [103]:

https://doi.org/10.3390/coatings16030355

Table 6. *Cont.*

Methods	Equations or Definitions	Parameters	Uses and Legends
Lissajous curve [150–152]	The Lissajous curve is mainly divided into two types: ① Elastic Lissajous curve (stress–strain curve) ② Viscous Lissajous curve (stress–strain rate curve)	① Elastic Lissajous curve: ·Strain-controlled mode: G'_M: minimum strain modulus, the tangent slope at $\gamma = 0$ G'_L: maximum strain modulus, the slope from $\gamma = \gamma_0$ to the origin	The viscoelastic state and degree of asphalt are judged according to the curve shape. Legend (Elastic Lissajous curve) [150]:
		·Stress-controlled mode: J'_M: minimum nonlinear compliance J'_L: maximum nonlinear compliance	
		② Viscous Lissajous curve ·Strain-controlled mode: η'_M: minimum strain rate viscosity, the tangent slope at $\dot{\gamma} = 0$ η'_L: maximum strain rate viscosity, the slope from $\dot{\gamma} = \dot{\gamma}_0$ to the origin	
		·Stress-controlled mode: Φ'_M: minimum fluidity Φ'_L: maximum fluidity	
Stress decomposition and strain decomposition [153,154]	① Nonlinear stress: $\tau(t) = \tau_{OB}(t) + \tau_{BO}(t)$ ② Nonlinear strain: $\gamma(t) = \gamma'(t) + \gamma''(t)$ ③ Chebyshev polynomials: ·Elastic: $\tau'(x) = \gamma_0 \sum_{n=0}^{n} e_n(\omega, \gamma_0) T_n(x)$ $\gamma'(t) - \tau_0 \sum_{n=0}^{n} c_n(\omega, \tau_0) T_n(x)\cdot$ Viscous: $\tau''(x) = \dot{\gamma}_0 \sum_{n=1}^{\infty} v_n(\omega, \gamma_0) T_n(y)$ $y''(t) = \tau_0 \sum_{n=1}^{n} f_n(\omega, \tau_0) T_n(x)$	$\tau_{OE}(x)$: apparent elastic stress, $x = \gamma/\gamma_0 = \sin\omega t$ $\tau_{EO}(y)$: apparent viscous stress, $x = \dot{\gamma}/\dot{\gamma}_0 = \cos\omega t$ γ': apparent elastic strain γ'': apparent viscous strain $e_n = G_n'(-1)^{(n-1)/2}$: n-order Chebyshev coefficient $v_n = G_n''/\omega$: n-order Chebyshev coefficient $c_n = J_n'$: n-order Chebyshev coefficient $f_n = n\omega J_n''$: n-order Chebyshev coefficient $T_n(x)$: n-order Chebyshev coefficient	The rheological behavior of the material was judged according to the Chebyshev coefficient (commonly used third-order Chebyshev coefficient). Legend [145]:
Constitutive model	Jeffreys model [155]: $\frac{\dot{\sigma}}{G_0} + \frac{\sigma}{\eta_0} = \left(\xi + \frac{\eta_\infty}{\eta_A}\right)\dot{\gamma} + \frac{\eta_\infty}{G_0}\dot{\gamma}$	ξ: constant η_∞: infinite shear viscosity G_0: shear modulus	The NLVE behavior under LAOS loading is described.
	Bergström-Boyce model [156]: $\dot{\gamma} - \dot{\gamma}_0(\overline{\lambda_B^v} - 1 + \xi)^c \left[R\frac{\tau}{\tau_{base}} - \overline{\tau}_{cu}\right]^m$	$\overline{\tau}_{cu}$: cut-off stress below which no flow will occur ξ: strain adjustment factor $\overline{\lambda_B^v}$: viscoelastic chain stretch	

Shan et al. [149] found that the matrix asphalt only has the first-order harmonic intensity in the LVE state, and the third-order and fifth-order harmonic relative intensity ratios appear in the NLVE state, and that the NLVE behavior of the asphalt is more significant with the increase in loading level. In the LAOS test, different control modes will have different results. For instance, in the strain-controlled mode, the third-order relative harmonic

 https://doi.org/10.3390/coatings16030355

strength (I1/I3) of asphalt increases with higher strain and frequency, and decreases with lower temperature. In other words, asphalt exhibits more pronounced NLVE behavior under high-strain, low-temperature, and high-frequency conditions [157–159]. However, under stress-controlled mode, the NLVE of asphalt becomes increasingly pronounced under high stress, high temperature, and low-frequency conditions [149,160]. For example, in the strain-controlled mode, the Lissajous curve exhibits substantial distortion under high stress, low frequency, and low temperature, indicating a more pronounced NLVE [157,158,161]. Conversely, under stress-control mode, asphalt exhibits enhanced nonlinear viscoelasticity under high stress, low frequency, and high temperature [149].

The stress decomposition and strain decomposition method is mainly used to judge the rheological properties of asphalt, such as stress (strain) softening (hardening) and shear thinning (thickening) [149,158]. There are many classical nonlinear viscoelastic constitutive models under LAOS loading [103,145,162–165], but few researchers have applied them to asphalt materials. Padmarekha et al. [152] developed a Frame-Invariant model to fit the stress–strain curves of matrix asphalt and rubber-powder-modified asphalt under LAOS loading. Diab et al. [156] used the Bergström-Boyce model to fit the stress–strain curves of asphalt in the LVE and NLVE regions. These models can effectively capture the NLVE behaviors exhibited by various asphalt types. However, as with the aforementioned NLVE analysis methods that rely on the LAOS test, the parameters in these models often lack physical significance or fail to correlate with pavement performance. In the study by Saboo et al. [159], it was found that I1/I3 was significantly correlated with the fatigue life Nf of asphalt, and a relationship independent of temperature, aging condition, and asphalt type was established. Therefore, it is feasible to establish a relationship between the evaluation index and road performance under the LAOS test, which can be further studied in the future.

4.3. Thixotropy

Due to the combined effects of vehicle load and the external environment, fatigue failure of asphalt pavement will occur during its service life, resulting in a series of pavement issues. In many early investigations on asphalt fatigue, researchers believed that asphalt fatigue is caused by damage and is an irreversible process. The fatigue factor, cumulative dissipated energy ratio, and other indicators are proposed to evaluate the fatigue characteristics of asphalt [120,166–168], and a series of models are used to predict the asphalt fatigue life [120,169,170]. In fact, changes in rheological parameters during asphalt fatigue damage are not solely due to damage, but also to the coupling effects of other factors, such as thixotropy and NLVE [171–175].

As depicted in Figure 5a, asphalt thixotropy refers to the orientation, extension, and unwinding of asphalt molecules under long-term loading, resulting in a decrease in viscosity. When the loading stops for a period of time, the asphalt molecules gradually return to the state before the load, and the viscosity gradually recovers [176].

The variation in viscoelastic parameters caused by thixotropy can be fully restored, but fatigue damage will decrease the effective area of asphalt, and the variation in viscoelastic parameters will be wholly or partially restored at a relatively low rate [177]. To enhance the precision of asphalt fatigue life prediction, researchers have delved deeper into examining the impact of thixotropy on asphalt's fatigue characteristics. The findings indicate that fatigue-life prediction models based on damage theory tend to overestimate the extent of asphalt damage. H. Di Benedetto et al. [178] took into account the influence of thixotropy when investigating asphalt fatigue characteristics. They divided the fatigue process into three stages (Figure 5b). During the time sweep test, reversible phenomena

https://doi.org/10.3390/coatings16030355

such as thixotropy or nonlinear viscoelasticity occur in the initial stage of asphalt, leading to a rapid decrease in the complex modulus, followed by stabilization (stage 1). As loading cycles increase, the asphalt develops small cracks due to fatigue, leading to a gradual decrease in the complex modulus (stage 2). Subsequently, these cracks grow and connect, forming larger cracks leading to asphalt failure, accompanied by a rapid drop in the complex modulus (stage 3). Shan [176] and Lv et al. [172] studied the effect of thixotropy on the fatigue characteristics of asphalt, and also believed that the decrease in modulus at the initial stage was caused by thixotropy and nonlinear viscoelasticity.

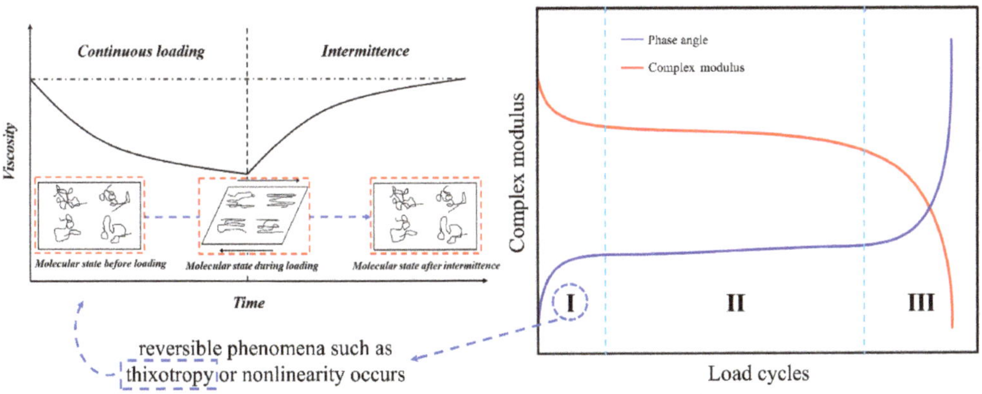

(a) Thixotropy of asphalt (b) Time sweep test

Figure 5. The variation of rheological parameters of asphalt during fatigue process.

5. Conclusions and Future Perspective

This study systematically reviews advances in the multiscale rheological behaviour of asphalt pavement materials, aiming to bridge the cognitive gap between their microstructure and macroscopic performance. Key consensus points, existing knowledge gaps, and priority areas for future research can be summarised as follows:

5.1. Conclusions

Current research has confirmed that bitumen constitutes a complex multiphase chemical continuum, with its macroscopic rheological properties rooted at the microscopic scale. The chemical composition, particularly the proportions and polarities of SARA components, fundamentally governs bitumen's viscoelasticity. Polar constituents (such as asphaltenes) primarily contribute elasticity, whilst non-polar components predominantly influence viscosity. Colloidal structure theories, notably the Yen-Mullins model, provide a robust framework for explaining asphaltene nanoaggregation and its performance implications. Regarding macroscopic characterisation, dynamic shear rheometer (DSR) testing within the linear viscoelastic (LVE) region has been standardised. Master curves constructed via the time–temperature superposition principle (TTSP), alongside various empirical algebraic and mechanical element models, effectively describe asphalt's linear response across broad temperature and frequency domains. Significant correlations exist between rheological parameters and microscopic morphologies such as the 'honeycomb structure' observed via atomic force microscopy (AFM), further substantiating the notion that microstructure governs macroscopic performance.

Despite the aforementioned consensus, several critical knowledge gaps continue to impede the establishment of a comprehensive structure-property relationship. Firstly, there

 https://doi.org/10.3390/coatings16030355

remains a lack of universal bridging models capable of quantitatively linking intermolecular interactions, colloidal structural indices, and measurable rheological parameters (such as |G*|, δ) across micro- and macro-scales. Secondly, the understanding of non-linear viscoelasticity (NLVE) remains far from mature. Most microstructure-property correlation studies remain confined to the linear viscoelastic (LVE) regime, whilst large-amplitude oscillatory shear (LAOS) testing methods—capable of authentically reflecting complex pavement loading conditions—and their data interpretation are still in their infancy. Furthermore, existing rheological models (e.g., empirical and fractional-derivative models) frequently suffer from limitations, including insufficient universality, ambiguous physical interpretations of parameters, and inadequate descriptions of extreme conditions. Finally, no definitive consensus has been reached regarding the chemical origins of microstructural features such as 'honeycomb structures' or their representativeness within the bulk matrix.

5.2. Future Perspective

Future research should focus on the following areas in order to address the aforementioned challenges:

(1) Developing deeply integrated multi-scale methodologies: There is an urgent need to combine advanced molecular simulations (such as coarse-grained molecular dynamics), high-resolution chemical imaging techniques (such as AFM-IR), and macroscopic rheological measurements to establish quantitative, verifiable models spanning molecular motion to colloidal network evolution, thereby bridging the theoretical gap between scales.

(2) Establish a universal model encompassing both linear and nonlinear behaviours: Focus on developing constitutive models that can accurately describe linear viscoelasticity across a wide temperature and frequency range while effectively characterising nonlinear responses under low-amplitude oscillatory stress (LAOS). Model parameters should possess clear physical significance and be directly linked to the chemical/colloidal properties of asphalt, thereby enhancing predictive capability and universality across different asphalt types, including various modified asphalts.

(3) Enhancing Fatigue Life Prediction Frameworks: Designing novel experimental protocols and analytical theories to isolate and quantify the coupled effects of reversible non-linear viscoelastic and thixotropic behaviour alongside genuine irreversible damage during fatigue processes, thereby developing more precise asphalt fatigue life prediction methodologies.

6. Patents

Patent: Training method, component content analysis method, and equipment for component content analysis (202511720449X).

Author Contributions: Conceptualization, Q.Z. and Z.C.; methodology, X.C.; validation, Q.Z., Q.L. and Y.Y.; formal analysis, X.C.; investigation, L.H.; resources, X.C.; writing—original draft preparation, Q.Z.; writing—review and editing, Z.C.; visualization, X.C.; supervision, Q.L.; project administration, J.G.; funding acquisition, L.H. and X.C.; All authors have read and agreed to the published version of the manuscript.

Funding: This work was sponsored by Chongqing Natural Science Foundation Project, China (NO. CSTB2025NSCQ-GPX0887), and the Natural Sciences Challenge Grant Program of Chongqing Jiaotong University (XJ2023000501).

Data Availability Statement: Data are contained within the article.

Acknowledgments: The authors gratefully acknowledge all financial support.

https://doi.org/10.3390/coatings16030355

Conflicts of Interest: The authors declare no conflicts of interest.

Appendix A

Appendix A.1

Table A1. Algebraic models.

Models	Equations	Parameters
Jongepier and Kuilman's model [179]	$H(\tau) = \frac{G_g}{\beta\sqrt{\pi}}\exp\left(-\left\{\frac{\ln\left(\frac{\tau}{\tau_m}\right)}{\beta}\right\}^2\right)$	$H(\tau)$: relaxation spectrum β: width of relaxation spectrum τ: relaxation time τ_m: mean relaxation time
Dobson's model [117]	$\log\omega_r^{-\sigma} = \log(G_r^{-\sigma} - 1) + \frac{20.5 - G_r^{-\sigma}}{230.3}$	$\omega_r = \eta_0\omega\alpha_T/G_g$ $G_r = \lvert G^*\rvert/G_g$ b: shear-susceptibility index
Dickinson and Witt's model [180]	$\log G_r = 0.5\left\{\log\omega_r - \left[\left(\log\omega_r\right)^2 + (2\beta)^2\right]^{0.5}\right\}$ $\delta(\omega_r) = 45\left\{1 - \log\omega_r\left[\left(\log\omega_r\right)^2 + (2\beta)^2\right]^{-0.5}\right\}$	$logG_g = 8.88 + 0.58\beta_{BW}$ $\beta_{BD} = logG_g - logG_x$
Christensen and Anderson (CA) model [181]	$G_g = \left[1 + \left(\omega_r\right)^{\frac{-\log 2}{R}}\right]^{\frac{-R}{\log 2}}$ $\delta = \frac{90}{1+\left(\omega_r\right)^{\frac{Log2}{R}}}$	$R = logG_g - logG_x$ Gg: Glass transition modulus (Pa), ω: Angular frequency (rad/s)
Fractional model [182]	$\lvert G^*\rvert = \eta_0\omega\left[\frac{\Pi_{i=1}^{m}\left(1+\left(\mu_i\omega\right)^2\right)}{\Pi_{i=1}^{n}\left(1+\left(\lambda_i\omega\right)^2\right)}\right]^{\frac{1}{2(n-m)}}$ $\delta = \frac{\pi}{2} + \frac{1}{(n-m)}\left[\sum_1^m \alpha tan(\mu_k\omega) - \sum_1^n \alpha tan(\lambda_k\omega)\right]$	μ_k, λ_k: relaxation time, $(\mu_k, \lambda_k > 0)$ n, m: numbers of relaxation time, $(n >> m)$ α: fourier transform of the Dirac delta function
Christensen, Anderson and Marasteanu (CAM) model [183]	$\lvert G^*\rvert = G_g\left[1 + \left(\frac{\omega_c}{\omega}\right)^v\right]^{\frac{-\omega}{v}}$ $\delta = \frac{90\omega}{\left[1+\left(\frac{\omega_c}{\omega}\right)^v\right]}$	$\nu = \log 2/R$
Modified Christensen, Anderson and Marasteanu model [102]	$\lvert G^*\rvert = G_e + \frac{G_g - G_e}{\left[1+\left(\frac{f_c}{f'}\right)^k\right]^{\frac{m_e}{k}}}$	$G_e = \lvert G^*\rvert(f \to 0)$ $G_g = \lvert G^*\rvert(f \to \infty)$
Al-Qadi and co-workers' model [184]	$\lvert G^*\rvert - G_g\left[1 - \frac{1}{\left[1+\left(\frac{\omega}{\omega_0}\right)^v\right]}\right]$ $\delta = \frac{90}{\left[1+\left(\frac{\omega}{\omega_0}\right)^v\right]^m}$	ω_0: the scale parameter v, m: dimensionless parameters
Polynomial model [185]	$\log\lvert G^*\rvert = A(\log f_r)^3 + B(\log f_r)^2 + C(\log f_r) + D$	f_r: reduced frequency (Hz), $\lvert G^*\rvert$: complex modulus (Pa)
Sigmoidal model [185]	$\log\lvert G^*\rvert = v - \frac{\alpha}{1+e^{\beta}+\gamma(\log(\omega_r))}$	v: lower asymptote α: the difference between the values of the upper and lower asymptote $\omega_r = 10^{\left(-\frac{\beta}{r}\right)}$

Appendix A.2

Table A2. Fractional derivative models.

Models	Equations	Parameters
Huet model [186]	$G^* = \frac{G_\infty}{1+o\left(i\omega\tau\right)^{-k}+\left(i\omega\tau\right)^{-h}}$, $(0 < h < k < 1)$	G_∞: limit of the complex modulus τ_m: mean relaxation time o: dimensionless constant
Huet-Sayegh (HS) model [187]	$G^* = G_0 + \frac{G_\infty - G_0}{1+o\left(i\omega\tau\right)^{-k}+\left(i\omega\tau\right)^{-h}}$	G_0: elastic modulus
Di Benedetto and Neifar (DBN) model [188]	$G^* = \left(\frac{1}{G_0} + \sum_{i=1}^{n}\frac{1}{G_i + i\omega\eta_i(T)}\right)^{-1}$	G_0: elastic modulus of the single spring $\eta_i(T)$: a viscosity function of the temperature
2S2P1D model [189]	$G^* = G_0 + \frac{G_\infty - G_0}{1+o\left(i\omega\tau\right)^{-k}+\left(i\omega\tau\right)^{-h}+\left(i\omega\tau\beta\right)^{-1}}$	o, β: dimensionless constan

https://doi.org/10.3390/coatings16030355

References

1. Shen, J. Rheology of asphalt materials (part 1). *Pet. Asph.* **1988**, *4*, 35–39.
2. Wang, Y.; Wang, W.; Wang, L. Understanding the relationships between rheology and chemistry of asphalt binders: A review. *Constr. Build. Mater.* **2022**, *329*, 127161. [CrossRef]
3. Petersen, J.C.; Robertson, R.E.; Branthaver, J.F.; Harnsberger, P.M.; Duvall, J.J.; Kim, S.S.; Anderson, D.A.; Christiansen, D.W.; Bahia, H.U. *Binder Characterization and Evaluation: Volume 1*; Report No. SHRP-A-367; Strategic Highway Research Program, National Research Council: Washington, DC, USA, 1994.
4. Yang, S. Research on Dynamic Viscoelastic Fractional Constitutive Model of Asphalt Binders and Mixtures. Ph.D. Thesis, Hunan University, Changsha, China, 2020.
5. Yan, X. *Rheology of Road Materials*; China Communications Press Co., Ltd.: Beijing, China, 2019.
6. Chen, Y.; Wu, Z.; Liang, J.; He, W.; Xiao, F. Method for optimizing key evaluation parameters of recycled asphalt: Based on rheology and chemistry. *J. Clean. Prod.* **2025**, *516*, 145844. [CrossRef]
7. Pipintakos, G.; Sreeram, A.; Mirwald, J.; Bhasin, A. Engineering bitumen for future asphalt pavements: A review of chemistry, structure and rheology. *Mater. Des.* **2024**, *244*, 113157. [CrossRef]
8. Bi, Y.; Chen, H.; Chen, Z.; Pei, J.; Zhang, J.; Luo, Z.; Wang, W.; Gao, J. A comprehensive review of rheological behaviors of asphalt binders, mastics, and mixtures from a generalized rheology perspective. *Fuel* **2025**, *393*, 134984. [CrossRef]
9. Liu, J.; Feng, Z.; Wang, Z.; Guo, W.; Jing, H.; Ren, S. Multiscale characterization of rubberized asphalt performance: Coupling rheology, adhesive behavior, and nanomorphology. *Constr. Build. Mater.* **2025**, *503*, 144548. [CrossRef]
10. Mortazavi, M.; Moulthrop, J.S. *The SHRP Materials Reference Library*; Binders; National Research Council: Washington, DC, USA, 1993.
11. Lesueur, D. The colloidal structure of bitumen: Consequences on the rheology and on the mechanisms of bitumen modification. *Adv. Colloid Interface Sci.* **2009**, *145*, 42–82. [CrossRef]
12. Domin, M.A.; Herod, A.A.; Kandiyoti, R.; Larsen, J.W.; Lazaro, M.-J.; Li, S.; Rahimi, P. A Comparative Study of Bitumen Molecular-Weight Distributions. *Energy Fuels* **1999**, *13*, 552–557. [CrossRef]
13. Bai, T.; Xiang, B.; Li, M.; Manica, R.; Liu, Q. Mechanism of modification of bitumen zeta potential by sodium citrate: Interactions between bitumen metal content and ligands. *Colloids Surf. A Physicochem. Eng. Asp.* **2022**, *653*, 130007. [CrossRef]
14. Robertson, R.E. *Chemical Properties of Asphalts and Their Relationship to Pavement Performance*; National Research Council: Washington, DC, USA, 1994.
15. Green, J.B.; Yu, S.; Pearson, C.D.; Reynolds, J.W. Analysis of sulfur compound types in asphalt. *Symp. Combust.* **1993**, *27*, 2085–2092. [CrossRef]
16. López, L.; Mónaco, S.L. Vanadium, nickel and sulfur in crude oils and source rocks and their relationship with biomarkers: Implications for the origin of crude oils in Venezuelan basins. *Org. Geochem.* **2017**, *104*, 53–68. [CrossRef]
17. Corbett, L.W. Composition of asphalt based on generic fractionation, using solvent deasphaltening, elution-adsorption chromatography, and densimetric characterization. *Anal. Chem.* **2002**, *41*, 576–579. [CrossRef]
18. Zhang, C.; Xu, T.; Shi, H.; Wang, L. Physicochemical and pyrolysis properties of SARA fractions separated from asphalt binder. *J. Therm. Anal. Calorim.* **2015**, *122*, 241–249. [CrossRef]
19. Michon, L.; Martin, D.; Planche, J.-P.; Hanquet, B. Estimation of average structural parameters of bitumens by 13C nuclear magnetic resonance spectroscopy. *Fuel* **1997**, *76*, 9–15. [CrossRef]
20. Speight, J. Petroleum Asphaltenes-Part 1: Asphaltenes, resins and the structure of petroleum. *Oil Gas Sci. Technol.* **2004**, *59*, 467–477. [CrossRef]
21. Mirwald, J.; Werkovits, S.; Camargo, I.; Maschauer, D.; Hofko, B.; Grothe, H. Investigating bitumen long-term-ageing in the laboratory by spectroscopic analysis of the SARA fractions. *Constr. Build. Mater.* **2020**, *258*, 119577. [CrossRef]
22. Sabbah, H.; Morrow, A.L.; Pomerantz, A.E.; Zare, R.N. Evidence for island structures as the dominant architecture of asphaltenes. *Energy Fuels* **2011**, *25*, 1597–1604. [CrossRef]
23. Robertson, R.E.; Branthaver, J.F.; Harnsberger, P.M.; Peterson, J.C.; Dorrence, S.M.; McKay, J.F.; Turner, F.; Pauli, T.; Huang, S.C.; Huh, J.D.; et al. *Fundamental Properties of Asphalts and Modified Asphalts, Volume I: Interpretive Report*; Federal Highway Administration: Washington, DC, USA, 2001.
24. Dealy, J.M. Rheological properties of oil sand bitumens. *Can. J. Chem. Eng.* **1979**, *57*, 677–683. [CrossRef]
25. Sultana, S.; Bhasin, A. Effect of chemical composition on rheology and mechanical properties of asphalt binder. *Constr. Build. Mater.* **2014**, *72*, 293–300. [CrossRef]
26. Eberhardsteiner, L.; Füssl, J.; Hofko, B.; Handle, F.; Hospodka, M.; Blab, R.; Grothe, H. Towards a microstructural model of bitumen ageing behaviour. *Int. J. Pavement Eng.* **2015**, *16*, 939–949. [CrossRef]

 https://doi.org/10.3390/coatings16030355

27. Mangiafico, S.; Di Benedetto, H.; Sauzéat, C.; Olard, F.; Pouget, S.; Planque, L. Effect of colloidal structure of bituminous binder blends on linear viscoelastic behaviour of mixtures containing Reclaimed Asphalt Pavement. *Mater. Des.* **2016**, *111*, 126–139. [CrossRef]
28. López-Paz, J.; Gracia-Fernández, C.; Gómez-Barreiro, S.; López-Beceiro, J.; Nebreda, J.; Artiaga, R. Study of bitumen crystallization by temperature-modulated differential scanning calorimetry and rheology. *J. Mater. Res.* **2012**, *27*, 1410–1416. [CrossRef]
29. Gawel, I.; Czechowski, F.; Bagińska, K. Study of wax isolated from bitumen. In Proceedings of the Eurasphalt & Eurobitume Congress, Strasbourg, France, 7–10 May 1996.
30. Lu, X.; Kalman, B.; Redelius, P. A new test method for determination of wax content in crude oils, residues and bitumens. *Fuel* **2008**, *87*, 1543–1551. [CrossRef]
31. Tan, Y.; Li, G.; Dan, L.; Lyu, H.; Meng, A. Research progress of bitumen microstructures and components. *J. Traffic Transp. Eng.* **2020**, *20*, 1–17.
32. Liu, H. Study On Wax-Reducing Modification and Ageing of Paving Asphalt. Ph.D. Thesis, China University of Petroleum (East China), Qingdao, China, 2012.
33. Soenen, H.; Besamusca, J.; Fischer, H.R.; Poulikakos, L.D.; Planche, J.-P.; Das, P.K.; Kringos, N.; Grenfell, J.R.A.; Lu, X.; Chailleux, E. Laboratory investigation of bitumen based on round robin DSC and AFM tests. *Mater. Struct.* **2014**, *47*, 1205–1220. [CrossRef]
34. Lu, X.; Langton, M.; Olofsson, P.; Redelius, P. Wax morphology in bitumen. *J. Mater. Sci.* **2005**, *40*, 1893–1900. [CrossRef]
35. Merusi, F.; Filippi, S.; Polacco, G. Effect of synthetic and functionalized waxes on bituminous binders: From the glassy state to the intermediate viscoelastic domain. *Constr. Build. Mater.* **2017**, *136*, 541–555. [CrossRef]
36. Polacco, G.; Filippi, S.; Paci, M.; Giuliani, F.; Merusi, F. Structural and rheological characterization of wax modified bitumens. *Fuel* **2012**, *95*, 407–416. [CrossRef]
37. Redelius, P.; Soenen, H. Relation between bitumen chemistry and performance. *Fuel* **2015**, *140*, 34–43. [CrossRef]
38. Guan, D.; Feng, S.; Zhang, L.; Shi, Q.; Zhao, S.; Xu, C. Mesoscale simulation for heavy petroleum system using structural unit and dissipative particle dynamics (SU–DPD) frameworks. *Energy Fuels* **2019**, *33*, 1049–1060. [CrossRef]
39. Jennings, P.W.; Pribanic, J.A.; Desando, M.A.; Raub, M.F.; Moats, R.; Smith, J.A.; Mendes, T.M.; McGrane, M.; Fanconi, B.; VanderHart, D.L.; et al. *Binder Characterization and Evaluation by Nuclear Magnetic Resonance Spectroscopy*; National Research Council: Washington, DC, USA, 1993.
40. Li, D.D.; Greenfield, M.L. Chemical compositions of improved model asphalt systems for molecular simulations. *Fuel* **2014**, *115*, 347–356. [CrossRef]
41. You, L.; Spyriouni, T.; Dai, Q.; You, Z.; Khanal, A. Experimental and molecular dynamics simulation study on thermal, transport, and rheological properties of asphalt. *Constr. Build. Mater.* **2020**, *265*, 120358. [CrossRef]
42. Kang, Y.; Zhou, D.; Wu, Q.; Liang, R.; Shangguan, S.; Liao, Z.; Wei, N. Molecular dynamics study on the glass forming process of asphalt. *Constr. Build. Mater.* **2019**, *214*, 430–440. [CrossRef]
43. De Oliveira, F.C.; Maia, J.M.; Tavares, F.W. Asphaltenes at the water-oil interface using DPD/COSMO-SAC. *Colloids Surf. A Physicochem. Eng. Asp.* **2021**, *625*, 126828. [CrossRef]
44. Li, G.; Han, M.; Tan, Y.; Meng, A.; Li, J.; Li, S. Research on bitumen molecule aggregation based on coarse-grained molecular dynamics. *Constr. Build. Mater.* **2020**, *263*, 120933. [CrossRef]
45. Aguilera-Mercado, B.M.; Herdes, C.; Murgich, J.; Müller, E.A. Mesoscopic Simulation of Aggregation of Asphaltene and Resin Molecules in Crude Oils. *Energy Fuels* **2005**, *20*, 327–338. [CrossRef]
46. Tang, J.; Wang, H. Coarse grained modeling of nanostructure and asphaltene aggregation in asphalt binder using dissipative particle dynamics. *Constr. Build. Mater.* **2022**, *314*, 125605. [CrossRef]
47. Yao, H.; Liu, J.; Xu, M.; Ji, J.; Dai, Q.; You, Z. Discussion on molecular dynamics (MD) simulations of the asphalt materials. *Adv. Colloid Interface Sci.* **2022**, *299*, 102565. [CrossRef]
48. Wang, H.; Zheng, W.; You, Z.; Ji, J.; Li, L.; Chen, Y. Research progress on compatibility evaluation methods of polymer modifiers and petroleum asphalt binders. *J. Traffic Transp. Eng.* **2023**, *23*, 8–26.
49. Cheng, D.; Little, D.N.; Lytton, R.L.; Holste, J.C. Surface Energy Measurement of Asphalt and Its Application to Predicting Fatigue and Healing in Asphalt Mixtures. *Transp. Res. Rec.* **2002**, *1810*, 44–53. [CrossRef]
50. Liu, B.; Long, J.; Ren, Q.; Li, B.; Shen, H.; Zhang, S. Preliminary Exploration for Supramolecular System of Tahe Asphaltene. *Acta Pet. Sin. (Pet. Process. Sect.)* **2017**, *33*, 16–24.
51. Bian, Y. Aggregation of Novelly Synthesized Phthalocyanines and Effect on Association of Heavy Petroleum Fractions. Ph.D. Thesis, China University of Petroleum (East China), Qingdao, China, 2015.
52. Xiao, S.; Li, M.; Li, Y. Characterization of functional groups and hydrogen bond of asphaltenes from petroleum and coal tar by infrared spectroscopy. *Pet. Process. Petrochem.* **2022**, *53*, 99–103.

 https://doi.org/10.3390/coatings16030355

53. Bukka, K.; Miller, J.D.; Hanson, F.V.; Misra, M.; Oblad, A. The influence of carboxylic acid content on bitumen viscosity. *Fuel* **1994**, *73*, 257–268. [CrossRef]
54. Steed, J.W.; Atwood, J.L. *Supramolecular Chemistry*; John Wiley & Sons: Chichester, UK, 2022.
55. Cho, D.-W.; Kim, K.; Lee, M.-J. Chemical model to explain asphalt binder and asphalt–aggregate interface behaviors. *Can. J. Civ. Eng.* **2010**, *37*, 45–53. [CrossRef]
56. Nellensteyn, F. The constitution of asphalt. *J. Inst. Pet. Technol.* **1924**, *10*, 311–323.
57. Mack, C. Colloid Chemistry of Asphalts. *J. Phys. Chem.* **1931**, *36*, 2901–2914. [CrossRef]
58. Pfeiffer, J.P.; Saal, R.N.J. Asphaltic Bitumen as Colloid System. *J. Phys. Chem.* **1940**, *44*, 139–149. [CrossRef]
59. Shan, L.; Zhang, E.; Liu, S.; Xu, H.; Tan, Y. Analysis of Microscopic Damage Mechanism of Asphalt Binder Through Atomic Force Microscopy (AFM). *China J. Highw. Transp.* **2020**, *33*, 171–177.
60. Gaestel, C.; Smadja, R.; Lamminan, K. Contribution à la connaissance des propriétés des bitumes routiers. *Rev. Gentile Routes Aérodromes* **1971**, *466*, 85–94.
61. Yu, X.; Burnham, N.A.; Tao, M. Surface microstructure of bitumen characterized by atomic force microscopy. *Adv. Colloid Interface Sci.* **2015**, *218*, 17–33. [CrossRef]
62. Li, Z.; Fa, C.; Zhao, H.; Zhang, Y.; Chen, H.; Xie, H. Investigation on evolution of bitumen composition and micro-structure during aging. *Constr. Build. Mater.* **2020**, *244*, 118322. [CrossRef]
63. Oyekunle, L.O. Certain Relationships between Chemical Composition and Properties of Petroleum Asphalts from Different Origin. *Oil Gas Sci. Technol.—Rev. d'IFP Energ. Nouv.* **2006**, *61*, 433–441. [CrossRef]
64. Li, J. Study on the Diffusion Behavior of Asphalt Rejuvenator and Influence Factors. Ph.D. Thesis, China University of Petroleum (East China), Qingdao, China, 2010.
65. Dwiggins, C., Jr. A small angle X-ray scattering study of the colloidal nature of petroleum. *J. Phys. Chem.* **1965**, *69*, 3500–3506. [CrossRef]
66. Ravey, J.; Ducouret, G.; Espinat, D. Asphaltene macrostructure by small angle neutron scattering. *Fuel* **1988**, *67*, 1560–1567. [CrossRef]
67. Overfield, R.; Sheu, E.; Sinha, S.; Liang, K. SANS study of asphaltene aggregation. *Fuel Sci. Technol. Int.* **1989**, *7*, 611–624. [CrossRef]
68. Bardon, C.; Barre, L.; Espinat, D.; Guille, V.; Li, M.H.; Lambard, J.; Ravey, J.; Rosenberg, E.; Zemb, T. The colloidal structure of crude oils and suspensions of asphaltenes and resins. *Fuel Sci. Technol. Int.* **1996**, *14*, 203–242. [CrossRef]
69. Mason, T.; Lin, M. Asphaltene nanoparticle aggregation in mixtures of incompatible crude oils. *Phys. Rev. E* **2003**, *67*, 050401. [CrossRef]
70. Yen, T.F. The colloidal aspect of a macrostructure of petroleum asphalt. *Fuel Sci. Technol. Int.* **1992**, *10*, 723–733. [CrossRef]
71. Shan, L.; Xie, R.; Wagner, N.J.; He, H.; Liu, Y. Microstructure of neat and SBS modified asphalt binder by small-angle neutron scattering. *Fuel* **2019**, *253*, 1589–1596. [CrossRef]
72. Dickie, J.P.; Yen, T.F. Macrostructures of the asphaltic fractions by various instrumental methods. *Anal. Chem.* **1967**, *39*, 1847–1852. [CrossRef]
73. Wang, Y.; Zhao, K. Different forms of asphaltene microstructures discovered in transmission electron microscopy. *J. Mater. Civ. Eng.* **2016**, *28*, 04016137. [CrossRef]
74. Mullins, O.; Sabbah, H.; Eyssautier, J.; Pomerantz, A.E.; Barré, L.; Andrews, A.B.; Ruiz-Morales, Y.; Mostowfi, F.; McFarlane, R.; Goual, L.; et al. Advances in Asphaltene Science and the Yen–Mullins Model. *Energy Fuels* **2012**, *26*, 3986–4003. [CrossRef]
75. Mullins, O.C. The Modified Yen Model. *Energy Fuels* **2010**, *24*, 2179–2207. [CrossRef]
76. Lesueur, D.; Gerard, J.F.; Claudy, P.; Letoffe, J.; Planche, J.; Martin, D. A structure-related model to describe asphalt linear viscoelasticity. *J. Rheol.* **1996**, *40*, 813–836. [CrossRef]
77. Zhang, J.; Sun, H.; Jiang, H.; Xu, X.; Liang, M.; Hou, Y.; Yao, Z. Experimental assessment of reclaimed bitumen and RAP asphalt mixtures incorporating a developed rejuvenator. *Constr. Build. Mater.* **2019**, *215*, 660–669. [CrossRef]
78. Filippelli, L.; De Santo, M.P.; Gentile, L.; Rossi, C.O. Quantitative evaluation of the restructuring effect of a warm mix additive on bitumen recycling production. *Road Mater. Pavement Des.* **2015**, *16*, 741–749. [CrossRef]
79. Mazumder, M.; Ahmed, R.; Ali, A.W.; Lee, S.-J. SEM and ESEM techniques used for analysis of asphalt binder and mixture: A state of the art review. *Constr. Build. Mater.* **2018**, *186*, 313–329. [CrossRef]
80. Das, P.K.; Baaj, H.; Tighe, S.; Kringos, N. Atomic force microscopy to investigate asphalt binders: A state-of-the-art review. *Road Mater. Pavement Des.* **2016**, *17*, 693–718. [CrossRef]
81. Loeber, L.; Sutton, O.; Morel, J.; Valleton, J.-M.; Muller, G. New direct observations of asphalts and asphalt binders by scanning electron microscopy and atomic force microscopy. *J. Microsc.* **1996**, *182*, 32–39. [CrossRef]

 https://doi.org/10.3390/coatings16030355

82. Masson, J.F.; Leblond, V.; Margeson, J. Bitumen morphologies by phase-detection atomic force microscopy. *J. Microsc.* **2006**, *221*, 17–29. [CrossRef] [PubMed]
83. Masson, J.F.; Leblond, V.; Margeson, J.; Bundalo-Perc, S. Low-temperature bitumen stiffness and viscous paraffinic nano- and micro-domains by cryogenic AFM and PDM. *J. Microsc.* **2007**, *227*, 191–202. [CrossRef]
84. Qin, Q.; Farrar, M.J.; Pauli, A.T.; Adams, J.J. Morphology, thermal analysis and rheology of Sasobit modified warm mix asphalt binders. *Fuel* **2014**, *115*, 416–425. [CrossRef]
85. Sourty, E.; Tamminga, A.; Michels, M.; Vellinga, W.-P.; Meijer, H. The microstructure of petroleum vacuum residue films for bituminous concrete: A microscopy approach. *J. Microsc.* **2011**, *241*, 132–146. [CrossRef]
86. Zhang, H.; Wang, H.; Yu, J. Effect of aging on morphology of organo-montmorillonite modified bitumen by atomic force microscopy. *J. Microsc.* **2011**, *242*, 37–45. [CrossRef] [PubMed]
87. Wu, S.-P.; Pang, L.; Mo, L.-T.; Chen, Y.-C.; Zhu, G.-J. Influence of aging on the evolution of structure, morphology and rheology of base and SBS modified bitumen. *Constr. Build. Mater.* **2009**, *23*, 1005–1010. [CrossRef]
88. Schmets, A.; Kringos, N.; Pauli, T.; Redelius, P.; Scarpas, T. On the existence of wax-induced phase separation in bitumen. *Int. J. Pavement Eng.* **2010**, *11*, 555–563. [CrossRef]
89. Das, P.K.; Kringos, N.; Wallqvist, V.; Birgisson, B. Micromechanical investigation of phase separation in bitumen by combining atomic force microscopy with differential scanning calorimetry results. *Road Mater. Pavement Des.* **2013**, *14*, 25–37. [CrossRef]
90. Pauli, A.T.; Grimes, R.W.; Beemer, A.G.; Turner, T.F.; Branthaver, J.F. Morphology of asphalts, asphalt fractions and model wax-doped asphalts studied by atomic force microscopy. *Int. J. Pavement Eng.* **2011**, *12*, 291–309. [CrossRef]
91. Allen, R.G.; Little, D.N.; Bhasin, A. Structural Characterization of Micromechanical Properties in Asphalt Using Atomic Force Microscopy. *J. Mater. Civ. Eng.* **2012**, *24*, 1317–1327. [CrossRef]
92. Dourado, E.R.; Simao, R.A.; Leite, L. Mechanical properties of asphalt binders evaluated by atomic force microscopy. *J. Microsc.* **2012**, *245*, 119–128. [CrossRef]
93. Allen, R.G.; Little, D.N.; Bhasin, A.; Glover, C.J. The effects of chemical composition on asphalt microstructure and their association to pavement performance. *Int. J. Pavement Eng.* **2014**, *15*, 9–22. [CrossRef]
94. Carrera, V.; Partal, P.; García-Morales, M.; Gallegos, C.; Páez, A. Influence of Bitumen Colloidal Nature on the Design of Isocyanate-Based Bituminous Products with Enhanced Rheological Properties. *Ind. Eng. Chem. Res.* **2009**, *48*, 8464–8470. [CrossRef]
95. Zhang, E.; Shan, L.; Qi, X.; Wang, X.; Fan, Y. Investigating the relationship between chemical composition and mechanical properties of asphalt binders using atomic force microscopy (AFM). *Constr. Build. Mater.* **2022**, *343*, 128001. [CrossRef]
96. Xing, C.; Liu, L.; Cui, Y.; Ding, D. Analysis of base bitumen chemical composition and aging behaviors via atomic force microscopy-based infrared spectroscopy. *Fuel* **2020**, *264*, 116845. [CrossRef]
97. Ganter, D.; Mielke, T.; Maier, M.; Lupascu, D.C. Bitumen rheology and the impact of rejuvenators. *Constr. Build. Mater.* **2019**, *222*, 414–423. [CrossRef]
98. Narayan, S.P.A.; Little, D.N.; Rajagopal, K.R. Nonlinear Viscoelastic Model for Describing the Response of Asphalt Binders within the Context of a Gibbs-Potential–Based Thermodynamic Framework. *J. Eng. Mech.* **2015**, *141*, 04014116. [CrossRef]
99. Kennedy, T.W.; Huber, G.A.; Harrigan, E.T.; Cominsky, R.J.; Hughes, C.S.; Von Quintus, H.; Moulthrop, J.S. *Superior Performing Asphalt Pavements (Superpave): The Product of the SHRP Asphalt Research Program*; National Research Council: Washington, DC, USA, 1994.
100. Zeiada, W.; Liu, H.; Ezzat, H.; Al-Khateeb, G.G.; Underwood, B.S.; Shanableh, A.; Samarai, M. Review of the Superpave performance grading system and recent developments in the performance-based test methods for asphalt binder characterization. *Constr. Build. Mater.* **2022**, *319*, 126063. [CrossRef]
101. Van Poel, C.D. A general system describing the visco-elastic properties of bitumens and its relation to routine test data. *J. Appl. Chem.* **1954**, *4*, 221–236. [CrossRef]
102. Zeng, M.; Bahia, H.U.; Zhai, H.; Anderson, M.R.; Turner, P. Rheological modeling of modified asphalt binders and mixtures. *J. Assoc. Asph. Paving Technol.* **2001**, *70*, 403–441.
103. Shan, L.; Qi, X.; Tan, Y.; Wang, D.; Wang, Y. Review of Nonlinear Viscoelasticity of Asphalt Binders under Oscillatory Shear Loading. *China J. Highw. Transp.* **2022**, *35*, 100–113.
104. Delgadillo, R.; Bahia, H.U.; Lakes, R. A nonlinear constitutive relationship for asphalt binders. *Mater. Struct.* **2012**, *45*, 457–473. [CrossRef]
105. Conte, T.; Chaouche, M. Rheological behavior of cement pastes under Large Amplitude Oscillatory Shear. *Cem. Concr. Res.* **2016**, *89*, 332–344. [CrossRef]
106. Goodrich, J.L. Asphalt and polymer modified asphalt properties related to the performance of asphalt concrete mixes (with discussion). *Proc. Assoc. Asph. Paving Technol.* **1988**, *57*, 116–175.

https://doi.org/10.3390/coatings16030355

107. Chailleux, E.; Ramond, G.; Such, C.H.; de La Roche, C. A mathematical-based master-curve construction method applied to complex modulus of bituminous materials. *Road Mater. Pavement Des.* **2006**, *7*, 75–92. [CrossRef]
108. Williams, M.L.; Landel, R.F.; Ferry, J.D. The Temperature Dependence of Relaxation Mechanisms in Amorphous Polymers and Other Glass-forming Liquids. *J. Am. Chem. Soc.* **1955**, *77*, 3701–3707. [CrossRef]
109. Ait-Kadi, A.; Brahimi, B.; Bousmina, M. Polymer blends for enhanced asphalt binders. *Polym. Eng. Sci.* **1996**, *36*, 1724–1733. [CrossRef]
110. Pellinen, T.K.; Witczak, M.W.; Bonaquist, R.F. Asphalt Mix Master Curve Construction Using Sigmoidal Fitting Function with Non-Linear Least Squares Optimization. In *Recent Advances in Materials Characterization and Modeling of Pavement Systems*; American Society of Civil Engineers: Reston, VA, USA, 2003.
111. Mirza, M.W.; Witczak, M.W. *Development of a Global Aging System for Short and Long Term Aging of Asphalt Cements (with Discussion)*; University of Maryland: College Park, MD, USA, 1995.
112. Levenberg, E. Smoothing Asphalt Concrete Complex Modulus Test Data. *J. Mater. Civ. Eng.* **2011**, *23*, 606–611. [CrossRef]
113. Chen, J.; Sun, Y.; Zhang, Y.; Xu, H.; Xu, Y. Modeling Method for Analysis of Dynamic Viscoelastic Behavior of Asphalt Mixture. *China J. Highw. Transp.* **2014**, *27*, 11–16.
114. Hu, Q.; Jiang, Y.; Li, J.; Fu, T. Fitting of relaxation modulus master curve for solid propellant and W.L.F. equation parameters. *J. Solid Rocket Technol.* **2003**, 46–48. [CrossRef]
115. Caracciolo, R.; Gasparetto, A.; Giovagnoni, M. Application of causality check and of the reduced variables method for experimental determination of Young's modulus of a viscoelastic material. *Mech. Mater.* **2001**, *33*, 693–703. [CrossRef]
116. Rouleau, L.; Deü, J.-F.; Legay, A.; Le Lay, F. Application of Kramers–Kronig relations to time–temperature superposition for viscoelastic materials. *Mech. Mater.* **2013**, *65*, 66–75. [CrossRef]
117. Yusoff, N.I.M.; Shaw, M.T.; Airey, G.D. Modelling the linear viscoelastic rheological properties of bituminous binders. *Constr. Build. Mater.* **2011**, *25*, 2171–2189. [CrossRef]
118. Zhu, J.; Lundberg, J.; He, L.; Li, G. Extending the Black diagram of bitumen to three dimensions. *Constr. Build. Mater.* **2022**, *349*, 128727. [CrossRef]
119. Peng, M.; Xu, Z. Research on nonlinear constitutive relationship of permanent deformation in asphalt pavements. *Sci. China Ser. G Phys. Mech. Astron.* **2006**, *49*, 671–682. [CrossRef]
120. Murali Krishnan, J.; Rajagopal, K.R. Review of the uses and modeling of bitumen from ancient to modern times. *Appl. Mech. Rev.* **2003**, *56*, 149–214. [CrossRef]
121. Krishnan, J.M.; Rajagopal, K.R. Thermodynamic Framework for the Constitutive Modeling of Asphalt Concrete: Theory and Applications. *J. Mater. Civ. Eng.* **2004**, *16*, 155–166. [CrossRef]
122. Darabi, M.K.; Abu Al-Rub, R.K.; Masad, E.A.; Huang, C.W.; Little, D.N. A thermo-viscoelastic–viscoplastic–viscodamage constitutive model for asphaltic materials. *Int. J. Solids Struct.* **2011**, *48*, 191–207. [CrossRef]
123. Yin, Y.; Zhang, X. Dynamic viscoelastic behavior of asphalt mixtures based on fractional derivative. *J. Cent. South Univ. (Sci. Technol.)* **2013**, *44*, 3891–3897.
124. Xu, Q.; Solaimanian, M. Modelling linear viscoelastic properties of asphalt concrete by the Huet–Sayegh model. *Int. J. Pavement Eng.* **2009**, *10*, 401–422. [CrossRef]
125. Shan, L.; Tan, Y.; Zhang, H.; Xu, Y. Analysis of Linear Viscoelastic Response Function Model for Asphalt Binders. *J. Mater. Civ. Eng.* **2016**, *28*, 04016010. [CrossRef]
126. Zanzotto, L.; Stastna, J.; Ho, S. Molecular weight distribution of regular asphalts from dynamic material functions. *Mater. Struct.* **1999**, *32*, 224–229. [CrossRef]
127. Themeli, A.; Chailleux, E.; Farcas, F.; Chazallon, C.; Migault, B.; Buisson, N. Molecular structure evolution of asphaltite-modified bitumens during ageing; Comparisons with equivalent petroleum bitumens. *Int. J. Pavement Res. Technol.* **2017**, *10*, 75–83. [CrossRef]
128. Pipintakos, G.; Lommaert, C.; Varveri, A.; Bergh, W.V.D. Do chemistry and rheology follow the same laboratory ageing trends in bitumen? *Mater. Struct.* **2022**, *55*, 146. [CrossRef]
129. Merusi, F.; Polacco, G.; Filippi, S.; Giuliani, F. Structural transitions and physical networks in wax-modified bitumens. *Road Mater. Pavement Des.* **2013**, *14*, 289–309. [CrossRef]
130. Wang, T.; Wang, J.; Hou, X.; Xiao, F. Effects of SARA fractions on low temperature properties of asphalt binders. *Road Mater. Pavement Des.* **2021**, *22*, 539–556. [CrossRef]
131. Siroma, R.S.; Nguyen, M.L.; Hornych, P.; Lorino, T.; Hung, Y.; Nicolaï, A.; Ziyani, L.; Chailleux, E. Molecular Agglomeration Index: Quantification of the Incidence of Asphaltene Molecular Agglomeration in Aged Asphalt Binders Through Rheological Measurements. *Transp. Res. Rec.* **2022**, *2676*, 89–103. [CrossRef]

 https://doi.org/10.3390/coatings16030355

132. Krolkral, K.; Haddadi, S.; Chailleux, E. Quantification of asphalt binder ageing from apparent molecular weight distributions using a new approximated analytical approach of the phase angle. *Road Mater. Pavement Des.* **2020**, *21*, 1045–1060. [CrossRef]
133. Paliukaitė, M.; Vaitkus, A.; Zofka, A. Influence of bitumen chemical composition and ageing on pavement performance. *Balt. J. Road Bridge Eng.* **2015**, *10*, 97–105. [CrossRef]
134. Weigel, S.; Stephan, D. Bitumen Characterization with Fourier Transform Infrared Spectroscopy and Multivariate Evaluation: Prediction of Various Physical and Chemical Parameters. *Energy Fuels* **2018**, *32*, 10437–10442. [CrossRef]
135. Yu, X.; Burnham, N.A.; Granados-Focil, S.; Tao, M. Bitumen's microstructures are correlated with its bulk thermal and rheological properties. *Fuel* **2019**, *254*, 115509. [CrossRef]
136. Oldham, D.; Hung, A.; Parast, M.M.; Fini, E.H. Investigating bitumen rejuvenation mechanisms using a coupled rheometry-morphology characterization approach. *Constr. Build. Mater.* **2018**, *159*, 37–45. [CrossRef]
137. Soenen, H.; Redelius, P. The effect of aromatic interactions on the elasticity of bituminous binders. *Rheol. Acta* **2014**, *53*, 741–754. [CrossRef]
138. Li, F.; Wang, Y.; Miljković, M.; Chan, K.M. Changes in the nanoscale asphaltene particles and relaxation spectra of asphalt binders during aging and rejuvenation. *Mater. Des.* **2022**, *219*, 110808. [CrossRef]
139. An, X.; Wang, R.; Kang, X.; Yue, J. A more accurate fatigue characterization of GO-modified asphalt binder considering non-linear viscoelastic behaviour and UV exposure effects. *Int. J. Fatigue* **2023**, *168*, 107396. [CrossRef]
140. Masad, E.; Somadevan, N. Microstructural finite-element analysis of influence of localized strain distribution on asphalt mix properties. *J. Eng. Mech.* **2002**, *128*, 1105–1114. [CrossRef]
141. Airey, G.D.; Rahimzadeh, B.; Collop, A.C. Linear rheological behavior of bituminous paving materials. *J. Mater. Civ. Eng.* **2004**, *16*, 212–220. [CrossRef]
142. Safaei, F.; Castorena, C. Material nonlinearity in asphalt binder fatigue testing and analysis. *Mater. Des.* **2017**, *133*, 376–389. [CrossRef]
143. Masad, E.A.; Huang, C.W.; D'Angelo, J.; Little, D.N. Characterization of asphalt binder resistance to permanent deformation based on nonlinear viscoelastic analysis of multiple stress creep recovery (MSCR) test. *Asph. Paving Technol. Assoc. Asph. Paving Technol.-Proc. Tech. Sess.* **2009**, *78*, 535–562.
144. Park, S.W.; Richard Kim, Y.; Schapery, R.A. A viscoelastic continuum damage model and its application to uniaxial behavior of asphalt concrete. *Mech. Mater.* **1996**, *24*, 241–255. [CrossRef]
145. Hyun, K.; Wilhelm, M.; Klein, C.O.; Cho, K.S.; Nam, J.G.; Ahn, K.H.; Lee, S.J.; Ewoldt, R.H.; McKinley, G.H. A review of nonlinear oscillatory shear tests: Analysis and application of large amplitude oscillatory shear (LAOS). *Prog. Polym. Sci.* **2011**, *36*, 1697–1753. [CrossRef]
146. Giacomin, A.J.; Bird, R.B.; Johnson, L.M.; Mix, A. Large-amplitude oscillatory shear flow from the corotational Maxwell model. *J. Non-Newton. Fluid Mech.* **2011**, *166*, 1081–1099. [CrossRef]
147. Yosick, J.A.; Giacomin, A.J.; Moldenaers, P. A kinetic network model for nonlinear flow behavior of molten plastics in both shear and extension. *J. Non-Newton. Fluid Mech.* **1997**, *70*, 103–123. [CrossRef]
148. Thompson, R.L.; Alicke, A.A.; de Souza Mendes, P.R. Model-based material functions for SAOS and LAOS analyses. *J. Non-Newton. Fluid Mech.* **2015**, *215*, 19–30. [CrossRef]
149. Shan, L.; He, H.; Wagner, N.J.; Li, Z. Nonlinear rheological behavior of bitumen under LAOS stress. *J. Rheol.* **2018**, *62*, 975–989. [CrossRef]
150. Läuger, J.; Stettin, H. Differences between stress and strain control in the non-linear behavior of complex fluids. *Rheol. Acta* **2010**, *49*, 909–930. [CrossRef]
151. Ewoldt, R.H.; Hosoi, A.E.; McKinley, G.H. New measures for characterizing nonlinear viscoelasticity in large amplitude oscillatory shear. *J. Rheol.* **2008**, *52*, 1427–1458. [CrossRef]
152. Padmarekha, A.; Chockalingam, K.; Saravanan, U.; Deshpande, A.P.; Krishnan, J.M. Large amplitude oscillatory shear of unmodified and modified bitumen. *Road Mater. Pavement Des.* **2013**, *14*, 12–24. [CrossRef]
153. Cho, K.S.; Hyun, K.; Ahn, K.H.; Lee, S.J. A geometrical interpretation of large amplitude oscillatory shear response. *J. Rheol.* **2005**, *49*, 747–758. [CrossRef]
154. Dimitriou, C.J.; Ewoldt, R.H.; McKinley, G.H. Describing and prescribing the constitutive response of yield stress fluids using large amplitude oscillatory shear stress (LAOStress). *J. Rheol.* **2013**, *57*, 27–70. [CrossRef]
155. Blackwell, B.C.; Ewoldt, R.H. A simple thixotropic–viscoelastic constitutive model produces unique signatures in large-amplitude oscillatory shear (LAOS). *J. Non-Newton. Fluid Mech.* **2014**, *208–209*, 27–41. [CrossRef]
156. Diab, A.; You, Z.; Adhikari, S.; Li, X. Modeling shear stress response of bituminous materials under small and large strains. *Constr. Build. Mater.* **2020**, *252*, 119133. [CrossRef]

 https://doi.org/10.3390/coatings16030355

157. Wang, Z. Research on Nonlinear Rheological Characteristics of Asphalt Under Laos Strain. Master's Thesis, Harbin Institute of Technology, Harbin, China, 2020.
158. Gulzar, S.; Underwood, B.S. Nonlinear Viscoelastic Response of Crumb Rubber Modified Asphalt Binder Under Large Strains. *Transp. Res. Rec.* **2020**, *2674*, 139–149. [CrossRef]
159. Saboo, N.; Sukhija, M.; Chaudhary, M. Relating asphalt binders response to LAS and LAOS tests at intermediate temperatures. *Mech. Time-Depend. Mater.* **2021**, *25*, 21–35. [CrossRef]
160. Shan, L.; Qi, X.; Duan, X.; Liu, S.; Chen, J. Effect of styrene-butadiene-styrene (SBS) on the rheological behavior of asphalt binders. *Constr. Build. Mater.* **2020**, *231*, 117076. [CrossRef]
161. Liang, R.; Jin, R.; Zhou, D.; Sun, W.; Kang, Y. Nonlinear rheological behaviors of epoxy asphalt binder compared to base asphalt binder and SBS modified asphalt binder at above ambient temperatures. *Constr. Build. Mater.* **2020**, *250*, 118755. [CrossRef]
162. Pearson, D.S.; Rochefort, W.E. Behavior of concentrated polystyrene solutions in large-amplitude oscillating shear fields. *J. Polym. Sci. Polym. Phys. Ed.* **1982**, *20*, 83–98. [CrossRef]
163. Helfand, E.; Pearson, D.S. Calculation of the nonlinear stress of polymers in oscillatory shear fields. *J. Polym. Sci. Polym. Phys. Ed.* **1982**, *20*, 1249–1258. [CrossRef]
164. Calin, A.; Balan, C.; Wilhelm, M. A New Approach to Determine the Nonlinear Parameter of the Giesekus Constitutive Model. *AIP Conf. Proc.* **2008**, *1027*, 1372–1374. [CrossRef]
165. Vittorias, I.; Parkinson, M.; Klimke, K.; Debbaut, B.; Wilhelm, M. Detection and quantification of industrial polyethylene branching topologies via Fourier-transform rheology, NMR and simulation using the Pom-pom model. *Rheol. Acta* **2007**, *46*, 321–340. [CrossRef]
166. Pronk, A.C.; Hopman, P.C. Energy dissipation: The leading factor of fatigue. The united states strategic highway research program. Sharing the benefits. In Proceedings of the Conference Organized by the Institution of Civil Engineers in Cooperation with U.S. Strategic Highway Research Program, London, UK, 29–31 October 1990.
167. Pronk, A. *Evaluation of the Dissipated Energy Concept for the Interpretation of Fatigue Measurements in the Crack Initiation Phase*; Road and Hydraulic Engineering Division: Delft, The Netherlands, 1995.
168. *AASHTO M320*; Standard Specification for Performance-Graded Asphalt Binder. American Association of State Highway and Transportation Officials: Washington, DC, USA, 2017.
169. Lee, H.-J.; Kim, Y.R. Viscoelastic Constitutive Model for Asphalt Concrete under Cyclic Loading. *J. Eng. Mech.* **1998**, *124*, 32–40. [CrossRef]
170. Si, Z.; Little, D.N.; Lytton, R.L. Characterization of Microdamage and Healing of Asphalt Concrete Mixtures. *J. Mater. Civ. Eng.* **2002**, *14*, 461–470. [CrossRef]
171. Shan, L.; Tan, Y.; Underwood, B.S.; Kim, Y.R. Separation of Thixotropy from Fatigue Process of Asphalt Binder. *Transp. Res. Rec.* **2011**, *2207*, 89–98. [CrossRef]
172. Lv, H.; Tan, Y.; Zhang, D.; Li, G. Differentiation of thixotropy from damage for accurately characterizing fatigue resistance of asphalt binder. *Constr. Build. Mater.* **2022**, *344*, 128225. [CrossRef]
173. Pérez-Jiménez, F.; Botella, R.; López-Montero, T.; Miró, R.; Martínez, A.H. Complexity of the behaviour of asphalt materials in cyclic testing. *Int. J. Fatigue* **2017**, *98*, 111–120. [CrossRef]
174. Pérez-Jiménez, F.E.; Botella, R.; Miró, R.; Martínez, A.H. Analysis of the thixotropic behavior and the deterioration process of bitumen in fatigue tests. *Constr. Build. Mater.* **2015**, *101*, 277–286. [CrossRef]
175. Pérez-Jiménez, F.E.; Botella, R.; Miró, R. Differentiating between damage and thixotropy in asphalt binder's fatigue tests. *Constr. Build. Mater.* **2012**, *31*, 212–219. [CrossRef]
176. Shan, L. Fatigue & Rheology Mechanism of Asphalt Binder Based on Viscoelastic Characteristic. Ph.D. Thesis, Harbin Institute of Technology, Harbin, China, 2010.
177. Lv, H.; Zhang, D.; Tan, Y.; Cai, L.; Liu, H. Development of a complex index to characterize the recovery evolution of asphalt binder after cyclic loading. *Int. J. Fatigue* **2022**, *156*, 106664. [CrossRef]
178. Di Benedetto, H.; De La Roche, C.; Baaj, H.; Pronk, A.; Lundström, R. Fatigue of bituminous mixtures. *Mater. Struct.* **2004**, *37*, 202–216. [CrossRef]
179. Jongepier, R.; Kuilman, B. The dynamic shear modulus of bitumens as a function of frequency and temperature. *Rheol. Acta* **1970**, *9*, 102–111. [CrossRef]
180. Dickinson, E.J.; Witt, H.P. The Dynamic Shear Modulus of Paving Asphalts as a Function of Frequency. *Trans. Soc. Rheol.* **1974**, *18*, 591–606. [CrossRef]
181. Christensen, D.W.; Anderson, D.A. Interpretation of dynamic mechanical test data for paving grade asphalt cements. *J. Assoc. Asph. Paving Technol.* **1992**, *61*, 67–116.
182. Šťastná, J.; Kee, D.D.; Powley, M.B. Complex Viscosity as a Generalized Response Function. *J. Rheol.* **1985**, *29*, 457–469. [CrossRef]

 https://doi.org/10.3390/coatings16030355

183. Marasteanu, M.; Anderson, D. Improved model for bitumen rheological characterization. In Proceedings of the Eurobitume Workshop on Performance Related Properties for Bituminous Binders, Luxembourg, 3–6 May 1999.
184. Elseifi, M.A.; Al-Qadi, I.L.; Flinstch, G.W.; Masson, J.-F. Viscoelastic Modeling of Straight Run and Modified Binders Using the Matching Function Approach. *Int. J. Pavement Eng.* **2002**, *3*, 53–61. [CrossRef]
185. Bonaquist, R.; Christensen, D.W. Practical Procedure for Developing Dynamic Modulus Master Curves for Pavement Structural Design. *Transp. Res. Rec.* **2005**, *1929*, 208–217. [CrossRef]
186. Huet, C. Étude par une méthode d'impédance du comportement viscoélastique des matériaux hydrocarbonés [Study of the Viscoelastic Behavior of Bituminous Mixes by Method of Impedance]. Ph.D. Thesis, Faculte des Sciences de Paris, Paris, France, 1963. (In French)
187. Sayegh, G. Viscoelastic properties of bituminous mixtures. In Proceedings of the International Conference on Structural Design of Asphalt Pavements, Ann Arbor, MI, USA, 7–11 August 1967.
188. Di Benedetto, H.; Mondher, N.; Sauzéat, C.; Olard, F. Three-dimensional Thermo-viscoplastic Behaviour of Bituminous Materials: The DBN Model. *Road Mater. Pavement Des.* **2007**, *8*, 285–315.
189. Olard, F.; Di Benedetto, H.; Eckmann, B.; Triquigneaux, J.P. Linear Viscoelastic Properties of Bituminous Binders and Mixtures at Low and Intermediate Temperatures. *Road Mater. Pavement Des.* **2003**, *4*, 77–107. [CrossRef]

https://doi.org/10.3390/coatings16030355

coatings

MDPI

Article

Effect of Hydrophobic Alkyl Chain Length on the Interfacial Adhesion Performance of Emulsified Asphalt–Aggregate Systems

Haijiao Xu [1], Pinhui Zhao [1,*], Wenyu Wang [1], Zhiyu Zhang [2], Jingtao Shi [3,*], Dongqi Yu [1], Hongbo Cui [1], Deshang Huang [1] and Jiawei Su [1]

[1] School of Transportation Engineering, Shandong Jianzhu University, Jinan 250101, China
[2] Beijing Zhengda Tongshun Inspection Technology Co., Ltd., Beijing 100195, China
[3] Petrochina Fuel Oil Co., Ltd. Research Institute, Beijing 100195, China
* Correspondence: zhaopinhui08@163.com (P.Z.); sjt0822@163.com (J.S.)

Abstract

To elucidate the mechanisms by which the hydrophobic hydrocarbon chain length of emulsifiers and the surface properties of aggregates influence the adhesive performance at the emulsified asphalt–aggregate interface, this study employed molecular dynamics simulations to construct interface models. Key parameters, including relative concentration, diffusion coefficients, and interfacial adhesion work, were systematically analysed to reveal the intrinsic effects of imidazoline-type emulsifier chain length and aggregate type on interfacial behaviour. The results indicate that increasing the hydrophobic chain length of the emulsifier suppresses the adsorption of emulsified asphalt at the aggregate interface. The diffusion coefficients of both emulsifier and asphalt molecules initially increase and subsequently decrease with chain length, with the non-polar asphalt components (aromatics and saturates) exhibiting greater sensitivity to chain length variations. Moderate extension of the hydrophobic chain enhances interfacial adhesion work, whereas exceeding the optimal chain length reverses this trend, weakening adhesion. Aggregate surface properties exert a significant influence on interfacial behaviour. Compared with the acidic SiO_2 (0 0 1) surface, the basic $CaCO_3$ (1 0 4) surface exhibits lower peak relative concentrations of emulsified asphalt, reduced sensitivity to variations in emulsifier chain length, lower molecular diffusion coefficients, and stronger interactions with asphalt molecules, resulting in superior interfacial adhesion. This study provides a molecular-level theoretical basis for the targeted design of emulsifier structures and the efficient adaptation of emulsified asphalt to different aggregate systems.

Keywords: molecular dynamics; emulsifier molecular structure; aggregate characteristics; adhesion performance

1. Introduction

Emulsified asphalt, as a low-carbon and energy-efficient material, has been widely applied in road engineering, offering both economic and environmental advantages, including ease of low-temperature construction, significant energy savings, and extended pavement service life [1–4]. It is a water-in-oil emulsion in which asphalt serves as the dispersed phase and water as the continuous phase, with emulsifiers adsorbing at the water–asphalt interface to stabilise asphalt dispersion at ambient temperature [5]. In practice, emulsified asphalt is typically mixed with mineral aggregates to produce asphalt mixtures for cold-mix, cold recycling, and surface treatment applications [6,7], with macroscopic properties such as mechanical strength, water stability, and high-temperature performance largely determined by the adhesion and coating effectiveness between the emulsified asphalt and aggregates [8].

At the microscopic level, the adhesive performance of the emulsified asphalt–aggregate interface is governed by a combination of interfacial properties and molecular interactions [9]. Among these, the role of the emulsifier at the interface is particularly critical. As a typical surfactant, the polar head group of the emulsifier interacts with water or active sites on the aggregate surface, while the hydrophobic hydrocarbon chain inserts into the asphalt phase, forming an ordered adsorption layer at the interface. This interfacial adsorption layer reduces the water–asphalt interfacial tension, facilitating asphalt wetting and spreading on the aggregate surface, while simultaneously enhancing interfacial stability through electrostatic repulsion and steric hindrance, thereby preventing detachment of the asphalt film in aqueous environments.

Beyond the emulsifier, aggregate type and its physicochemical surface properties also play a crucial role in the adhesive performance of emulsified asphalt [10,11]. Different aggregates exhibit significant variations in chemical composition, surface charge characteristics, and polar functional groups, which directly influence their interfacial interactions with emulsified asphalt [12]. Studies have shown that carbonate aggregates, such as limestone, exhibit markedly higher interfacial adhesion strength with asphalt than acidic aggregates, such as granite or quartz, which are primarily composed of SiO_2 [13,14]. Under ambient conditions, the influence of aggregate type on adhesion is notably greater than that of asphalt type [15]. Moreover, under humid or combined heat–moisture conditions, acidic aggregates are more prone to debonding and stripping than basic aggregates [16], further highlighting the significant impact of aggregate surface properties on emulsified asphalt adhesion.

Molecular dynamics (MD) simulation provides an efficient approach to investigate the dynamic evolution of materials at the molecular scale and has been widely applied in the field of road materials. This method not only enables the study of how modifiers affect asphalt viscosity, interfacial adhesion, and the interaction mechanisms between asphalt and aggregates [17,18], but also allows detailed analysis of molecular diffusion and distribution at interfaces, thereby elucidating the microscopic mechanisms underlying adhesion [19]. In recent years, numerous studies have employed MD simulations to explore interfacial behaviour in emulsified asphalt systems. Liu et al. [20] used MD to examine the adhesion mechanisms and characteristics of emulsified asphalt residue–aggregate interfaces under varying emulsifier contents. Zhu et al. [21] investigated the effect of cationic emulsifier alkyl chain length on the demulsification of cationic emulsified asphalt on $CaCO_3$ surfaces. Kong et al. [22,23] combined MD modelling and macroscopic experiments to study mechanisms at aggregate interfaces, quantifying the role of phenyl functional groups at the molecular level and assessing the influence of hydrocarbon chain length on adhesion between emulsified asphalt and acidic aggregates. Zhang et al. [24] employed MD simulations to examine asphalt–basalt and asphalt–steel slag interfacial adhesion. Meng et al. [25] provided atomic-scale insights into asphalt–aggregate adhesion and failure behaviour. Wang et al. [26] constructed MD models of emulsified asphalt–aggregate interfaces with varying water contents to investigate the effect of moisture on adhesion. Sun et al. [27] used MD to study molecular interactions between matrix and aged asphalt binders with silica and calcite. Collectively, these studies confirm that emulsified asphalt adhesion to aggregates is closely related to emulsifier molecular structure and aggregate surface properties, with molecular dynamics simulation serving as a critical tool for probing their relationships and underlying mechanisms at the microscopic level.

https://doi.org/10.3390/coatings16030330

In view of the aforementioned background, this study employed molecular dynamics (MD) simulations to systematically investigate the influence of hydrophobic hydrocarbon chain length of imidazoline-type emulsifiers on the adhesion and adsorption behaviour of emulsified asphalt on $CaCO_3$ (1 0 4) and SiO_2 (0 0 1) surfaces. The findings provide theoretical guidance for the optimisation of emulsified asphalt systems and hold significant practical value for enhancing the quality of road engineering construction.

2. Materials and Methods

2.1. Base Asphalt Molecular Model

Asphalt is a complex multicomponent mixture composed of tens of thousands of hydrocarbons with varying molecular weights and structural characteristics [28,29]. In recent years, extensive research has been conducted both domestically and internationally to achieve precise microscopic characterisation of asphalt composition and structure. The most widely used molecular models of asphalt include the average molecular model, the three-component model, and the four-component model. In this study, based on Corbett's SARA fractionation method, representative compounds from the four fractions—saturates, aromatics, resins, and asphaltenes—were selected as the fundamental molecular units for constructing the asphalt molecular model [30–32].

Using Materials Studio (MS), representative molecules of the four asphalt fractions were constructed as fundamental molecular units, and their initial geometries were optimised to verify structural plausibility. Based on this, the Construction tool in the Amorphous Cell (AC) module of MS was employed to fill the four types of molecules into a cubic simulation box with periodic boundary conditions according to the predefined component ratios (Table 1), thereby establishing the asphalt molecular model (Figure 1). It should be noted that the molecular model in this study was entirely constructed based on literature data and the SARA fractionation method, without direct experimental verification via microscopic imaging techniques such as AFM, SEM, or TEM. The plausibility of the model was confirmed through geometry optimization and comparison with reported molecular characteristics in the literature.

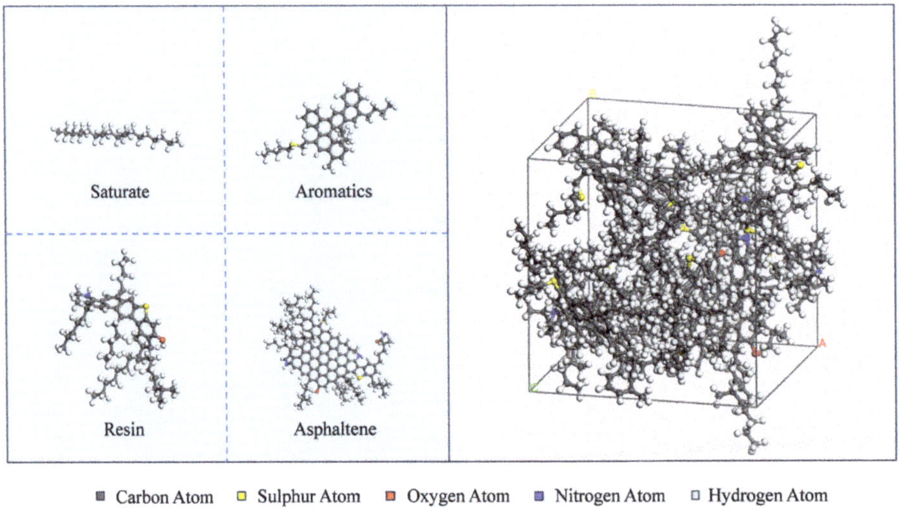

Figure 1. Molecular model of four-component asphalt.

https://doi.org/10.3390/coatings16030330

Table 1. Number of molecules in the asphalt molecular model.

Molecular Name	Molecular Formula	Molar Mass (g/mol^{-1})	Number Added	Proportion (%)
Saturate	$C_{22}H_{46}$	310.61	5	7.3
Aromatics	$C_{46}H_{50}S$	634.966	7	20.8
Resin	$C_{59}H_{85}NOS$	856.395	3	12
Asphaltene	$C_{149}H_{177}N_3O_2S_2$	2106.19	1	9.9

2.2. Emulsified Asphalt Molecular Model

The interfacial behaviour between emulsified asphalt and aggregates is largely governed by the action of emulsifiers. Emulsifier molecules contain both hydrophilic and lipophilic groups, which preferentially adsorb at the asphalt–water interface to form a stable interfacial adsorption layer. Under the influence of this layer, the asphalt phase is dispersed as droplets within the continuous aqueous phase, allowing the emulsified asphalt to come into contact with mineral aggregate surfaces. Upon contact, the emulsifier molecules act as a critical mediator linking the asphalt phase, aqueous phase, and aggregate surface; their interfacial configuration and molecular orientation directly influence the spreading, wetting, and adhesion of asphalt on the aggregate surface, thereby determining the interfacial interaction characteristics and adhesion stability of the emulsified asphalt–aggregate system.

Building on previous studies, this work selected long-chain aliphatic imidazoline-type emulsifiers to construct molecular models [33]. These emulsifiers are synthesised via amidation–cyclisation of tetraethylenepentamine with organic acids. Their hydrophilic groups (polar imidazoline rings) form hydrogen bonds with water molecules, enhancing interfacial compatibility, while the hydrophobic groups (long-chain alkanes) interact via van der Waals forces with the hydrophobic structures of asphalt molecules, anchoring the emulsifier at the surface of asphalt droplets. To distinguish molecular structural variations, the C_nN_m nomenclature is adopted, where n denotes the number of carbon atoms in the long alkyl chain and m the number of nitrogen atoms in the hydrophilic group; the corresponding emulsified asphalt system is referred to as the C_nN_m emulsified asphalt system. Molecular models of the long-chain aliphatic imidazoline-type emulsifiers with different alkyl chain lengths, together with water molecules, are shown in Figure 2.

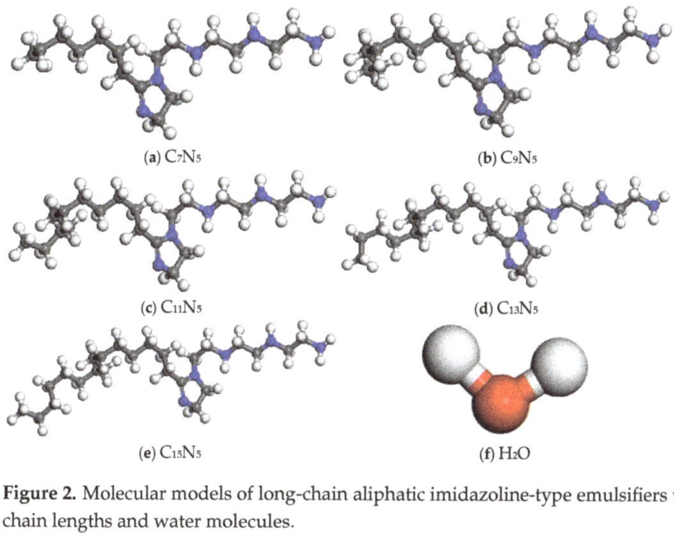

(a) C_7N_5 (b) C_9N_5 (c) $C_{11}N_5$ (d) $C_{13}N_5$ (e) $C_{15}N_5$ (f) H_2O

Figure 2. Molecular models of long-chain aliphatic imidazoline-type emulsifiers with varying alkyl chain lengths and water molecules.

https://doi.org/10.3390/coatings16030330

Based on the aforementioned molecular models of imidazoline-type emulsifiers and water, the emulsified asphalt molecular model was constructed using the Amorphous Cell (AC) module in Materials Studio (MS). The model was configured with a molar ratio of emulsifier to water of 15:85 [34], while the molar ratio of the emulsifier–water solution to asphalt was set at 5:5. A cubic simulation box with periodic boundary conditions was generated to represent the infinite extent of the real system, ensuring that the simulation results are representative. The detailed composition of the emulsifier–water solution molecular model is listed in Table 2.

Table 2. Model molecular composition of emulsifier aqueous solutions.

Composition	Molecular Name	Molecular Formula	Molar Mass (g/mol^{-1})	Number Added
C_7N_5	Emulsifier Water molecule	$C_{16}H_{35}N_5$ H_2O	297.491 18.015	5 502
C_9N_5	Emulsifier Water molecule	$C_{18}H_{39}N_5$ H_2O	325.545 18.015	5 502
$C_{11}N_5$	Emulsifier Water molecule	$C_{20}H_{43}N_5$ H_2O	353.599 18.015	5 502
$C_{13}N_5$	Emulsifier Water molecule	$C_{22}H_{47}N_5$ H_2O	381.653 18.015	4 502
$C_{15}N_5$	Emulsifier Water molecule	$C_{24}H_{51}N_5$ H_2O	409.707 18.015	4 502

The optimisation procedure for the emulsified asphalt molecular model and the parameters used in the molecular dynamics simulations were kept consistent with those of the asphalt molecular model, thereby enhancing the reliability and consistency of the results. The schematic diagram of the molecular model structure of $C_{11}N_5$ emulsified asphalt, constructed based on the number of molecules in Tables 1 and 2, is shown in Figure 3.

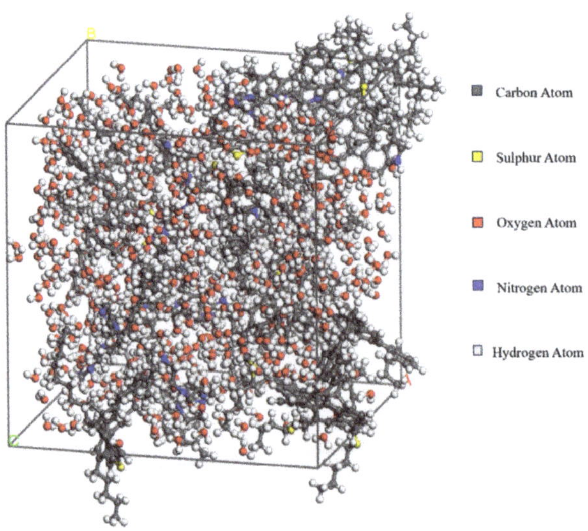

Figure 3. Molecular model of $C_{11}N_5$ emulsified asphalt.

https://doi.org/10.3390/coatings16030330

2.3. Aggregate Molecular Model

In road engineering materials, mineral aggregates are complex inorganic mixtures with diverse sources, primarily composed of oxides such as SiO_2, CaO, and Fe_2O_3. Given the complexity and variability of actual aggregate mineral compositions, molecular dynamics simulations typically employ single-mineral crystal models to represent aggregates equivalently, enabling analysis of their interfacial interactions with asphalt systems. Based on existing studies and model applicability, this work selected SiO_2 and $CaCO_3$ as representative acidic and basic aggregates, respectively, and employed the widely used SiO_2 (0 0 1) and $CaCO_3$ (1 0 4) crystal planes to construct aggregate surface models for molecular dynamics analysis of emulsified asphalt–aggregate interfacial behaviour. The lattice parameters of the aggregate unit cells used are listed in Table 3.

Table 3. Lattice Constants of $CaCO_3$ and SiO_2.

Aggregate	Lattice Constant					
	a/Å	b/Å	c/Å	α/°	β/°	γ/°
$CaCO_3$	4.99	4.99	17.061	90	90	120
SiO_2	4.193	4.193	5.4052	90	90	120

During the construction of the aggregate models, the Cut tool in Materials Studio (MS) was used to cleave the mineral unit cells to generate specific crystal planes, with the number of layers set to control model thickness. It should be noted that cleavage of the SiO_2 crystal surface produces unsaturated atomic structures. If used directly in molecular dynamics simulations, these may undergo unrealistic interactions with water molecules or oxygen atoms, compromising the accurate representation of quartz aggregate interfacial behaviour under service conditions. Therefore, it is necessary to repair the broken bonds on the SiO_2 crystal surface to obtain a structurally stable and chemically reasonable aggregate surface model.

On this basis, the constructed crystal plane models were periodically extended in the x and y directions to achieve the model dimensions required for interfacial studies. A 30 Å vacuum layer was then introduced perpendicular to the crystal plane to mitigate the influence of periodic boundary conditions on the interface system, converting the original two-dimensional crystal plane into a three-dimensional supercell suitable for molecular dynamics simulations. The final aggregate unit cell molecular models are shown in Figure 4.

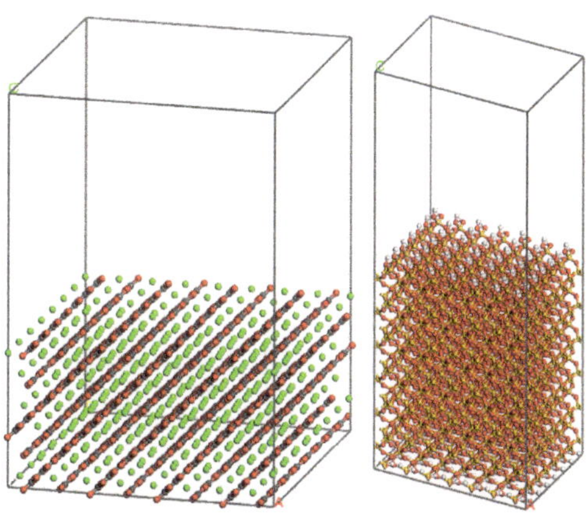

(a) $CaCO_3$ (0 1 4) (b) SiO_2 (0 0 1)

Figure 4. Molecular model of $C_{11}N_5$ emulsified asphalt.

 https://doi.org/10.3390/coatings16030330

2.4. Emulsified Asphalt–Aggregate Interfacial Model

Based on the emulsified asphalt molecular models and aggregate unit cell models constructed in Sections 2.1–2.3, the emulsified asphalt–aggregate interface models were assembled using a "lower aggregate–upper emulsified asphalt" spatial stacking approach. To mitigate potential interference from periodic boundary conditions on interfacial microstructure and molecular dynamics behaviour, a 50 Å vacuum layer was added above the emulsified asphalt layer. A schematic of the final emulsified asphalt–aggregate interface molecular model is shown in Figure 5.

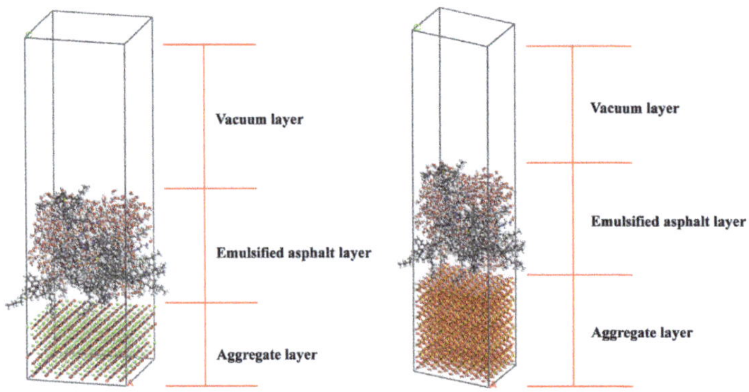

(a)$CaCO_3$(1 0 4) (b)SiO_2(0 0 1)

Figure 5. Molecular model of the emulsified asphalt–aggregate interface.

2.5. Simulation Parameters and Validation

2.5.1. Parameter Settings

Molecular simulation techniques can reveal the static structures and dynamic diffusion behaviours of substances at the microscopic scale, with atomic model construction and the choice of force field being critical [35]. The COMPASS force field, widely used in molecular simulations, accurately describes various interactions within molecular systems, making it well suited for simulating molecular aggregation and geometry optimisation during initial model construction [36]. Accordingly, the COMPASS III force field was employed throughout all simulations in this study.

It should be noted that periodic boundary conditions (PBCs) were imposed in the x, y, and z directions to minimize the boundary effects on the simulation results.

The constructed base asphalt molecular model was first subjected to geometry optimisation to eliminate unreasonable initial configurations. During geometry optimisation, the system temperature was set to 298 K, the non-bonded interaction cutoff radius was 15.5 Å, and energy minimisation was carried out using the conjugate gradient algorithm. The Coulombic electrostatic interactions and van der Waals interactions were calculated using the Ewald summation method and the Atom-Based method, respectively. After geometry optimisation, a 300 ps molecular dynamics pre-equilibration was performed to ensure the stability and reliability of the base molecular structure. The time step was set to 1 fs, and the simulation was conducted under the NPT ensemble. During the simulation, the Andersen thermostat and Berendsen barostat were employed to maintain the system temperature and pressure at 298 K and 1.0×10^{-4} GPa, respectively.

For the constructed emulsified asphalt/aggregate interface model, geometry optimisation was likewise performed using the same parameters and computational settings as those applied to the base asphalt model. Subsequently, pre-equilibration, annealing,

 https://doi.org/10.3390/coatings16030330

and final equilibration were carried out in sequence. First, a 200 ps pre-equilibration was conducted under the NPT ensemble to eliminate the influence of the initial configuration. Then, the system was subjected to five heating–cooling annealing cycles between 298 K and 500 K to release internal stresses and promote structural relaxation. After annealing, a 1000 ps final equilibrium simulation was performed under the NPT ensemble to ensure that the system reached a stable state. The pre-equilibration and final equilibration processes of the emulsified asphalt/aggregate system adopted the same simulation parameters and settings as those used for the base asphalt molecular pre-equilibration.

2.5.2. Model Validation

To validate the constructed asphalt molecular model, density and glass transition temperature were selected as key evaluation indicators. Density is a critical physical parameter reflecting asphalt's brittleness and thermal stability [37], while T_g denotes the characteristic temperature at which asphalt transitions from a glassy to a viscoelastic state, strongly correlating with its low-temperature mechanical behaviour and in-service stability, thus serving as an important criterion for assessing the reasonableness and accuracy of the molecular model [38]. Based on this approach, molecular dynamics simulations were conducted to examine the evolution of density over simulation time (Figure 6a) and to determine the model's glass transition temperature (Figure 6b). As shown in Figure 6a, density fluctuations of the asphalt model gradually converged after 50 ps and stabilised at an equilibrium value of 1.011 g·cm^{-3}, within the typical range of 1.01–1.04 g·cm^{-3} for real asphalt materials, indicating that the system had essentially reached thermodynamic equilibrium after 300 ps of dynamics. From Figure 6b, the glass transition temperature of the model was determined to be 253.3 K, falling within the temperature range of 238.15 K to 258.15 K, which is typical for asphalt transitioning from the glassy to viscoelastic state. These results demonstrate that the constructed asphalt molecular model is both reasonable and reliable, providing a robust foundation for subsequent construction of emulsified asphalt systems and simulation of interfacial interactions.

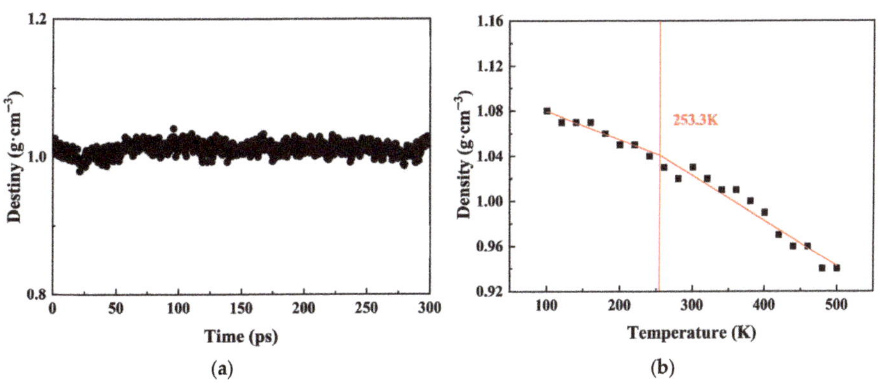

Figure 6. (**a**) Density evolution of the asphalt molecular model over time and (**b**) glass transition temperature of the asphalt molecular model.

Based on the validated reliability of the asphalt molecular model through density and glass transition temperature, it is essential to further verify the thermodynamic equilibrium of the constructed emulsified asphalt–aggregate interface model to ensure the scientific validity of subsequent molecular dynamics simulations. In MD simulations, system temperature and total energy are the primary indicators of equilibrium, with their stability

 https://doi.org/10.3390/coatings16030330

directly determining the credibility of interfacial behaviour analyses. Figure 7 presents the evolution of temperature and total energy of the interface model over a 1000 ps simulation. As shown, the system's total energy converges clearly without significant drift, while the temperature fluctuates slightly around the set value. This indicates that, following 1000 ps of equilibration, the system has reached a stable thermodynamic equilibrium, confirming the model's reliability and providing a solid foundation for quantitative analysis of interfacial adhesion, adsorption, and diffusion behaviours.

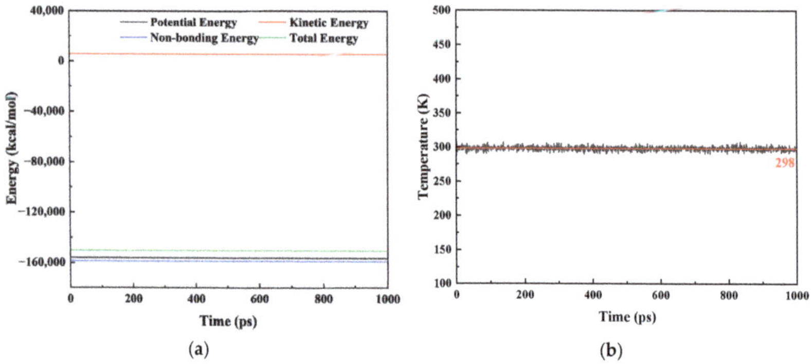

Figure 7. Temporal evolution of (**a**) total energy and (**b**) temperature in the emulsified asphalt–aggregate molecular model.

3. Results

3.1. Effect of Hydrophobic Hydrocarbon Chain Length on the Interfacial Distribution of Emulsified Asphalt–Aggregate Systems

Relative molecular concentration is an important parameter for characterising the spatial distribution of molecules within the interfacial region of the model. The position of the peak in the concentration profile quantitatively reflects the degree of molecular enrichment in that region. By analysing the number density of molecules along different spatial coordinates, this parameter provides a direct visualisation of molecular aggregation and spatial distribution at the interface, offering foundational data for further analysis of interfacial interaction mechanisms. The relative molecular concentration is calculated using Equations (1) and (2):

$$C_A = \frac{L_A}{D_A} \tag{1}$$

$$D_A = \frac{M_A \times n_A}{N_a \times V} \tag{2}$$

In the equations, C_A denotes the relative concentration, while D_A, L_A, M_A, and N_a represent the actual concentration, the local density, molecular weight, and number of molecules of species A, respectively. N_a and V correspond to Avogadro's constant and the system volume.

In this study, molecular dynamics simulations were used to analyse the relative molecular concentration distribution of emulsified asphalt molecules along the z-axis (i.e., the vertical direction of the simulation box) in the emulsified asphalt–aggregate systems. Five time points—200 ps, 400 ps, 600 ps, 800 ps, and 1000 ps—were selected to systematically examine the spatiotemporal evolution of asphalt molecules across different interface models. Based on these results, the effects of the hydrophobic hydrocarbon chain length of imidazoline-type emulsifiers and aggregate surface properties on the distribution behaviour of emulsified asphalt molecules

https://doi.org/10.3390/coatings16030330

were further evaluated. The resulting relative molecular concentration profiles are presented in Figures 8 and 9, with the boxes indicating the positions of the curve peaks.

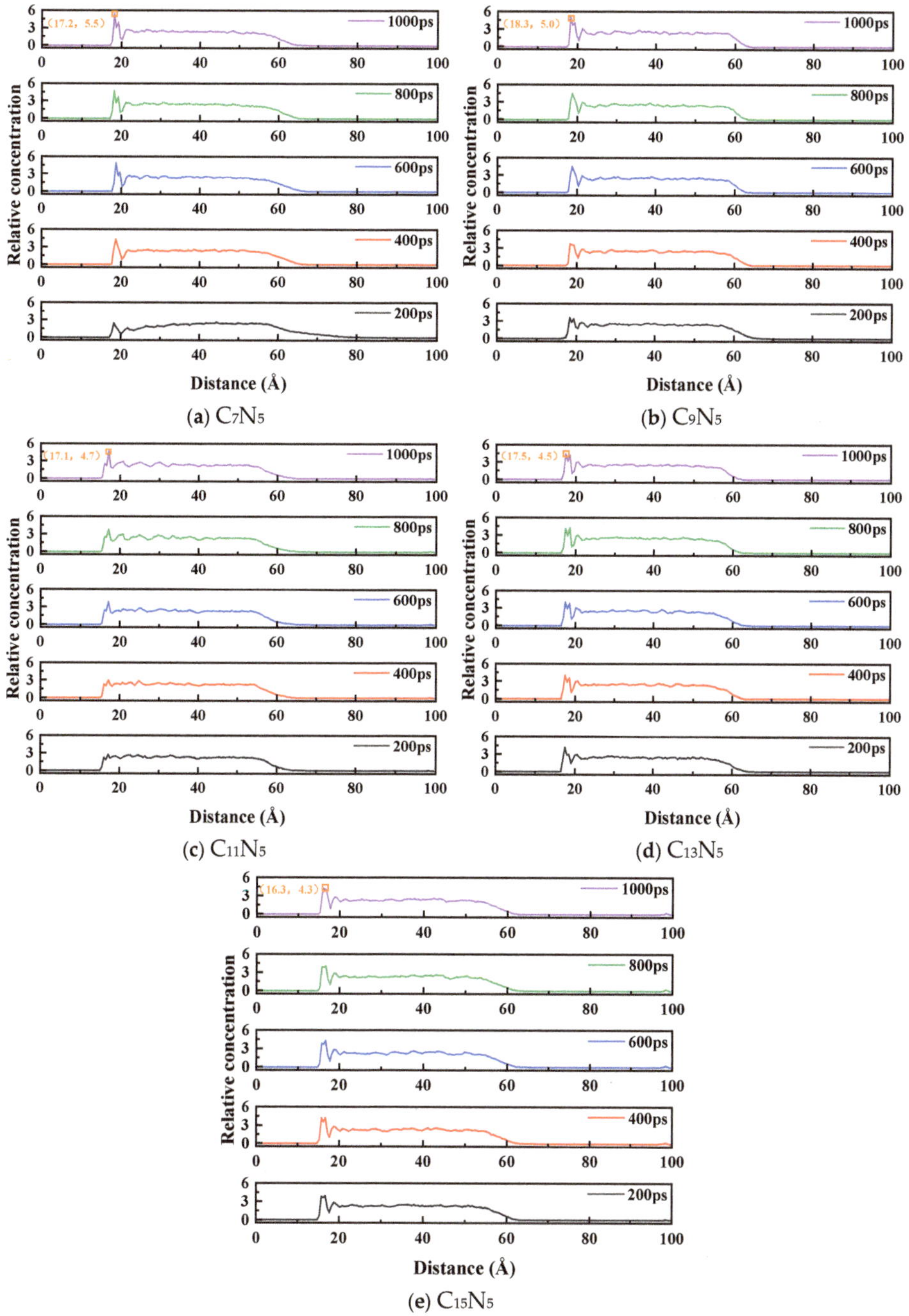

Figure 8. Relative concentration distribution of emulsified asphalt molecules along the z-axis on the $CaCO_3$ (1 0 4) surface.

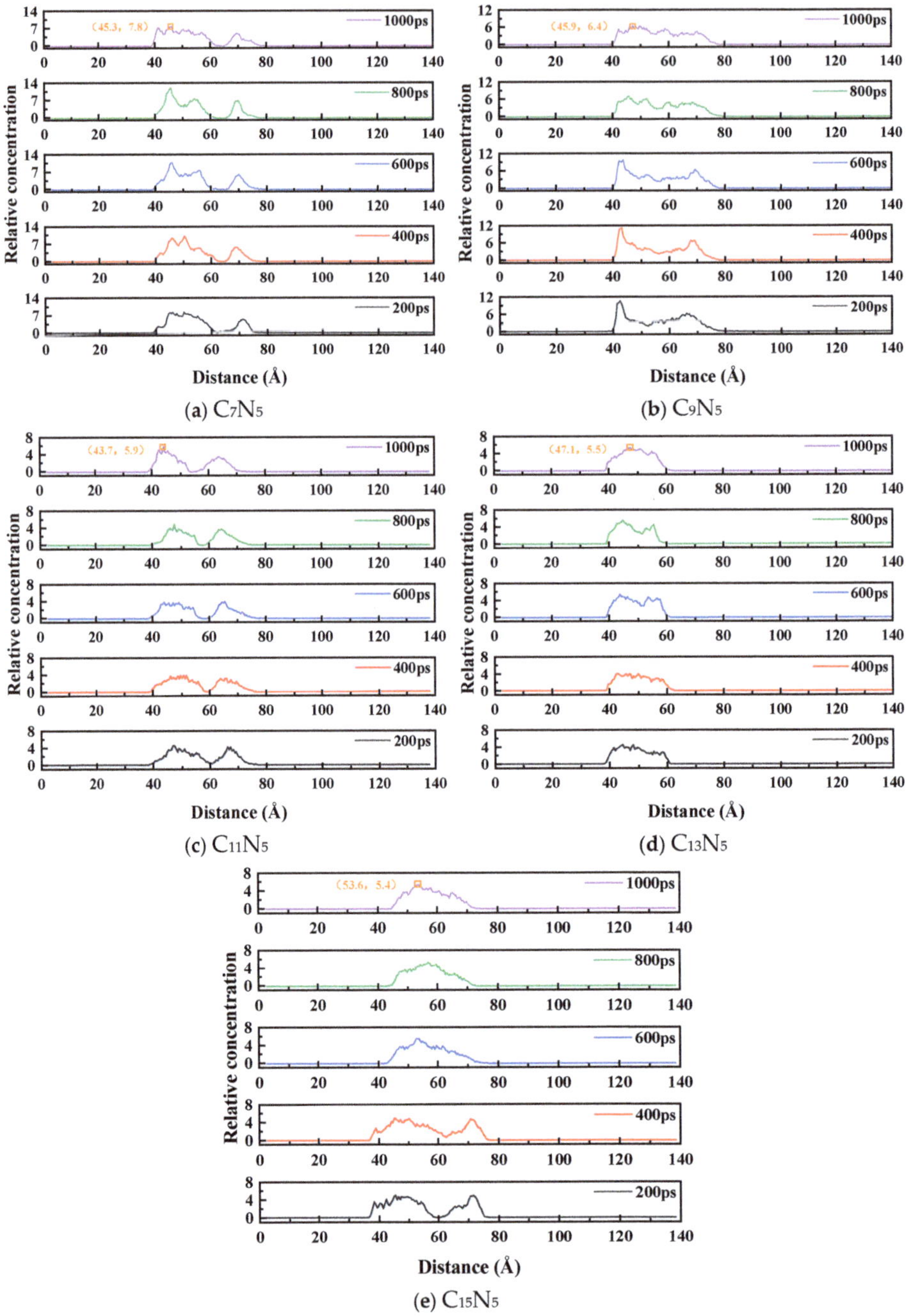

Figure 9. Relative concentration distribution of emulsified asphalt molecules along the z-axis on the SiO_2 (0 0 1) surface.

Analysis of Figure 8 shows that the vertical coordinate range of 0–15 Å corresponds to the $CaCO_3$ (1 0 4) aggregate region, where the relative concentration of emulsified asphalt

 https://doi.org/10.3390/coatings16030330

is zero. Within the range of hydrophobic hydrocarbon chain lengths studied, the relative concentration of emulsified asphalt molecules near the $CaCO_3$ (1 0 4) surface increases significantly, forming a single concentration peak. This indicates that the active structures in emulsified asphalt preferentially interact with the $CaCO_3$ (1 0 4) surface, resulting in adsorption and the formation of a stable monolayer. As the vertical distance from the aggregate surface increases, the relative concentration gradually stabilises, reflecting the reduced influence of aggregate surface functional groups and the transition of molecules to a freely distributed state.

Moreover, the adsorption peak of emulsified asphalt on the $CaCO_3$ (1 0 4) surface decreases progressively with increasing emulsifier hydrophobic chain length, indicating a negative correlation between asphalt adsorption and chain length on this basic aggregate. This behaviour arises because long-chain emulsifier molecules tend to form micellar structures through intermolecular hydrophobic interactions, reducing the effective contact probability of individual emulsifier molecules with the aggregate surface and thus lowering asphalt adsorption. Among the five emulsifier systems examined, the C_7N_5 emulsified asphalt system exhibits the highest concentration peak, reaching 5.5 at a vertical coordinate of 17.2 Å. When the hydrophobic alkyl chain length of the emulsifier increases from C_7 to C_9, the adsorption concentration peak decreases most significantly, with a reduction of approximately 9.1%. On average, for every additional two carbon atoms in the hydrophobic chain, the corresponding adsorption peak decreases by about 6.0%.

Figure 9 shows the relative concentration profiles of emulsified asphalt molecules on the SiO_2 (0 0 1) surface. As observed, the vertical coordinate range of 0–38 Å corresponds to the SiO_2 (0 0 1) aggregate region, where no asphalt molecules are present and the relative concentration is zero. For emulsified asphalt systems with different chain lengths, distinct concentration peaks appear on the SiO_2 (0 0 1) surface. Consistent with the trend observed on the $CaCO_3$ (1 0 4) surface, the peak relative concentration of emulsified asphalt on the SiO_2 (0 0 1) surface decreases with increasing hydrophobic chain length of the emulsifier, dropping from 7.8 for the C_7N_5 system to 5.4 for the $C_{15}N_5$ system. On average, the adsorption peak decreases by approximately 8.6% for every two-carbon increase in the hydrophobic chain length. These results further confirm the influence of the hydrophobic hydrocarbon chain length of imidazoline-type emulsifiers on the adsorption behaviour of emulsified asphalt at the aggregate interface, and indicate that this effect is consistent across aggregates with different surface characteristics.

By comparing the relative concentration distributions of emulsified asphalt molecules on $CaCO_3$ (1 0 4) and SiO_2 (0 0 1) surfaces, it is evident that, for systems with the same hydrophobic hydrocarbon chain length, both the spatial distribution patterns and adsorption intensities differ significantly between aggregates with distinct surface characteristics. Specifically, the z-axis concentration profile of asphalt on the $CaCO_3$ (1 0 4) surface exhibits a clearly stabilised concentration region, whereas the concentration distribution on the SiO_2 (0 0 1) surface shows more pronounced oscillatory behaviour. Quantitative analysis further indicates that the peak relative concentrations on the $CaCO_3$ (1 0 4) surface are considerably lower than those on SiO_2, and the average decrease in peak concentration with increasing emulsifier chain length is smaller on $CaCO_3$ (1 0 4) than on SiO_2.

Mechanistically, for acidic aggregates such as SiO_2 (0 0 1), van der Waals interactions are the primary force governing the adsorption of emulsified asphalt at the interface. Increasing the hydrophobic hydrocarbon chain length of the emulsifier significantly strengthens both intermolecular and molecule–aggregate van der Waals interactions, rendering the SiO_2 surface highly sensitive to changes in chain length. In contrast, $CaCO_3$ (1 0 4), as a basic aggregate, exhibits a more complex interaction mechanism with asphalt molecules,

 https://doi.org/10.3390/coatings16030330

potentially involving a combination of electrostatic forces, hydrogen bonding, and other interactions, resulting in a comparatively moderate response to variations in emulsifier chain length. This distinction further elucidates the coupled influence of aggregate surface chemistry and emulsifier molecular structure on the adsorption behaviour at the asphalt–aggregate interface.

Upon contact with the aggregate surface, the polar head groups of emulsifier molecules interact with functional groups on the aggregate. Driven by these interfacial forces, the molecular arrangement within the emulsified asphalt system undergoes reorganisation, causing active asphalt molecules to separate from the originally stable colloidal dispersion and orientationally adsorb onto the aggregate surface, completing the asphalt demulsification process. Analysis of the relative concentration distributions of emulsified asphalt on $CaCO_3$ (1 0 4) and SiO_2 (0 0 1) surfaces indicates that by 800 ps, both the peak concentrations and their spatial positions have stabilised, demonstrating that the demulsification process is essentially complete [14]. By 1000 ps, asphalt demulsification is fully achieved, forming a structurally stable adsorption layer at the emulsified asphalt–aggregate interface, providing a microscopic structural basis for the subsequent enhancement of interfacial bonding strength.

Figures 10 and 11 illustrate the microscopic distribution of the four asphalt components on $CaCO_3$ (1 0 4) and SiO_2 (0 0 1) surfaces after demulsification. The results show that all four asphalt components exhibit distinct concentration peaks on both aggregate surfaces, confirming that demulsification is complete and that the asphalt molecules have successfully adhered to the aggregates. Mechanistically, the conjugated benzene rings in the aromatic fraction can undergo π–π stacking with conjugated groups or polar sites on the aggregate surface, enabling directional adsorption. Saturates, such as small-chain alkanes, can aggregate around the hydrophobic tails of emulsifier molecules via hydrophobic interactions and migrate with the emulsifier toward the aggregate surface. Polar functional groups in resins and asphaltenes (e.g., hydroxyl and carboxyl groups) can form hydrogen bonds, electrostatic interactions, or other intermolecular forces with either the polar head groups of the emulsifier or active sites on the aggregate surface, such as hydroxyl or carbonyl groups. These mechanisms collectively allow effective adsorption of all asphalt components, though the spatial positions, intensities, and shapes of the concentration peaks differ significantly among components. Furthermore, both $CaCO_3$ (1 0 4) and SiO_2 (0 0 1) are hydrophilic mineral aggregates, enabling water molecules to adopt various stable conformations at the interface. This indicates that, regardless of the hydrophobic chain length of the emulsifier, water molecules readily orient and form a continuous interfacial layer on the aggregate surface. The MD simulation results (Figures 10 and 11) further confirm that water molecules stably adsorb on both aggregates, with their distribution closely linked to the chemical characteristics of the aggregate surfaces.

3.2. *Effect of Hydrophobic Hydrocarbon Chain Length on the Interfacial Diffusion Behaviour of Emulsified Asphalt–Aggregate Systems*

The diffusion coefficient is a key parameter characterising the dynamic activity of molecules within the interfacial region and plays a crucial role in studies of interfacial mechanisms at the molecular scale. Its value is typically derived from the Einstein diffusion equation and is closely related to molecular mobility: a larger diffusion coefficient indicates higher interfacial dynamic activity and faster molecular migration. The Einstein relation describes the diffusion behavior of particles in Brownian motion. It estimates the self-diffusion coefficient through the mean square displacement of particles over time. The diffusion coefficient is calculated using the following equation:

 https://doi.org/10.3390/coatings16030330

$$D = \lim_{t \to \infty} \frac{1}{6Nt} < \sum_{i=1}^{N} \mid r_i(t) - r_i(0) \mid^2 > \approx \frac{k}{6} \tag{3}$$

In the equation, D denotes the molecular diffusion coefficient, N is the number of molecules in the simulation system, t represents the elapsed time, $<\cdot>$ denotes the ensemble average (or time average), $r_i(t)$ and $r_i(0)$ are the positions of particle i at times t and 0, respectively, and k is the slope of the mean square displacement.

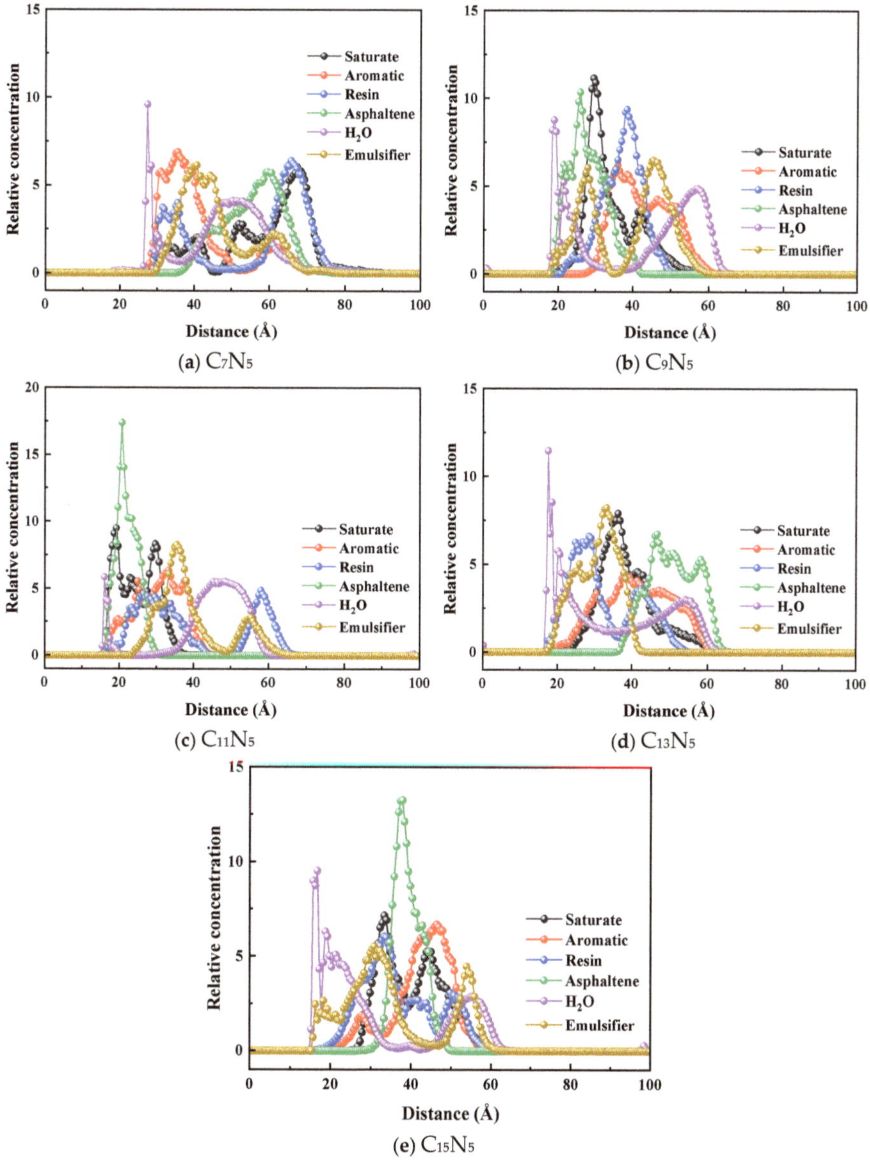

Figure 10. Relative concentration distribution of components along the z-axis on the $CaCO_3$ (1 0 4) surface.

https://doi.org/10.3390/coatings16030330

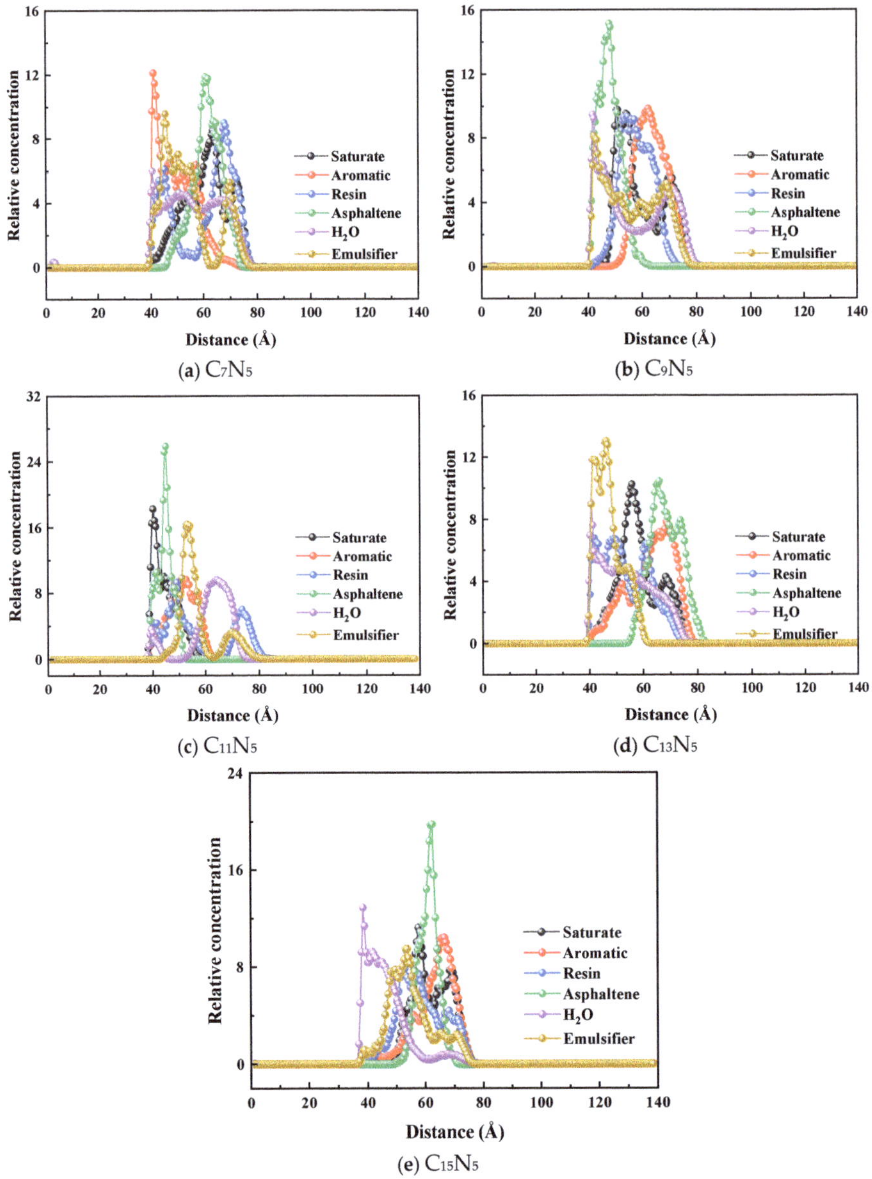

Figure 11. Relative concentration distribution of components along the z-axis on the SiO_2 (0 0 1) surface.

In the emulsified asphalt–aggregate interfacial system, the variation in diffusion coefficients for emulsified asphalt and emulsifier molecules is shown in Figure 12. As illustrated, both types of molecules exhibit the same trend: when the emulsifier's hydrophobic hydrocarbon chain length ranges from 7 to 13 carbon atoms, the diffusion coefficients increase with chain length; however, when the chain length reaches 15 carbon atoms, the diffusion coefficients decrease.

https://doi.org/10.3390/coatings16030330

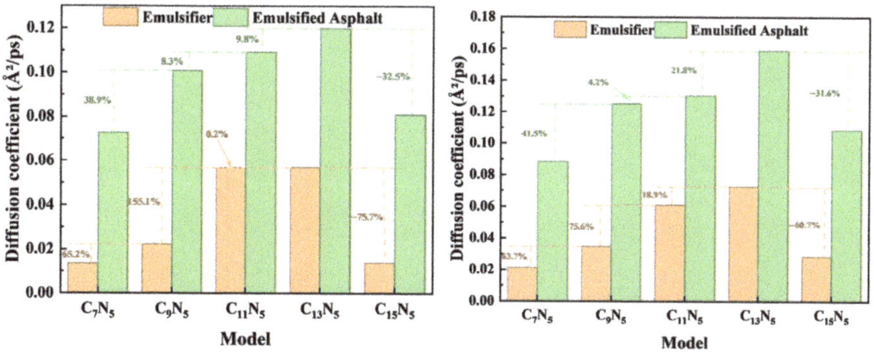

(a) (b)

Figure 12. Diffusion coefficients of emulsified asphalt molecules and emulsifier molecules on the (**a**) $CaCO_3$(1 0 4) and (**b**) SiO_2(0 0 1) aggregate surfaces.

The hydrophobic chain length of the emulsifier is a key factor affecting its hydrophilic–lipophilic balance (HLB). As the chain length increases, the interaction between the hydrophobic segments of the emulsifier and asphalt molecules becomes stronger, which facilitates the formation of a more ordered and stable interfacial structure at the water–asphalt interface. This structural enhancement improves local molecular mobility, leading to an increase in diffusion behavior. However, when the hydrophobic chain length becomes excessively long, the molecular chains tend to undergo folding and entanglement. This strengthens the van der Waals interactions between the emulsifier and asphalt molecules or the aggregate surface, thereby increasing interfacial adsorption constraints. Consequently, molecular motion is restricted, resulting in a decrease in the diffusion coefficients of both the emulsifier and the emulsified asphalt. Additionally, the difference between the diffusion coefficients of emulsified asphalt and the emulsifier on the same aggregate surface reflects the influence of chain length on the overall mobility of asphalt molecules. Furthermore, the self-diffusion coefficients of both emulsified asphalt and the emulsifier on the $CaCO_3$ (1 0 4) surface are generally lower than those on SiO_2 (0 0 1), indicating more restricted molecular motion, lower migration rates, and reduced mobility on the basic aggregate surface. This phenomenon highlights the effect of aggregate surface chemistry and interfacial interaction strength on asphalt diffusion behaviour. Specifically, the $CaCO_3$ (1 0 4) surface contains abundant Ca^{2+} active sites, which strongly interact electrostatically and coordinatively with the polar head groups of the emulsifier and polar asphalt components, significantly enhancing interfacial binding and limiting molecular mobility. In contrast, the SiO_2 (0 0 1) surface is primarily composed of Si–OH and Si–O–Si units, with weaker surface polarity, resulting in lower interfacial interaction with emulsifier and asphalt molecules and allowing greater molecular freedom, which is reflected in higher diffusion coefficients.

Further quantitative analysis of the diffusion coefficients reveals notable differences in sensitivity to hydrophobic hydrocarbon chain length between emulsifier and emulsified asphalt molecules. When the emulsifier's hydrophobic chain length increases from 9 to 11 carbon atoms, the diffusion coefficient of the emulsifier exhibits the most pronounced change: rising by over 100% on the $CaCO_3$ (1 0 4) surface and by 75.6% on the SiO_2 (0 0 1) surface. This indicates that, within this chain length range, the mobility of emulsifier molecules is particularly sensitive to chain length. For emulsified asphalt molecules, the change in diffusion coefficients is also primarily concentrated within the 9–11-carbon-atom chain length range, with increases exceeding 38% on both aggregate

 https://doi.org/10.3390/coatings16030330

surfaces. These findings demonstrate that structural variations in emulsifier molecules with shorter hydrophobic chains have a more significant impact on the overall migration ability of asphalt molecules.

In summary, the hydrophobic hydrocarbon chain length of the emulsifier and the chemical characteristics of the aggregate surface have a significant influence on the molecular mobility of the individual asphalt components. To further elucidate this mechanism, a targeted analysis of the diffusion coefficients of the four asphalt components and water molecules was conducted. The variation in diffusion coefficients for the asphalt components and water molecules is presented in Figure 13.

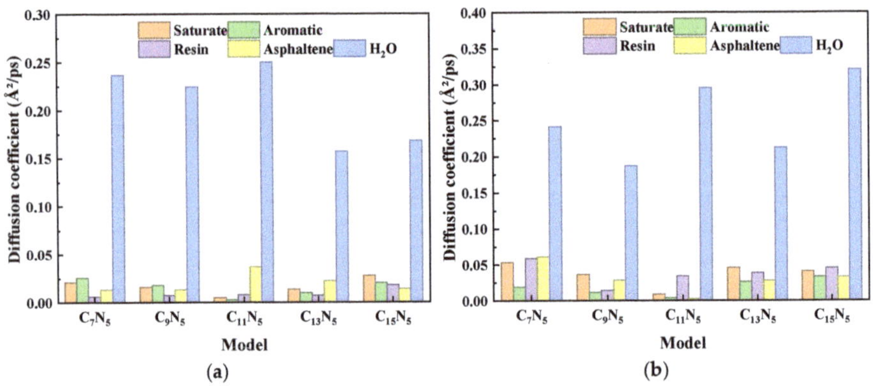

Figure 13. Diffusion coefficients of the four asphalt components and water molecules on the (**a**) $CaCO_3$ (1 0 4) surface and (**b**) SiO_2 (0 0 1) surface.

The analysis of diffusion coefficients in Figure 13 indicates that differences in the surface characteristics of $CaCO_3$ (1 0 4) and SiO_2 (0 0 1) aggregates significantly affect the mobility of polar asphalt components (asphaltenes and resins), whereas the motion of non-polar components (aromatics and saturates) is primarily governed by the hydrophobic hydrocarbon chain length of the emulsifier.

Specifically, the diffusion coefficients of asphaltenes and resins on the two aggregate surfaces exhibit opposite trends, likely reflecting the contrasting chemical properties of the aggregates. In contrast, the diffusion of aromatics and saturates is relatively insensitive to the aggregate surface type but shows a non-monotonic trend with increasing emulsifier chain length, first decreasing and then increasing. This confirms that the mobility of non-polar components is mainly controlled by the emulsifier's hydrocarbon chain length. Within a certain chain length range, longer emulsifier chains can markedly hinder the free diffusion of non-polar molecules, thereby affecting the microscopic packing density in the interfacial region. In both aggregate systems examined, the $C_{11}N_5$ emulsifier corresponds to an inflection point in the diffusion coefficients of saturates, aromatics, and asphaltenes, suggesting that this chain length represents a critical threshold for influencing the diffusion behaviour of non-polar and asphaltene molecules. Additionally, water molecules exhibit relatively high diffusion coefficients in both interfacial systems, indicating rapid mobility throughout the system. In comparison, emulsified asphalt and emulsifier molecules, due to their larger molecular size and stronger intermolecular interactions, experience significantly restricted free motion.

https://doi.org/10.3390/coatings16030330

3.3. Effect of Hydrophobic Hydrocarbon Chain Length on the Interfacial Adhesion Performance of Emulsified Asphalt–Aggregate Systems

Adhesion work is a key parameter for quantitatively characterising the interfacial bonding strength between two phases. In an emulsified asphalt–aggregate system, a positive adhesion work indicates that emulsified asphalt can spontaneously adsorb onto the aggregate surface, with higher values corresponding to stronger interfacial adsorption. Conversely, a negative adhesion work implies that emulsified asphalt cannot form a stable adsorption layer on the aggregate, and spontaneous interfacial adsorption is unlikely. In this study, adhesion work was calculated according to Equation (4).

$$W_{adhesion} = \frac{E_{bind}}{A} = \frac{-E_{inter}}{A} = \frac{-[E_{total} - (E_A + E_B)]}{A} \times K \quad (4)$$

In the equation, E_{bind} represents the binding energy of the A–B interface; E_{inter} is the interaction energy between A and B, with negative values indicating attraction and positive values indicating repulsion; E_{total} denotes the total free energy of the fully relaxed A–B system; E_A and E_B are the free energies of the fully relaxed A and B models, respectively; K is a conversion factor, taken as 659.

Based on the above equation, the interfacial adhesion work for the $CaCO_3$ (1 0 4) and SiO_2 (0 0 1) surfaces was calculated, as shown in Figure 14.

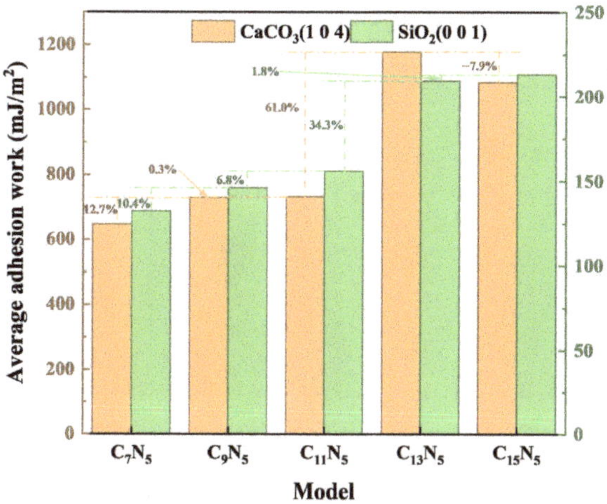

Figure 14. Variation in interfacial adhesion work of emulsified asphalt–aggregate systems with emulsifier hydrocarbon chain length.

Figure 14 shows that the adhesion work for both aggregate interfaces is positive, indicating that the interfacial adhesion of emulsified asphalt to the aggregates occurs spontaneously. For the $CaCO_3$ (1 0 4) surface, adhesion work initially increases with the emulsifier's hydrophobic hydrocarbon chain length but decreases when the chain exceeds 13 carbon atoms. In contrast, adhesion work on the SiO_2 (0 0 1) surface continues to rise steadily with increasing hydrocarbon chain length. These results indicate that moderately extending the emulsifier's hydrophobic chain promotes the formation of a dense, stable adsorption layer at the interface, thereby enhancing interfacial adhesion performance. Further quantitative analysis reveals that when the hydrocarbon chain length increases from 11 to 13 carbon atoms, the adhesion work reaches its maximum increase for both surfaces, at 61.0% for $CaCO_3$ (1 0 4) and 34.3% for SiO_2 (0 0 1). Notably, when the chain

 https://doi.org/10.3390/coatings16030330

length reaches 15 carbon atoms, adhesion work on SiO_2 (0 0 1) continues to increase, but by only 1.8%, whereas that on $CaCO_3$ (1 0 4) decreases by 7.9%. These findings indicate that beyond 13 carbon atoms, further chain elongation contributes little to enhancing interfacial adhesion and may even reduce adhesion strength, particularly on basic aggregates. At the molecular level, short-chain emulsifiers have shorter tails, resulting in low adsorption density, limited hydrophobic interactions, and consequently low work of adhesion. In contrast, the tails of long-chain emulsifier (C_{15}) is excessively long, increasing molecular flexibility and making them prone to coiling or entanglement. This leads to a decrease in adsorption arrangement density and restricted diffusion, hindering an increase in the work of adhesion. The tail length of the C_{13} emulsifier ensures sufficient hydrophobic contact with the surface without causing steric hindrance due to excessive length. Simultaneously, its highest diffusion coefficient enables the molecules to rapidly reach and uniformly cover the surface, thereby forming the most stable adsorption layer and minimizing the interfacial free energy. Furthermore, the hydrophilic head group of the C_{13} emulsifier can form effective electrostatic or hydrogen-bonding interactions with both acidic and alkaline aggregate surfaces, allowing it to exhibit relatively strong adhesion under different surface chemical environments. Moreover, under identical emulsifier chain lengths, the adhesion work on $CaCO_3$ (1 0 4) consistently exceeds that on SiO_2 (0 0 1), demonstrating that emulsified asphalt exhibits stronger interfacial adhesion with basic aggregates, in agreement with previous studies [14].

Based on the adhesion work calculations, the contributions of electrostatic and van der Waals interactions were further decomposed, as shown in Figure 15, illustrating the relative roles and dominant mechanisms of these interactions in interfacial adhesion.

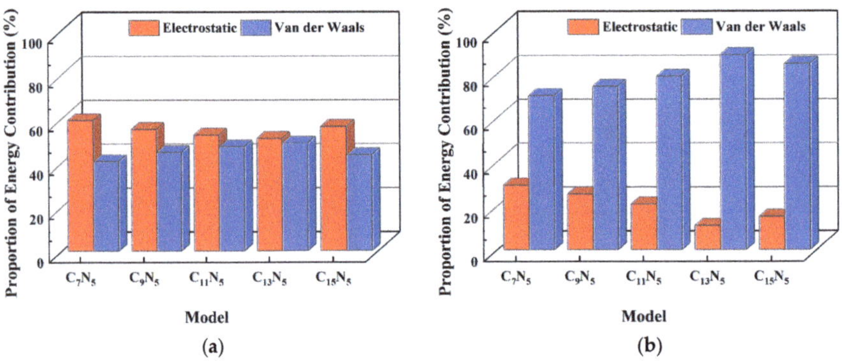

Figure 15. Contribution ratios of interaction energies for the (**a**) $CaCO_3$ (1 0 4) and (**b**) SiO_2 (0 0 1) systems.

Analysis of Figure 15 indicates that, for both aggregate surfaces, the contribution of van der Waals interactions to adhesion work initially increases and then decreases as the emulsifier's hydrophobic chain lengthens, whereas the contribution from electrostatic interactions exhibits the opposite trend. This suggests that extending the hydrophobic chain of imidazoline-type emulsifiers can enhance van der Waals interactions to a certain extent; however, beyond an optimal chain length, emulsifier molecules tend to aggregate and entangle within the asphalt phase and at the interface, preventing uniform spreading on the aggregate surface and consequently reducing the van der Waals contribution. Overall, the variation in adhesion work with hydrophobic chain length closely follows the trend of van der Waals contribution. The effect of imidazoline-type emulsifier chain length

 https://doi.org/10.3390/coatings16030330

on interfacial adhesion is fundamentally mediated through modulation of van der Waals interactions: as the hydrophobic chain length increases, intermolecular van der Waals forces are strengthened, promoting the formation of a dense and stable interfacial adsorption structure and thereby enhancing interfacial adhesion performance.

Further analysis reveals that on the $CaCO_3$ (1 0 4) basic aggregate surface, electrostatic interactions contribute over 50% of the interfacial adhesion work, serving as the dominant force governing adhesion. In contrast, on the SiO_2 (0 0 1) acidic aggregate surface, van der Waals interactions account for more than 80% of the adhesion contribution. This observation highlights a fundamental difference in the microscopic adhesion mechanisms between these two representative aggregate crystal planes. For the basic $CaCO_3$ (1 0 4) surface, electrostatic forces—including ionic coordination and electrostatic adsorption—constitute the primary source of interfacial bonding strength, with van der Waals forces playing only a secondary, supportive role. Conversely, the interfacial adhesion of the acidic SiO_2 (0 0 1) surface relies predominantly on van der Waals interactions between the imidazoline-type emulsifier and the aggregate surface.

Analysis of the energy contribution proportions reveals, at a mechanistic level, the fundamental reason why basic aggregates exhibit higher adhesion work than acidic aggregates: electrostatic interactions—dominated by Coulombic forces and ionic coordination—belong to the category of strong interactions, whereas van der Waals forces are weak physical interactions. Consequently, in basic aggregate systems governed by strong electrostatic interactions, the adhesion work of emulsified asphalt is higher, and interfacial adhesion performance is superior. In conjunction with the findings on hydrophobic chain length, optimizing chain length is crucial for enhancing adhesion on van der Waals-dominated acidic aggregates, while it has limited impact on electrostatically dominated basic aggregates. Therefore, different emulsifier molecular design strategies are required for the two aggregate types to achieve optimal interfacial adhesion performance.

4. Discussion

Within the carbon chain length range from C_7 to C_{15}, the adsorption peak shows a continuous decreasing trend. This is attributed to the increase in molecular occupied area caused by the lengthening of the carbon chain, which reduces the saturated adsorption capacity per unit surface area. However, the mere decrease in adsorption capacity is insufficient to evaluate the performance of different chain lengths, and a comprehensive consideration of adsorption quality and efficiency is still required. Both the diffusion coefficient and the work of adhesion reach their maximum values at C_{13}. This synergistic peak reveals that C_{13} achieves the optimal balance between kinetics and thermodynamics. Specifically, the largest diffusion coefficient indicates that C_{13} molecules experience the lowest migration resistance in the bulk phase and can migrate to the interfacial region at the fastest rate. Meanwhile, the maximum work of adhesion demonstrates that the adsorption layer formed by C_{13} at the interface is the most stable and exhibits the strongest anti-peeling ability.

The chain length of C_{13} avoids the loose adsorption layer caused by insufficient binding force in short-chain molecules, and also overcomes the increased mass transfer resistance and reduced utilization of effective adsorption sites resulting from molecular curling and steric hindrance in overly long carbon chains. Therefore, although C_{13} does not exhibit the highest adsorption peak, it trades moderate adsorption capacity for the optimal interfacial bonding strength and fastest adsorption rate, making it the critical chain length in this system that achieves the best synergy between efficient mass transport and strong adsorption.

https://doi.org/10.3390/coatings16030330

This study systematically investigates the adsorption behavior, diffusion characteristics, and adhesion properties of emulsifiers with chain lengths ranging from C_7 to C_{15} on different aggregate surfaces, elucidating the influence mechanism of imidazoline emulsifiers on the interfacial adhesion performance of emulsified asphalt/aggregate from a molecular perspective. The results reveal the critical regulatory role of emulsifier tail chain length on the emulsified asphalt/aggregate interfacial interaction, providing a theoretical basis for optimizing interfacial molecular arrangement and adhesion performance. From the perspective of road engineering applications, the adhesion performance between emulsified asphalt and aggregate is a crucial factor affecting the initial strength, interface stability, and durability of pavement structures. Based on the findings of this study, selecting emulsifiers with appropriate tail chain lengths enables the formation of highly stable adsorption layers on different types of aggregate surfaces, thereby optimizing the interfacial performance of emulsified asphalt mixtures and enhancing pavement stripping resistance and long-term service life. This research not only provides molecular-level support for an in-depth understanding of the emulsified asphalt–aggregate interface mechanism but also offers a scientific basis for the formulation design and construction process optimization of emulsified asphalt materials in road engineering.

5. Conclusions

This study, based on molecular dynamics simulations and key characterization parameters including relative concentration, diffusion coefficient, and interfacial adhesion work, systematically investigated the influence of imidazoline-type emulsifier molecular structure on the adhesion performance of the emulsified asphalt–aggregate interface, and further analyzed the effect of aggregate surface properties on the interfacial adhesion mechanism. The main conclusions are as follows:

(1) The hydrophobic alkyl chain length of imidazoline-type emulsifiers exerts a significant regulatory effect on interfacial behavior: increasing the hydrophobic chain length suppresses the adsorption amount of emulsified asphalt at the interface. Hydrophobic chains of appropriate length enhance molecular mobility, thereby improving the interfacial adhesion performance between emulsified asphalt and aggregates. When the chain length exceeds the optimal value (C_{13}), both molecular diffusion rates and interfacial adhesion performance are weakened. In addition, the diffusion behaviors of nonpolar components (aromatics and saturates) respond noticeably to changes in alkyl chain length. Based on the combined analysis of adsorption behaviour, diffusion coefficient, and adhesion work, the interfacial performance reaches the optimum at an alkyl chain length of C_{13}.

(2) The surface properties of aggregates significantly influence interfacial adhesion: compared with the acidic SiO_2 (0 0 1) aggregate, the $CaCO_3$ (1 0 4) basic aggregate exhibits overall lower relative concentration peaks of emulsified asphalt, with peak values less sensitive to variations in the hydrophobic alkyl chain length of the emulsifier. Molecular mobility is lower, adhesion work is higher, and interfacial adhesion performance is superior. These observations collectively indicate strong interactions between emulsified asphalt and the surface of basic aggregates.

(3) The dominant energetic mechanism of interfacial adhesion depends on the type of aggregate: the adhesion of emulsified asphalt on the acidic SiO_2 (0 0 1) aggregate is primarily governed by van der Waals interactions, whereas the interfacial adhesion on the basic $CaCO_3$ (1 0 4) aggregate is dominated by strong electrostatic interactions.

https://doi.org/10.3390/coatings16030330

Author Contributions: Writing—original draft preparation, H.X.; writing—review and editing, P.Z.; data curation, Z.Z., J.S. (Jingtao Shi), W.W., D.Y., H.C., D.H. and J.S. (Jiawei Su). All authors have read and agreed to the published version of the manuscript.

Funding: This research was funded by the Natural Science Foundation of Shandong (ZR2021ME189) and the National Natural Science Foundation of China (51608511).

Institutional Review Board Statement: Not applicable.

Informed Consent Statement: Not applicable.

Data Availability Statement: The original contributions presented in this study are included in the article. Further inquiries can be directed to the corresponding author.

Conflicts of Interest: Author Zhiyu Zhang was employed by the company Beijing Zhengda Tongshun Inspection Technology Co., Ltd. Author Jingtao Shi was employed by the company Petrochina Fuel Oil Co., Ltd. Research Institute. The remaining authors declare that the research was conducted in the absence of any commercial or financial relationships that could be construed as a potential conflict of interest.

References

1. Blaauw, S.A.; Maina, J.W. Life Cycle Inventory for Pavements—A Case Study of South Africa. *Transp. Eng.* **2021**, *3*, 100049. [CrossRef]
2. Zhao, P.H.; Liang, H.Y.; Wu, W.X.; Yang, Y.S.; Liu, Y.M.; Li, C.; Meng, W.K.; Zhang, R.; Song, X.Q.; Wang, C.L.; et al. Rheological properties of high-asphalt-content emulsified asphalt. *Constr. Build. Mater.* **2024**, *419*, 135511. [CrossRef]
3. Zhang, H.; Wang, Z.; Wang, Q. Quantitative evaluation of cement emulsified asphalt mortar and aggregate adhesion performance with dynamic mechanical analysis. *Constr. Build. Mater.* **2020**, *262*, 120043. [CrossRef]
4. Wang, Y.Y.; Leng, Z.; Li, X.; Hu, C.C. Cold recycling of reclaimed asphalt pavement towards improved engineering performance. *J. Clean. Prod.* **2018**, *171*, 1031–1038. [CrossRef]
5. Liu, J.; Zheng, X.G.; Li, T.J.; Wang, C.P.; Yanxu, Z.; Ying, L.; Liangwei, L.; Wei, S.; Zheng, X.; Li, T.; et al. Mechanical properties and micromorphology of redispersible emulsified asphalt powder modified cement mortar. *Chem. Ind. Eng. Prog.* **2022**, *41*, 2015–2021. [CrossRef]
6. Chu, F.S.; Li, Q.S.; Sun, C.J.; Luo, Y.; Li, J.C.; Feng, J.R.; Zhang, H.L. Investigation on the pavement performance and application of cold mix emulsified asphalt mixture in the surface layer. *J. Clean. Prod.* **2025**, *507*, 145542. [CrossRef]
7. Huang, W.Q.; Cao, M.M.; Xiao, L.L.; Li, J.; Zhu, M. Experimental study on the fatigue performance of emulsified asphalt cold recycled mixtures. *Constr. Build. Mater.* **2023**, *369*, 130607. [CrossRef]
8. Shi, C.Y.; Ge, J.G.; Yu, H.A.; Qian, G.P.; Zhou, H.Y.; Ma, Y.T.; Nian, T.F.; Yao, D.; Wang, Y.; Zhong, Y.X. Interfacial adhesion properties and debonding mechanisms in rejuvenated asphalt mixtures. *Constr. Build. Mater.* **2024**, *425*, 135973. [CrossRef]
9. Tang, Y.J.; Fu, Z.; Raos, G.; Ma, F.; Zhao, P.; Hou, Y.J. Molecular dynamics simulation of adhesion at the asphalt-aggregate interface: A review. *Surf. Interfaces* **2024**, *44*, 103706. [CrossRef]
10. Tan, W.; Zhu, S.X.; Kong, L.Y.; Peng, Y.; Xu, L.; Fu, Y.G. Influence of aggregate chemical composition on the demulsification rate of emulsified asphalt. *Front. Mater.* **2022**, *9*, 1079431. [CrossRef]
11. Chen, L.X.; Li, J.L.; Xiang, H.; Zhang, J.; Feng, E.L.; Kong, L. Effect of Aggregate Type on Asphalt-Aggregate Adhesion and Its Quantitative Characterization. *Materials* **2025**, *18*, 3696. [CrossRef]
12. Wang, H.N.; Xu, H.; Wang, J.F.; Ying, J.Z.; Shen, S.; Zhang, S. Influence of aggregate physicochemical properties on interfacialadhesion performance of asphalt-aggregate. *J. Chang Univ. (Nat. Sci. Ed.)* **2022**, *42*, 52–61.
13. Yin, Y.P.; Chen, H.X.; Kuang, D.L.; Song, L.F.; Wang, L. Effect of chemical composition of aggregate on interfacial adhesion property between aggregate and asphalt. *Constr. Build. Mater.* **2017**, *146*, 231–237. [CrossRef]
14. Zhu, S.X.; Kong, L.Y.; Peng, Y.; Chen, Y.; Zhao, T.H.; Jian, O.Y.; Zhao, P.H.; Sheng, X.; Li, Z. Mechanisms of interface electrostatic potential induced asphalt-aggregate adhesion. *Constr. Build. Mater.* **2024**, *438*, 137255. [CrossRef]
15. Lin, M.; Lei, Y.; Li, P.; Zhang, Q. Interface adhesion performance and adhesion mechanism between graphene/rubber composite modified asphalt and aggregate. *Chem. Ind. Eng. Prog.* **2025**, *44*, 991–1002. [CrossRef]
16. Cui, S.; Blackman, B.R.K.; Kinloch, A.J.; Taylor, A.C. Durability of asphalt mixtures: Effect of aggregate type and adhesion promoters. *Int. J. Adhes. Adhes.* **2014**, *54*, 100–111. [CrossRef]
17. Yao, X.G.; Li, C.X.; Xu, T. Interfacial adhesive behaviors between SBS modified bitumen and aggregate using molecular dynamics simulation. *Surf. Interfaces* **2022**, *33*, 102245. [CrossRef]

https://doi.org/10.3390/coatings16030330

18. Chu, L.; Luo, L.; Fwa, T.F. Effects of aggregate mineral surface anisotropy on asphalt-aggregate interfacial bonding using molecular dynamics (MD) simulation. *Constr. Build. Mater.* **2019**, *225*, 1–12. [CrossRef]
19. Qu, X.; Liu, Q.; Guo, M.; Wang, D.W.; Oeser, M. Study on the effect of aging on physical properties of asphalt binder from a microscale perspective. *Constr. Build. Mater.* **2018**, *187*, 718–729. [CrossRef]
20. Liu, K.; Yang, Q.; Qiu, X.; Xu, W.; Xiao, S.; Gu, Y.; Ye, Y. An investigation toward adhesion characteristics of emulsified asphalt residue–aggregate interface through MD simulation. *Constr. Build. Mater.* **2024**, *438*, 137251. [CrossRef]
21. Zhu, S.X.; Kong, L.Y.; Zhao, P.H.; Peng, Y.; Feng, B.; Zeng, Q.L.; Yang, M.Z.; Zhang, H. Effect of alkyl chain length on the demulsification process of cationic emulsified asphalt on a $CaCO_3$ surface. *Constr. Build. Mater.* **2024**, *417*, 135154. [CrossRef]
22. Kong, L.Y.; Zhu, S.X.; Peng, Y.; Zhao, P.H.; Ouyang, J.; Shao, Q.; Wang, W.N. Effect of phenyl functional groups on the adsorption behaviour of dodecyl anionic emulsifiers on oxides on the aggregate surface. *Int. J. Pavement Eng.* **2024**, *25*, 2303662. [CrossRef]
23. Kong, L.Y.; Zhu, S.X.; Chen, Y.; Peng, Y.; Zhao, P.H.; Ouyang, J.; Feng, B. Quantifying the influence of lipophilic carbon chain length on adhesion between emulsified asphalt and acidic aggregates. *Int. J. Pavement Eng.* **2024**, *25*, 2376219. [CrossRef]
24. Zhang, Z.; Zheng, X.G.; Li, J.Y.; Xu, G.; Tan, L.J. Mechanism of reinforced interfacial adhesion between steel slag and highly devulcanized waste rubber modified asphalt and its influence on the volume stability in steel slag asphalt mixture. *Constr. Build. Mater.* **2024**, *447*, 138129. [CrossRef]
25. Meng, X.L.; Li, X.Y.; Yan, N.H.; Li, W.B. Research on Dynamic Adhesion and Failure Characteristics of Asphalt-AggregateInterface Based on Molecular Dynamics. *Mater. Rep.* **2025**, *39*, 97–104.
26. Wang, S.H.; Chen, Y.; Wang, L.H.; Cui, N.X.; Li, C.F.; Sun, S.F. Study on Water Damage of Asphalt-Aggregate Based on Molecular Dynamics. *Materials* **2025**, *18*, 209. [CrossRef] [PubMed]
27. Sun, W.; Wang, H. Moisture effect on nanostructure and adhesion energy of asphalt on aggregate surface: A molecular dynamics study. *Appl. Surf. Sci.* **2020**, *510*, 145435. [CrossRef]
28. Artok, L.; Su, Y.; Hirose, Y.; Hosokawa, M.; Murata, S.; Nomura, M. Structure and Reactivity of Petroleum-Derived Asphaltene. *Energy Fuels* **1999**, *13*, 287–296. [CrossRef]
29. Wiehe, I.A.; Liang, K.S. Asphaltenes, resins, and other petroleum macromolecules. *Fluid Phase Equilibria* **1996**, *117*, 201–210. [CrossRef]
30. Zhang, L.Q.; Greenfield, M.L. Effects of polymer modification on properties and microstructure of model asphalt systems. *Energy Fuels* **2008**, *22*, 3363–3375. [CrossRef]
31. Li, D.D.; Greenfield, M.L. Chemical compositions of improved model asphalt systems for molecular simulations. *Fuel* **2014**, *115*, 347–356. [CrossRef]
32. Yao, H.; Dai, Q.L.; You, Z.P. Molecular dynamics simulation of physicochemical properties of the asphalt model. *Fuel* **2016**, *164*, 83–93. [CrossRef]
33. Zhao, P.H.; Ren, R.B.; Zhou, H.; Ouyang, J.; Li, Z.G.; Sun, D.C. Preparation and properties of imidazoline surfactant as additive for warm mix asphalt. *Constr. Build. Mater.* **2021**, *273*, 121692. [CrossRef]
34. Kong, L.Y.; Zhu, S.X.; Quan, X.J.; Peng, Y. Effect of phenyl functional group on the demulsification process of dodecyl anion emulsified asphalt. *Constr. Build. Mater.* **2022**, *354*, 129196. [CrossRef]
35. Liao, R.J.; Li, B.F.; Cheng, L.; Yang, L.J.; Zhao, X.; Hao, J. Review on Numerical Simulation and Experimental Study of Gas Transport Mechanisms in Oil-paper Insulation. *High Volt. Eng.* **2025**, *51*, 3940–3957. [CrossRef]
36. Liu, Y.Y.; Li, F.Q.; Liu, D.; Wang, J.T.; Luo, X. Molecular simulation study on the interfacial properties of recycled asphalt-aggregate at the nanoscale. *Chem. Ind. Eng. Prog.* **2025**, *44*, 4302–4310. [CrossRef]
37. Zhang, L.Q.; Greenfield, M.L. Analyzing properties of model asphalts using molecular simulation. *Energy Fuels* **2007**, *21*, 1712–1716. [CrossRef]
38. Lesueur, D. The colloidal structure of bitumen: Consequences on the rheology and on the mechanisms of bitumen modification. *Adv. Colloid Interface Sci.* **2009**, *145*, 42–82. [CrossRef]

https://doi.org/10.3390/coatings16030330

coatings

MDPI

Article

Utilization of Water-Based Drill Cuttings Stabilized by a Novel Composite Stabilizer for Pavement Base Applications

Shucheng Tan [1,2], Hua Wen [3,*], Hua Tang [1,2], Wentao Fu [1,2], Xiaoyan Guo [1,2], Biaotian Bai [1,2], Jiujiang Wu [3,4] and Xiaoyu Tan [3]

1 CNPC Chuanqing Drilling Engineering Company Limited, Chengdu 610051, China; tansc_sc@cnpc.com.cn (S.T.); tanghua_sy@cnpc.com.cn (H.T.); fuwt_sc@cnpc.com.cn (W.F.); guoxiaoyan@cnpc.com.cn (X.G.); baibt_sy@cnpc.com.cn (B.B.)
2 CCDC SHUYU Engineering Construction Co., Ltd., Chengdu 610084, China
3 School of Civil Engineering and Architecture, Southwest University of Science and Technology, Mianyang 621010, China; jwu766@uwo.ca (J.W.); txiaoyu1016@163.com (X.T.)
4 Department of Civil and Environmental Engineering, Western University, London, ON N6A 5B9, Canada
* Correspondence: wenhua189@163.com

Abstract

Water-based drill cuttings generated during onshore natural gas development are complex solid wastes that may pose environmental risks if improperly managed. This study evaluates the feasibility of reutilizing water-based drill cuttings as pavement base materials after stabilization using a novel composite stabilizer composed of cement, stabilizer liquid agent, and water-reducing powder (CLP stabilizer). Mix proportion optimization was conducted through compaction and 7-day unconfined compressive strength tests, followed by evaluation of road performance, including strength, compressive rebound modulus, water stability, and temperature shrinkage, with stabilized powder stabilized soil as a control. Microstructural characteristics were analyzed using X-ray diffraction and scanning electron microscopy, and environmental safety was assessed through heavy metal leaching tests and background soil investigation. The results show that the optimal mixture ratio of curing agent (5% cement + 2% liquid stabilizer + 8% superplasticizer powder) satisfies the strength requirement for pre-drilling road bases, exhibiting superior performance compared to the control group. When the stabilizer dosage reaches 9%, the 7-day unconfined compressive strength achieves a maximum of 3.38 MPa, representing a 51% increase over the control group. At a stabilizer dosage of 12%, the splitting tensile strength reaches a peak value of 0.901 MPa, showing a 60.3% improvement. These results indicate enhanced deformation resistance, water stability, and reduced temperature shrinkage rates. Microstructural analysis indicates that the formation of calcium silicate hydrate (C-S-H) gel and ettringite (AFt phase) leads to a denser structure and enhanced durability. Heavy metal concentrations comply with relevant standards, demonstrating controllable environmental risks and supporting sustainable pavement base application.

Keywords: water-based drill cuttings; CLP stabilizer; pavement base; road performance; environmental safety; microstructural mechanism

1. Introduction

Large quantities of water-based drill cuttings are generated during natural gas exploration and development. These cuttings are typically complex in composition and may contain heavy metals and residual chemical additives. If improperly managed, they can contaminate soil and groundwater, posing risks to surrounding ecosystems and public health [1]. Therefore, the resource utilization of water-based drill cuttings after harmless treatment, particularly in pavement base engineering, represents a promising approach to mitigating environmental pressure while alleviating shortages of conventional construction materials, such as sand and gravel [2].

Driven by infrastructure demands (roads, foundations) and material shortages, soil solidification technology has rapidly developed and been widely adopted. Qin et al. [3] used three components, namely, cement clinker, sodium bisulfate, and Na_2SiF_6, as the main materials of the stabilizer to chemically reinforce the coastal silt. The 28-day strength of the consolidated soil was verified by the unconfined compressive strength test. The test results show that the activity of the slag in the stabilizer is fully stimulated, and the consolidation effect on the silt is very significant. In the study, Babatunde et al. [4] used gum Arabic as a sustainable soil stabilizer and employed three different contents of biopolymers to investigate the effects of gum Arabic on the rheological behavior, strength, and stiffness characteristics of treated sand. The compressive strength of the treated specimens was significantly enhanced after 28 days of curing. Li et al. [5] used the method of ionic soil stabilizer (ISS)—cement synergistic solidification in the study to solidify coastal pickled dispersed soil. The results showed that the solidification effect was the best when the ISS content was 5%. Zhong et al. [6] use high-content slag soil to prepare controllable low-strength filling materials to treat and effectively utilize the engineering sludge generated by urban development projects, achieving engineering performance reconstruction and resource utilization.

Scholars have conducted extensive research on the harmless disposal and resource utilization of water-based drill cuttings, providing crucial insights for their high-value applications in the fields of building materials and cementitious materials, such as Hu et al. [7]. In this paper, the feasibility of water-based drilling cuttings as building materials additives after treatment is discussed, and the reuse path of water-based drilling cuttings in specific industrial processes is studied, which provides a solid scientific theoretical support for the sustainable management of drilling cuttings. Xie et al. and Liuyang et al. [8,9] successfully achieved the secondary pyrolysis resource utilization of oil-based drilling waste ash by mixing drilling cuttings containing Cr(VI) pollutants with bauxite in different ratios to prepare proppants, providing an effective solution for the disposal of heavy metal-contaminated drilling cuttings. The study by Liu et al. [1] demonstrated that processed water-based drilling cuttings exhibit high calcium content properties, enabling partial substitution for conventional Portland cement. They proposed the preparation of highly reactive supplementary cementitious materials through mechanical activation methods, effectively achieving high-value resource recovery from water-based drilling cuttings. Liu et al. [10] demonstrated that combining calcined drilling cuttings with waste glass not only compensates for the low silicon content in water-based drilling cuttings but also significantly accelerates cement hydration, thereby expanding the application prospects of drilling cuttings in the cementitious materials field.

The application of solid waste materials as stabilizing agents in road engineering has also been extensively explored. Tanyıldızı et al. [11] evaluated the stabilizing effect of waste hazelnut shell ash (WHSA) on expansive soil. The experiments showed that a WHSA content of 2% to 10% could significantly reduce the plasticity, expansion, and contraction of soil. The environmentally friendly and low-cost characteristics could effectively improve the engineering properties of expansive soil and are suitable for road subgrade reinforcement. Qiu et al. [12] replaced cement with calcium carbide residue, blast furnace slag, and fly ash, and combined polypropylene fiber to reinforce the silt powder soil in Dongying

 https://doi.org/10.3390/coatings16040406

City, significantly improving the mechanical properties of the stabilized soil. Sun et al. [13] mixed fly ash (FA) and granulated blast furnace slag (GBFS) in a certain proportion as the basic building material for low-carbon pavement, which can effectively replace traditional materials and promote the utilization of waste materials in pavement applications.

Despite these advances, road base materials remain predominantly reliant on conventional cementitious systems such as Portland cement, quicklime, industrial slags, and natural aggregates. Developing new soil stabilizer systems with improved soil compatibility and environmental performance has become critical for addressing resource constraints and ecological pressures in road engineering. Jia et al. [14] prepared subgrade materials from iron tailings containing fly ash and lime and improved their mechanical strength and frost resistance and reduced the dry shrinkage rate through ionic soil stabilizers (ISS). The results show that when the ISS is 0.67%, the 7-day unconfined compressive strength increases by 195.5%. Pushpakumara and Mendis [15] treated highly plastic clayey soil with different ratios of rice husk ash (RHA) and lime. The results show that when the ratio is 10% rice husk ash + 20% lime, the unconfined compressive strength of the stabilized soil increases by 54.05%, and the friction angle value in the shear strength parameter increases by 60.48%. Lei et al. [16] and Hou et al. [17] successfully synthesized a multifunctional polymer soil stabilizer (MPSS) by copolymerizing modified nano-zinc oxide with acrylic acid and used it as an effective additive to prevent soil erosion. The results showed that the stabilizer exhibited excellent water retention and plant compatibility.

Although substantial progress has been made in soil stabilization and solid waste utilization, significant technical challenges remain in treating high-water-content water-based drilling cuttings associated with oil and gas extraction. Conventional stabilizers often fail to achieve efficient and economical solidification due to the high moisture content, complex chemical composition, and potential heavy metal contamination of these materials. To address these limitations, this study evaluates the feasibility of reutilizing water-based drill cuttings as pavement base materials through stabilization with a novel composite system composed of cement, stabilizer liquid agent, and water-reducing powder (CLP stabilizer). At first, water-based drill cuttings collected from an engineering site were first characterized in terms of their fundamental physical properties. The optimal stabilizer proportion was determined through mix design, compaction tests, and 7-day unconfined compressive strength tests, followed by stabilization treatment of the drill cuttings. Subsequently, the road performance of the stabilized materials under different curing ages and stabilizer dosages was evaluated through unconfined compressive strength tests, splitting tensile strength tests, compressive resilient modulus tests, temperature shrinkage tests, and water stability tests, thereby identifying the optimal solidifying agent content. Finally, microstructural evolution, pore morphology, and mineral composition of both stabilized drill cuttings and powder-stabilized soil were analyzed. Heavy metal concentrations in leachate and surrounding soil were examined to assess the environmental compatibility and practical applicability of the stabilized material.

2. Material Analysis and Mix Design

2.1. *Analysis of Basic Properties of Raw Materials*

2.1.1. Water-Based Drill Cuttings

The water-based drill cuttings are taken from the engineering cuttings of a drilling well with a depth of 500 m to 2000 m in Suining, Sichuan Province. Their lithology is grayish-brown, in a moist state, with an unpleasant smell, and formed unevenly sized blocky bodies after absorbing water. In the early stage of the experiment, the moisture content of the on-site soil samples was analyzed first to clarify the composition of the soil

 https://doi.org/10.3390/coatings16040406

samples in this area. The phase analysis of soil samples was conducted by X-ray diffraction (XRD) detection [18], and the detection results are shown in Figure 1.

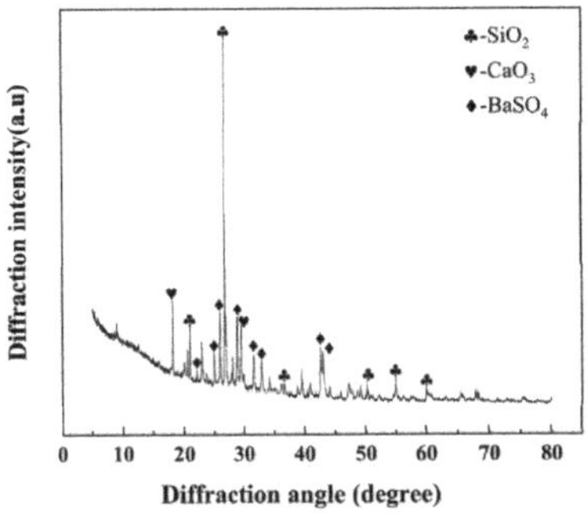

Figure 1. XRD pattern of water-based drill cuttings.

It can be seen that the mineral composition of water-based drill cuttings mainly includes quartz, barite, calcite, etc., and it also contains a certain proportion of cuttings particles. The composition of water-based drill cuttings was detected by an X-ray fluorescence spectrometer. The chemical composition and content are shown in Table 1. The heavy metal detection results of water-based drill cuttings samples are shown in Table 2. The samples mainly contain mercury, arsenic, copper, petroleum, chlorides, and other substances, and the content of each substance is far below the standard limit, which meets the relevant requirements of the "Integrated wastewater discharge standard" (GB 8978-1996) [19] and "Environmental quality standards for surface water" (GB 3838-2002) [20].

Table 1. Chemical components and contents of water-based drilling cuttings (%).

Material Name	SiO_2	BaO	SO_3	Al_2O_3	CaO	Fe_2O_3	K_2O	Cl	Na_2O	MgO	Other
Content	39.6	20.33	13.81	10.55	4.43	3.28	2.74	1.91	1.41	1.14	0.8

Table 2. Heavy metal content detected in water-based drill cuttings (mg/L).

Surveillance Project	Detection Value	Standard Limit	Surveillance Project	Detection Value	Standard Limit
PH	8.22	6~9	Pb	ND [1]	1.0
Hg	0.00026	0.05	Zn	ND [1]	2.0
As	0.00710	0.5	COD	53.8	100
Cd	ND [1]	0.1	Petroleum products	0.06	5.0
Cu	0.01	0.5	Chloride	0.64	10
Ni	ND [1]	1.0	-	-	-

[1] "ND" indicates a test result below the detection limit.

According to the "Highway geotechnical test code" (JTG 3430-2020) [21], the basic physical property indicators of water-based drill cuttings were determined, and the results are shown in Table 3 and Figure 2. The particle analysis results show that the soil sample is a fine-grained soil with uniform particle distribution but poor gradation. It is determined to be a low liquid limit clay through plasticity diagram analysis.

https://doi.org/10.3390/coatings16040406

Table 3. Basic physical properties of soil samples.

Material Name	Natural Water Content (%)	Optimum Moisture Content (%)	Maximum Dry Density (g/cm^3)	Liquid Limit (%)	Plastic Limit (%)	Plasticity Index	Specific Gravity
Water-based drill cuttings	39.6	20.33	13.81	10.55	4.43	3.28	2.74

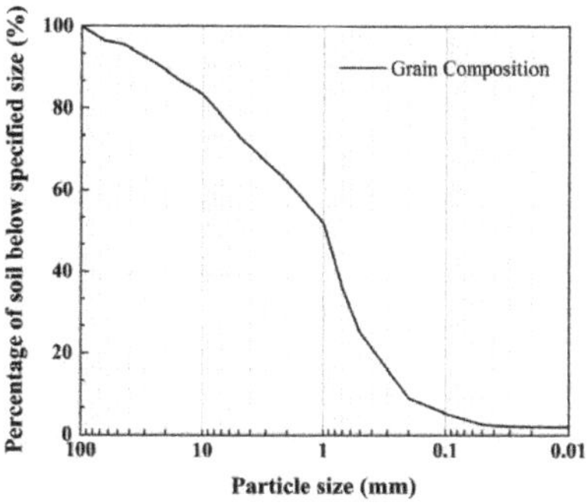

Figure 2. Particle size curve.

2.1.2. Stabilizing Agent Materials

Cement

This experiment selected P.O42.5R ordinary Portland cement produced by Hongshi Cement Co., Ltd. in Jiangyou City, Sichuan Province. Its chemical composition and main physical property indicators are as Table 4.

Table 4. Chemical components of P.O42.5R cement (%).

Material Name	CaO	SiO_2	Al_2O_3	Fe_2O_3	MgO	SO_3	Na_2O	K_2O
C ontent	61.34	23.61	6.45	3.76	1.92	2.17	0.33	0.42

Basic Properties of Stabilizers

This experiment employed a novel soil stabilizer independently developed by Southwest University of Science and Technology. The stabilizer consists of stabilizing liquid, water-reducing powder, and cement (as shown in Figure 3), abbreviated as CLP stabilizer, which is mixed with cement in a specified proportion. The stabilizing liquid agent comprises 20–35% aqueous epoxy resin, 4–8% isocyanates, 0.1–2% lignin fibers, 15–23% neutral water glass, 10–18% acrylamide, 0.8–3% sodium polycarboxylate, 1–2% anionic fibers, 0.2–0.4% magnesium oxide, and 150–200% water. The water-reducing powder is a gray powdery solid made up of 20–41% potassium chloride, 4–10% sodium sulfide, 3–7% magnesium chloride, 16–26% calcium chloride, 3–5% polyanionic cellulose, and 0.3–2% desulfurized ash. Its core function lies in enhancing the density of the compacted soil, reducing the soil's water permeability and simultaneously strengthening the soil's compressive strength and water resistance.

2.2. Mix Proportion Design of Water-Based Drill Cuttings Stabilized Soil

2.2.1. Preliminary Selection of Mix Proportion

Taking the optimal moisture content, maximum dry density, and 7-day unconfined compressive strength of the specimens as evaluation indicators, 9 sets of stabilizer mix ratio

 https://doi.org/10.3390/coatings16040406

schemes, as shown in Table 5, were designed according to relevant test specifications [22], aiming to explore the influence law of cement, stabilizer, and water-reducing powder dosage on the performance of stabilized soil.

(**a**) Stabilizer liquid agent. (**b**) Water-reducing powder agent.

Figure 3. CLP stabilizer.

Table 5. Mix ratio design.

Test Number	Cement Content (%)	Stabilizer Liquid Agent (%)	Water Reducing Powder (%)
GH1	4	0	0
GH2	5	0	0
GH3	6	0	0
GH4	4	1	6
GH5	5	1	6
GH6	6	1	6
GH7	4	2	8
GH8	5	2	8
GH9	6	2	8

2.2.2. Determine the Maximum Dry Density and the Optimal Moisture Content

The maximum dry density and the optimal moisture content of the stabilized soil are determined using the heavy compaction (Class II-2) method. After the curing of the specimens was completed, a 7-day unconfined compressive strength test was conducted to evaluate the maximum ability of the material to resist axial compressive stress.

Three test samples are prepared for each experimental configuration, with the final test results analyzed using the average of valid data [23]. The experimental results are shown in Figure 4. When only cement is added, the maximum dry density of stabilized soil is relatively low, while the optimal moisture content is relatively high. The 7-day unconfined compressive strength of GH1, GH2, and GH3 groups is 1.12 MPa, 1.18 MPa, and 1.27 MPa, respectively. Under constant cement dosage, as the CLP stabilizer dosage increases, the optimal moisture content shows a downward trend, whereas the maximum dry density first increases and then decreases. However, the 7-day unconfined compressive strength increases in all cases (comparisons between GH4 and GH7, GH5 and GH8, GH6 and GH9). When the cement dosage is 5%, stabilizer dosage is 2%, and superplasticizer dosage is 8%, the maximum dry density reaches 1.919 g/cm^3, the optimal moisture content is 17.6%, and the 7-day unconfined compressive strength reaches its maximum value of 2.60 MPa. The compressive strength of stabilized soil with the addition of stabilizer is significantly higher than that of pure cement-stabilized soil, indicating a linear relationship between the stabilizer dosage and the 7-day unconfined compressive strength. The strength

https://doi.org/10.3390/coatings16040406

growth rate of stabilized soil is the fastest when the stabilizer dosage is increased from 0% to 1% stabilizer liquid agent plus 6% water reducing powder. As the stabilizer dosage continues to increase, the strength growth rate gradually slows down. This phenomenon suggests that a critical dosage of CLP stabilizer exists in the modification process of water-based drilling cuttings. Once this critical value is exceeded, the strength of the stabilized soil becomes difficult to further improve.

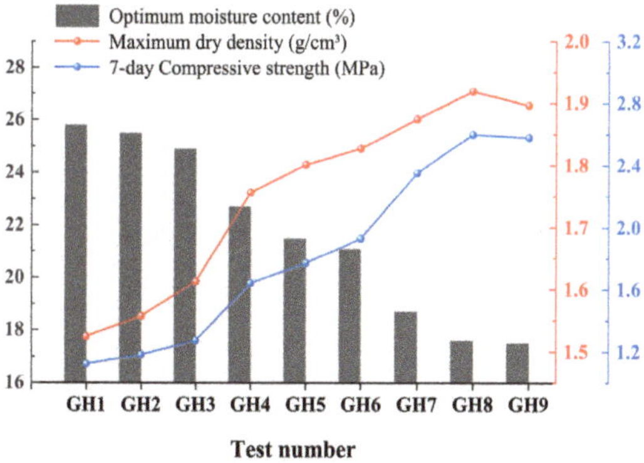

Figure 4. Test results of samples with different mix ratios.

2.2.3. Optimized Design of Mix Proportion

Based on the analysis of the mix proportion test, compaction test, and 7-day unconfined compressive strength test results, it can be known that the stabilizer liquid agent and water-reducing powder agent have the most significant impact on the strength of water-based drill cuttings stabilized soil, while excessive cement content will have an adverse effect on the stability of the stabilized soil. According to the technical requirements of "Application standard of soil stabilizer" (CJJ/T 286-2018) [24] and "Technical guidelines for construction of highway roadbases" (JTG/T F20-2015) [25], the 7-day unconfined compressive strength of the road base in the pre-drilling project should reach 2.5 MPa to 3.0 MPa. Taking into account factors such as strength performance, economy, and construction requirements comprehensively, the final optimal mix ratio was determined to be 5% cement + 2% stabilizer liquid agent + 8% water-reducing powder.

3. Experimental Study on the Performance of Water-Based Drill Cuttings Stabilized Soil Roads

Using the powdery soil from the rear hill of Southwest University of Science and Technology as the control group, this study compared the road performance differences between powdery stabilized soil and water-based drill cuttings stabilized soil, with 10 test groups designed as shown in Table 6. The CLP stabilizer dosage ratio is 5% cement + 2% stabilizer + 8% water-reducing powder, where the percentages represent the proportion of each component relative to the total soil sample volume.

3.1. Unconfined Compressive Strength Test

To study the variation law of the unconfined compressive strength of water-based drill cuttings stabilized soil under different factors, the soil sample ratio was determined based on the optimal mix ratio in the previous text. According to the specification requirements [26], four standard curing ages of 7 days, 14 days, 28 days, and 90 days

 https://doi.org/10.3390/coatings16040406

are set. On the last day of each curing age, the specimens are immersed in water for 24 h, and then the unconfined compressive strength test is carried out.

Table 6. The dosage scheme of the soil sample solidification test.

Soil Sample Classification	Test Number	Soil Sample Dosage (%)	Stabilizer Dosage (%)
Powdery soil	FGH1	100	0
	FGH2	100	3
	FGH3	100	6
	FGH4	100	9
	FGH5	100	12
Water-based drill cuttings	ZGH1	100	0
	ZGH2	100	3
	ZGH3	100	6
	ZGH4	100	9
	ZGH5	100	12

The final failure mode of stabilized soil is shown in Figure 5a. During the test, the failure process of water-based drill cuttings stabilized soil presented typical brittle failure characteristics, which are specifically divided into the integrity stage, the crack propagation stage, and the fracture failure stage. The test results are shown in Figure 5b. Under the same curing age and dosage of stabilizer, the strength of water-based drill cuttings stabilized soil is much higher than that of powdery stabilized soil. When the dosage of stabilizer is 9%, the strength reaches the maximum value. After that, with the increase in the dosage of stabilizer, the strength of the stabilized soil decreases. With the extension of the curing period, the strength of the stabilized soil continues to increase. This occurs because when water-based drilling cuttings fully interact with CLP stabilizer, the hydration reaction of the mixture produces increasing amounts of substances such as alunite, hydrated calcium aluminate, and hydrated calcium silicate as the stabilizer dosage increases. Consequently, the unconfined compressive strength of water-based drill cuttings stabilized soil improves, while its cementation effect, filling capacity, and soil particle compaction capability become more pronounced.

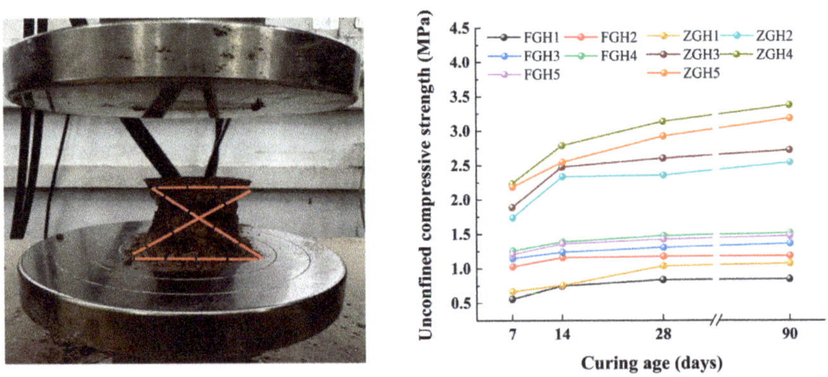

(**a**) Failure mode of stabilized soil specimen. (**b**) Curve of unconfined compressive strength with age.

Figure 5. The final failure mode and test results of stabilized soil.

Among them, the growth rate is the fastest from 7 days to 14 days. At 90 days, the strength of water-based drill cuttings stabilized soil reaches 3.38 MPa, with an increase rate of 51%. In addition, the strength increase in water-based drill cuttings during the

https://doi.org/10.3390/coatings16040406

curing period is significantly higher than that of powdery stabilized soil. It can be seen from this that a high dosage of stabilizer can more effectively enhance the compressive performance of stabilized soil, and its enhancing effect is more prominent compared to that of a low dosage of stabilizer. Therefore, effectively controlling the dosage of stabilizer and optimizing the curing process to promote the full progress of hydration reaction are of great significance for further enhancing the compressive strength of the stabilized soil.

3.2. Splitting Tensile Strength Test

This test prepared samples in accordance with relevant regulations and conducted splitting tensile strength tests at different ages [27]. A 6.35 mm pressure strip was used to load at a rate of 1 mm/min until the specimen reached the ultimate stress and failed. The maximum pressure value at the time of specimen failure was recorded, and then the splitting strength was calculated. The failure mode of the specimen is shown in Figure 6a.

(**a**) Failure mode of the specimen. (**b**) Influence of curing age on the splitting strength.

Figure 6. The final failure mode and test results of stabilized soil.

The variation laws of splitting tensile strength of water-based drill cuttings stabilized soil and powdery stabilized soil at different curing ages are shown in Figure 6b. Based on the variation law of the unconfined compressive strength of different specimens with age in the previous text, it can be known that the development trend of the splitting tensile strength is similar to that of the unconfined compressive strength. Whether it is water-based drill cuttings stabilized soil or powdery stabilized soil, with the increase in age, within the period of 7 days to 28 days, the strength growth curve is relatively steep, and the splitting strength increases significantly. During the period of 28 days to 90 days, the growth trend of the curve gradually slows down, but it still maintains growth. For the powdering stabilized soil without stabilizer, the splitting strengths at the ages of 7 days, 14 days, 28 days, and 90 days were 0.124 MPa, 0.146 MPa, 0.149 MPa, and 0.153 MPa, respectively. The splitting strengths of water-based drill cuttings stabilized soil without stabilizer reached 0.253 MPa, 0.268 MPa, 0.269 MPa, and 0.271 MPa, respectively, during the corresponding curing time. This indicates that when the CLP stabilizer is not added, the overall splitting tensile strength of the stabilized soil is relatively low with the increase in age. However, after the addition of stabilizers, although the splitting strength of powdery stabilized soil has increased, it is still far lower than that of water-based drill cuttings stabilized soil. It can be seen that the curing age has a very significant effect on the improvement of splitting strength.

As can be seen from Figure 7, the splitting tensile strength gradually increases with the increase in stabilizer dosage and then tends to stabilize. Under the condition that the filler

 https://doi.org/10.3390/coatings16040406

is water-based drill cuttings, when the curing age was extended from 7 days to 90 days, and the dosage of CLP stabilizer was 3%, the splitting strength increased from 0.478 MPa to 0.641 MPa, with an increase of 34.1%. When the dosage was 6%, the splitting strength increased from 0.541 MPa to 0.795 MPa, with an increase of 47.0%. When the dosage was 9%, the splitting strength increased from 0.553 MPa to 0.852 MPa, with an increase of 54.1%. When the dosage was 12%, the splitting strength increased from 0.572 MPa to 0.901 MPa, with an increase of 60.3%. This indicates that the dosage of the stabilizer has a significant impact on the improvement of the splitting strength.

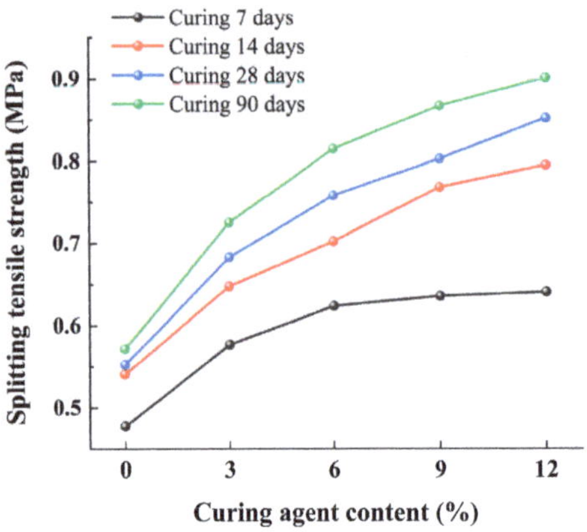

Figure 7. Relationship between stabilizer content and splitting tensile strength.

Based on the above data analysis, it can be known that after the water-based drill cuttings filler is added with a stabilizer, a series of physical and chemical reactions will occur inside the stabilized soil during the mixing and compaction process, promoting the formation of cementing substances between the mixtures. These cementing substances enhance the agglomeration of stabilized soil, making the soil structure more compact. Meanwhile, the ettringite crystals produced by the cement in the CLP stabilizer exert a mechanical interlocking effect [28]. However, excessive stabilizers will generate a considerable amount of heat, causing local expansion or cracking of the soil and reducing the toughness of the soil particles. Therefore, the selection of an appropriate amount of stabilizer plays a significant role in enhancing the splitting strength of water-based drill cuttings stabilized soil, and it plays a dominant role in the later strength improvement.

3.3. Compressive Rebound Modulus Test

The test results obtained by the method shown in Figure 8a are presented in Figure 8b. When subjected to external forces, the anti-deformation ability of water-based drill cuttings stabilized soil is significantly better than that of powdery stabilized soil [29]. During the curing process, the internal restraint effect of water-based drill cuttings stabilized soil is enhanced, specifically manifested as a reduction in elastic deformation and an increase in the rebound modulus. At the 7-day curing age, there is no significant difference in the rebound modulus between the stabilized soil without the addition of stabilizer and that with the addition of stabilizer. This might be due to the fact that the early hydration reaction had not been fully carried out, the internal framework structure of the material had not been fully formed, and the strengthening effect of the stabilizer had not been fully manifested. However, as the curing period extends, especially after 28 days or longer, the role of the stabilizer gradually becomes prominent, making the internal structure of the stabilized soil denser and the cementing

 https://doi.org/10.3390/coatings16040406

firmer, and the gap in the rebound modulus between the two gradually widens. This result fully demonstrates that the addition of stabilizers can effectively increase the compressive rebound modulus of stabilized soil and enhance its ability to resist deformation.

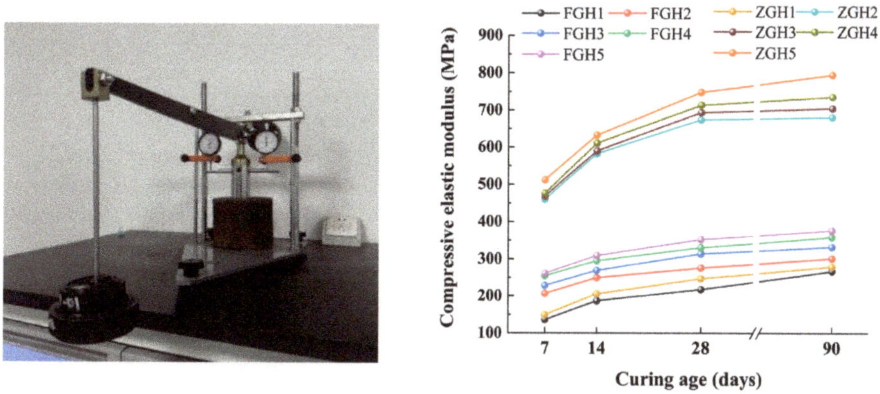

(**a**) Lever pressure gauge. (**b**) Relationship between compressive resilience modulus and age.

Figure 8. Test and result of compression and rebound.

Figure 9 presents the relationship curves between unit pressure and rebound deformation of four groups of stabilized soil when no stabilizer is added and when the stabilizer content is 12%, as well as the relationship graph of the rebound modulus of stabilized soil with age. Under the action of vertical loads, as the curing time increases, the rebound deformation of the specimens gradually decreases. This indicates that the water-based drill cuttings cured subgrade structure with the addition of stabilizer has a stronger elastic deformation capacity and a higher compressive strength of the subgrade.

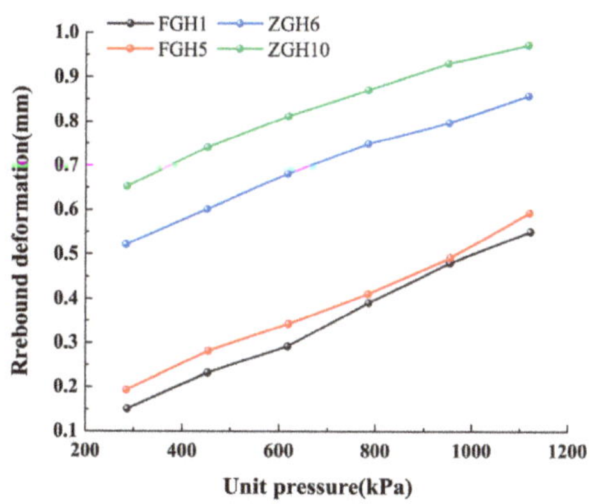

Figure 9. Relationship between unit pressure and rebound deformation.

3.4. Water Stability Test

To study the water stability of water-based drill cuttings stabilized soil and powdery stabilized soil under different ratios [30], and to verify the curing effect of the stabilizer at the same time, two groups of specimens were prepared according to the test scheme shown in Table 7. A group of specimens was cured under standard curing conditions (temperature 20 °C $\pm$ 2 °C, humidity $\geq$ 95%), and then unconfined compressive strength

https://doi.org/10.3390/coatings16040406

tests were carried out. Another group of specimens was cured under the same standard curing conditions until the last day and then taken out and immersed in water for 24 h before undergoing an unconfined compressive strength test.

Table 7. Water stability test scheme.

Maintenance Days	7	14	28
R_0	7 days standard curing	14 days standard curing	28 days standard curing
R_W	6 days standard curing + 1 day soaking	13 days standard curing + 1 day soaking	27 days standard curing + 1 day soaking

The test results are calculated as shown in Figure 10. Under different curing ages, the CLP stabilizer reacts with water-based drill cuttings, generating insoluble hydration products such as calcium silicate hydrate with "hydrophobicity". These reaction products keep increasing, filling the pores of the stabilized soil, reducing the influence of water on the stabilized soil, enhancing the compactness of the stabilized soil, and preventing it from collapsing when exposed to water. The water stability of water-based drill cuttings stabilized soil continuously improves with the increase in stabilizer dosage and age. The water stability coefficient shows an upward trend and gradually stabilizes, ultimately meeting the relevant specification requirements in the "Application standard of soil stabilizer" (CJJ/T 286-2018) [24]. However, the test results of powdery stabilized soil are exactly the opposite. The electrostatic effect carried by water molecules replaces the adsorption force on the surface layer of powder-stabilized soil particles, resulting in a decrease in the compactness between particles, making the soil loose and weak, with poor water stability, and failing to meet the performance requirements for road use.

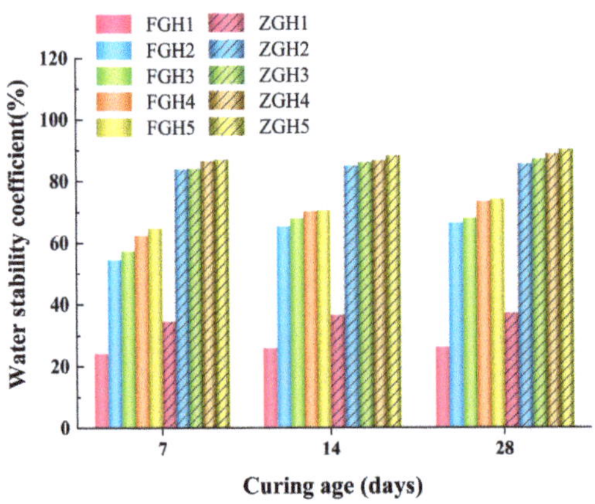

Figure 10. Water stability coefficient of stabilized soil at different curing ages.

3.5. Temperature Shrinkage Performance Test

In order to evaluate the crack resistance capacity of subgrade fill materials, small beam specimens with a density of 96% and dimensions of 50 mm × 50 mm × 200 mm are prepared in accordance with relevant test procedures. After drying the specimens to a constant weight, they were standard-cured for 7 days. The temperature shrinkage performance test was carried out in a high- and low-temperature alternating test chamber with a gradient of 10 °C within the temperature range of 40 °C to −30 °C [31], as shown in Figure 11.

https://doi.org/10.3390/coatings16040406

Figure 11. Temperature shrink test specimen.

The curve of the temperature shrinkage coefficient varying with the temperature range drawn from the test results of the temperature shrinkage test is shown in Figure 12.

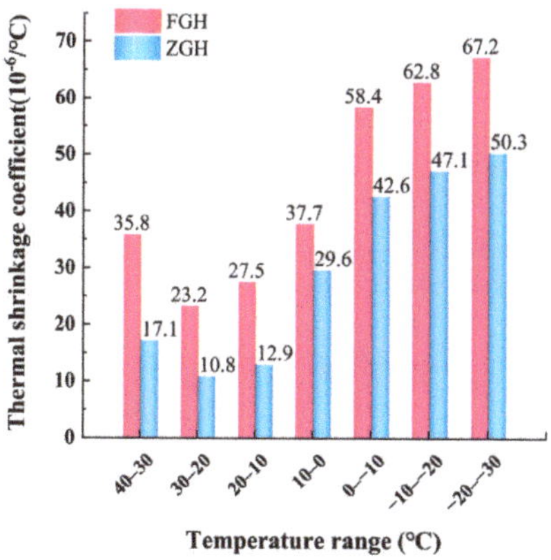

Figure 12. Comparison of the temperature shrinkage performance of the two stabilized soils.

During the process, when the temperature drops from 40 °C to −20 °C, the temperature shrinkage performance of water-based drill cuttings stabilized soil is superior to that of powdery stabilized soil. Specifically, it is characterized by a lower temperature shrinkage coefficient, smaller shrinkage deformation, stronger resistance to temperature shrinkage, and the ability to effectively reduce volume shrinkage caused by temperature changes.

The temperature shrinkage coefficients of water-based drill cuttings stabilized soil and powdery stabilized soil show a similar pattern with temperature variation: when the temperature drops from 40 °C to 0 °C, the temperature shrinkage coefficients fluctuate within a certain range, and the overall variation amplitude is relatively small. When the temperature further drops to −30 °C, the coefficient of temperature shrinkage rises rapidly. This is mainly because when the temperature drops below freezing point, water freezes into ice and expands in volume, and the coefficient of temperature contraction of ice is relatively large. In addition, the effect of the stabilizer weakens the influence of temperature changes on water-based drill cuttings stabilized soil. Its active groups form a randomly distributed grid structure through chemical reactions, effectively inhibiting the severe shrinkage caused by the drop in temperature.

Under the same temperature conditions, the temperature shrinkage coefficient of water-based drill cuttings stabilized soil is significantly lower than that of powdery stabilized soil, indicating that it has a stronger resistance to temperature shrinkage deformation and can more effectively resist the shrinkage effect caused by temperature changes. This is mainly because the smaller the coefficient of temperature shrinkage, the smaller the shrinkage deformation of the material when the temperature changes, thereby reducing the structural stress concentration caused by temperature differences and lowering the risk of cracks and damage.

https://doi.org/10.3390/coatings16040406

4. Microscopic Mechanism and Environmental Impact Analysis

4.1. Microscopic Mechanism Analysis

4.1.1. XRD Analysis

The DMAX-1400 X-ray diffractometer (Rigaku, Tokyo, Japan) was employed to analyze the phase composition and relative crystalline content of both powdery stabilized soil and water-based drilling cuttings stabilized soil [18]. Using MDI Jade 6.0 software to interpret the diffraction patterns, the phase identification was performed by cross-referencing peak positions and relative intensities against the ICDD/PDF standard diffraction database, ultimately yielding the phase composition and relative content results of the specimens.

The mineral composition and relative content of powdery stabilized soil are shown in Figure 13a. The primary mineral component is quartz (SiO_2), along with feldspar and calcite ($CaCO_3$). After adding the stabilizer, the characteristic diffraction peaks of these minerals did not disappear. In a strongly alkaline environment, the hydration reactions in the stabilizer break, generating amorphous calcium silicate hydrate (C-S-H) gel. However, no characteristic peaks of ettringite (AFt phase) and $Ca(OH)_2$ were detected in the XRD pattern, indicating minimal formation of hydration reaction products such as C-S-H gel, ettringite (AFt phase), and $Ca(OH)_2$. This resulted in a loose structure and insufficient cementation of the cured soil, ultimately leading to adverse effects, including low strength, poor hydraulic stability, and susceptibility to shrinkage and cracking. This phenomenon is attributed to the fact that SiO_2 and CaO in the powdery stabilized soil fail to fully dissociate during the hydration process of the solidifying agent, resulting in insufficient OH^- concentration and the inability to effectively release Ca^{2+} and silicate ions, thereby inhibiting the formation of calcium silicate hydrate, calcium aluminate hydrate, and $Ca(OH)_2$ crystals.

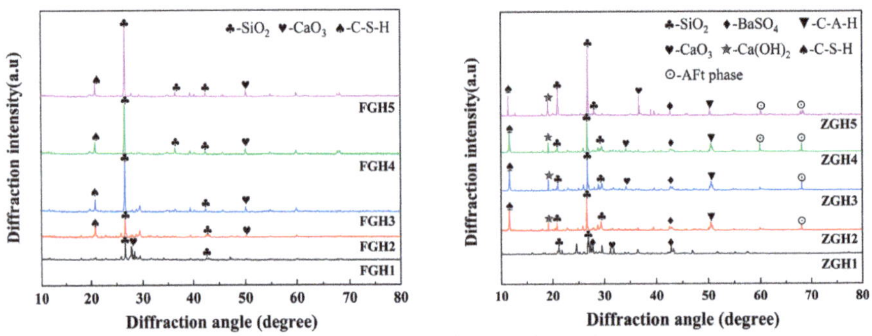

(a) FGH. (b) ZGH.

Figure 13. XRD results of different stabilized soils.

The XRD analysis of water-based drill cuttings stabilized soil (ZGH) is shown in Figure 13b. The addition of the solidifying agent triggered a series of physical and chemical changes, including hydration reaction, filling effect, carbonization, micro-expansion phenomenon, and ion exchange. In CLP stabilizer, cement clinker rapidly hydrates upon contact with water, forming $Ca(OH)_2$ and C-S-H gel while releasing high concentrations of Ca^{2+}, OH^-, and AlO_2^- ions. These ions react with water to form needle-like calcite crystals. In water-based drilling cuttings, active Al_2O_3 and SiO_2 are activated under high alkaline conditions, releasing Al^{3+} ions that serve as an aluminum source for calcite formation, thereby further promoting extensive calcite precipitation [32]. The gel encapsulates internal particles within the structure, forming an interlocking network that enhances the

 https://doi.org/10.3390/coatings16040406

compressive strength, splitting tensile strength, and deformation resistance of the structure. The ettringite (AFt phase) fills the pores, increasing structural density, reducing structural shrinkage, and improving water stability.

With increased curing agent dosage, the hydration products increase, leading to enhanced diffraction peak intensities for $CaCO_3$, ettringite (AFt phase), hydrated calcium silicate (C-S-H), and hydrated calcium aluminate (C-A-H). The resulting structure demonstrates superior mechanical properties and durability.

The increase in the dosage of stabilizer is positively correlated with the characteristic peak intensity of gelling products and swelling products. The main expansive substance, ettringite (AFt phase), forms stable calcium hydroxide by adsorbing calcium ions in the aqueous phase. This process not only provides an alkaline environment for the continuous formation of C-A-H and C-S-H gels but also its volume expansion effect, which can bridge the particle gaps and promote the formation of a dense structure within the material. This dual mechanism, through the synergistic effect of pore filling and volume micro-expansion, enhances the mechanical and durability performance indicators of the curing system.

4.1.2. Microscopic Scanning Electron Microscopy Analysis

To explore the mechanism of action and microstructure changes of CLP stabilizer on water-based drill cuttings stabilized soil and powdery stabilized soil, referring to relevant research methods [33,34], the morphology, distribution density, and growth status of the products were qualitatively assessed using a TM-4000 Zeiss scanning electron microscope (Hitachi, Tokyo, Japan) at magnification levels of 1000 times, 2500 times, and 5000 times.

As shown in Figure 14, at 1000 times magnification, the particles of powder stabilized soil are in a discrete accumulation state with clear boundaries between particles. A large number of connected large pores exist, and no continuous cementing structure is formed, resulting in a loose and porous structure overall. This is the main reason for the low strength, poor water stability, and insufficient bearing capacity of the stabilized powder soil. At 2500 times magnification, only a small amount of flaky $Ca(OH)_2$ crystals and trace flocculent C-S-H gel can be observed. The cementing products are sparsely distributed and fail to form effective wrapping and overlapping between particles. At 5000 times magnification, there is no obvious interwoven network structure at the particle interface. The hydration products are extremely scarce and unevenly distributed, which cannot achieve effective cementation between particles and pore filling. On the whole, the powder stabilized soil group has a single type of hydration products, an extremely low yield, and poor structural compactness.

At 1000 times magnification, the surface of soil particles in the water-based drill cuttings stabilized soil is completely covered by a large amount of flocculent C-S-H gel. The gel bonds discrete particles tightly into an integral whole, the number and size of pores are significantly reduced, and the structural integrity is greatly improved. At 2500 times magnification, abundant punctate C-A-H gel and rod-like ettringite (AFt phase) are generated in the system. The gel and acicular crystals interpenetrate and overlap with each other, initially forming a three-dimensional framework structure with a remarkable pore-filling effect. At 5000 times magnification, C-S-H gel, C-A-H gel, and ettringite (AFt phase) crystals coexist densely and interweave sufficiently, filling micropores and microcracks completely with no obvious connected pores, resulting in a dense and continuous microstructure.

 https://doi.org/10.3390/coatings16040406

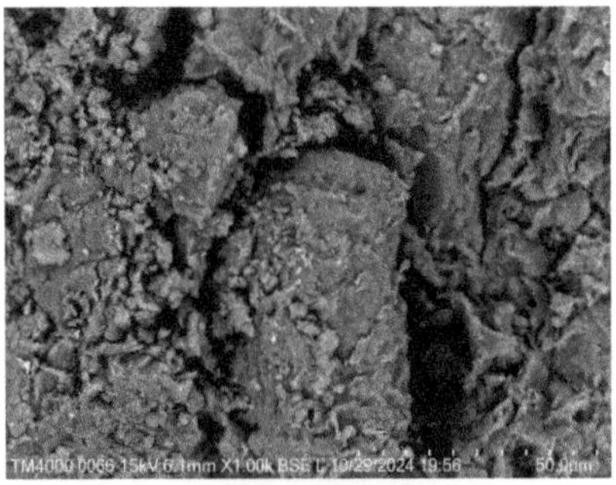

(**a**) FGH, 1000 times.

(**b**) ZGH, 1000 times.

(**c**) FGH, 2500 times.

(**d**) ZGH, 2500 times.

(**e**) FGH, 5000 times.

(**f**) ZGH, 5000 times.

Figure 14. SEM morphology comparison of powder-stabilized soil and water-based drill bit-stabilized soil.

Based on the above analysis results, the amount of hydration products in the ZGH group is much higher than that in the FGH group. The main reasons are as follows: the flocculent C-S-H gel, as the dominant cementing phase, is produced in a large quantity, which provides bonding force and structural strength for particles inside the system; a great number of rod-like ettringite (AFt phase) crystals are formed, which exert the effects of pore filling and micro-expansion, thus improving the compactness and deformation resistance of the structure; the punctate C-A-H gel assists in cementation and enhances the toughness of the matrix; and the generated $CaCO_3$ crystals fill the pores and strengthen the chemical stability of the structure. The products of different reactions act synergistically, transforming the structure of stabilized soil from loose to dense and significantly improving the mechanical properties and road performance of the material.

https://doi.org/10.3390/coatings16040406

4.2. Environmental Impact Analysis

To evaluate whether water-based cuttings resource utilization meets environmental requirements, the pH levels of leachate from drilling cuttings and surrounding soil were measured, and metal content was analyzed in accordance with the "Leaching Method for Toxicity of Solid Wastes-Sulfuric Acid and Nitric Acid Method" (HJ/T299-2007) [35]. Comparative analysis was conducted against the basic water quality requirements for surface water subcategories specified in the national standard "Surface Water Environmental Quality Standards" (GB3838-2002) [20] to assess compliance with environmental protection standards for water-based cuttings leachate.

The test samples were dried at 105 °C, ground, and sieved through a 200-mesh sieve. Pre-treatment was performed according to the microwave digestion method specified in Table 8. After microwave digestion, the digestion solution was diluted with ultrapure water to a final volume of 50 mL for testing.

Table 8. Pre-treatment methods of microwave digestion (unit: mL).

Pre-Treatment with HNO_3	Pre-Treatment Conditions	HNO_3	HCI	HF	H_2O_2	Conditions for Acid Reflux
5	Heated at 165 °C for 20 min	6	3	3	1	Acid removal at 175 °C for 35 min

4.2.1. Analysis of Water-Based Cuttings Leachate

The test results are shown in Table 9. The pH value of the water-based drill cuttings leachate was measured to be 8.64. This is because the main curing component in the stabilizer is an alkaline substance. No heavy metals, Co, V, or Pb, were detected in the leachate. The contents of As, Ba, Mn, Fe, Ni, Zn, Cr, Cu, Cd, Ag, and Be all comply with the first-level standards in Table 3 of the "Technical specification for co-processing of solid waste in cement kiln" (GN30760-2024) [36], the "Integrated wastewater discharge standard" (GB 8978-1996) [19], and the "Technical specifications of water-based drilling waste treatment and resource utilization for onshore oil & natural gas exploitation" (SY/T 7466-2020) [37].

Table 9. Test results of heavy metals and petroleum leaching solution in water-based drilling cuttings.

Surveillance Project	Water-Based Cuttings	GB30760-2024 Table 3	GB8978 Level 1 Standard	SY/T7466-2020 Table 1
pH	8.64	-	6~9	6~9
As	0.05	0.1	0.5	0.5
Ba	0.85	-	-	10.0
Mn	<0.01	1.0	2.0	-
Fe	0.22	-	-	-
Co	ND [1]	-	-	-
Ni	<0.02	0.2	1.0	1.0
Zn	<0.01	1.0	2.0	2.0
V	ND [1]	-	-	-
Cr	0.07	0.2	1.5	1.5
Cu	0.03	1.0	0.5	0.5
Pb	ND [1]	0.3	1.0	1.0
Cd	<0.01	0.03	0.1	0.1
Ag	<0.01	-	0.5	0.5
Be	<0.004	-	0.005	-
Petroleum	0.07	-	5	5

[1] "ND" indicates a test result below the detection limit.

https://doi.org/10.3390/coatings16040406

Therefore, the structural layer materials containing rock cuttings can withstand the infiltration and erosion of rainwater for a long time. After the leachate seeps into the soil, it will not cause the accumulation of heavy metals or the adsorption of organic toxins, nor will it inhibit the activity of microorganisms or damage soil fertility. It has no significant impact on the heavy metal content in groundwater or surface water quality.

4.2.2. Background Soil Heavy Metal Analysis

Most well sites are surrounded by farmland, and the background soil around the well sites is also one of the raw materials for structural layer materials. To assess the environmental pollution risks during the utilization of water-based drill cuttings stabilized soil structural layer materials, it is necessary to conduct detection and analysis of the background soil around the well site [38]. The detection results of heavy metal content in the background soil around the well site are shown in Table 10.

Table 10. Test results of heavy metals and petroleum leaching solution in water-based drilling cuttings.

Surveillance Project	Background Soil Surface	Background Soil Layer	GB15618-2018 Stricter Value in Filter	GB36600-2018 Stricter Value in Filter	DB51/2978-2023 Stricter Value in Filter
As	10.24	6.75	20	20	20
Hg	0.108	0.038	1.0	8	8
Ba	508	636	-	-	2766
Fe	35,674	28,013	-	-	-
Mn	822	788	-	-	3593
Co	16.0	15.6	-	20	20
Ni	51.9	39.2	190	150	150
Zn	126.4	83.3	300	-	-
V	93.4	77.2	-	165	-
Cr	75.6	56.8	250	150	1202
Cu	26.9	22.8	100	2000	2000
Sb	1.20	1.06	-	20	20
Pb	30.6	25.8	170	400	-
Cd	ND [1]	ND [1]	0.6	20	20
Ag	0.9	0.9	-	-	-
Se	ND [1]	ND [1]	-	-	243
Be	1.57	1.46	-	15	15
Tl	ND [1]	ND [1]	-	-	1.0
Ti	ND [1]	ND [1]	-	-	-

[1] "ND" indicates a test result below the detection limit.

According to Table 10, in terms of the total amount of heavy metals, no heavy metals, Cd, Se, T1, or Ti, were detected in the surface and middle layers of the background soil around the well site. Among them, the detection results of As, Hg, Ba, Fe, Mn, Co, Ni, Zn, V, Cr, Cu, Sb, Pb, Cd, Ag, and Be were all relatively low. The test results do not exceed the "Soil environmental quality risk control standard for soil contamination of agricultural land" (GB15618-2018) [39] or the "Risk control standard for soil contamination of development land in Sichuan Province" (DB51 2978-2023) [40]. The limit values were stipulated. This indicates that under extreme conditions, the content of heavy metals in the background soil around the field remains at a relatively low level. When the concentration of heavy metals in the soil is equal to or lower than the standard limit, the impact on human health can be ignored. If the limit is exceeded, it may pose health risks. Further detailed

https://doi.org/10.3390/coatings16040406

investigations and risk assessments are required to clarify the extent of pollution and the degree of risk [41].

The research finds that the content of heavy metals in the soil structure layer of the well site area has significant spatial heterogeneity, and the distribution varies greatly in different regions. After treating water-based drill cuttings with improved curing technology and using them as pavement base materials, during the curing process, heavy metals are fixed in a stable matrix, and their migration ability is significantly reduced, making them less likely to seep into groundwater or spread to the surrounding background soil due to precipitation under long-term environmental effects. Therefore, this treatment method effectively reduces the risk of heavy metal pollution, and the environmental risk is within a controllable range. Meanwhile, the application of solidified fillers has enhanced the resource utilization rate of drilling waste, achieving the reduction, harmless treatment, and resource utilization of waste.

5. Conclusions

This study systematically investigated the engineering feasibility of utilizing water-based drill cuttings from Suining, Sichuan Province, as pavement base materials through mix design optimization, performance evaluation, microstructural analysis, and environmental assessment. The main conclusions are summarized as follows:

(1) The water-based drill cuttings exhibit an optimum moisture content of 20.2%, a maximum dry density of 1.91 g/cm^3, a liquid limit of 34.8%, a plastic limit of 15.5%, a plasticity index of 19.3%, and a specific gravity of 2.96, classifying them as low liquid limit soil suitable for subgrade filling. The optimal mix proportion determined by orthogonal design is 5% cement + 2% CLP stabilizer + 8% water-reducing powder. Under this condition, the 7-day unconfined compressive strength of specimens increased by 51%; when the stabilizer dosage is 12%, the splitting tensile strength increased by 60.3%. Excellent mechanical properties, deformation resistance, and hydraulic stability all meet the technical requirements for pre-drilling engineering pavement base layers.

(2) Microstructural analyses using XRD and SEM revealed that hydration of the CLP-stabilized drill cuttings produces cementitious gels (C-A-H and C-S-H) and expansive ettringite (AFt). The flocculent gels interwoven with needle-like AFt form a dense three-dimensional network structure, enhancing particle bonding, reducing pore space, and improving overall compactness and mechanical strength. The rational properties of stabilized soil are, therefore, significantly improved.

(3) The heavy metal concentrations in the surrounding background soil meet the soil pollution risk control standards for agricultural and construction land. Stabilization treatment effectively immobilized heavy metals within the matrix, reducing their leaching potential and controlling environmental risks, thereby achieving the reduction, harmless treatment, and resource utilization of drilling waste.

This study focused on evaluating the road performance of water-based drill cuttings stabilized soil for application in pre-drilling engineering pavement bases and medium- to light-traffic highways. Long-term durability performance under freeze–thaw and wet–dry cycles was not investigated, and its applicability to heavy-traffic highway base courses was not further assessed. Future research should, therefore, examine long-term durability and optimize mix design to verify its suitability for higher-grade pavement structures.

Author Contributions: S.T.: project administration and conceptualization. H.W.: writing—original draft and supervision. H.T.: resources and data curation. W.F.: visualization; X.G.: investigation and

https://doi.org/10.3390/coatings16040406

formal analysis. B.B.: methodology. J.W.: writing—review and editing. X.T.: writing—original draft and visualization. All authors have read and agreed to the published version of the manuscript.

Funding: This study was funded by the CNPC Key Core Technology Research Project: Research on Key Technologies and Equipment for the Resource Utilization of Drilling waste. (No. 2025ZG73).

Institutional Review Board Statement: Not applicable.

Informed Consent Statement: Not applicable.

Data Availability Statement: The data presented in this study is available upon request from the corresponding author.

Acknowledgments: The authors gratefully acknowledge the anonymous reviewers and the editor for their constructive comments and valuable suggestions, which have significantly improved the quality and clarity of this manuscript.

Conflicts of Interest: Authors Shucheng Tan, Hua Tang, Wentao Fu, Xiaoyan Guo, Biaotian Bai were employed by the company CNPC Chuanqing Drilling Engineering Company Limited and CCDC SHUYU Engineering Construction Co., Ltd. The remaining authors declare that the research was conducted in the absence of any commercial or financial relationships that could be construed as a potential conflict of interest.

References

1. Liu, Y.C.; Yang, J.; Chen, M.Y.; Yan, X.; Wang, Z.M. Water-based drill cuttings, a shale gas extraction waste as supplementary cementitious material and optimization. *Constr. Build. Mater.* **2022**, *346*, 128419. [CrossRef]
2. Zhou, X.; Wang, G.; Tang, G.; Jia, C.; Zhang, H. Microbial Synergistic Active Magnesium Oxide Solidification of Water-Based Drilling Cuttings and Analysis of Its Durability Strength Loss Mechanism. *ACS Sustain. Chem. Eng.* **2024**, *12*, 11968–11979. [CrossRef]
3. Qin, P.F. Experimental study on reinforcement treatment of coastal silty soft soil. *Sci. Rep.* **2026**, *16*, 7688. [CrossRef]
4. Babatunde, Q.O.; Byun, Y.H. Soil Stabilization Using Gum Arabic Biopolymer. *Biopolymers* **2026**, *117*, e70077. [CrossRef]
5. Li, G.; Ren, Z.F.; Liu, J.K.; Chen, H.; Wen, Z.; Fu, C.Q. Enhancing coastal saline dispersive soil with ionic stabilizer-cement synergy: Mechanisms and performance. *Case Stud. Constr. Mater.* **2025**, *23*, e05519. [CrossRef]
6. Zhong, H.J.; Zhao, J.H.; Li, H.D.; Wang, N.N.; Zhao, X.; Zhang, L.; Fei, Y.L.; Quan, J. Eco-friendly controlled low strength material by construction muck: A systematic study from rheological properties, mechanical performance and microstructure. *J. Build. Eng.* **2025**, *112*, 113981. [CrossRef]
7. Hu, Y.S.; Mu, S.Q.; Zhang, J.J.; Li, Q.B. Regional Distribution, Properties, Treatment Technologies, and Resource Utilization of Oil-Based Drilling Cuttings: A Review. *Chemosphere* **2022**, *308*, 136145. [CrossRef]
8. Xie, P.; Wu, H.Y.; Yang, H.; Zhang, Y.; Xia, S.B. Utilization of Cr-Contaminated Oil-Based Drilling-Cuttings Ash in Preparation of Bauxite-Based Proppants. *Environ. Eng. Sci.* **2022**, *39*, 73–80. [CrossRef]
9. Liuyang, X.Y.; Yang, H.; Huang, S.Y.; Zhang, Y.; Xia, S.B. Resource Utilization of Secondary Pyrolysis Oil-Based Drilling Cuttings Ash for Removing Cr (VI) contaminants: Adsorption properties, kinetics and mechanism. *J. Environ. Chem. Eng.* **2020**, *8*, 104474. [CrossRef]
10. Liu, Y.C.; Ma, E.R.; Chen, M.Y.; Wu, J.W.; Yang, J.; Wang, Z.M.; Yang, L.; Li, H. Efficient mechanical activation for preparing highly active supplementary cementitious materials: High-value resource recovery of water-based drilling cuttings. *Process Saf. Environ. Prot.* **2026**, *208*, 108436. [CrossRef]
11. Tanyıldızı, M.; Nasıroğlu, S.; Zeybek, A.; Gökalp, İ. Sustainable use of waste hazelnut shell ash in road construction to stabilize expansive subgrad. *Constr. Build. Mater.* **2024**, *432*, 136663. [CrossRef]
12. Qiu, X.; Yang, J.; Wu, Y.; Liu, Q. Effect of Fiber Content on Mechanical Properties of Fiber-Reinforced CGF All-Solid-Waste Binder-stabilized soil. *Materials* **2024**, *17*, 388. [CrossRef] [PubMed]
13. Sun, Y.F.; Yang, Y.; Niu, X.; Yang, J.; Deng, K.; Song, S.; Shi, X.; Wen, Z.; Wang, Y. Pavement Performance of Fine-Grained Soil Stabilized by Fly Ash and Granulated Blast Furnace Slag-Based Geopolymer as Road Base Course Material. *J. Mater. Civ. Eng.* **2024**, *36*, 04024167. [CrossRef]
14. Jia, J.; Lu, X.; Zhu, J.; Wang, J.; Zhang, L.; Cheng, X. Effect of ionic soil stabilizer on the properties of lime and fly ash stabilized iron tailings as pavement base. *Constr. Build. Mater.* **2024**, *450*, 138558. [CrossRef]

 https://doi.org/10.3390/coatings16040406

15. Pushpakumara, J.H.B.; Mendis, W.S.W. Suitability of Rice Husk Ash (RHA) with lime as a soil stabilizer in geotechnical applications. *Int. J. Geo-Eng.* **2022**, *13*, 4. [CrossRef]
16. Lei, B.F.; Zhang, X.C.; Fan, H.H.; Wu, S.J.; Zhao, C.Z.; Ni, W.B.; Liu, C.H. A Strategy to Optimize the Mechanical Properties and Microstructure of Loess by Nano-Modified Soil Stabilizer. *Materials* **2025**, *18*, 4435. [CrossRef]
17. Hou, J.J.; Chen, Z.X.; Wang, W.Q.; Guo, H.N.; Liu, M.Q.; Guo, T.T.; Wang, J.L.; Mi, J.L.; Lei, Z.Q. Preparation of modified nano-zinc oxide-based multifunctional polymer soil stabilizer materials for controlling soil erosion and environmental remediation. *Surf. Interfaces* **2025**, *73*, 107630. [CrossRef]
18. Tabet, W.E.; Cerato, A.B.; Elwood Madden, A.S.; Jentoft, R.E. Characterization of Hydration Products' Formation and Strength Development in Cement-Stabilized Kaolinite Using TG and XRD. *J. Mater. Civ. Eng.* **2018**, *30*, 04018261. [CrossRef]
19. *GB 8978-1996*; Integrated Wastewater Discharge Standard. Ministry of Ecology and Environment of the People's Republic of China: Beijing, China, 1996.
20. *GB 3838-2002*; Environmental Quality Standards for Surface Water. Ministry of Ecology and Environment of the People's Republic of China: Beijing, China, 2002.
21. *JTG 3430-2020*; Highway Geotechnical Test Code. Ministry of Transport of the People's Republic of China: Beijing, China, 2020.
22. Wang, Z.; Li, H.; Ye, X.; Luo, W.; Zhang, B.; Hu, A.; Wu, L. Research on layered steel fiber reinforced concrete mix ratio design based on orthogonal test. *Coatings* **2024**, *14*, 697. [CrossRef]
23. Xiao, Z.A.; Guo, M.L.; Liang, J.R.; Yin, Z.Y.; Chen, X.Y.; Du, Z.H.; Xu, X.Y. Mechanical Behavior and Microscopic Mechanism of Solidified Sulfate Saline Soil under the Synergistic Effect of Temperature and Salt Content. *J. Cold Reg. Eng.* **2026**, *40*, 04026011. [CrossRef]
24. *CJJ/T 286-2018*; Application Standard of Soil Stabilizer. Ministry of Housing and Urban-Rural Construction of the People's Republic of China: Beijing, China, 2018.
25. *JTG/T F20-2015*; Technical Guidelines for Construction of Highway Roadbases. Ministry of Transport of the People's Republic of China: Beijing, China, 2015.
26. Feng, M.J.; Wang, D.Q.; Zhang, H.B.; Ma, H.F.; Huang, X.M.; Gao, W.B.; Jiang, Q.Q. Volume stability evaluation and life prediction of steel slag water-stabilised materials. *Road Mater. Pavement Des.* **2026**, *27*, 345–365. [CrossRef]
27. Wang, H.W.; Zhang, J.H.; Dong, L.J.; Zentar, R.; Shi, Y.; Sun, D.Y. Enhancing tensile strength and microstructural properties of solidified high-moisture dredged sediment through solid waste-based binder stabilization and basalt fiber reinforcement. *Case Stud. Constr. Mater.* **2026**, *24*, e05752. [CrossRef]
28. Wang, C.Q.; Liu, K.; Huang, D.M. Property of concrete made of recycled shale gas drilling cuttings. *Environ. Sci. Pollut. Res. Int.* **2021**, *29*, 2098–2106. [CrossRef]
29. Cheng, Y.H.; Zhu, B.L.; Yang, S.H.; Tong, B.Q. Design of concrete mix proportion based on particle packing voidage and test research on compressive strength and elastic modulus of concrete. *Materials* **2021**, *14*, 623. [CrossRef]
30. Wu, D.Z.; Chen, M.L.; Chen, K.Y.; Zhang, Z.L.; Cai, Q.M. An Experimental Study on the Effect of Desulfurized Gypsum on Geopolymer-stabilized soil. *Int. J. Geomech.* **2025**, *25*, 04025091. [CrossRef]
31. Hu, J.F.; Huang, Y.Y.; Ge, R.Y.; Wang, J.M.; Lu, W.L. Out-of-plane compression performance of unidirectionally arrayed short fiber reinforced honeycomb structures: Experimental and numerical analysis. *Compos. Struct.* **2026**, *378*, 119935. [CrossRef]
32. Wang, C.Q.; Mei, X.D.; Zhang, C.; Liu, D.S.; Xu, F.L. Mechanism study on co-processing of water-based drilling cuttings and phosphogypsum in non-autoclaved aerated concrete. *Environ. Sci. Pollut. Res. Int.* **2020**, *27*, 23364–23368. [CrossRef]
33. Aboutabikh, M.; Soliman, A.M.; El Naggar, M.H. Properties of cementitious material incorporating treated oil sands drill cuttings waste. *Constr. Build. Mater.* **2016**, *111*, 751–757. [CrossRef]
34. Liu, D.; Wang, C.; Mei, X.; Zhang, C. Environmental performance, mechanical and microstructure analysis of non-fired bricks containing water-based drilling cuttings of shale gas. *Constr. Build. Mater.* **2018**, *183*, 215–225. [CrossRef]
35. *HJ/T299-2007*; Leaching Method for Toxicity of Solid Wastes-Sulfuric Acid and Nitric Acid Method. Ministry of Ecology and Environment of the People's Republic of China: Beijing, China, 2007.
36. *GB/T 30760-2024*; Technical Specification for Co-Processing of Solid Waste in Cement Kiln. National Market Supervision Administration, National Standardization Administration Committee: Beijing, China, 2024.
37. *SY/T 7466-2020*; Technical Specifications of Water-Based Drilling Waste Treatment and Resource Utilization for Onshore Oil & Natural Gas Exploitation. Industry Standard of Petroleum and Natural Gas of the People's Republic of China: Beijing, China, 2020.
38. Zuo, W.P.; Gu, H.L.; Liao, Q.P.; Chao, J.; Qi, C.C.; Liu, J.Q.; Lv, Y.C.; Tian, C.; Lin, Z. Predicting heavy metal distribution coefficient in global soil via machine learning: The effect of mineral heterogeneity. *Environ. Res.* **2026**, *295*, 123957. [CrossRef] [PubMed]

 https://doi.org/10.3390/coatings16040406

39. *GB 15618-2018*; Soil Environmental Quality Risk Control Standard for Soil Contamination of Agricultural Land. Ministry of Ecology and Environment of the People's Republic of China: Beijing, China, 2018.
40. *DB51 2978-2023*; Risk Control Standard for Soil Contamination of Development Land in Sichuan Province. Sichuan Provincial Department of Ecology and Environment, Sichuan Provincial Market Supervision Administration: Chengdu, China, 2023.
41. Mkhonza, N.P.; Zondo, S.; Vilakazi, S. Toxic heavy metals distribution in urban soils of Africa: A systematic review. *Environ. Monit. Assess.* **2026**, *198*, 205. [CrossRef] [PubMed]

 https://doi.org/10.3390/coatings16040406

coatings

Article

Mechanism of Immersion Crushing on Alkali-Silica Reaction (ASR) in Glass Mortar

Kai Yan [1], Yuanbo Song [1,2], Lianfang Sun [1,2,*], Qian Zhang [1,2] and Tianfeng Yuan [1,2]

[1] Key Laboratory of Building Structural Retrofitting and Underground Space Engineering, Ministry of Education, Shandong Jianzhu University, Jinan 250101, China
[2] School of Civil Engineering, Shandong Jianzhu University, Jinan 250101, China
[*] Correspondence: sunlianfang@sdjzu.edu.cn

Abstract: The utilization of waste glass as an aggregate in cement-based materials provides both environmental and economic benefits, but the alkali-silica reaction (ASR) caused by the reactive silica in glass aggregates is a significant challenge for its application. This study investigates the impact of different crushing methods on the ASR of glass aggregate mortar, with a focus on the effect of immersion crushing using calcium chloride ($CaCl_2$) solution. Glass aggregates were prepared using conventional crushing, water immersion crushing, and $CaCl_2$ immersion crushing methods. The ASR expansion and compressive strength of the mortar were evaluated through accelerated ASR tests, compressive strength testing, and microstructural analysis using SEM/EDS and mercury intrusion porosimetry (MIP). Results show that immersion crushing significantly mitigated ASR expansion and the associated loss in compressive strength. The $CaCl_2$ immersion method yielded the most pronounced effect. Compared with conventional crushing, it reduced the ASR expansion by approximately 45% and improved the compressive strengths by approximately 20%. Microstructural analysis revealed that the $CaCl_2$ treatment led to a higher Ca/Si ratio in the ASR gel, which reduced the gel's water-absorbing swelling ability and consequently suppressed ASR-induced expansion. Additionally, the $CaCl_2$ immersion crushing method resulted in the smallest changes in porosity and pore size distribution. These findings provide a theoretical basis for the safe use of waste glass in cement-based materials and contribute to the promotion of resource recycling in the construction industry.

Keywords: waste glass aggregate; mortar; alkali-silica reaction; immersion crushing; microstructure

1. Introduction

Concrete is the most widely used construction material in the world today. Its sustainable development faces dual challenges of resource depletion and solid waste disposal [1]. Replacing natural aggregates in concrete with waste materials is one of the key approaches to achieving a green transformation in the construction industry [2,3]. Waste glass, a typical urban solid waste, is produced in large quantities each year. It is chemically stable, difficult to degrade naturally, and landfilling consumes valuable land resources [4]. Therefore, crushing waste glass for use as fine aggregate or powder in concrete can not only reduce environmental burdens but also alleviate the excessive exploitation of natural sand resources, offering significant environmental and economic benefits [2,3].

However, the amorphous silica in glass aggregates is highly chemically reactive and tends to undergo an alkali-silica reaction (ASR) with alkali metal ions (e.g., K^+, Na^+) in the pore solution [5,6]. The ASR produces alkali-silica gel, which absorbs water and swells in humid environments, generating substantial internal stress that eventually leads to cracking, spalling, and deterioration of the mechanical properties of concrete [7]. This potential durability issue has severely limited the large-scale application of waste glass in concrete structures.

In recent years, various methods have been proposed to mitigate the ASR of glass aggregates. These include: (1) using low-alkali cement or incorporating supplementary cementitious materials (such as fly ash, silica fume) to lower the pore solution alkalinity and optimize pore structure [8,9]; (2) applying surface treatments to glass aggregates, such as annealing the glass particles at high temperatures, which could remove the residual cracks [10]; and (3) limiting the replacement ratio of glass aggregates [11]. Nevertheless, these approaches often suffer from high costs, complex processes, or limited effectiveness. Therefore, it is crucial to explore innovative strategies that address ASR from the perspective of the aggregate's intrinsic properties and provide a fundamental solution to this problem.

Notably, the reactivity of ASR is closely related to the interfacial characteristics of glass aggregates. Studies have shown that microcracks generated during glass crushing serve as primary sites for ASR initiation [10,12]. In conventional crushing, the newly formed crack surfaces are directly exposed to air, and their surface chemistry depends on the bulk composition of the glass. This indicates that the aggregate preparation method plays a decisive role in the ASR process. If the chemical environment of microcrack surfaces can be modified during this critical stage, it may be possible to alter the subsequent ASR pathway.

From the perspective of ASR mechanisms, two typical types of products and reaction zones are generally observed [13]. On the aggregate surface, where calcium hydroxide ($Ca(OH)_2$) from the cement paste is enriched, dissolved silicate species preferentially combine with calcium ions (Ca^{2+}) to form calcium–silicate–hydrate (C–S–H) layers with high Ca/Si ratios [13]. This dense layer is not expansive and can act as a semipermeable membrane that selectively restricts ion migration [14], thereby slowing down the reaction. Long-term observations have shown that the thickness of this reaction layer tends to stabilize at later stages [15]. In contrast, within the internal microcracks of the aggregate, where $Ca(OH)_2$ is scarce, the dissolved silica primarily reacts with sodium and potassium ions to form ASR gels with low Ca/Si ratios. These gels exhibit strong hygroscopic swelling behavior [16]. The limited calcium ions diffusing into cracks are insufficient to promote effective gel carbonation. SEM/EDS analyses have confirmed that the Ca/Si ratio of gels inside cracks is significantly lower than that on the aggregate surface [12,16], and low Ca/Si ratios have been shown to markedly increase the swelling capacity of ASR gels [17]. This results in a self-accelerating cycle of "microcrack propagation $\rightarrow$ intensified ASR $\rightarrow$ expansion-induced new cracking" [10,12].

Based on the above analysis, calcium ions (Ca^{2+}) play a central role in regulating the properties of ASR gels. Inspired by this understanding, our research group has innovatively proposed a pretreatment method involving immersed crushing of glass aggregates in a calcium-ion-rich solution. Preliminary studies have demonstrated that this method can significantly reduce ASR-induced expansion [18]. However, the underlying micro-mechanisms—particularly the regulation of gel chemistry within internal cracks—remain unclear.

This study aims to systematically elucidate the mechanism by which immersed crushing, especially in a $CaCl_2$ solution medium, mitigates ASR, from the macroscopic to the microscopic scale. Glass aggregates were prepared via the immersed crushing process, and a multi-scale analytical approach was employed. At the macroscopic level, the evolution of mortar expansion and compressive strength was monitored to evaluate ASR inhibition. At the mesoscopic level, changes in pore structure (porosity and pore size distribution) during ASR were analyzed. At the microscopic level, SEM/EDS was used to characterize the chemical composition (particularly the Ca/Si ratio) and morphology of ASR gels to reveal the underlying mechanism. The findings of this study will provide new theoretical insights and practical pathways for the safe utilization of waste glass in cement-based materials, promoting higher-value recycling of glass solid waste.

The novelty of this study lies in the innovative application of immersion crushing in a calcium-rich solution ($CaCl_2$) as a pretreatment method for waste glass aggregates. Unlike conventional approaches that focus on modifying the cementitious matrix or limiting aggregate content, this method directly targets the reactive microcracks within the glass aggregates at the moment of their formation. By pre-enriching the crack surfaces with Ca^{2+} ions, we fundamentally alter the chemical pathway of ASR gel formation in the most vulnerable regions—internal cracks—leading to the formation of less expansive, high-Ca/Si gels. This work provides the first multi-scale evidence linking the crushing medium chemistry to the microchemical properties of ASR gels and the resulting macroscopic expansion behavior, offering a novel and practical strategy for mitigating ASR in glass-containing cementitious materials.

2. Materials and Methods

2.1. Materials

Ordinary Portland cement (P·O 42.5) was used, and its physical and chemical properties are shown in Tables 1 and 2. Natural river sand was used as the inactive fine aggregate, with a particle size range of 0–1.18 mm. Waste glass particles, with a particle size range of 1.18–4.75 mm, were used as the active fine aggregate, and their physical and chemical properties are also shown in Tables 1 and 2. $CaCl_2$ was used as an analytical-grade reagent, and pure water was used.

Table 1. Physical properties of glass particles and cement.

Parameter	Specific Gravity	Specific Surface Area	Water Absorption	Initial Setting Time	Final Setting Time
Sand	2.65	-	1.65%	-	-
Glass	2.49	-	0.26%	-	-
Cement	3.2	3621 cm^2/g	-	75 min	320 min

Table 2. Chemical properties of glass particles and cement.

Content (%)	CaO	Al_2O_3	SiO_2	Fe_2O_3	MgO	SO_3	Na_2O
Glass	9.49	2.82	67.13	0.01	1.83	0.01	14.01
Cement	61.94	4.73	20.72	3.72	3.07	2.31	0.01

2.2. Experimental Procedures

2.2.1. Glass Aggregate Preparation

Glass aggregates were prepared using three different crushing methods: (1) Conventional crushing control group (air crushing); (2) Immersion crushing control group (crushed in pure water); (3) Immersion crushing experimental group (crushed in a 1.0 mol/L $CaCl_2$ solution). The corresponding glass aggregates were labeled as Control, IC-W, and IC-C, respectively. The preparation process is illustrated in Figure 1a. After being washed and dried, the prepared glass bottles were crushed with a hammer while immersed in water or calcium solution. After crushing, all aggregates were sieved to obtain particles within the 1.18–4.75 mm size range to ensure consistent grading and eliminate the influence of particle size distribution on the test results. Previous studies have indicated that whether the particles are cleaned after immersion crushing—i.e., whether residual solution remains on the particle surfaces after soaking—has no significant effect on the macroscopic ASR expansion [18]. Therefore, this study focused on one scenario, in which the obtained glass particles were rinsed with running water, dried, and then used for subsequent experiments.

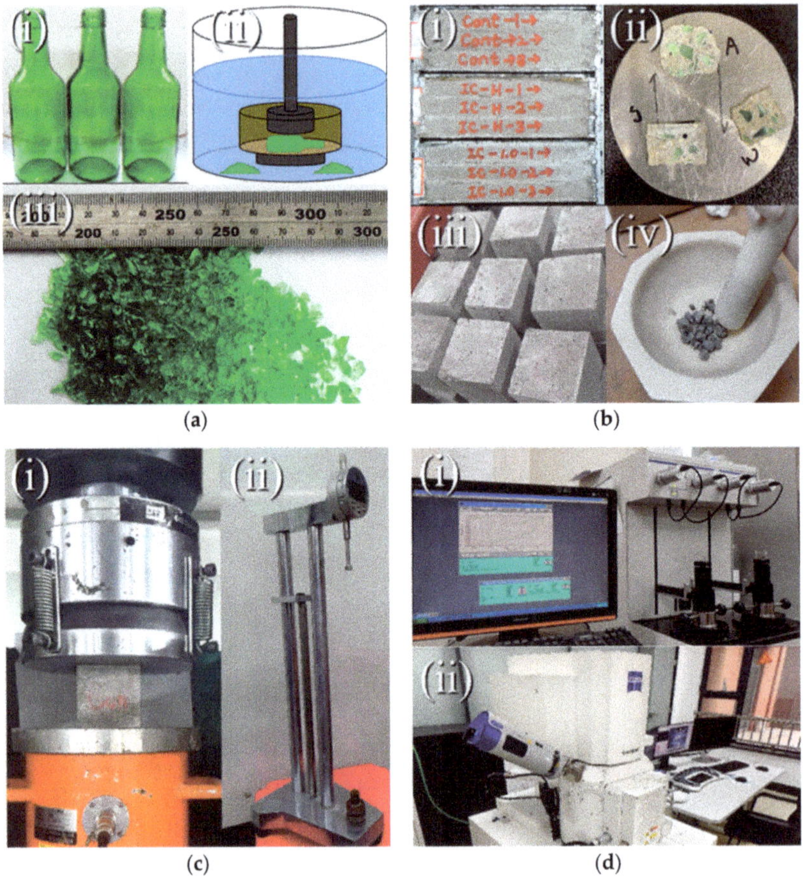

Figure 1. Schematic diagram of the experimental procedure: (**a**(**i**)–**a**(**iii**)) aggregate preparation; (**b**(**i**)–**b**(**iv**)) specimen preparation; (**c**(**i**),**c**(**ii**)) test setup for compressive and AST test; (**d**(**i**),**d**(**ii**)) microscopic characterization.

2.2.2. Alkali-Silica Reaction (ASR) Test

To quantitatively evaluate the ASR activity of different glass aggregates, mortar bar expansion tests were conducted following the accelerated test method specified in ASTM C1260. The mortar bar specimens (25 mm × 25 mm × 285 mm) were prepared with mix proportions shown in Table 3. The sieve analysis of the sand and glass aggregate is shown in Figure 2. Glass aggregates replaced 35% of natural sand by mass, a substitution level sufficient to induce noticeable ASR expansion while maintaining adequate workability [18]. The bulk density of sand and glass aggregate was 1560 kg/m^3. Three groups of specimens were prepared, each containing three mortar bars corresponding to the aggregate types described in Section 2.2.1, as illustrated in Figure 1b(i).

Table 3. Mix proportions of mortar specimens.

Water	Cement	Sand	Glass
0.47	1	1.45	0.8

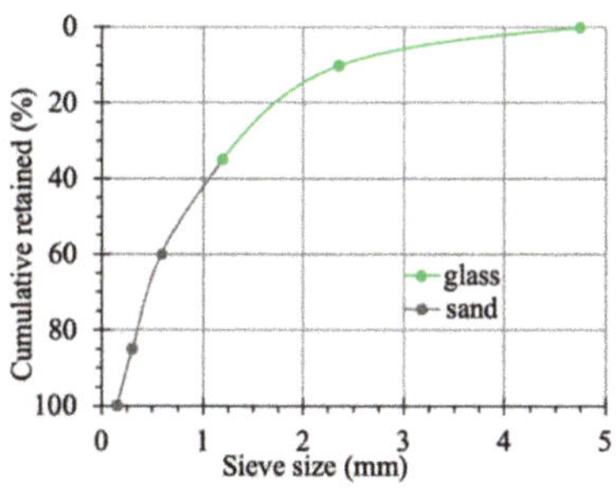

Figure 2. Sieve analysis of sand and glass aggregate.

After 24 h of casting, the specimens were demolded and cured in water at 80 °C for 24 h. Subsequently, they were immersed in a 1.0 mol/L NaOH solution at 80 °C to accelerate ASR development. The change in specimen length was periodically measured using a comparator, as shown in Figure 1c(ii). The average length change of three specimens from each group was recorded as the ASR expansion ratio, which served as the key macroscopic indicator of ASR severity.

2.2.3. Compressive Strength Test

To investigate the degradation of mechanical performance due to ASR and the mitigating effects of different aggregates, compressive strength tests were conducted under both standard and accelerated ASR curing conditions. Mortar cubes (70.7 mm × 70.7 mm × 70.7 mm) were prepared using the same mix proportions as the mortar bars (Table 3). Two batches of specimens were produced, each consisting of three groups (corresponding to the three aggregate types used in Section 2.2.1), and each group contained three specimens, as shown in Figure 1b(iii).

The first batch was cured under standard conditions (temperature 20 ± 2 °C, relative humidity ≥ 95%) for 28 days. The second batch was immersed in a 1.0 mol/L NaOH solution at 80 °C for 28 days to simulate ASR-induced degradation. After curing, compressive strength tests were carried out using a universal testing machine (Figure 1c(i)). The average strength of the three specimens in each group was recorded as the compressive strength of that group. Comparing the strength loss between the two curing conditions provided insight into the extent of ASR-induced structural damage.

2.2.4. Microstructural Analysis

To elucidate the characteristics of ASR products and the underlying mechanism of immersed crushing from a microscopic perspective, samples were collected from the mortar bars after completion of the ASR test. Thin slices (approximately 2.0 mm thick) were prepared from the midsection of the bars using a low-speed diamond saw (Isomet 1000, Buehler Ltd., Lake Bluff, IL, USA), as shown in Figure 1b(ii), and immediately immersed in anhydrous ethanol for 24 h to terminate hydration. Then, the slices were dried at 60 °C for another 24 h, and their surfaces were polished using sandpaper before performing the SEM and EDX analysis. To enhance the electrical conductivity at the surface, the specimen was coated with platinum. Afterward, the microstructure of the glass–paste interfacial zones was observed using a scanning electron microscope (SEM) equipped with energy-dispersive X-ray spectroscopy (EDS) (Hitachi S-4800, Tokyo, Japan). The SEM images were captured at an accelerating voltage of 20 kV. Point analyses were conducted on selected ASR gel products (Figure 1d(ii)) to determine their local chemical composition. Particular attention was paid to comparing the Ca/Si ratios of ASR gels within glass aggregate cracks across different groups, as this parameter serves as key evidence for validating the study's central hypothesis.

2.2.5. Pore Structure Analysis

The expansive stress of ASR gels leads to the initiation and propagation of microcracks in mortar, thereby altering its pore structure. After compressive strength testing, representative samples (approximately 5 mm × 5 mm × 5 mm) were extracted from the specimens, as shown in Figure 1b(iv), and immersed in anhydrous ethanol to stop hydration. Pore structure analysis was conducted using mercury intrusion porosimetry (MIP, AutoPore IV 9500, NC, USA), as shown in Figure 1d(i), to obtain the total porosity and pore size distribution data. By comparing the pore structure evolution under different curing conditions, the ASR-induced damage could be quantitatively characterized at the mesoscopic level and correlated with macroscopic mechanical performance.

3. Results and Discussion

3.1. *Macro ASR Expansion*

Figure 3 illustrates the effect of glass aggregate crushing methods on ASR in mortar bars. As shown in Figure 3a, the ASR expansion rate of mortar bars with different aggregates increased over time, but the rate of expansion gradually slowed down. This was primarily due to the formation of reaction products on the aggregate surface, which slowed the ingress of corrosive ions. The crushing method significantly affected the ASR expansion rate. The conventional crushing group exhibited the highest expansion rate, followed by the underwater crushing group and the $CaCl_2$ immersion crushing group. The underwater crushing group showed a 24% reduction in expansion compared to the conventional crushing group (Figure 3b), mainly because underwater crushing significantly reduced the surface defects [19] and lowered the glass's tensile strength [20]. Fewer surface defects mean a reduction in ASR reaction sites, while a lower tensile strength may allow residual cracks to develop more readily into through-going fractures during the crushing process, thereby decreasing the formation of residual cracks and subsequently mitigating ASR expansion. The $CaCl_2$ immersion crushing group showed the lowest expansion rate, reducing ASR expansion by approximately 45% compared to the conventional crushing group (Figure 3b), indicating the presence of additional influencing mechanisms beyond those in the underwater crushing process.

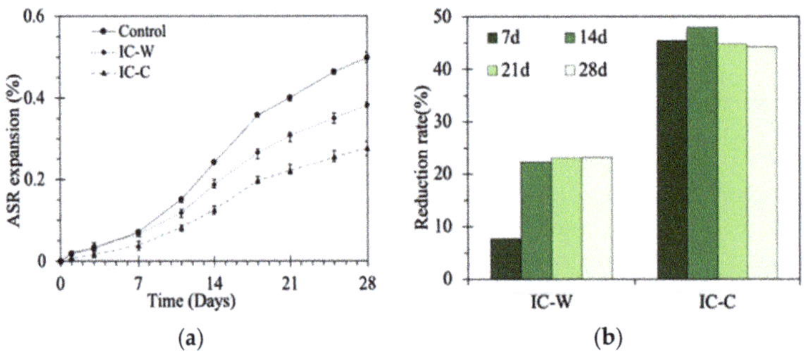

Figure 3. Effect of crushing process on ASR: (**a**) ASR expansion; (**b**) ASR expansion reduction rate.

Figure 4 shows the surface morphology of the mortar bars after 28 days of accelerated ASR testing, as observed with a handheld digital microscope, with the observation areas located in the middle of each bar. As shown in Figure 4, large macrocracks were visible on the surface of the mortar bars containing conventionally crushed glass, with the widest cracks; fewer cracks were observed in the underwater crushed glass mortar bars; and the $CaCl_2$ immersion crushed glass mortar bars had the best surface morphology, with only a few fine cracks. These surface morphological changes align with the macro ASR expansion trends.

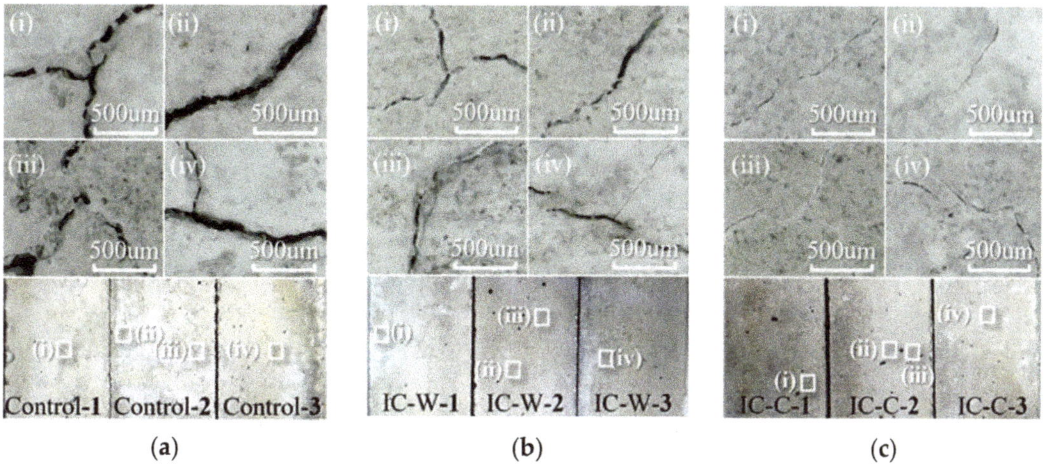

Figure 4. Appearance of mortar bars after ASR for 28 days: (**a(i)**–**a(iv)**) Control; (**b(i)**–**b(iv)**) IC-W; (**c(i)**–**c(iv)**) IC-C.

3.2. Compressive Strength

Figure 5 presents the compressive strength of mortar specimens with glass aggregates under different curing conditions at 28 days. Under standard curing conditions, the compressive strength of all groups was approximately 45 MPa, indicating that the preparation method of glass aggregates had no significant effect on the mortar's compressive strength. In contrast, under the accelerated ASR curing conditions, the compressive strength of all specimens significantly decreased, with the conventional crushing group experiencing the greatest reduction, down to approximately 35 MPa. The underwater crushing group and the $CaCl_2$ immersion crushing group had compressive strengths of approximately 38 MPa and 42 MPa, respectively, which represented improvements of 8.5% and 20% compared to the conventional crushing group. This was mainly due to the absence of ASR reactions under standard curing conditions. In contrast, the faster ASR curing conditions triggered ASR reactions, leading to expansion and damage, which caused a decrease in strength. As shown in Figures 3 and 4, the $CaCl_2$ immersion crushing group exhibited the best performance in suppressing macro-expansion and crack formation, leading to the smallest reduction in compressive strength. Conversely, the conventional crushing exhibited the greatest expansion and cracking, resulting in the most significant decrease in strength. The compressive strength results of each group are consistent with the ASR expansion trends in Section 3.1.

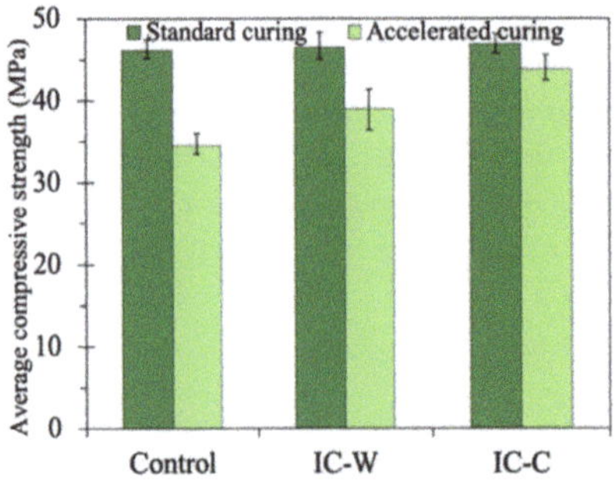

Figure 5. Compressive strength at day 28 with different curing conditions.

3.3. Microstructural Morphology

Figure 6 presents SEM images of products formed inside and outside the cracks of glass aggregates after 28 days, along with EDS elemental distribution maps for precise localization and the chemical analysis of reaction products. The corresponding chemical compositions are summarized in Table 4. In all groups (Control, IC-W, and IC-C), a dense reaction layer was observed on the glass surface, while products within internal cracks were more porous and exhibited drying shrinkage cracks—particularly in the Control and IC-W groups. These features correspond to the typical morphology of high-Ca/Si C–S–H layers on aggregate surfaces and low-Ca/Si ASR gels inside cracks [13].

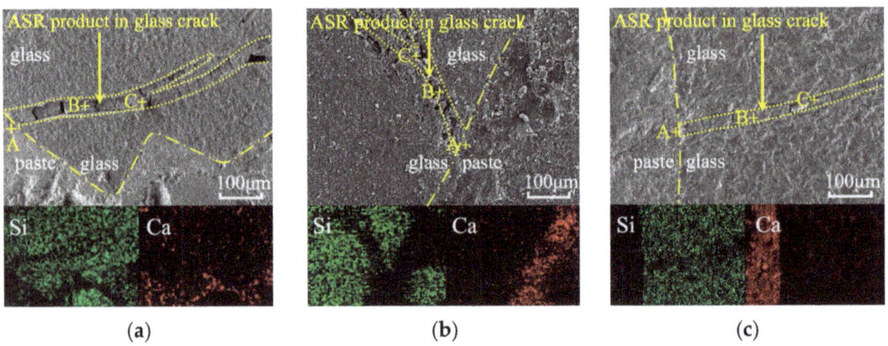

Figure 6. SEM-EDS analysis of specimens incorporating glass aggregates: (**a**) Control; (**b**) IC-W; (**c**) IC-C.

Table 4. Chemical composition of reaction products.

Figure	Point	Composition (%)					
		Si	Ca	Na	K	Ca/Si	(Na + K)/Si
Figure 6a	A	17.14	61.97	19.79	1.09	3.62	1.22
Figure 6a	B	47.35	16.55	34.50	1.61	0.35	0.76
Figure 6a	C	53.42	17.25	28.51	0.81	0.32	0.55
Figure 6b	A	15.46	68.83	13.36	2.35	4.45	1.02
Figure 6b	B	55.45	21.55	20.60	2.40	0.39	0.41
Figure 6b	C	56.24	16.33	25.82	1.61	0.29	0.49
Figure 6c	A	19.03	57.45	23.11	0.40	3.02	1.24
Figure 6c	B	47.78	26.11	23.41	2.70	0.55	0.55
Figure 6c	C	51.50	21.76	23.74	3.01	0.42	0.52

EDS point analyses (Table 4) revealed that products near crack openings possessed high Ca/Si ratios, confirming their identification as C–S–H gels. This occurs because the swelling ASR gel within cracks migrates outward and becomes enriched in Ca^{2+} at the surface. In contrast, products deeper within the cracks exhibited low Ca/Si ratios, characteristic of expansive ASR gels.

A key finding of this study is that the ASR gels within cracks in the IC-C group had significantly higher Ca/Si ratios than those in the Control and IC-W groups. This indicates that during crushing in the $CaCl_2$ solution, the freshly formed crack surfaces instantly adsorbed large amounts of Ca^{2+}. These pre-adsorbed ions later served as local calcium sources during ASR, promoting the formation of high-Ca/Si reaction products. According to Gholizadeh-Vayghan et al. [17], the swelling capacity of ASR gels decreases markedly with increasing Ca/Si ratio. Therefore, the generation of higher-Ca/Si gels in the IC-C group fundamentally reduces the swelling potential of ASR gels, providing the microscopic mechanism for its superior ASR inhibition.

3.4. Pore Structure

Figure 7 shows the total porosity and pore size distribution of specimens with different crushing methods and curing conditions after 28 days. As seen in Figure 7, under standard curing conditions, all groups exhibited a total porosity of approximately 17%. Under accelerated ASR curing conditions, the total porosity of all groups increased to varying degrees, with the highest porosity observed in the conventional crushing group, followed by the underwater crushing group and the $CaCl_2$ liquid immersion crushing group. This trend corresponds with the decrease in compressive strength observed in Section 3.2, as higher porosity results in lower compressive strength [21].

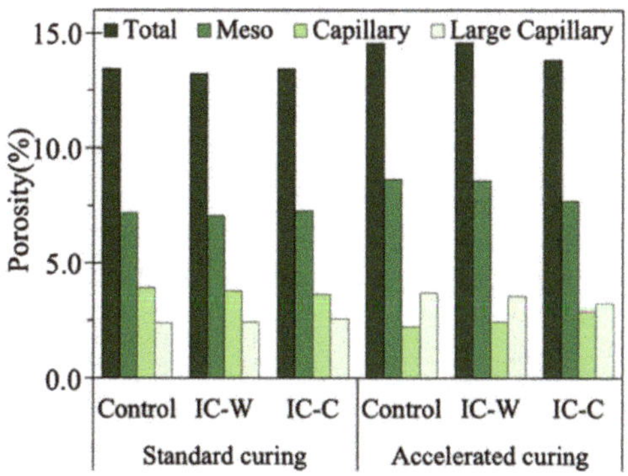

Figure 7. Porosity and pore size distribution at day 28 with different curing conditions.

Based on pore size, the pores in the mortar were classified into three categories: mesopores (less than 50 nm), capillary pores (50–1000 nm), and macropores (greater than 1000 nm) [22]. Under standard curing conditions, the pore size distribution in all groups was essentially unaffected by the crushing method. However, under accelerated ASR curing conditions, the proportion of capillary pores decreased significantly, while the proportion of mesopores and macropores increased to varying extents. Among these, the specimens from the $CaCl_2$ liquid immersion crushing group showed the lowest proportion of mesopores and macropores and the highest proportion of capillary pores. The shift in pore size distribution, characterized by an increase in mesopores and macropores, is primarily attributed to the formation of microcracks induced by ASR expansion. While the deposition of ASR gel products might have a minor, localized filling effect on some smaller pores, the MIP data conclusively shows that this potential effect is overwhelmingly offset by the net increase in harmful larger pores due to cracking.

4. Conclusions

- The immersion crushing effectively suppressed ASR and the associated deterioration in the performance of glass-containing mortar. Compared with conventional crushing, immersion crushing in water and $CaCl_2$ solution reduced the 14-day ASR expansion of mortar by approximately 24% and 45%, respectively, and significantly mitigated the loss of compressive strength under ASR conditions. Macroscopic observations confirmed that mortars incorporating glass aggregates crushed in $CaCl_2$ solution exhibited almost no visible surface cracks after ASR exposure.
- ASR degrades the pore structure of mortar, whereas immersion crushing effectively alleviates this degradation. ASR increased total porosity and shifted the pore size distribution toward more harmful mesopores and macropores. Specimens subjected to immersion crushing in $CaCl_2$ solution (IC-C group) showed the smallest increases in porosity and in the proportion of harmful pores, indicating the least internal structural damage.
- The microchemical mechanism underlying ASR suppression by immersion crushing—particularly in $CaCl_2$ solution—was revealed. In glass aggregates crushed in a $CaCl_2$ medium, the ASR gel formed along internal microcracks exhibited a higher Ca/Si ratio. This resulted from the pre-enrichment of Ca^{2+} on the surfaces of newly generated cracks during crushing, which altered the gel formation pathway and composition at the reaction source. Consequently, low-expansivity, high-Ca/Si-ratio gels were produced, fundamentally inhibiting ASR-induced expansion and damage.

Author Contributions: Conceptualization, Y.S. and L.S.; methodology, Y.S. and L.S.; investigation, L.S.; writing—original draft preparation, Y.S.; writing—review and editing, Q.Z. and T.Y.; supervision, K.Y.; funding acquisition, L.S. All authors have read and agreed to the published version of the manuscript.

Funding: This work was supported by the National Natural Science Foundation of China (Nos. 52378520, 52038006), the Natural Science Foundation of Shandong Province (No. ZR2025QC515), and the research fund of Shandong Jianzhu University.

Institutional Review Board Statement: Not applicable.

Informed Consent Statement: Not applicable.

Data Availability Statement: Data will be made available on request.

Conflicts of Interest: The authors declare no conflicts of interest.

References

1. Al-Kheetan, M.J. Properties of lightweight pedestrian paving blocks incorporating wheat straw: Micro-to macro-scale investigation. *Results Eng.* **2022**, *16*, 100758. [CrossRef]
2. Rakshvir, M.; Barai, S.V. Studies on recycled aggregates-based concrete. *Waste Manag. Res.* **2006**, *24*, 225–233. [CrossRef]
3. Patel, D.; Shrivastava, R.; Tiwari, R.P.; Yadav, R.K. Properties of cement mortar in substitution with waste fine glass powder and environmental impact study. *J. Build. Eng.* **2020**, *27*, 100940. [CrossRef]
4. Abdel-Shafy, H.I.; Mansour, M.S.M. Solid waste issue: Sources, composition, disposal, recycling, and valorization. *Egypt. J. Pet.* **2018**, *27*, 1275–1290. [CrossRef]
5. Topçu, I.B.; Boğa, A.R.; Bilir, T. Alkali–silica reactions of mortars produced by using waste glass as fine aggregate and admixtures such as fly ash and Li_2CO_3. *Waste Manag.* **2008**, *28*, 878–884. [CrossRef]

6. Du, H.; Tan, K.H. Use of waste glass as sand in mortar: Part II–Alkali–silica reaction and mitigation methods. *Cem. Concr. Compos.* **2013**, *35*, 118–126. [CrossRef]
7. Rajabipour, F.; Giannini, E.; Dunant, C.; Ideker, J.H.; Thomas, M.D. Alkali–silica reaction: Current understanding of the reaction mechanisms and the knowledge gaps. *Cem. Concr. Res.* **2015**, *76*, 130–146. [CrossRef]
8. Cota, F.P.; Melo, C.C.D.; Panzera, T.H.; Araújo, A.G.; Borges, P.H.R.; Scarpa, F. Mechanical properties and ASR evaluation of concrete tiles with waste glass aggregate. *Sustain. Cities Soc.* **2015**, *16*, 49–56. [CrossRef]
9. Berry, M.; Stephens, J.; Cross, D. Performance of 100% Fly Ash Concrete with Recycled Glass Aggregate. *ACI Mater. J.* **2011**, *108*, 378–384. [CrossRef]
10. Maraghechi, H.; Fischer, G.; Rajabipour, F. The role of residual cracks on alkali silica reactivity of recycled glass aggregates. *Cem. Concr. Compos.* **2012**, *34*, 41–47. [CrossRef]
11. Ali, E.E.; Al-Tersawy, S.H. Recycled glass as a partial replacement for fine aggregate in self compacting concrete. *Constr. Build. Mater.* **2012**, *35*, 785–791. [CrossRef]
12. Rajabipour, F.; Maraghechi, H.; Fischer, G. Investigating the alkali-silica reaction of recycled glass aggregates in concrete materials. *J. Mater. Civ. Eng.* **2010**, *22*, 1201–1208. [CrossRef]
13. Idir, R.; Cyr, M.; Tagnit-Hamou, A. Pozzolanic properties of fine and coarse color-mixed glass cullet. *Cem. Concr. Compos.* **2011**, *33*, 19–29. [CrossRef]
14. Ichikawa, T.; Miura, M. Modified model of alkali-silica reaction. *Cem. Concr. Res.* **2007**, *37*, 1291–1297. [CrossRef]
15. Zheng, K. Pozzolanic reaction of glass powder and its role in controlling alkali–silica reaction. *Cem. Concr. Compos.* **2016**, *67*, 30–38. [CrossRef]
16. Liaudat, J.; Carol, I.; López, C.M.; Leemann, A. ASR expansions at the level of a single glass-cement paste interface: Experimental results and proposal of a reaction-expansion mechanism. *Constr. Build. Mater.* **2019**, *218*, 108–118. [CrossRef]
17. Gholizadeh-Vayghan, A.; Rajabipour, F. Quantifying the swelling properties of alkali-silica reaction (ASR) gels as a function of their composition. *J. Am. Ceram. Soc.* **2017**, *100*, 3801–3818. [CrossRef]
18. Sun, L.; Kim, M.; Doh, J.H.; Zi, G. A novel method of crushing glass aggregates to reduce the alkali-silica reaction. *KSCE J. Civ. Eng.* **2021**, *25*, 4763–4770. [CrossRef]
19. Kokura, K.; Tomozawa, M.; MacCrone, R.K. Defect formation in SiO_2 glass during fracture. *J. Non-Cryst. Solids* **1989**, *111*, 269–276. [CrossRef]
20. Stockdale, G.F.; Tooley, F.V.; Ying, C.W. Changes in the tensile strength of glass caused by water immersion treatment. *J. Am. Ceram. Soc.* **1951**, *34*, 116–121. [CrossRef]
21. Chen, X.; Wu, S.; Zhou, J. Influence of porosity on compressive and tensile strength of cement mortar. *Constr. Build. Mater.* **2013**, *40*, 869–874. [CrossRef]
22. Care, S. Effect of temperature on porosity and on chloride diffusion in cement pastes. *Constr. Build. Mater.* **2008**, *22*, 1560–1573. [CrossRef]

coatings

MDPI

Article

Numerical Modeling of the Relationship Between Mechanical Properties and Void Morphology in Porous Asphalt Mixtures

Hongchang Wang [1,2,*], Haozhe Wu [1] and Congying Liu [1]

[1] School of Civil Engineering, Nanjing Forestry University, Nanjing 210037, China; whz18066035916@163.com (H.W.); 13852154483@163.com (C.L.)

[2] Jiangsu Highway Intelligent Detection and Low-Carbon Maintenance Engineering Research Center, Nanjing 210037, China

* Correspondence: seuwhc@126.com; Tel.: +86-025-85427747

Highlights

What are the main findings?

- The Monte Carlo method and ABAQUS finite element software are used to build a three-dimensional finite element model of voids, and the relationship between the pore morphology, distribution, and mechanical properties of the porous asphalt mixture is studied.
- The correlation between the stress–strain characteristics of the mesostructure and the void morphology was studied, and the stress characteristics of the aggregate in the porous asphalt mixture and the damaging trend in the mesostructure were summarized.
- The relationship between the strength of the porous asphalt mixture and the void ratio was studied.

What is the implication of the main finding?

- A microvoid model of porous asphalt mixture with optimal mechanical properties is proposed.

Abstract

Traditional macroscopic test methods (e.g., uniaxial compression and indirect tensile strength tests) cannot accurately describe the internal microstructure and its influence on the mechanical properties of asphalt mixtures from a microscopic perspective. With the advancement of digital image processing (DIP) techniques and numerical simulation methods, relatively complete workflows for microstructure characterization and mesostructural evaluation of composite materials have been established. In this study, a three-dimensional finite element model incorporating voids was developed using the Monte Carlo method and ABAQUS software to investigate the relationships between void morphology, distribution, and the mechanical properties of porous asphalt mixtures. By varying void size and shape in the model, correlations between mesostructural stress–strain characteristics and void morphology were derived. The stress distribution around aggregates and damage initiation trends in the mesostructure were summarized. The relationship between the anticipated strength (numerically assessed) of porous asphalt mixture and the void ratio was established through simulations of models with varying void ratios. Finally, a micromechanical model for porous asphalt mixture with optimized mechanical performance is proposed, featuring an icosahedral void shape, a void diameter range of 3–4 mm, and a void ratio of 18%.

 https://doi.org/10.3390/coatings16020214

Keywords: porous asphalt; void morphology; microstructural modeling; finite element analysis; mechanical performance

1. Introduction

Currently, research on porous asphalt mixtures predominantly focuses on macroscopic scales, with mechanical properties often evaluated through laboratory tests [1–4]. While essential, such tests are costly and limited in their ability to characterize detailed void information within specimens. For instance, macroscopic approaches typically correlate void ratio with mixture performance [5–8], whereas microscopic analyses can evaluate the influence of void shape, size, and distribution patterns on mechanical properties. Therefore, investigating the effect of voids on mechanical properties requires a fine-scale (micro-/mesoscale) perspective to characterize void features and their influence mechanisms. Two primary approaches exist for fine-scale modeling of asphalt mixtures: image-based models and computer-generated models [9–11]. Image-based models are constructed directly from images of real mixtures, accurately capturing the geometry and spatial distribution of aggregates, asphalt matrix, and voids. While offering high realism, these methods are costly and time-consuming, and their accuracy depends heavily on advanced imaging and processing techniques [12–16]. Recent studies, such as those employing digital image processing (DIP) of warm mix asphalt modified with nanocolemanite and nanoulexite minerals, demonstrate the critical role of DIP in evaluating microstructural effects on performance [17,18]. Computer-generated models offer a more economical and efficient approach by using algorithms to generate aggregate particles within a virtual asphalt matrix in a random or rule-based manner. This facilitates rapid model generation and iterative optimization. However, ensuring that generated geometries accurately represent real mixtures remains a challenge, requiring robust and flexible algorithms. For example, Qian et al. [12] employed 3D image reconstruction to analyze void space. Zhao et al. [13] developed methods to identify and characterize connected voids. Zhu et al. [14] used CT imaging to model AC-13 mixtures and validate void parameters. Xu et al. [15,16] emphasized that 3D void morphology evaluation can reflect actual conditions, largely independent of compaction methods. Zhao et al. [19] constructed a 2D FE model from CT slices to simulate indirect tensile tests, analyzing aggregate content effects. Wang et al. [20] improved the Zwikker–Kosten acoustic model by incorporating additional microstructural factors, optimizing porous asphalt concrete (PAC) for sound absorption. Kollmann et al. [21] combined X-ray tomography and DIP to reconstruct asphalt microstructure. Wang and Yang et al. [22,23] performed laboratory studies linking microscopic pore features to mixture performance using CT images. These studies represent significant progress in image-based modeling. Meanwhile, research on direct computer-generated models is also advancing rapidly. For instance, Cao et al. [24] used a random aggregate FE model to predict the dynamic modulus of asphalt concrete, showing good agreement with experimental results. Ma et al. [25,26] proposed the "Occupancy and Removal Method (ORM)" and a rhombic deformation framework to efficiently generate models with high aggregate content. Wang Zishuo [27] used laser scanning to capture aggregate geometry and employed a Bubble Pack algorithm in PFC3D to create discrete element models, where voids were represented by removing spheres from a regular mortar arrangement. Chen et al. [28] employed a Monte Carlo stochastic aggregate model to investigate the effects of specimen size, coarse aggregate area ratio, and porosity on the nominal tensile strength of concrete through uniaxial tensile simulations. Zhang et al. [29] developed a random aggregate

https://doi.org/10.3390/coatings16020214

placement model in ABAQUS to simulate asphalt concrete mesostructures, demonstrating good agreement with the experimental results. Wang and Yuan [30] applied Monte Carlo simulations to analyze variability in asphalt pavement mechanical responses, validating the method's reliability for probabilistic analysis. Cao et al. [31] integrated random aggregate modeling with ABAQUS to predict dynamic modulus, showing improved accuracy over traditional analytical methods.

The fine structure of asphalt mixtures directly influences their macroscopic deformation and strength behavior, justifying a shift from macroscale to micro-/mesoscale investigations. Therefore, developing a fine-scale model that incorporates void mechanics—i.e., the effect of voids on the surrounding matrix—is crucial for analyzing porous asphalt mixtures. This study employs the general-purpose finite element software ABAQUS (2016) to establish a three-dimensional micromechanical model of a porous asphalt mixture. By comparing models with different void shapes, sizes, and ratios, we analyze their stress–strain responses, revealing damage initiation mechanisms and force transmission characteristics within the microstructure. Furthermore, the modeling approach combines the Monte Carlo method for stochastic void generation with ABAQUS for mechanical simulation, a methodology that has been increasingly applied in asphalt research [28–30].

The primary objectives of this study are: (1) to develop a fine-scale 3D finite element model incorporating stochastic voids using Monte Carlo and ABAQUS; (2) to systematically investigate the effects of void shape, size, and ratio on the deformation and mechanical properties of porous asphalt mixtures; and (3) to identify an optimal void configuration that minimizes stress concentration and maximizes mechanical performance. The paper is organized as follows: Section 2 details the materials and modeling methodology; Section 3 presents the simulation results regarding deformation; Section 4 analyzes mechanical properties; Section 5 provides a comprehensive discussion comparing results with existing literature; and Section 6 summarizes the conclusions and practical implications.

2. Materials and Methods

2.1. Model Conceptualization and Scale Definition

In this study, the terms microscale refer to the scale of individual constituents (asphalt mastic, aggregates, and voids), mesoscale to the representative volume element (RVE) containing multiple voids and matrix, and macroscale to the homogenized mixture behavior. Our focus is on the mesoscale, where void morphology directly influences local stress fields. The matrix is modeled as a homogeneous linear elastic material representing the asphalt mastic (including fine aggregates and binder), with properties summarized in Table 1. Aggregate particles are not explicitly modeled in this phase to isolate the effect of void morphology; instead, the matrix properties implicitly account for the composite effect of mastic and aggregate stiffness. This simplification is common in initial void morphology studies [21,27].

Table 1. Material parameters for the finite element model.

Material	Modulus of Elasticity (MPa)	Poisson's Ratio
Matrix (Asphalt mastic)	1400	0.35
Voids (Numerical representation)	1×10^{-6}	0.01

 https://doi.org/10.3390/coatings16020214

2.2. Numerical Modeling Framework

2.2.1. Stochastic Void Generation Using Monte Carlo Method

Monte Carlo methods, also known as stochastic simulation methods, are employed to generate random void distributions within the matrix. The steps include: (1) defining the stochastic process for void placement; (2) optimizing the algorithm for efficiency; (3) sampling from prescribed probability distributions for void location and size; and (4) constructing the 3D geometry [31,32].

Common void shapes investigated include spheres, icosahedrons, dodecahedrons, and hexahedrons. While icosahedral voids may seem idealized, they serve as a geometric benchmark to study the effect of increasing facet count and reduced sharpness compared to common polyhedral approximations of irregular voids. The generation algorithm was implemented in Python (3.6), producing center coordinates and dimensions for each void, which were then imported into ABAQUS via scripts.

2.2.2. Finite Element Model Setup in ABAQUS

A representative volume element (RVE) of size 50 mm $\times$ 50 mm $\times$ 25 mm was used for uniaxial compression simulation. The matrix was assigned linear elastic properties (Table 1). Voids were modeled as empty spaces by assigning extremely low stiffness values ($E \approx 1 \times 10^{-6}$ MPa, $\nu \approx 0.01$) to the void elements, a common numerical technique to represent negligible load-bearing capacity while maintaining mesh connectivity. This approach avoids convergence issues associated with modeling true voids/empty spaces in Abaqus Standard. The boundary conditions simulated uniaxial compression: the bottom surface fixed in the Y-direction, and a displacement-controlled load applied on the top surface. The mesh was generated using a global size of 2 mm after a sensitivity analysis.

2.2.3. Model Variants and Simulation Plan

Models were generated to investigate three variables:

(1) Void shape: sphere, icosahedron, dodecahedron, hexahedron (constant void ratio ~20%, equivalent diameter range 4–6 mm).
(2) Void size: diameter ranges 3–4, 4–5, 5–6, 6–7, 7–8, 8–9 mm (icosahedron shape, constant void ratio ~20%).
(3) Void ratio: 18%, 19%, 20%, 21%, 22% (icosahedron shape, diameter range 3–4 mm).

For each configuration, uniaxial compression was simulated to extract stress, strain, and displacement fields.

2.2.4. Validation of Modeling Approach

To enhance credibility, the modeling methodology was partially validated by comparing stress distributions with analytical solutions for simple void arrangements and with prior numerical studies [17,22]. While full experimental validation is beyond this paper's scope, the consistent trends observed in the literature lend support to the approach.

2.3. Image-Based Microstructure Characterization

Although this study primarily uses computer-generated models, we acknowledge the importance of digital image processing (DIP) for real microstructure quantification. For instance, techniques like those used in [21,23] could provide actual void shape distributions to inform model parameters. Future work will integrate CT-based void morphology statistics into the stochastic generation process.

Figure 1 shows a simulated 3D random spherical void model (RVE: 50 $\times$ 50 $\times$ 25 mm) with void diameters ranging from 3 to 9 mm and a void ratio of 20%. Figure 2 shows the

https://doi.org/10.3390/coatings16020214

meshed model. The global mesh size was 2 mm, ensuring at least 2–3 elements across smaller voids.

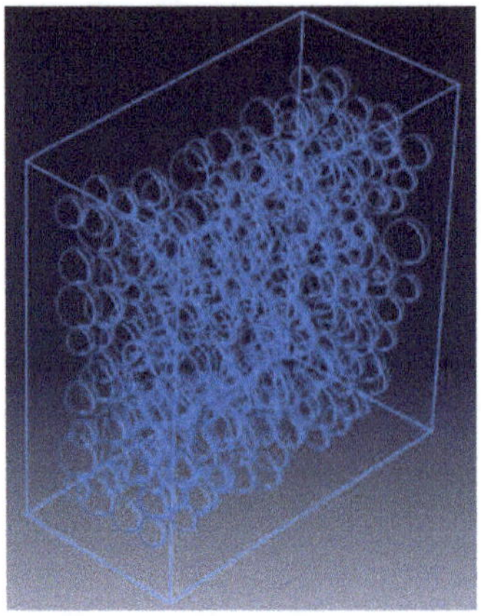

Figure 1. 3D random sphere void model.

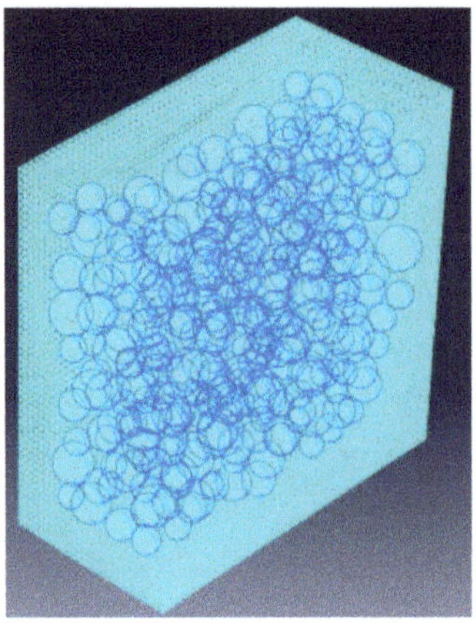

Figure 2. Divided three-dimensional random sphere void model.

3. Results: Deformation Analysis in Relation to Void Patterns

3.1. Effect of Void Shape on Deformation

The porous asphalt mixture contains a large number of individual voids to form a complex spatial network structure system, and the mechanical properties of porous asphalt mixtures are deeply affected by this. Differences in the size, shape, and number of voids make the internal composition of porous asphalt mixtures vary widely, which can also be called porous asphalt mixtures have different void patterns, but different void patterns will have different degrees of impact on the performance of asphalt mixtures.

First, four models with the same gradation and void shapes of sphere, n-octahedron, n-dodecahedron, and n-hexahedron are established, and the X-direction strain cloud diagrams of each model after the loading calculation are shown in Figure 3, and the values of each strain are shown in Figure 4.

From Figure 4, it can be found that the tensile strain is less than the compressive strain for the four different void shape models in the X- and Y-directions. Starting from the

 https://doi.org/10.3390/coatings16020214

icosahedron, as the number of polyhedral prisms decreases, the compressive strain in the Y-direction also increases with decreasing number of edges in the void model, in which the compressive strain increases about 20 times from the icosahedron to the dodecahedron. To deeply analyze the trend of the strain, the maximum values of the tensile and compressive strains in the X-, Y-, and Z-directions were regressed and are shown in Figure 5.

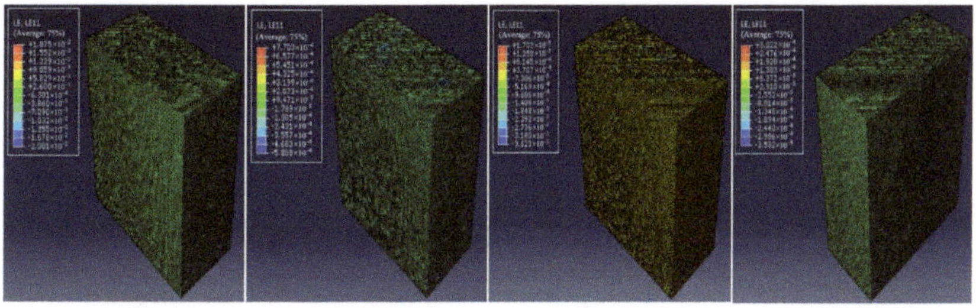

(**a**) sphere (**b**) icosahedron (**c**) dodecahedron (**d**) hexahedron

Figure 3. X-direction strain cloud images of four different void shape models.

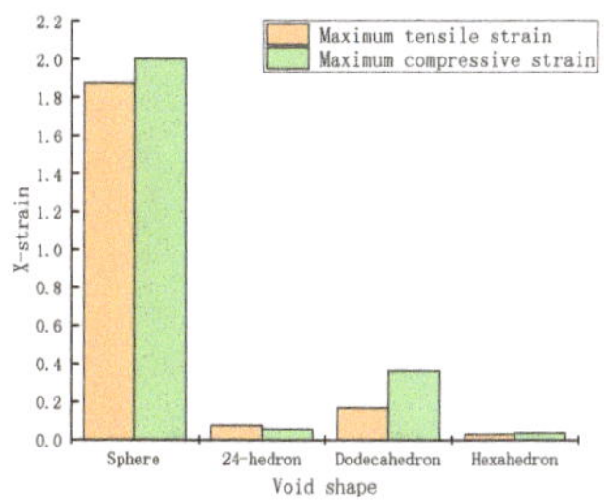

(**a**) X-directional strain

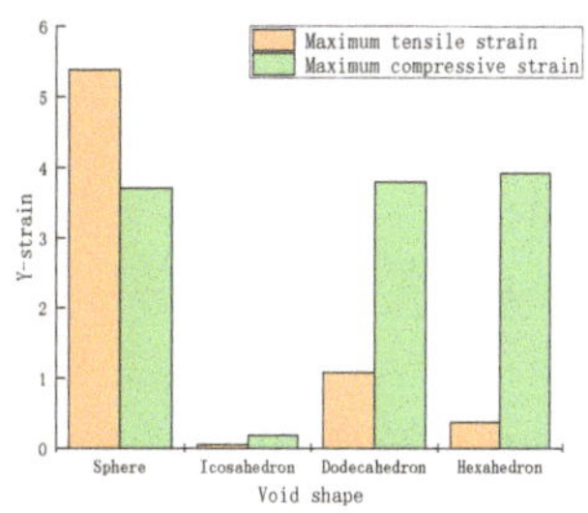

(**b**) Y-direction strain

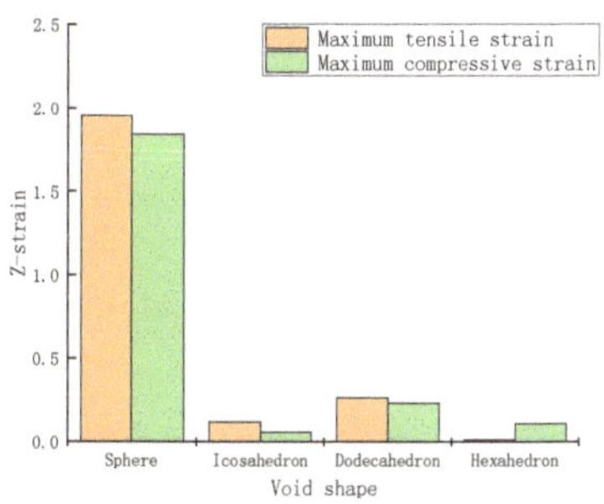

(**c**) Z-direction strain

Figure 4. X, Y, Z strain histograms of four different void shape models.

From the above comparison results, it is easy to observe that the maximum tensile and compressive strains in the Y-direction, along with the maximum compressive strains in the Z-direction, increase as the number of polyhedral prisms decreases. The remaining maximum tensile and compressive strains in the X-direction and the maximum tensile stress in the Z-direction exhibit a decreasing trend. However, since the above maximum strain values only refer to the changes at each point in the model, to see the changes in the overall model, it is necessary to use the highest displacement vectors of the three models

 https://doi.org/10.3390/coatings16020214

with different void shapes to characterize the total vertical changes in the models under standard axial loads. As an example, the displacement vectors in the top Y-direction of the icosahedron, dodecahedron, and hexahedron void models are shown in Figure 6.

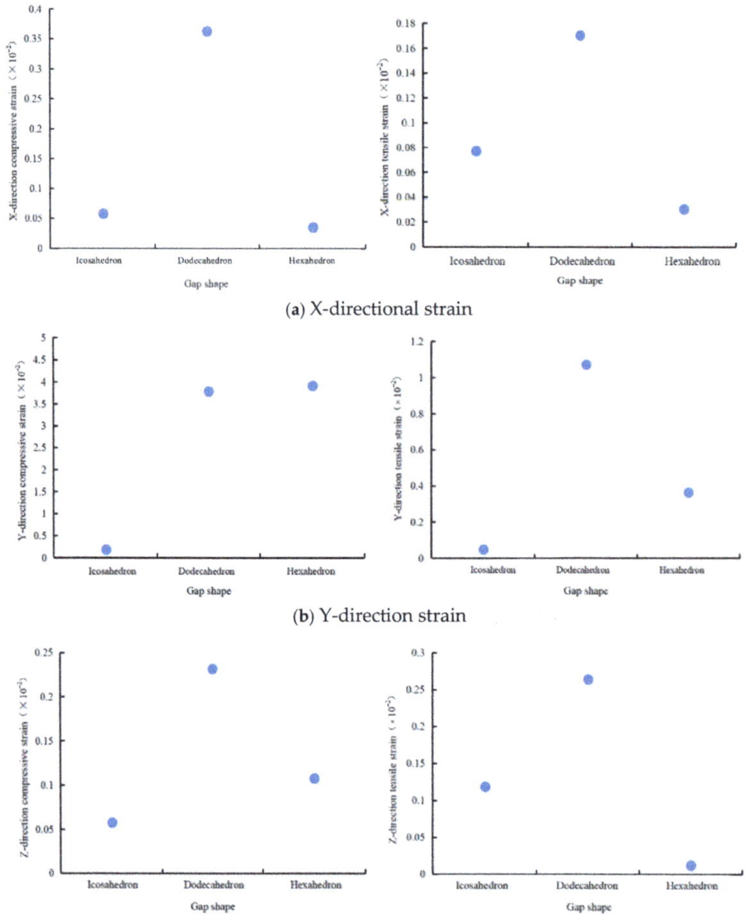

(**a**) X-directional strain

(**b**) Y-direction strain

(**c**) Z-direction strain

Figure 5. Individual strain cases for the icosahedral-hexahedral model.

The different positions indicated along the horizontal position of the tip indicate the unevenness of the change in the pavement. By observation, it is observed that regardless of the shape of the voids, the change at the two ends is always smaller than that in the central part. However, mainly because the test simulates a uniaxial situation, both the left and right sides are constrained and, consequently, their deformation is also affected. From the icosahedron to the hexahedron, the trend of the change in the displacement in the Y-direction of the tip is roughly similar for all models as the horizontal distance from the tip increases. Starting from the icosahedron, the magnitude of the deformation slowly increases as the number of polyhedral void prongs decreases, with the hexahedral void model having the largest deformation. Therefore, it can be inferred that the deformation of the fine-view void model increases as the number of polyhedral prongs decreases.

https://doi.org/10.3390/coatings16020214

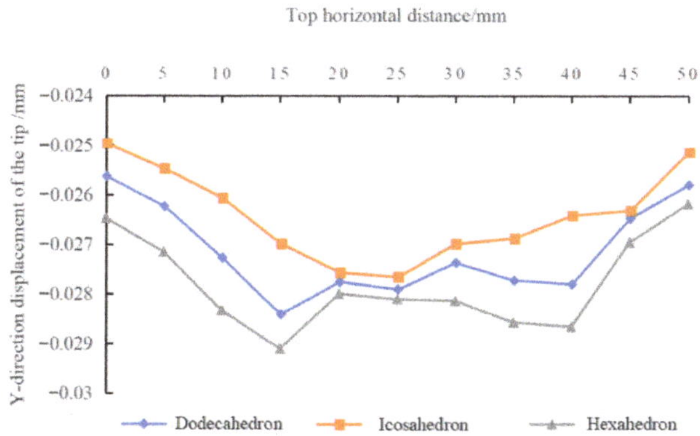

Figure 6. Sum of the displacement vectors in the Y-direction of the top three models of the void shape.

Again, to make the pattern more obvious, the icosahedron and hexahedron void models were taken separately to compare the cloud plots at the places of maximum strain. As seen in Figure 7, the strains at the edges of the voids, as well as between the voids and the voids, are somewhat larger than elsewhere for both the icosahedron and the hexahedron. In addition, comparing Figure 7a,b still shows that the tip strain of the voids increases a lot due to the fact that the corners of the ortho-hexahedron voids are a little more sharply defined, which indicates that the tip responds sharply to the mechanical properties of the porous asphalt mixtures with a considerable effect.

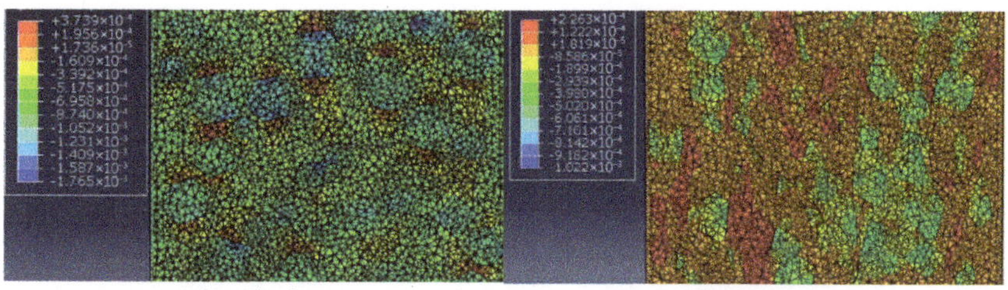

(**a**) icosahedron (**b**) hexahedron

Figure 7. Partial strain cloud diagrams of tetrahedral and hexahedral void models.

The path A-A of the modeled icosahedral void is established, and the X-, Y-, and Z-direction strains along this path are shown in Figure 8. Comparing the path A-A plot of the icosahedral void with the strains along this path, it can be seen that, at the two adjacent voids, the X-, Y-, and Z-direction strains all show peaks. It can also be analyzed that the entire icosahedral model produces the maximum Y-directional strain, followed by the X-directional strain, and the minimum Z-directional strain.

Observation of Figure 8 reveals that, first, the X-direction tensile strain and Y-direction compressive strain vary linearly in each range where the path intersects with each void. Second, the tensile strain of the X-direction at 30 mm and the compressive strain of the Y-direction at 18 mm of the path reach the maximum value, and the strain extremes appear at the intersection of the path and the gap edge, which belongs to the case where the two gap corners are adjacent to each other. This implies that the region adjacent to the gap is highly strained and susceptible to damage such as cracking.

https://doi.org/10.3390/coatings16020214

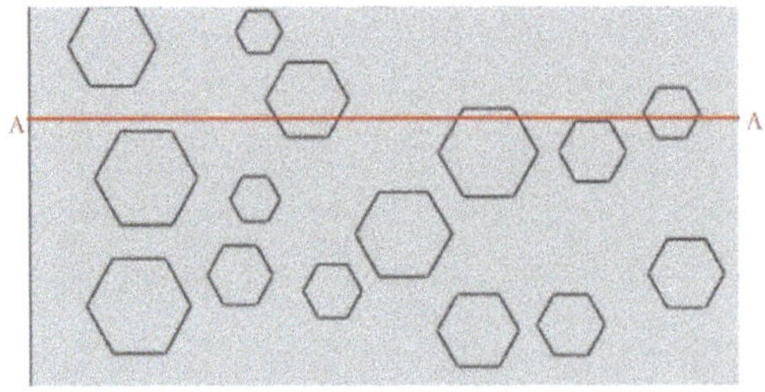

(a) Void model path A-A

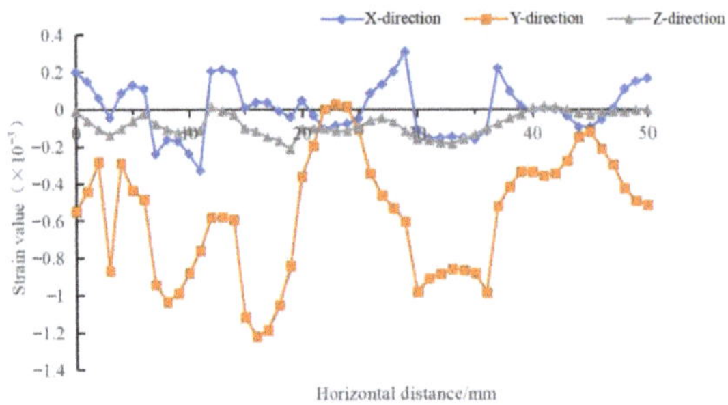

(b) X, Y, Z strain of path A-A

Figure 8. Twenty-tetrahedral void model path A-A and the strain on this path.

3.2. *Effect of Void Size on Deformation*

Six models with void diameters ranging from 3 to 4 mm, 4 to 5 mm, 5 to 6 mm, 6 to 7 mm, 7 to 8 mm, and 8 to 9 mm were established, and then the strains under standard axial loads were recorded, and strain cloud diagrams of the Y-direction of various models (see Figure 9) and the various strains (see Figure 10) were recorded.

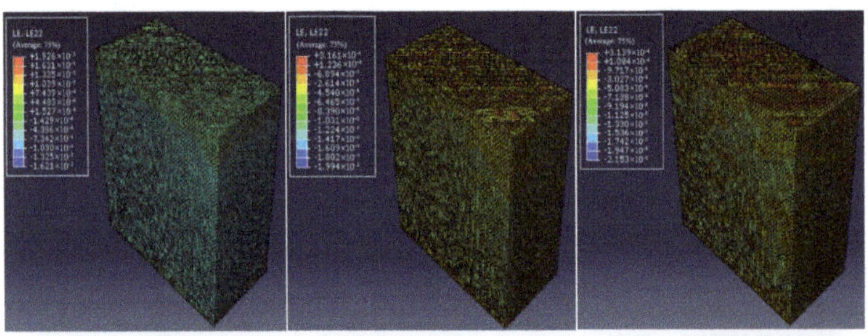

(a) Gap diameter 3~4 mm (b) Gap diameter 4~5 mm (c) Gap diameter 5~6 mm

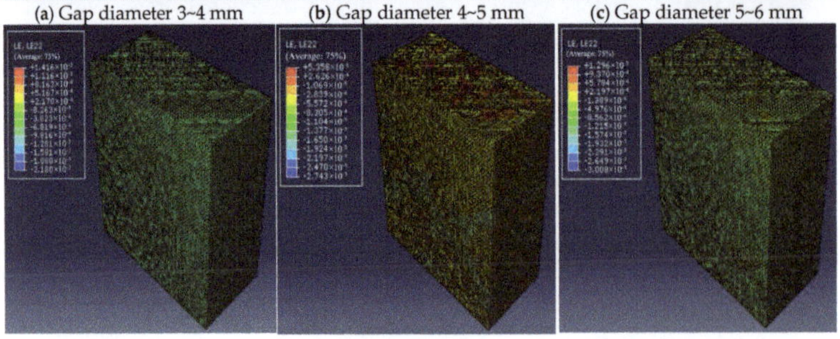

(d) Gap diameter 6~7 mm (e) Gap diameter 7~8 mm (f) Gap diameter 8~9 mm

Figure 9. Y-direction strain cloud diagrams of six models with different gap diameters.

 https://doi.org/10.3390/coatings16020214

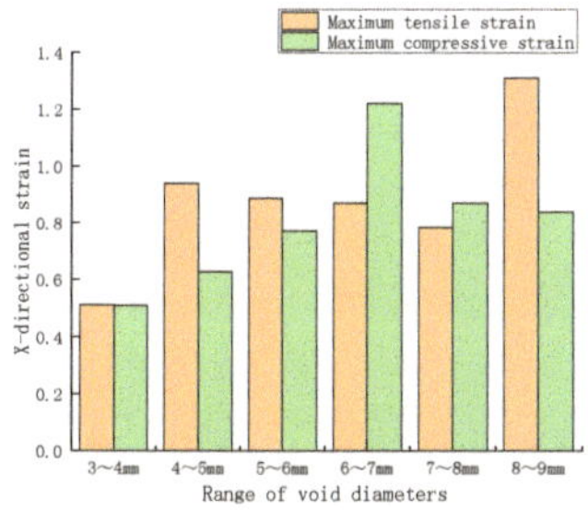

(**a**) X-direction strain histogram

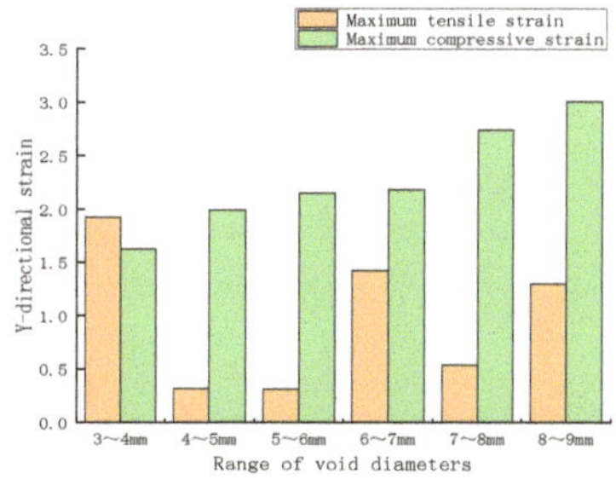

(**b**) Y-direction strain histogram

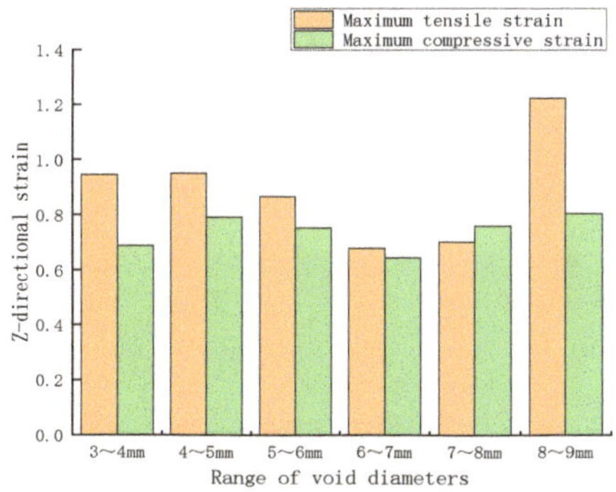

(**c**) Z-direction strain histogram

Figure 10. Histogram of maximum strain in X-, Y-, and Z-directions of models with different gap ranges.

Comparing (a), (b), and (c) in Figure 10, it is easy to see that the overall trend of X- and Y-directional strains increases with the increase in the void diameter. Among them, the compressive strain in the X-direction removes the compressive strain of the model with a gap range of 6~7 mm, and the rest of them are steadily increasing with an increase in the gap diameter range. As the diameter of the voids increases, the effective bearing area continuously decreases, the stability of the skeleton constantly drops, and the bonding performance of the asphalt mortar declines, resulting in a linear decrease in the material's resistance to compression. Therefore, the compressive strain steadily rises. The tensile strain in the Y-direction is much smaller than the compressive strain, which is also due to the fact that the asphalt mixture model is subjected to the action of the vertical direction load. The Z-direction deformation, except for the tensile deformation in the model with a gap range of 8~9 mm, is a little higher, and the rest of them are similar in size. The tensile and compressive strains are similar in size, and the rest of them are similar in value. The Z-direction strain, except for the 8~9 mm gap range model, which has a slightly higher value of tensile strain, the rest of the tensile and compressive strains are similar in size, without an obvious pattern of change.

3.3. *Effect of Void Ratio on Deformation*

The void ratio is modeled as 18%, 19%, 20%, 21%, and 22%, and the strain results are shown in Figures 11 and 12.

From Figure 12, it can be seen that the effect of the size of the void ratio on the X-direction strain is more obvious, and the X-direction strain increases linearly with the

 https://doi.org/10.3390/coatings16020214

increase in the void ratio; then we look at the compressive strain of the Y-direction, the maximum compressive strain of the model with 18% void ratio is the smallest, which is 1.842×10^{-3}, and the maximum compressive strain is the model with a void ratio of 19%, which is 4.058×10^{-3}, and more than doubled; finally, the effect of the void ratio on the maximum compressive strain of the Z-direction is not very large, and the magnitudes are similar for models with different void ratios. Figure 13 depicts the vertical deformation plots (also called top displacement vector sums) for different models of the void ratio.

The results show that the vertical deformation of the model increases with the increase in void ratio, and the vertical deformation of the model also increases from −0.026 to −0.031 when the void ratio increases from 18% to 22%, and the trend of the vertical deformation at the top of the different void ratios is more or less the same, except that the change in the sum of the top displacement vectors of the models of the remaining void ratios starts to fluctuate abnormally after the void ratio increases to 21%. Therefore, it can be concluded that the deformation of porous asphalt mixtures increases with increasing void ratio, and the 18% void ratio model has the smallest vertical deformation but the best mechanical properties.

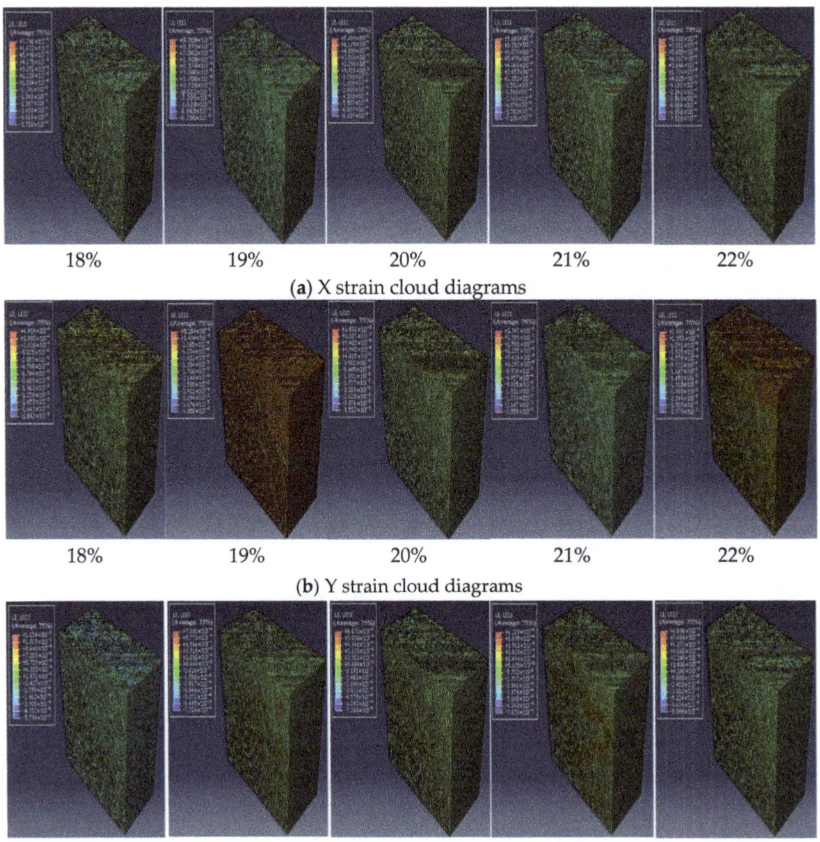

18% 19% 20% 21% 22%

(**a**) X strain cloud diagrams

18% 19% 20% 21% 22%

(**b**) Y strain cloud diagrams

18% 19% 20% 21% 22%

(**c**) Z strain cloud diagrams

Figure 11. X, Y, and Z strain cloud diagrams of models with porosity of 18%, 19%, 20%, 21%, 22%.

 https://doi.org/10.3390/coatings16020214

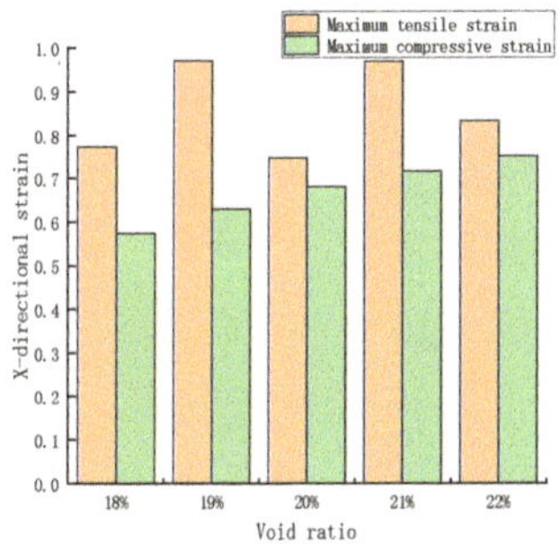

(a) X-direction strain histogram

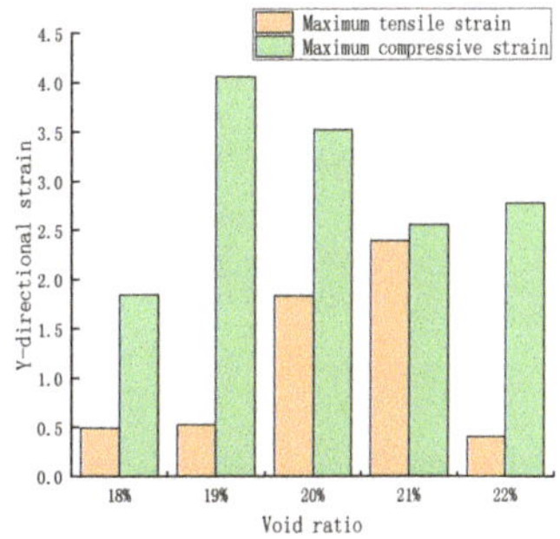

(b) Y-direction strain histogram

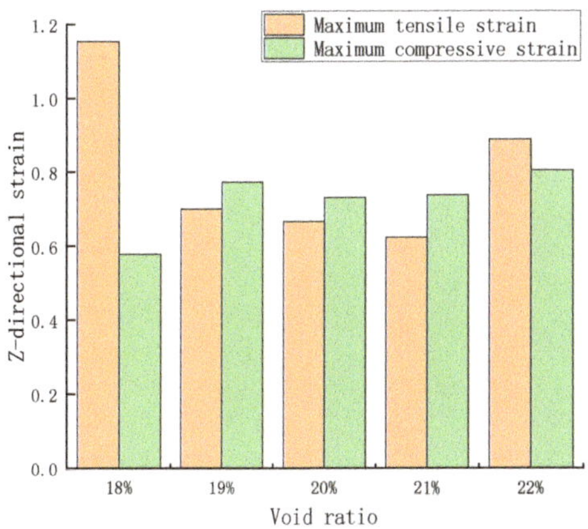

(**c**) Z-direction strain histogram

Figure 12. X, Y, and Z strain histograms of six sets of models in the void range.

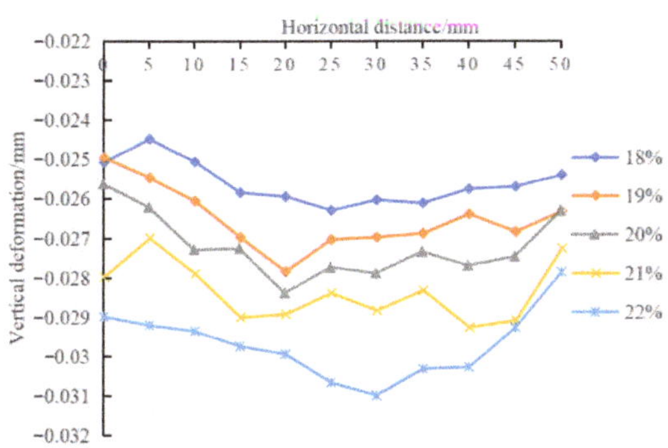

Figure 13. Top vertical deformation map with different void ratios.

4. Results: Mechanical Properties in Relation to Void Morphology

4.1. Effect of Void Shape on Mechanical Properties

The simulation study is carried out for four different shapes of void models with the same gradation. The void shapes are spherical, icosahedral, dodecahedral, and hexahe-

https://doi.org/10.3390/coatings16020214

dral, and their anisotropic stress clouds are shown in Figure 14, and the values of stress magnitude for each model are shown in Figure 15.

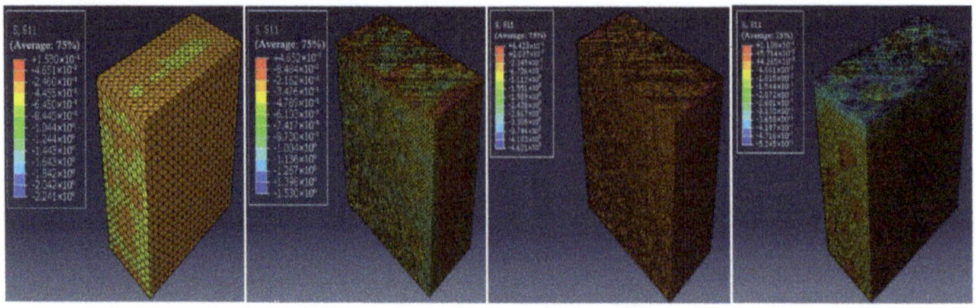

(**a**) Sphere (**b**) Icosahedron (**c**) Dodecahedron (**d**) Hexahedron

Figure 14. X-direction stress cloud images of four different void shape models.

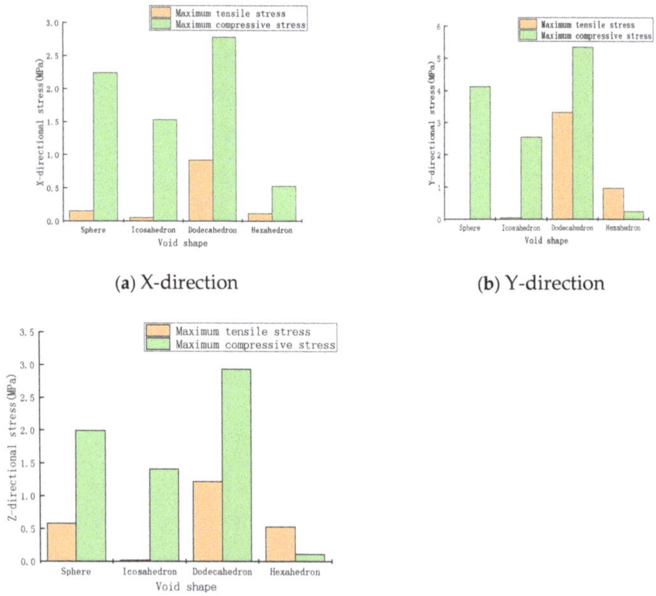

(**a**) X-direction

(**b**) Y-direction

(**c**) Z-direction

Figure 15. The histogram of the maximum stress in the X-, Y-, and Z-directions of four different void shape models.

In Figure 14, it can be seen that the stress extremes appear in the middle of the neighboring voids, and the closer the voids are to each other, the stronger the stress response. In Figure 15, the maximum pressure value in the Y-direction appears in the void model with a dodecahedron shape, with a value of 5.349 MPa, and the minimum value appears in the void model with an icosahedron shape, with a value of 2.552 MPa. Obviously, both 5.349 MPa and 2.552 MPa are much larger than the standard axial load of 0.7 MPa. This also indicates that the analysis of porous asphalt mixtures from a detailed point of view has a strong influence on the response to stress. Perspective analysis has a great influence on the mechanical properties of porous asphalt mixtures. Figure 16 depicts the compressive stress in the X-direction and its comparison results for each model with different void shapes.

https://doi.org/10.3390/coatings16020214

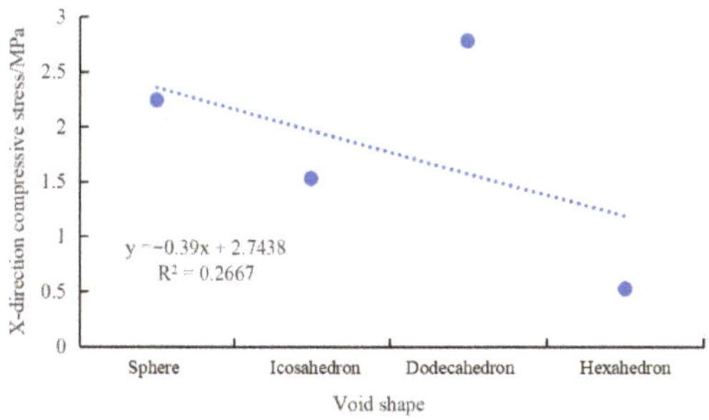

Figure 16. The relationship between the void shape and compressive force of the X-direction stress and its regression results.

Combining Figures 15 and 16, it is not difficult to find that porous asphalt mixtures have a significant mechanical response to the change in void shape. Specifically, as the number of void edges decreases, the compressive stress in the X-direction of different models of the shape of the void shows a linear decreasing trend, while the change in the compressive stress in the Y-direction is also quite significant. These observations further confirm the important influence of the shape of the void on the mechanical properties of porous asphalt mixtures.

To investigate in depth the specific effects of void shapes on the mechanical properties of porous asphalt mixtures, we chose an icosahedral void model as the research object and set specific research paths on it. These paths include horizontal paths A′-A′ shown in Figure 17 and vertical paths B′-B′ shown in Figure 18.

Looking at and analyzing Figure 17 carefully, it is not difficult to find that the stress extremes always appear at the locations where paths A-A are in contact with each void, in which the Y-directional compressive stress at the adjacent corners of two different voids even reaches 1.416 MPa. The remaining paths are in contact with each void range, and the changes in X- and Y-directional stresses are mostly manifested as the linear compressive stress.

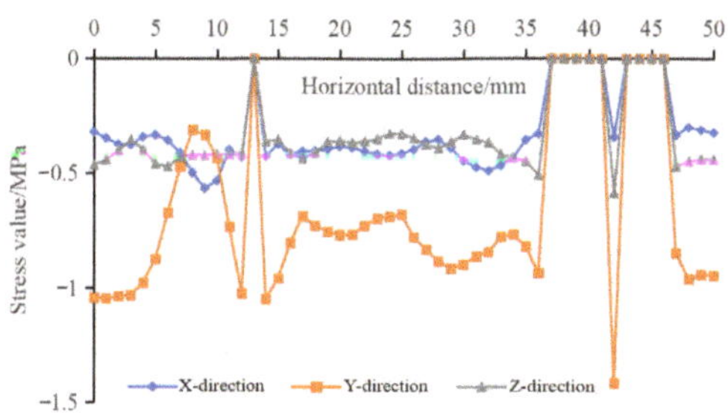

Figure 17. Twenty-tetrahedral void model path A′-A′ and stress on this path.

The comparative analysis of Figure 18 shows that, first, as in the analysis of paths A-A above, the stress extremes on paths B-B all occur at the locations where the paths are in contact with each void edge. Second, the stresses on the path B-B in both the x-direction and the y-direction exhibit linear compressive stresses. Obviously, the damage-prone areas in porous asphalt mixtures are the sharp edges of the voids.

https://doi.org/10.3390/coatings16020214

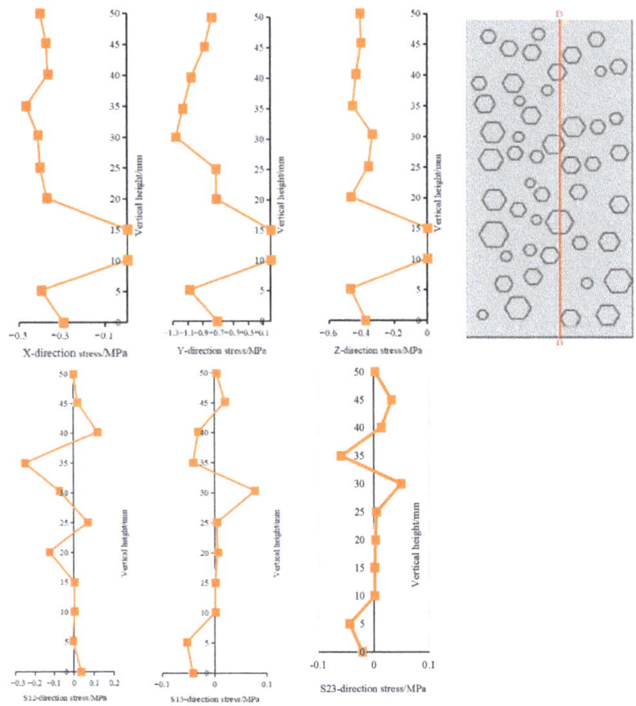

Figure 18. Stress in all directions on path B-B.

4.2. *Effect of Void Size on Mechanical Properties*

The X-direction stress clouds for the icosahedral void model with different sizes are shown in Figure 19, and the stress data in each direction are shown in Figure 20.

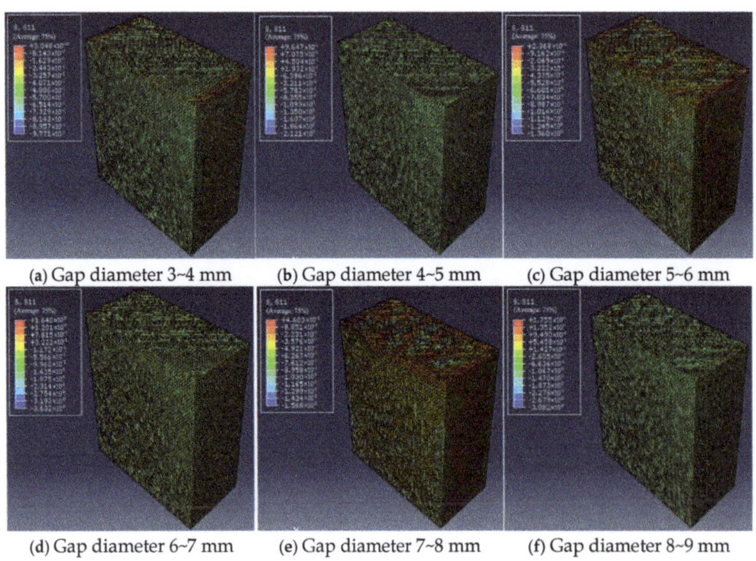

Figure 19. The stress cloud diagram in the x-direction of a tetrahedral void model with the same void ratio and different diameter ranges.

https://doi.org/10.3390/coatings16020214

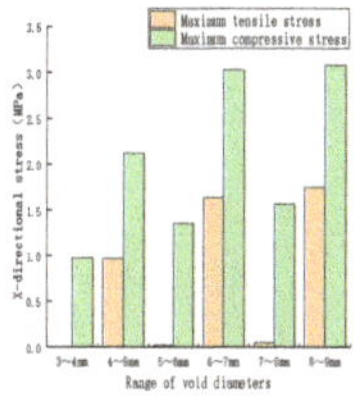

(a) X-direction stress diagram

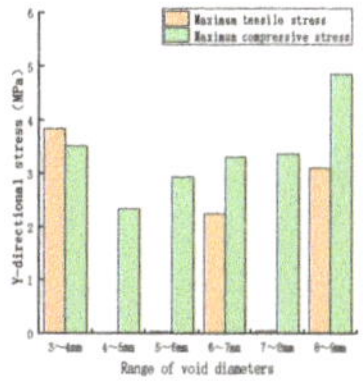

(b) Y-direction stress diagram

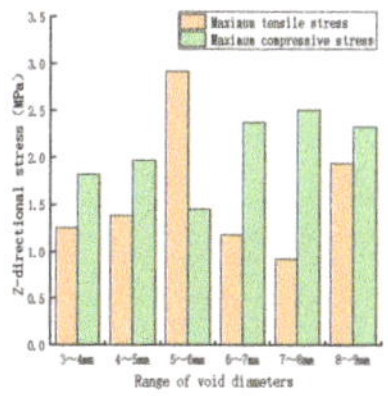

(c) Z-direction stress diagram

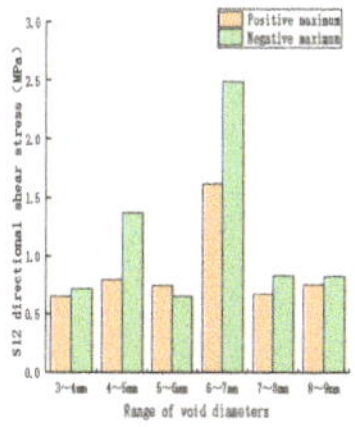

(d) S12 shear stress plots

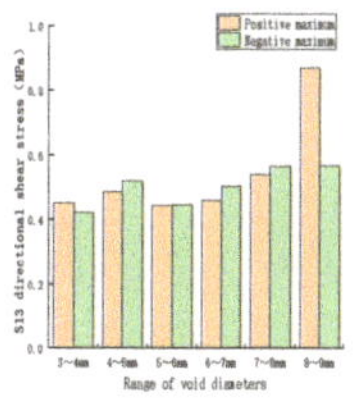

(e) S13 shear stress plots

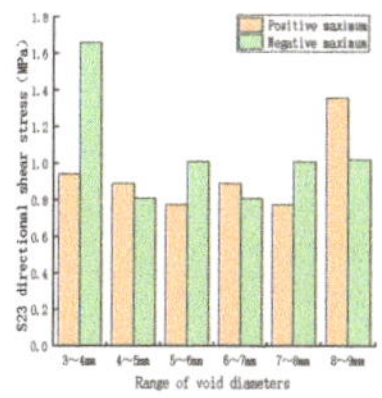

(f) S23 shear stress plots

Figure 20. Histograms of the X-, Y-, and Z-directions and maximum shear stress of the tetrahedral. void models with different diameter ranges.

Combining Figures 19 and 20, it is found that the Y-directional compressive stress increases with a larger range of void diameters. Therefore, the model with a 3~4 mm void diameter range has the best stress.

4.3. Effect of Void Ratio on Mechanical Properties

The shear stress clouds for the models with voids of 18%, 19%, 20%, 21%, and 22% are shown in Figure 21, and the stresses for each model are shown in Figure 22.

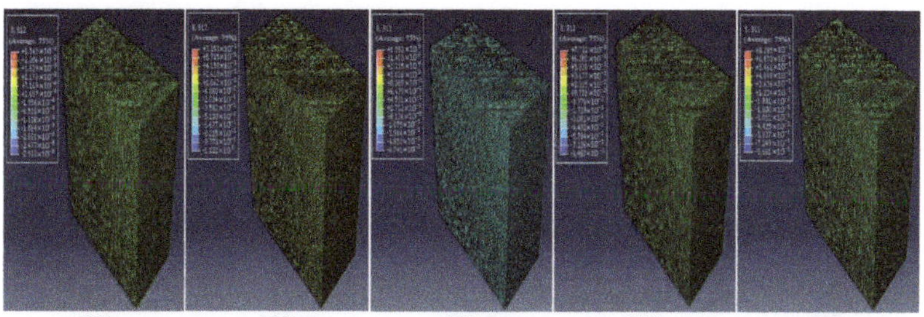

18% 19% 20% 21% 22%

(a) Shear stress clouds in the S12 direction

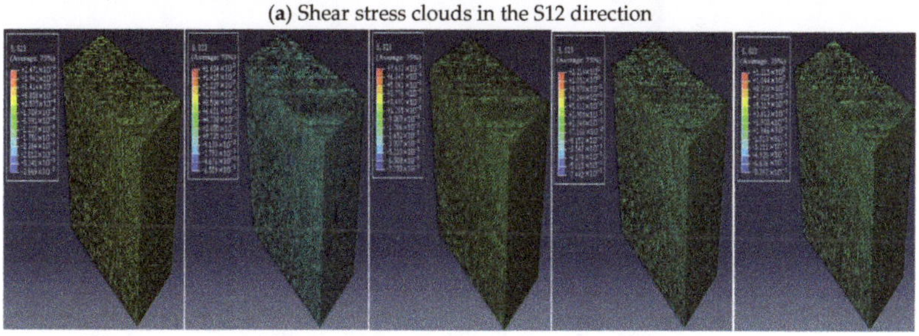

18% 19% 20% 21% 22%

(b) Shear stress clouds in the S13 direction

Figure 21. *Cont.*

 https://doi.org/10.3390/coatings16020214

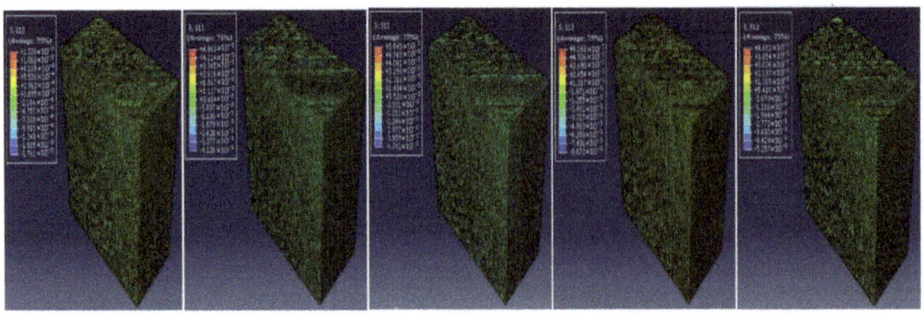

18% 19% 20% 21% 22%

(**c**) Shear stress clouds in the S23 direction

Figure 21. Shear stress contours of models with different porosity.

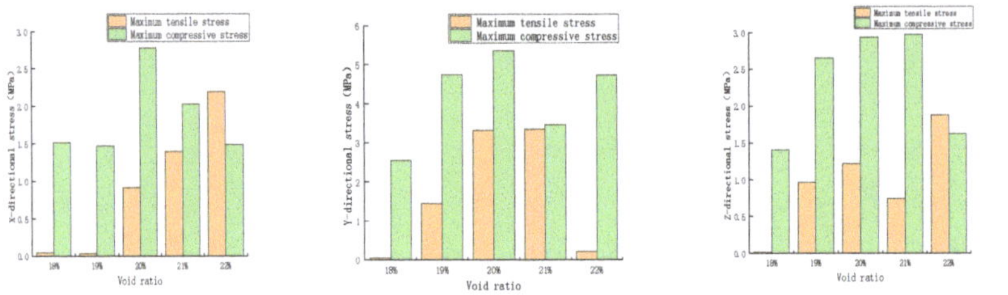

(**a**) X-direction stress diagram (**b**) Y-direction stress diagram (**c**) Z-direction stress diagram

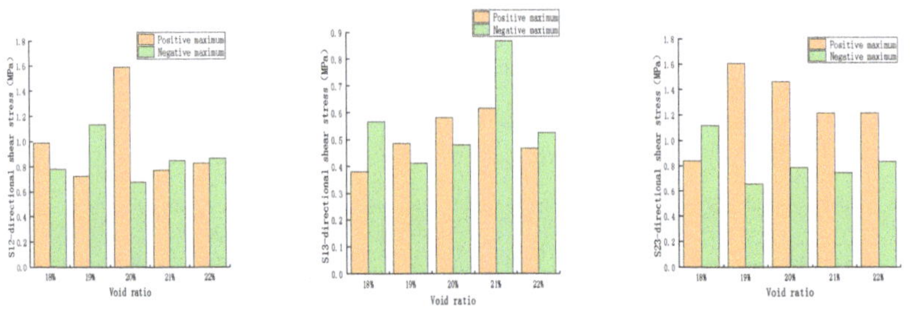

(**d**) S12 shear stress plots (**e**) S13 shear stress plots (**f**) S23 shear stress plots

Figure 22. Histogram of the X-, Y-, and Z-directions, and maximum shear stress of models with different porosities.

Via a comprehensive analysis of the above graphs, it can be found that the 18% void ratio model has the least force in all directions, i.e., the best mechanical properties.

5. Discussion

This study numerically investigated the mesoscale mechanical response of porous asphalt mixtures as a function of void morphology. The key finding—that an icosahedral void shape with a diameter of 3–4 mm and 18% void ratio yields optimal mechanical performance—can be discussed in light of existing literature.

First, regarding the finding of the optimal void shape (icosahedron), although precise control of void geometry is challenging in actual production, this result highlights the importance of reducing sharp edges and corners in order to alleviate stress concentration. This aligns with the conclusions of numerous studies that emphasize the beneficial effect of void roundness on durability [13,21]. The icosahedron, with its larger number of faces and

 https://doi.org/10.3390/coatings16020214

more uniform curvature, can be considered as an "idealized" approximation of a smooth polyhedral void.

Second, concerning the optimal void size (3–4 mm) and void ratio (18%), our simulations indicate that smaller voids can reduce local strain concentration while maintaining a relatively high void ratio. This agrees with the phenomenon often observed in experimental studies, where smaller interconnected voids provide a better balance between drainage capacity and strength [5,9]. However, the actual void size distribution in real mixtures is continuous. Future models should consider realistic distributions obtained from CT images [12,15].

Third, the Monte Carlo + ABAQUS modeling approach adopted in this study shows consistent predictive trends with other similar studies [24,25], such as the decrease in stiffness with increasing void ratio. Nevertheless, our work further quantifies the coupled influence of shape and size, providing more detailed guidance for microstructural design.

Limitations: The model simplifies the matrix as linear elastic, neglecting the viscoelastic and damage evolution behavior of asphalt. The void shapes are regular geometric bodies, differing from real irregular shapes. The aggregate–mortar interface is not explicitly modeled. These simplifications may affect quantitative accuracy, but the qualitative trends should align with the underlying physical mechanisms. While direct shape control is infeasible, mix design and compaction practices that promote aggregate rearrangement and mastic flow (e.g., using rounded aggregates, optimal binder viscosity) can indirectly encourage the formation of less angular, more favorable void networks.

Practical Implications: The findings suggest that during mix design and compaction processes, efforts should be made to avoid creating sharp and large-sized voids. Gradation optimization should be pursued to promote the formation of a more rounded, uniformly distributed, small-void structure, thereby enhancing the mixture's resistance to deformation and damage. For example, adjusting aggregate angularity or using modified asphalt to improve coating may indirectly foster more favorable void morphology.

Future Research: Future studies could integrate CT scanning to acquire statistical parameters of real void morphology, driving more realistic stochastic models; introduce viscoelastoplastic constitutive models to simulate time-dependent behavior; and conduct parallel experimental validation, particularly using techniques like digital image correlation to measure local strain fields.

6. Conclusions

(1) Among the four different void shapes modeled, the maximum strain values increased as the number of void edges decreased. Tensile strains were consistently lower than compressive strains. The compressive strain in the Y-direction increased notably with fewer edges, while tensile and compressive strains in the X- and Y-directions exhibited linear trends, aligned with stress concentration effects at sharper void corners. Reducing void sharpness (i.e., promoting more rounded, less angular void geometries) is beneficial for mechanical performance.

(2) Both tensile and compressive strains in the X-direction increased with void ratio. The compressive strain in the Y-direction also rose with void ratio, whereas the tensile strain in the Y-direction changed only slightly over the 18%–22% range, indicating a weaker sensitivity to void ratio.

(3) The model with icosahedral voids, a void diameter range of 3–4 mm, and a void ratio of 18% exhibited the most favorable mechanical performance in terms of minimized stress concentration, lower overall deformation, and higher stiffness retention. This

https://doi.org/10.3390/coatings16020214

configuration represents a micromechanical design goal for enhancing the durability of porous asphalt mixtures.

Practical Implications and Future Work: The findings provide a mesoscale rationale for material design: reducing void sharpness and controlling void size distribution can improve mechanical performance without sacrificing drainage capacity. For industry, this implies that attention to aggregate shape, compaction uniformity, and binder properties that promote smoother void walls could yield longer-lasting porous pavements. Future studies should integrate real void morphology from CT scans, incorporate viscoelastic and damage models, and validate predictions through coordinated experimental testing.

Author Contributions: H.W. (Hongchang Wang): conceptualization, formal analysis, project administration, writing—review and editing; H.W. (Haozhe Wu): data analysis, methodology, writing—original draft, C.L.: visualization, writing—original draft. All authors have read and agreed to the published version of the manuscript.

Funding: This research received no external funding.

Institutional Review Board Statement: Not applicable.

Informed Consent Statement: Not applicable.

Data Availability Statement: All data, models, and codes generated or used in this study are included in the submitted manuscript.

Acknowledgments: The authors would like to express sincere gratitude to all organizations that supported this research.

Conflicts of Interest: The authors declare no conflicts of interest.

References

1. Ramadhansyah, P.J.; Masri, K.A.; Mangi, S.A.; Yusak, M.M.; Mashros, N.; Warid, M.M.; Satar, M.K.; Haziman, W.M. Strength and Porosity of Porous Concrete Pavement Containing Nano Black Rice Husk Ash. *IOP Conf. Ser. Mater. Sci. Eng.* **2020**, *712*, 012037. [CrossRef]
2. Ma, X.; Li, Q.; Cui, Y.C.; Ni, A.Q. Performance of porous asphalt mixture with various additives. *Int. J. Pavement Eng.* **2018**, *19*, 355–361. [CrossRef]
3. Wu, H.; Chu, Y.; Gu, X.; Tang, J. Multilevel Characterization of Bonding Strength of High Viscosity Modified Asphalt Mortar and Its Correlation with Failure Forms of Porous Mixtures. *Build. Struct.* **2023**, *53*, 1390–1398.
4. Yan, B.; Yao, B. Research on Aggregate Skeleton Optimization of Porous Asphalt Mixture Based on Discrete Element Method. *Highway* **2025**, *70*, 14–21.
5. Gu, H. Research on the Influence of Coarse Aggregate Texture on the Road Performance of Steel Slag Porous Asphalt Mixture. *Hunan Commun. Sci. Technol.* **2023**, *49*, 65–68+73.
6. Ma, X.; Hu, X.Q.; Zhou, P.S. Internal structure changes within porous asphalt mixture with different saturation under vertical repeated load. *Constr. Build. Mater.* **2023**, *372*, 130727. [CrossRef]
7. Li, J.; He, Z.; Guan, Z. Construction of a physical model for the void ratio of mineral aggregates in porous asphalt mixtures. *J. Harbin Inst. Technol.* **2022**, *54*, 139–147.
8. Ji, X.; Wang, H. Microscopic Void Characteristics Laws of Porous Asphalt Mixture Based on Digital Image Processing Technology. *J. China Foreign Highw.* **2018**, *38*, 257–261.
9. Radzi, N.A.; Masri, K.A.; Ramadhansyah, P.J.; Jasni, N.E.; Arshad, A.K.; Ahmad, J.; Mashros, N.; Yaacob, H. Stability and Resilient Modulus of Porous Asphalt Incorporating Steel Fiber. *IOP Conf. Ser. Mater. Sci. Eng.* **2020**, *712*, 012027. [CrossRef]
10. Shahnewaz, S.M.; Masri, K.A.; Ghani, N.A.; Jaya, R.P.; Choo, C.S.; Giannakopoulou, P.P.; Rogkala, A.; Lampropoulou, P.; Petrounias, P. Porous asphalt mixtures enriched with bamboo fibers as a new approach for future sustainable construction. *Constr. Build. Mater.* **2023**, *407*, 133456. [CrossRef]
11. Yang, X.; You, Z.; Wang, Z.; Dai, Q. Review on heterogeneous model reconstruction of stone-based composites in numerical simulation. *Constr. Build. Mater.* **2016**, *117*, 229–243. [CrossRef]

https://doi.org/10.3390/coatings16020214

12. Qian, N.; Wang, D.; Li, D.; Shi, L. Three-dimensional mesoscopic permeability of porous asphalt mixture. *Constr. Build. Mater.* **2020**, *236*, 117430. [CrossRef]
13. Zhao, Y.; Wang, X.; Jiang, J.; Zhou, L. Characterization of interconnectivity, size distribution and uniformity of air voids in porous asphalt concrete using X-ray CT scanning images. *Constr. Build. Mater.* **2019**, *213*, 182–193. [CrossRef]
14. Zhu, R.; Kang, A.; Zhang, Y.; Zhi, W.; Zhu, L. Feasibility Study of Discrete Element Numerical Simulation Technology-Based Microscopic Model for Asphalt Mixtures. *Jiangsu Build. Mater.* **2023**, *6*, 31–34.
15. Xu, H.; Shi, H.; Tan, Y. Investigation and Characterization of 3DVoid Mesostructures in Asphalt Mixtures. *China J. Highw. Transp.* **2020**, *33*, 210–220.
16. Xu, H.; Shi, H.; Tan, Y. Review on Genome Research on Microscopic Structural Characteristics of Asphalt Concrete. *Chin. J. Nat.* **2020**, *42*, 249–261.
17. Kutuk-Sert, T.; Ozturk, M.; Kutuk, S. Digital image processing of warm mix asphalt enriched with nanocolemanite and nanoulexite minerals. *Constr. Build. Mater.* **2023**, *399*, 132542. [CrossRef]
18. Kutuk, S.; Kutuk-Sert, T. An examination of nanoparticle colemanite mineral added warm mix asphalt. *Constr. Build. Mater.* **2020**, *243*, 118252. [CrossRef]
19. Zhao, Y.; Jiang, L.; Jiang, J.; Ni, F. Accuracy Improvement for Two-Dimensional Finite-Element Modeling while Considering Asphalt Mixture Meso-Structure Characteristics in Indirect Tensile Test Simulation. *J. Mater. Civ. Eng.* **2020**, *32*, 04020275. [CrossRef]
20. Wang, Z.; Xie, J.; Gao, L.; Liu, M.; Liu, Y. Improvement of acoustic model and structural optimization design of porous asphalt concrete based on meso-structure research. *Constr. Build. Mater.* **2020**, *265*, 120327. [CrossRef]
21. Kollmann, J.; Lu, G.; Liu, P.; Xing, Q.; Wang, D.; Oeser, M.; Leischner, S. Parameter optimisation of a 2D finite element model to investigate the microstructural fracture behaviour of asphalt mixtures. *Theor. Appl. Fract. Mech.* **2019**, *103*, 102319. [CrossRef]
22. Wang, D.; Liu, P.; Xu, H.; Kollmann, J.; Oeser, M. Evaluation of the polishing resistance characteristics of fine and coarse aggregate for asphalt pavement using Wehner/Schulze test. *Constr. Build. Mater.* **2018**, *163*, 742–750. [CrossRef]
23. Yang, B.; Li, H.; Zhang, H.; Xie, N.; Zhou, H. Laboratorial investigation on effects of microscopic void characteristics on properties of porous asphalt mixture. *Constr. Build. Mater.* **2019**, *213*, 434–446. [CrossRef]
24. Cao, P.; Jin, F.; Changjun, Z.; Feng, D. Investigation on statistical characteristics of asphalt concrete dynamic moduli with random aggregate distribution model. *Constr. Build. Mater.* **2017**, *148*, 723–733. [CrossRef]
25. Ma, H.; Xu, W.; Li, Y. Random aggregate model for mesoscopic structures and mechanical analysis of fully-graded concrete. *Comput. Struct.* **2016**, *177*, 103–113. [CrossRef]
26. Ma, H.; Song, L.; Xu, W. A novel numerical scheme for random parameterized convex aggregate models with a high-volume fraction of aggregates in concrete-like granular materials. *Comput. Struct.* **2018**, *209*, 57–64. [CrossRef]
27. Wang, Z. Material Design and Performance Simulation Evaluationof UTAC-10 Based on Discrete Element Method. Master's Thesis, South China University of Technology, Guangzhou, China, 2019.
28. Chen, C.; Zhang, Q.; Keer, L.M.; Yao, Y.; Huang, Y. The multi-factor effect of tensile strength of concrete in numerical simulation based on the Monte Carlo random aggregate distribution. *Constr. Build. Mater.* **2018**, *165*, 585–595. [CrossRef]
29. Zhang, F.; Teng, X.; Binbin, L. Meso mesoscopic model of asphalt concrete based on random aggregate placement. *J. Railw. Sci. Eng.* **2019**, *16*, 1216–1223.
30. Yuan, P.; Wang, D. Monte Carlo Simulation Analysis of Variability of Mechanical Responses for Asphalt Pavement. *J. South China Univ. Technol. (Nat. Sci. Ed.)* **2008**, *36*, 55–63.
31. Cao, P.; Jin, F.; Feng, D.; Zhou, C.; Hu, W. Prediction on dynamic modulus of asphalt concrete with random aggregate modeling methods and viscoelastic theory. *Constr. Build. Mater.* **2016**, *125*, 987–997. [CrossRef]
32. Xu, Z.J. *Monte Carlo Methodology*; Shanghai Science & Technology Publishers: Shanghai, China, 1985.

https://doi.org/10.3390/coatings16020214

coatings

Article

Mesoscale Mechanical Analysis of Concrete Based on a 3D Random Aggregate Model

Shuaishuai Wei [1], Huan Zhang [2,*], Ding Wang [3,4], Xuchun Wang [1] and Mengdi Cao [2]

[1] School of Civil Engineering, Qingdao University of Technology, Qingdao 266520, China; shuaishuai_wei@163.com (S.W.); xuchunwang_qut@163.com (X.W.)
[2] Shandong Transportation Institute, Jinan 250102, China; 13188871310@163.com
[3] School of Mechanical and Electrical Engineering, China University of Mining and Technology, Beijing 100083, China; dingwang_qut@163.com
[4] State Key Laboratory for Tunnel Engineering, China University of Mining and Technology, Beijing 100083, China
* Correspondence: zhanghuan095@163.com

Abstract: The shape, size, and interfacial transition zone (ITZ) of aggregates significantly impact the nonlinear mechanical behavior of concrete. This study investigates concrete's mechanical response and damage mechanisms by developing a three-dimensional, three-phase mesoscale model comprising coarse aggregates, mortar, and ITZ to explore the compressive performance of concrete. A method for simulating the random distribution of aggregates based on three-dimensional grid partitioning is proposed, where the value of each grid point represents the maximum aggregate radius that can be accommodated if the point serves as the aggregate center. Aggregates are generated by randomly selecting grid points that meet specific conditions, avoiding overlapping distributions and significantly improving computational efficiency as the generation progresses. This model effectively enhances the precision and efficiency of aggregate distribution and provides a reliable tool for studying the random distribution characteristics of aggregates in concrete. Additionally, an efficient discrete element model (DEM) was established based on this mesoscale model to simulate the compressive behavior of concrete, including failure modes and stress–strain curves. The effects of aggregate shape and maximum aggregate size on the uniaxial compressive failure behavior of concrete specimens were investigated. Aggregate shape has a particular influence on the compressive strength of concrete, and the compressive strength decreases with an increase in maximum aggregate size. Combined with existing experimental results, the proposed mesoscale model demonstrates high reliability in analyzing the compressive performance of concrete, providing valuable insights for further research on the mechanical properties of concrete.

Keywords: aggregate shape; maximum aggregate size; DEM; mesoscopic simulation; compressive strength

1. Introduction

Concrete is a typical heterogeneous multiphase composite material composed of coarse aggregates, mortar, and the interfacial transition zone (ITZ) between them. Due to its excellent mechanical properties, durability, and engineering adaptability, concrete is widely used in various civil engineering applications [1–3]. However, significant physical and mechanical differences exist among the internal components of concrete, especially regarding the geometry, size distribution, and spatial arrangement of aggregates, which greatly influence the overall mechanical response and failure mechanism of concrete [4,5]. Therefore, accurately characterizing the spatial distribution of aggregates and conducting multiscale mechanical behavior analysis based on this is key to understanding the nonlinear mechanical properties of concrete materials. Mesoscale numerical simulation of concrete holds significant theoretical and practical importance [6].

In recent years, with the continuous advancement of computational mechanics and material modeling technologies, mesoscale numerical simulation methods have received widespread attention in the mechanical study of concrete [7–9]. Particularly at the two-dimensional (2D) level, various random aggregate generation methods have been proposed, such as the Monte Carlo method, digital image-based techniques, and Voronoi diagram-based algorithms [10–13]. These approaches have achieved notable progress in simulating aggregate shape, size distribution, and spatial occupation characteristics, thereby supporting 2D mesoscale mechanical analyses to a certain extent. The Voronoi-based methods enable the generation of polygonal aggregates resembling real shapes observed in concrete sections [10], while digital image methods can reconstruct actual microstructures from thin-section images or CT scans [11], thus ensuring more accurate spatial distributions. Monte Carlo simulations provide a statistical framework for controlling gradation and area fractions [12]. However, due to the inherent limitations of 2D models in representing the actual spatial distribution and mechanical interactions of aggregates, they are unable to accurately capture the complex three-dimensional (3D) behavior of aggregates and the propagation paths of cracks. This restricts their applicability in realistically simulating the failure processes of concrete. Consequently, 3D mesoscale models have emerged as essential tools for investigating the heterogeneous properties and failure mechanisms of concrete. In 3D models, coarse aggregates, mortar, and ITZ are typically treated as separate physical phases. By accurately simulating their geometrical configurations, mechanical properties, and interactions, these models enable more realistic predictions of concrete's mechanical response and damage evolution under loading.

Although three-dimensional mesoscale modeling can more realistically reflect the internal structure and mechanical response of concrete, the generation of randomly distributed aggregates in 3D still faces numerous challenges [3,14–16]. Existing studies mainly adopt the following types of methods for 3D aggregate modeling: (1) sphere packing methods, which achieve aggregate filling by controlling the particle size distribution and enforcing non-overlapping constraints [17,18]; (2) image reconstruction-based 3D visualization techniques, such as CT image reconstruction or digital voxel models, which are suitable for simulating realistic aggregate morphologies [19,20]; and (3) geometric modeling methods based on mathematical shape descriptions (e.g., polyhedra, ellipsoids) to approximate the actual shape of aggregates [8,21,22]. The work of Yang and Wang [17] introduced an improved dynamic packing algorithm for polydisperse spherical particles, significantly enhancing packing efficiency. Wang et al. [19] used high-resolution CT scanning and segmentation algorithms to extract realistic 3D aggregate shapes, which were then meshed for finite element analysis. These methods each have their strengths and weaknesses in terms of modeling accuracy, but they commonly suffer from high computational costs, low modeling efficiency, and difficulties in overlap control—issues that become particularly significant in large-scale or high-volume-fraction simulations. Furthermore, to ensure the stability and convergence of subsequent numerical simulations, the aggregate generation process must balance spatial uniformity and physical realism, which further increases the complexity of modeling. Therefore, achieving efficient and automated 3D aggregate generation and model construction while maintaining physical realism remains one of the key challenges in the current field of mesoscale concrete modeling.

Based on the development of three-dimensional mesoscale models, researchers have gradually applied them to analyze the mechanical behavior of concrete, particularly in studying crack evolution, failure mechanisms, and strength prediction [23–26]. Common approaches include the finite element method (FEM), finite difference method (FDM), and phase-field method, which, under the assumption of continuum mechanics, can effectively simulate internal stress distributions and macroscopic responses [27–30]. However, concrete is inherently a discontinuous and heterogeneous material, and its failure process often involves particle-level cracking, sliding, and rearrangement. Therefore, traditional continuum-based methods exhibit limitations in capturing nonlinear failure behavior. For instance, crack paths cannot naturally emerge, local stress concentrations may lead to numerical instability, and it remains difficult to accurately model damage evolution in the ITZ region—especially in post-peak and large-deformation stages. Continuum-based FEM tends to smear cracks and underestimate strain localization [28], whereas the incorporation of discontinuity-enhanced formulations has been emphasized to better capture crack initiation and branching [29]. To overcome these limitations, increasing attention has been given to the discrete element method (DEM) for modeling and analyzing the mesoscale structure of concrete [31–33]. DEM treats the material as an assembly of rigid or elastic particles, and by defining contact interactions between particles, it can explicitly simulate crack initiation, propagation, and coalescence, making it inherently suitable for modeling discontinuous failure processes. Moreover, DEM excels in handling large deformations, particle interactions, and nonlinear mechanical properties, making it particularly effective in studying the influence of aggregate shape, interface strength, and particle distribution on overall mechanical performance. Therefore, integrating DEM into an efficient 3D aggregate generation method holds great potential for significantly improving the accuracy and stability of concrete mechanical simulations and provides a robust numerical tool for uncovering failure mechanisms and guiding structural design optimization.

Over the last decade, the typical take-and-place method involving overlap detection has become well developed [34–38]. Despite substantial efforts to improve the efficiency of aggregate placement, overlap detection remains complex and time-consuming, especially in 3D cases. As described in the recent literature, conventional 2D random aggregate concrete samples generated using this method require several hours, while 3D methods may take dozens of hours [35–38]. This study addresses some challenges associated with traditional methods for generating random aggregates in concrete and proposes a novel algorithm with significant advantages. Traditional methods typically require constant calculation of the distance relationships between newly generated aggregates and existing ones. As more aggregates are generated, the algorithm may become redundant, leading to a significant decline in efficiency, mainly when dealing with densely packed composite materials. The random aggregate generation algorithm proposed in this study performs exceptionally well in addressing these issues. The main advantage of this algorithm is that the speed of generating individual new aggregates gradually increases as the number of aggregates grows. By employing an innovative approach, the occupied regions of already-generated aggregates are excluded from subsequent calculations for new aggregates, effectively avoiding overlaps between newly generated and existing aggregates. This strategy ensures that the overall generated aggregates remain non-overlapping and significantly enhances computational efficiency. The effectiveness of this method was validated through discrete element numerical simulations and physical experiments. Additionally, the study investigated the effects of aggregate shape and particle size on the uniaxial compressive strength of concrete specimens.

2. Materials and Methods

Herein, this paper proposes a method for generating randomly distributed irregular concrete aggregates, including a 3D grid partitioning method, a random ellipsoid generation method, an irregular polyhedron generation method, an optimization method for irregular polyhedra, and a method for generating ITZs.

2.1. Three-Dimensional Grid Partitioning

This paper proposes an original method for partitioning 3D meshes [39,40]. The method calculates a value d for each mesh, where d can be interpreted as the radius of the most enormous inscribed sphere. The core of the method involves dividing the unoccupied regions into 3D meshes and assigning values to all mesh regions. The maximum value, $d_{\max}$, represents the radius of the most enormous inscribed sphere, and the major axis of the generated aggregate is constrained not to exceed $d_{\max}$. After generating new aggregates, the mesh values for regions not occupied by aggregates must be recalculated.

The specific steps are as follows: We assume the entire computational domain initially consists of cement mortar before the first aggregate generation. A boundary layer is added around the computational domain to serve as the boundary conditions, and its value is set to 0. For each cement mortar mesh, the d-value is calculated. All cement mortar meshes are initialized with d = 0.5. If all meshes within a distance of d + 0.5 from a given cement mortar mesh also consist of cement mortar, the d-value is incremented by 0.5; otherwise, the d-value remains unchanged. This process is repeated until d-values for all cement mortar meshes are calculated, resulting in the distribution of d-values across the computational domain. The specific process is illustrated in Figure 1, and Figure 2 presents d-value distributions for some simple meshes.

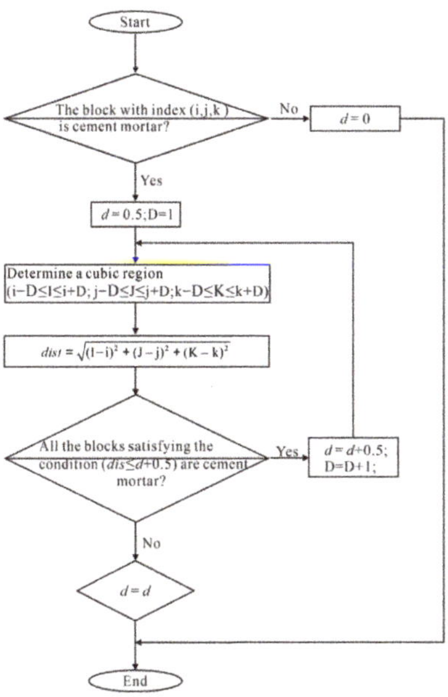

Figure 1. Flowchart of d-value distribution. D represents half the side length of the cubic computational domain. I, J, and K denote the coordinates of the small cubes within the computational domain in the x, y, and z directions, respectively.

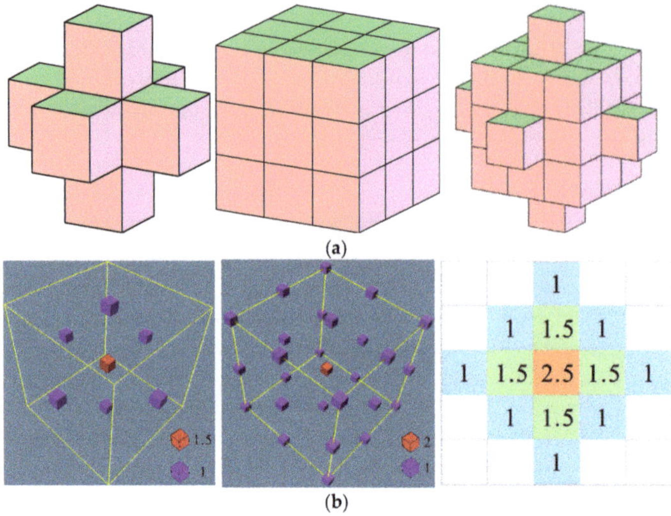

Figure 2. Simple cement mortar model and d-value distribution: (**a**) schematic diagram of a simple model, (**b**) d-value distribution map, where different colors represent the corresponding *d*-values of each grid element shown in (**a**).

2.2. Arbitrary Distribution Ellipsoid Generation Method

This paper adopts an aggregate generation strategy of first placing large-sized aggregates, followed by small-sized aggregates. This approach better fills the gaps between coarse aggregates, enhancing the compactness of the concrete. It ensures a uniform distribution of all components in the concrete, thereby improving its workability and performance. The subsequent addition of small-sized aggregates further fills the voids, reducing the porosity and enhancing the compactness and durability of the concrete.

The maximum value in the d-value distribution corresponds to the maximum length of the ellipsoid's major axis. Ellipsoids with larger major axes are generated first to achieve a more reasonable random aggregate distribution. Regions where the d-value is greater than or equal to a are identified. These regions can accommodate the ellipsoid. The center point and orientation angle of the ellipsoid are randomly selected within these regions. The parametric equation of the ellipsoid in spherical coordinates is

$$\begin{aligned} x(\theta,\phi) &= rx + a\cos(\theta)\sin(\phi) \\ y(\theta,\phi) &= ry + b\sin(\theta)\sin(\phi) \\ z(\theta,\phi) &= rz + c\cos(\phi) \end{aligned} \tag{1}$$

where (rx, ry, rz) are the coordinates of the ellipsoid's center. a, b, and c are the semi-axis lengths of the ellipsoid along the three coordinate axes. θ is the polar angle, and ϕ is the azimuthal angle.

The rotation matrices R_x, R_y, and R_z are obtained by rotating around the x, y, and z axes, respectively. They are represented as follows:

$$R_x = \begin{bmatrix} 1 & 0 & 0 \\ 0 & \cos(\theta_x) & -\sin(\theta_x) \\ 0 & \sin(\theta_x) & \cos(\theta_x) \end{bmatrix} \tag{2}$$

$$R_y = \begin{bmatrix} \cos(\theta_y) & 0 & \sin(\theta_y) \\ 0 & 1 & 0 \\ -\sin(\theta_y) & 0 & \cos(\theta_y) \end{bmatrix} \tag{3}$$

$$R_z = \begin{bmatrix} \cos(\theta_z) & -\sin(\theta_z) & 0 \\ \sin(\theta_z) & \cos(\theta_z) & 0 \\ 0 & 0 & 1 \end{bmatrix} \tag{4}$$

The overall rotation matrix is

$$R = R_z R_y R_x \tag{5}$$

The coordinates of any point (x, y, z) inside the ellipsoid are transformed into new coordinates after being rotated by the rotation matrix R. The specific expression is as follows:

$$\begin{bmatrix} x' \\ y' \\ z' \end{bmatrix} = R \begin{bmatrix} x \\ y \\ z \end{bmatrix} \tag{6}$$

2.3. Generation Method for Irregular Polyhedra

One of the most frequently occurring sequences in nature is the Fibonacci sequence. With further exploration, it was discovered that this sequence leads to the golden ratio [41,42]. The Golden Section Spiral method has been widely applied and validated in science and engineering, demonstrating a strong theoretical foundation and practical effectiveness. Due to its ability to generate points with excellent uniformity and distribution properties, the Golden Section Spiral method is extensively used in computer graphics, scientific computation, and materials science. Its solid theoretical basis has been verified and applied in numerous studies.

The Golden Section Spiral method is highly extensible and can generate points on spheres, ellipsoids, and other geometric shapes. Its application is not limited to specific geometries. Simple yet effective, this method can produce large-scale point sets with minimal computation, avoiding complex iterative or optimization processes. It is particularly adept at generating uniformly distributed points on spherical or ellipsoidal surfaces. Uniform distribution ensures equal intervals between points, avoiding clustering or sparse regions.

To prevent collinearity or coplanarity, this paper utilizes the Golden Section Spiral method to generate 2N uniformly distributed points on the surface of an ellipsoid. N points are randomly selected from these to construct irregular polyhedra, ensuring randomness in the polyhedron's shape.

For the i-th point generated using the Golden Section Spiral method on the surface of the ellipsoid, its polar and azimuthal angles can be expressed as follows:

$$\begin{aligned} \phi_i &= arc\cos\left(1 - \tfrac{2(i-0.5)}{N}\right) \\ \theta_i &= \pi(1+\sqrt{5})(i-0.5) \end{aligned} \tag{7}$$

The expression for mapping points onto the surface of an ellipsoid is

$$(X_i, Y_i, Z_i) = (a\sin(\phi_i)\cos(\theta_i), b\sin(\phi_i)\sin(\theta_i), c\cos(\phi_i)) \tag{8}$$

This generates a 3D irregular aggregate (Figure 3) comprising 10 vertices.

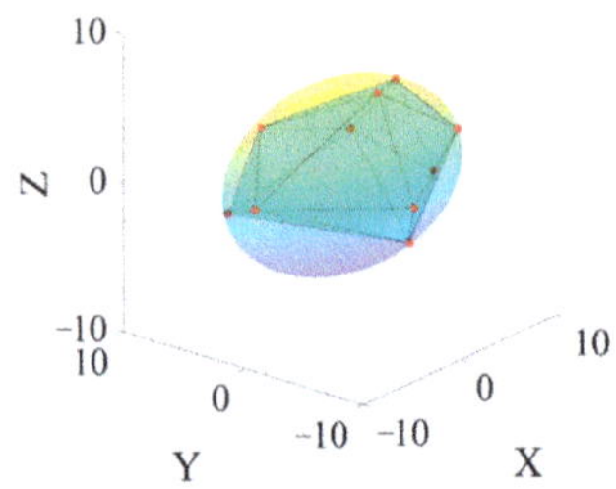

Figure 3. Three-dimensional irregular aggregate.

2.4. *Optimization of Irregular Polyhedra*

Although the above method can generate irregular aggregates, the resulting aggregates require optimization before they can be used for numerical calculations. Three aspects of optimization are performed here: eliminating flatness, removing triangular faces with very short edges, and eliminating triangular faces with extremely small interior angles.

Flat polyhedra, especially those with complex shapes [43,44], may have overlapping vertices or edges. They may also have intersecting or overlapping faces on the same plane. Aggregates with very short edges can affect computational efficiency. Extremely small interior angles can also reduce simulation accuracy. This study further optimizes the generated aggregates to address this issue.

First, sphericity describes the degree of flatness to avoid obtaining flat-shaped aggregates. Sphericity B is defined as [45]

$$B = \sqrt[3]{\frac{L_i L_s}{L_l^2}} \tag{9}$$

where L_l, L_i, and L_s represent the longest, intermediate, and shortest particle dimensions, respectively. Considering that planar aggregates usually have very low sphericity, the minimum sphericity is set to 0.4. Therefore, only aggregates with a sphericity greater than 0.4 can be generated.

Next, edge length optimization is performed by calculating the length of each edge of the aggregate. If the shortest edge in a triangular face is smaller than the predefined minimum length L_{min}, one of the vertices of that edge is removed (Figure 4). L_{min} is set as one tenth of the aggregate's maximum edge length. The new aggregate, defined by the remaining vertices, is further optimized until all edge lengths exceed L_{min}.

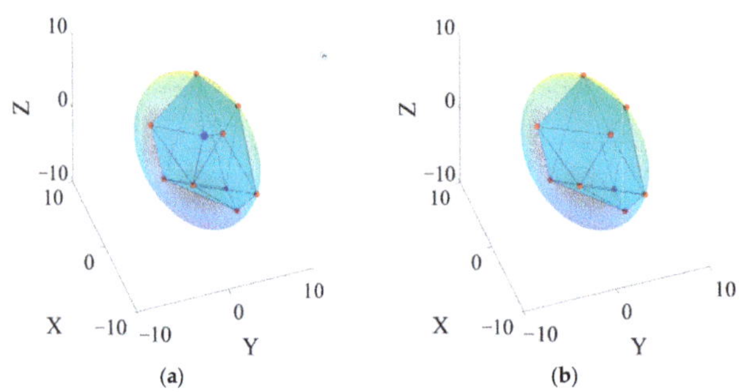

Figure 4. Optimization of triangles with shortest edges: (**a**) aggregate before optimization; (**b**) aggregate after optimization. The blue dots represent the vertices to be removed, and the red dots indicate the vertices to be retained.

Finally, interior angle optimization is carried out by calculating the interior angles of each triangular face of the aggregate. If the smallest interior angle is less than the predefined value β (set to 10° here), the other two interior angles of the triangle are compared. The vertex corresponding to the larger interior angle is removed (Figure 5). The new aggregate, formed by the remaining vertices, undergoes further optimization until all interior angles exceed β. After the geometric optimization of all aggregates, the resulting aggregates are well-shaped and free from overlaps.

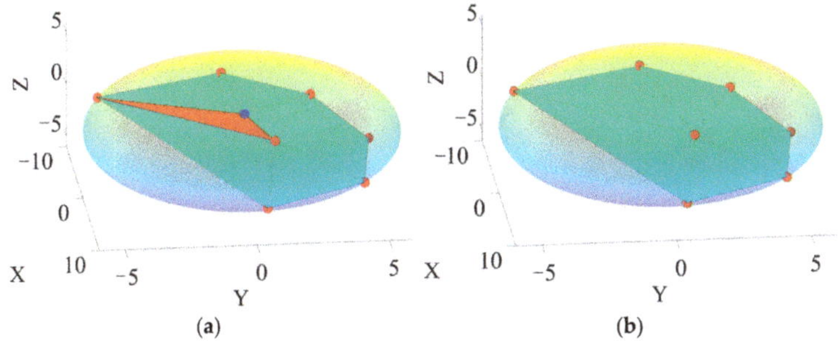

Figure 5. Optimization of triangles with minimum interior angles: (**a**) aggregate before optimization; (**b**) aggregate after optimization. The blue dots represent the vertices to be removed, and the red dots indicate the vertices to be retained.

2.5. Generate Bonding Interface

First, calculate the centroid of the polyhedron. The centroid is the average of all vertex coordinates:

$$(centroid_x, centroid_y, centroid_z) = \left(\frac{1}{N}\sum_{i=1}^{N} X_i, \frac{1}{N}\sum_{i=1}^{N} Y_i, \frac{1}{N}\sum_{i=1}^{N} Z_i\right) \tag{10}$$

where $(centroid_x, centroid_y, centroid_z)$ represents the centroid coordinates, N is the number of vertices, and X_i, Y_i, and Z_i are the coordinates of the i-th vertex.

$$d_i = \sqrt{(X_i - centroid_x)^2 + (Y_i - centroid_y)^2 + (Z_i - centroid_z)^2} \tag{11}$$

Next, calculate the distance d_i from each vertex to the centroid:

Assuming the thickness of the bonding interface is l, the distance from the shrunken new vertex to the centroid is

$$d_i' = d_i - l \tag{12}$$

The coordinates of the new shrunken vertex are

$$\begin{aligned} X_i' &= centroid_x + (X_i - centroid_x)\tfrac{d_i'}{d_i} \\ Y_i' &= centroid_y + (Y_i - centroid_y)\tfrac{d_i'}{d_i} \\ Z_i' &= centroid_y + (Z_i - centroid_z)\tfrac{d_i'}{d_i} \end{aligned} \tag{13}$$

Figure 6 shows the bonding interface constructed in this study. In the figure, the red points represent the original vertices of the irregular polyhedron randomly selected on the ellipsoidal surface. In contrast, the blue points represent the vertices of the shrunken irregular polyhedron. The region between the two polyhedra constitutes the bonding interface. Since the bonding interface is generated using a shrinking method, DD = H + 2l, ensuring sufficient space for the generation of aggregates and the bonding interface. Figure 7 illustrates the flowchart for generating three-dimensional randomly distributed irregular aggregates. Figure 8 displays the three-dimensional randomly distributed irregular aggregates, with different colors representing aggregates of different particle sizes.

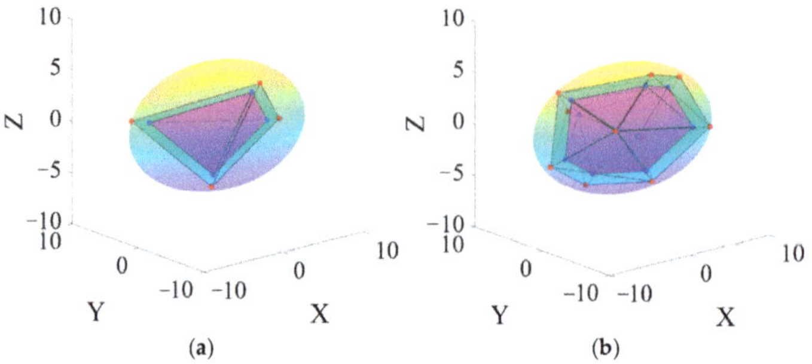

Figure 6. Schematic diagram of the 3D bonding interface: (**a**) randomly select 4 points; (**b**) randomly select 10 points. Red points are the original polyhedron vertices on the ellipsoidal surface, and blue points are the vertices of the shrunken polyhedron. The bonding interface lies between the two polyhedra.

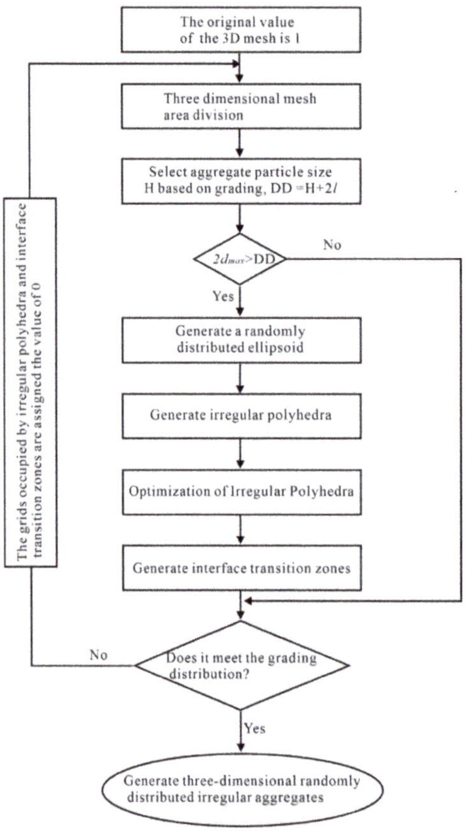

Figure 7. Flowchart for generating 3D randomly distributed irregular aggregates. $d_{\max}$ represents the maximum value after dividing the 3D grid region, and l represents the thickness of the bonding interface.

Figure 8. Three-dimensional randomly distributed irregular aggregates. Different colors represent aggregates of different particle sizes.

In traditional methods for generating 3D random aggregates, the spatial relationship between a new aggregate and the existing aggregates must be determined [36,46]. As the number of generated aggregates increases, the time required to generate new aggregates also grows. The random aggregate generation model proposed in this study adopts a three-dimensional grid partitioning method. Each grid cell's value represents the maximum aggregate radius that can be accommodated if the grid point is used as the aggregate center. By randomly selecting grid points that meet the conditions, aggregates are generated, ensuring a reasonable distribution of aggregates. The innovation of the model lies in the fact that when a new aggregate is generated, the regions occupied by existing aggregates are excluded from further grid partitioning. This approach avoids overlap between new and existing aggregates and significantly reduces the time required to generate new aggregates as the number of aggregates increases. Twenty irregular aggregates with a particle size of 20 mm are randomly generated within a model of dimensions 75 mm × 75 mm × 75 mm (Figure 9). Figure 10 shows the time required to generate each aggregate. The figure indicates that as the number of generated aggregates increases, the time required to generate new aggregates decreases, with a noticeable reduction in the time required for the initial stages of aggregate generation.

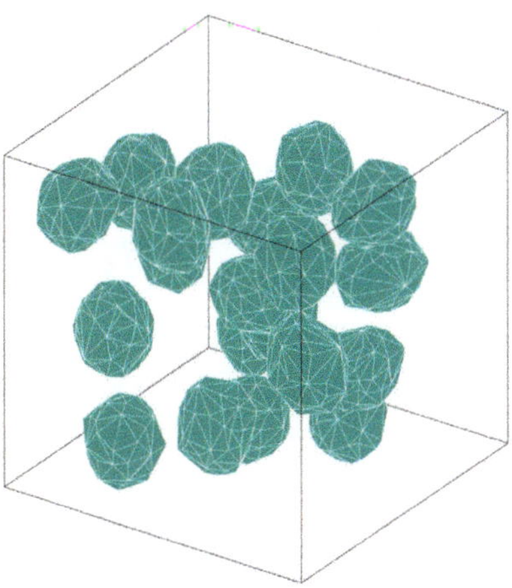

Figure 9. Randomly distributed irregular aggregates of the same size.

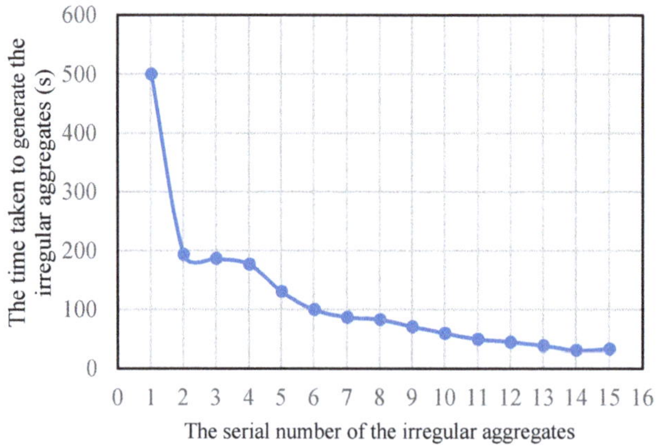

Figure 10. The time taken to generate irregular aggregates of the same size.

3. Discrete Element Method

The particle flow method is founded on Newton's second law and the force–displacement law. The computational process revolves around two core iterative steps: applying Newton's second law to particle elements and applying the force–displacement law to contact points. Newton's second law primarily determines each particle element's motion state and updates its state variables accordingly. The force–displacement law updates contact state variables based on the relative motion at each contact point, thereby updating contact forces, as illustrated in Figure 11.

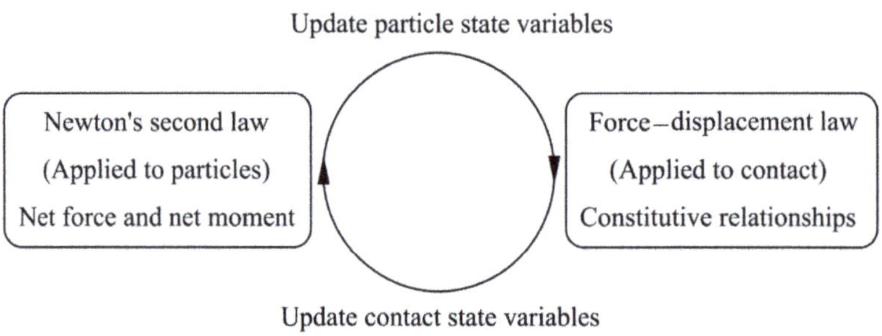

Figure 11. Diagram of the PFC computational iteration process.

Additionally, Newton's second law does not apply to the movement of walls, as their motion is generally predefined. In PFC simulations, the force–displacement law only considers the contact interactions between particle elements and walls during calculations.

3.1. Force–Displacement Law

The force–displacement law is central to the computational process of PFC simulations. By introducing the normal stiffness and tangential stiffness parameters, PFC effectively resolves the relationship between force and displacement for two contacting elements. The normal vector of the contacting elements determines the exact position of the contact point. In particle flow simulations, the contact forms are mainly categorized into "particle–particle" and "particle–wall" interactions, both crucial for describing the dynamic behavior of the particle flow system, as shown in Figure 12. In this figure, $R^{(1)}$, $R^{(2)}$ represents the radii of the particles, $x^{(1)}$ and $x^{(2)}$ denote the centers of the particles, x_w stands for the center of rotation and displacement of the wall, and g_c indicates the distance between two elements. When g_c is negative, it signifies the contact amount (or overlap), denoted as δ.

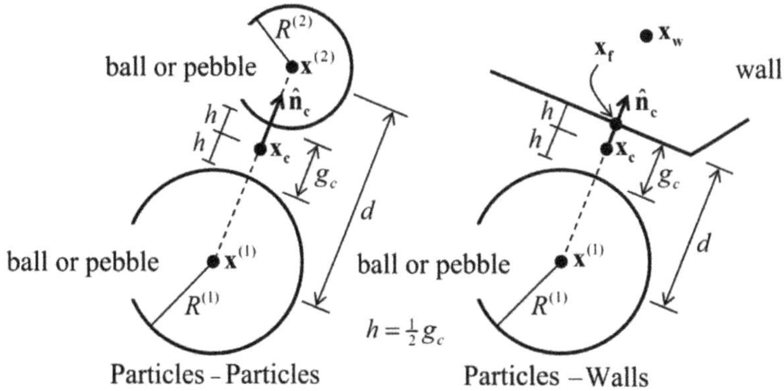

Figure 12. Forms of element contact.

In applying the force–displacement law to specific calculations, the contact force F_c between two elements is divided into normal and tangential components. The normal force is proportional to the overlap amount, denoted as

$$F_{cf} = k_n U \tag{14}$$

where F_{cf} is the normal force, k_n is the normal stiffness, and U is the overlap amount.

The tangential force calculation is more complex due to its dependence on the motion and loading history of the particle elements. It is usually expressed in incremental form. When the relative displacement between two entities is ΔU, the incremental tangential force at the i-th time step can be expressed as

$$\Delta F_{cs}^{i} = k_n \Delta U \tag{15}$$

where ΔF_{cs}^{i} is the incremental tangential force at the i-th step and k_n is the tangential stiffness.

Considering the tangential force from the previous time step, the total tangential force is given by

$$F_{cs} = F_{cs}^{i-1} + \Delta F_{cs}^{i} \tag{16}$$

These equations can be used to calculate the total contact force and contact moment of the contact elements in each computational time step.

3.2. Laws of Motion

In the particle flow model, the motion of particle elements is determined by the forces and moments acting on them. These forces and moments reflect the overall motion through the particle's linear and angular velocities. The motion equations for particles consist of two main components: the linear motion equation, resulting from the net force, and the rotational motion equation, resulting from the net moment.

The linear motion equation is expressed as

$$\begin{bmatrix} F_x \\ F_y \\ F_z - mg \end{bmatrix} = m \begin{bmatrix} a_x \\ a_y \\ a_z \end{bmatrix} \tag{17}$$

where F_x, F_y, and F_z are the net forces in the x, y, and z directions, respectively; m represents the mass of the particle; a_x, a_y, and a_z are the accelerations in the x, y, and z directions, respectively; and g denotes gravitational acceleration.

The rotational motion equation is expressed as

$$\begin{bmatrix} M_x \\ M_y \\ M_z \end{bmatrix} = \begin{bmatrix} I_{xx} & I_{xy} & I_{xz} \\ I_{yx} & I_{yy} & I_{yz} \\ I_{zx} & I_{zy} & I_{zz} \end{bmatrix} \cdot \begin{bmatrix} a_x \\ a_y \\ a_z \end{bmatrix} \tag{18}$$

where M_x, M_y, and M_z represent the net moments in the x, y, and z directions, respectively; I_{xx}, I_{yy}, and I_{zz} are the elements of the inertia matrix; and a_x, a_y, and a_z are the angular accelerations in the x, y, and z directions, respectively.

3.3. Constitutive Models in PFC

In PFC, the description of macroscopic material constitutive properties relies on the definition of particle micro-contact models and their parameters. PFC includes various micro-contact models, categorized into three main types: contact stiffness models, sliding models, and bonding models. This article uses bonding models for numerical simulation.

The bonding model simulates the cohesive or cementitious behavior between particles by introducing additional bonding forces. This model effectively describes strong connections between particles due to chemical bonds, cementitious materials, or other bonding mechanisms. It is particularly suitable for simulating non-loose materials with significant bonding characteristics, such as rocks and concrete. Depending on the type of bonding between particles, the bonding model can be classified into linear contact bonding models and linear parallel bonding models.

In the contact bonding model, the contact range between particles is restricted, approximating a point-contact state. This contact mode can be considered as a spring with constant normal and tangential stiffness at the contact point, allowing force transmission but not moment transmission. Conversely, the parallel bonding model has broader mechanical properties, permitting the transmission of both forces and moments between particles, as illustrated in Figure 13a. In this model, the contact area between adjacent particle elements is assumed to form a circular or rectangular bond material, with the constitutive relationship shown in Figure 13b. This relationship comprises linear elements, damping elements, and contact bonds.

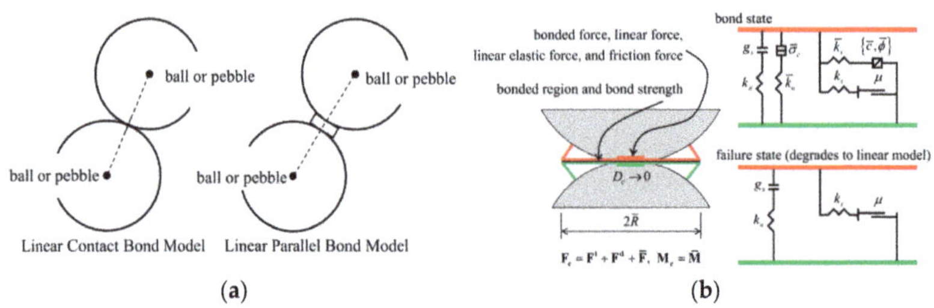

Figure 13. Comprehensive schematic diagram of the structure and constitutive relationship for the bonded model: (**a**) schematic diagram of the bonded model; (**b**) schematic diagram of the constitutive relationship for the parallel bonded model.

The model's characteristics are defined by five parameters: normal stiffness, tangential stiffness, normal strength, tangential strength, and bond radius. These parameters collectively define the linear elastic deformation, friction, and viscous behavior between particle elements. Adjusting these parameters allows for a more accurate simulation of the mechanical properties of rock materials, enhancing the accuracy and reliability of the simulations.

4. Results and Discussion

4.1. Physical Experiments and Numerical Simulations

The main materials used for the actual concrete specimens included cement, fine aggregates, coarse aggregates, water, and admixtures. The cement used was ordinary Portland cement. The fine aggregate was natural river sand, and the coarse aggregate was crushed stone (Figure 14a). Tap water was used. The admixture was PCA-I type high-efficiency water reducer. The concrete mix proportions were configured according to design requirements. The concrete strength grade used in the experiments was C30, with the specific mix proportions provided in Table 1. It should be noted that the concrete mix components in this study were not scaled down to match the characteristic dimensions. For all specimens, the maximum size of the coarse aggregate was approximately 30 mm. The mass contents of the cement and water were 340 kg/m^3 and 178 kg/m^3, respectively. The water-to-cement ratio was approximately 0.52. The total mass content of the concrete (Sum) was 2278 kg/m^3. The prepared concrete mix was poured into standard cubic molds with dimensions of 150 mm × 150 mm × 150 mm. To ensure that the specimens were compact and uniform, tamping rods or vibrating tables were used to consolidate the concrete during pouring. After pouring, the surface was leveled with a scraper, and a damp cloth was placed over the specimens after initial setting to prevent water evaporation. After demolding, the specimens were immediately placed in a curing room under standard conditions with a temperature of 20 ± 2 °C and a relative humidity of no less than 95%. The curing duration was 28 days, during which the specimens were kept moist (Figure 14b). Compressive strength tests were performed on the concrete specimens using a hydraulic pressure testing machine (Figure 14c). The loading rate of the testing machine was 0.5 MPa/s, ensuring uniform and stable loading. The ultimate load values for the three specimens were 753.75 kN, 778.5 kN, and 747 kN, respectively (Figure 15). The average compressive strength was calculated to be 33.8 MPa.

(a) (b) (c)

Figure 14. Compressive strength test of standard concrete specimens: (**a**) Aggregates; (**b**) concrete specimens; (**c**) hydraulic testing machine.

Table 1. Concrete mixture proportions utilized.

Aggregate 0–5 mm (kg/m^3)	Aggregate 5–10 mm (kg/m^3)	Aggregate 10–30 mm (kg/m^3)	Cement (kg/m^3)	Water (kg/m^3)	Admixture (kg/m^3)	w/c Ratio	Sum (kg/m^3)	Compressive Strength fc (Mpa)
780	180	800	340	178	11.20	0.52	2289.20	33.8

Figure 15. Concrete specimen after uniaxial compressive strength test.

At the mesoscopic scale, concrete is typically considered a three-phase composite material consisting of aggregate particles, a mortar matrix, and an interfacial transition zone (ITZ) between the first two phases. Evaluating the composite behavior of concrete at the mesoscale requires generating a random aggregate structure in which the shape, size, and distribution of the aggregate particles statistically resemble those of real concrete. Additionally, aggregates, including fine and coarse aggregates, usually account for 60%–80% of concrete's volume, significantly influencing its performance, mix proportions, and cost-efficiency. For simplicity in this study, fine aggregates and cement paste are combined into the mortar matrix. Coarse aggregates make up approximately 40%–50% of the concrete's volume [47]. The shape of aggregate particles depends on the type of aggregate. Gravel aggregates are generally rounded, while crushed-stone aggregates are angular [34]. Modeling with spherical aggregates can satisfy the grading distribution in real concrete but lacks accuracy in representing aggregate morphology. Using ellipsoidal aggregates for modeling can meet three parameters of real concrete—grading, aspect ratio, and angular distribution—but still falls short of capturing the aggregate's true morphology. Modeling with polygonal aggregates can satisfy four parameters: grading, aspect ratio, angular distribution, and roundness. Grading refers to determining the particle size distribution of aggregates, often expressed as a cumulative percentage passing through a series of sieve sizes. This can be represented through formulas, tables, or charts. In practice, one of the most well-known and widely accepted aggregate distributions is the one proposed by Fuller. The Fuller curve-based aggregate grading provides optimal aggregate packing, generating more fine aggregate particles to maximize concrete density and macroscopic strength. The grading curve formula is

$$P = 100\sqrt{\frac{D}{D_{\max}}} \tag{19}$$

where P is the cumulative percentage passing a sieve with aperture diameter D, and $D_{\max}$ is the maximum size of aggregate particles.

In order to determine the mechanical parameters, especially the compressive strengths, a cubic concrete sample size of 150 mm × 150 mm × 150 mm was established, as presented in Figure 16. The concrete mix composition used in the numerical model was the same as that listed in Table 1. To ensure that the 3D three-phase random aggregate model accurately reflected the material composition and physical characteristics of the concrete used in the experiments, the mass-based mix proportions provided in Table 1 were systematically converted into volume fractions and mechanical input parameters required by the model. First, the volume fraction of coarse aggregates was derived by combining their mass proportion with the known densities of the constituent materials, and this value was used to define the aggregate content in the 3D model. Second, based on the actual composition of the concrete, the model was divided into three distinct phases: coarse aggregates, the mortar matrix, and the interfacial transition zone (ITZ). The shapes and sizes of the aggregates were randomly generated to match the maximum particle size and grading of the crushed stone used in the experiments. Finally, the mechanical properties of each phase—including Young's modulus, Poisson's ratio, and compressive strength—as well as the ITZ thickness, were determined based on the target strength grade of the experimental mix, the published literature, and known material characteristics. This parameterization ensures consistency between the numerical model and the experimental conditions in terms of both material composition and mechanical behavior, thereby providing a reliable basis for subsequent simulation analyses and result comparisons. The Fuller's curve was utilized to describe the aggregate size distribution. The numbers of random aggregates within the particle size ranges of 5–10 mm and 10–30 mm are 1227 and 258, respectively. The major mechanical parameters of the three meso components used in the simulations are provided in Table 2. It is to be noted that Young's modulus, Poisson's ratio, the compressive strength of the aggregate particles, and the mortar matrix are obtained from the tests, while the mechanical properties of the ITZs are given with reasonable assumption [48]. The real typical thickness of the ITZ is about 0.05 mm; however, because of the limitation in computational capacity, it is impossible to use such small elements to model the real thickness of the ITZ.

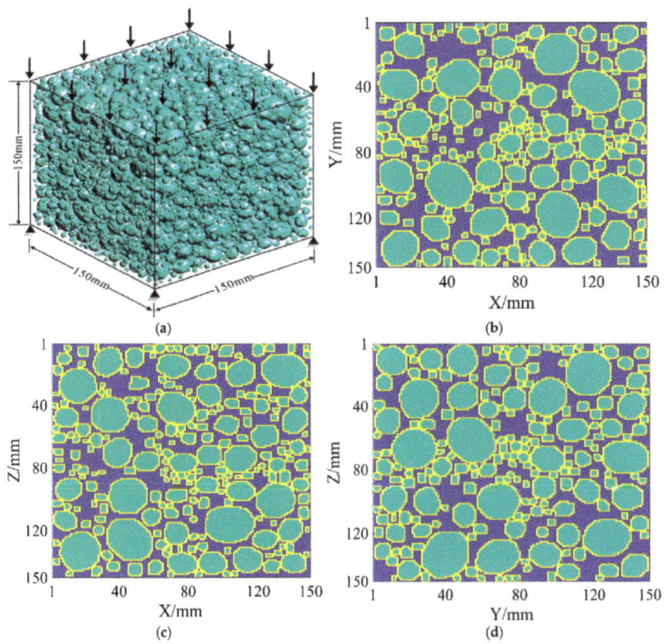

Figure 16. Modeling of the proposed mesoscale numerical model: (**a**) uniaxial compressive simulation of concrete samples with dimensions of 150 mm × 150 mm × 150 mm, the arrows indicate the applied force on the top surface; (**b**) XY section of the concrete model; (**c**) XZ section of the concrete model; (**d**) YZ section of the concrete model. Green, yellow, and blue represent aggregate, ITZ, and mortar, respectively.

Table 2. Mechanical parameters of the meso components.

	Mortar Matrix	ITZ	Aggregate Particle
Young's modulus E (GPa)	32.5	26	70
Poisson ratio v	0.2	0.22	0.16
Compressive strength fc (MPa)	31.6	23.9	80

In the concrete model generated, green, blue, and red spherical particles represent mortar, aggregate, and ITZ, respectively (Figure 17). It is worth noting that due to experimental limitations, it is challenging to measure the mechanical properties of the ITZ directly. In numerical simulations, the mechanical properties of the ITZ are typically described as weaker mechanical parameters of the mortar matrix, with the ITZ assumed to range between 30% and 80% of the mortar matrix's strength [49,50].

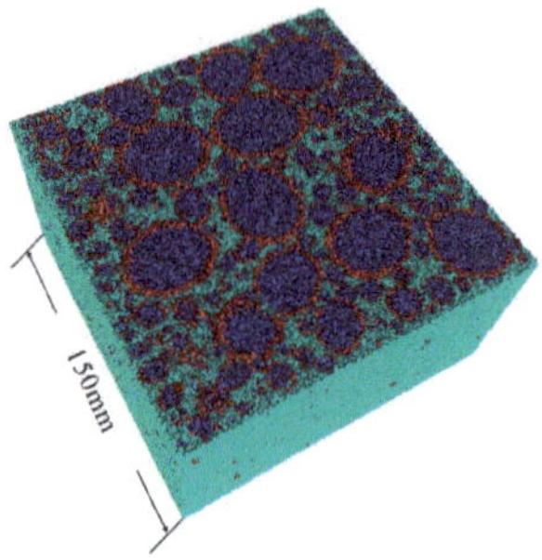

Figure 17. DEM three-dimensional numerical model. Blue, red, and green represent aggregate, ITZ, and mortar, respectively.

In this study, concrete is treated as a three-phase composite material consisting of a mortar matrix, coarse aggregates, and an ITZ. The physical and mechanical properties of each phase are summarized in Table 2. The elastic modulus, Poisson's ratio, and compressive strength of the mortar matrix and aggregates were obtained from standard specimen tests. Since the ITZ is difficult to isolate and test directly in experiments, its mechanical parameters were adopted from the existing literature [49,50], and its strength was set to approximately 75% of that of the mortar. Regarding boundary conditions, uniaxial compression was applied by imposing vertical displacement on the top and bottom loading plates. The bottom plate was fully fixed in all directions, while the top plate was allowed to move only in the vertical direction. The lateral sides of the specimen were left unconstrained, employing free boundary conditions to better replicate realistic loading conditions. To meet the requirements of quasi-static loading, a time-stepping algorithm was used for the simulation, with the total number of steps reaching several tens of thousands. During each time step, the forces and displacements between particles were updated. The simulation was terminated when the axial stress dropped to 80% of its peak value [51]. For model validation, the simulated specimen's dimensions (150 mm × 150 mm × 150 mm), mix proportions (Table 1), maximum aggregate size (30 mm), and loading rate were all consistent with the experimental setup. The coarse aggregate used was natural crushed stone, with an elastic modulus set to 70 GPa, which is consistent with the properties of hard limestone rocks [48]. Figure 18a presents the simulation results under uniaxial compression with an ultimate strength of 31.1 MPa, corresponding to the experimental result of 33.8 MPa. The simulation results show good consistency with the experimental results. In this study, the mechanical parameters listed in Table 2 were adopted in the subsequent simulations. Figure 18b shows the numerical failure mode. Under uniaxial compression loading, the interfacial transition zone (ITZ) is the weakest phase in concrete, where numerous microcracks develop. Microcracks extending into the mortar form interfacial cracks, which eventually connect to form macroscopic cracks; this progression represents a typical feature of brittle failure in concrete. The simulated failure mode is consistent with the results observed in laboratory experiments. Furthermore, the results of this study were compared with those obtained from an existing two-dimensional (2D) mesoscale finite element model [52]. Although the 2D model also incorporated randomly distributed coarse aggregates to evaluate the mechanical behavior of concrete, and its resulting stress–strain curve (Figure 18a) exhibits a similar macroscopic trend to our numerical results, it is inherently limited to in-plane mechanical behavior and thus fails to accurately capture the three-dimensional characteristics of crack propagation. In contrast, the 3D mesoscale model developed in this study not only accurately reproduces the stochastic distribution of aggregate sizes but also fully captures the spatial distribution of cracks within the specimen volume. The predicted strain localization zones and crack paths show good agreement with experimentally observed fracture morphologies. In the 2D model, cracks tend to concentrate along the weakest plane, and their inclination and continuity are restricted by the simplified treatment of the thickness direction, making it difficult to reflect the complex through-thickness failure modes. Overall, although 2D simulations offer certain advantages in computational efficiency and in capturing general mechanical trends, the 3D mesoscale model provides more realistic and detailed representations of crack evolution paths, spatial failure mechanisms, and size effects. This makes it a more reliable numerical tool for investigating concrete fracture mechanisms and for applications at the engineering scale.

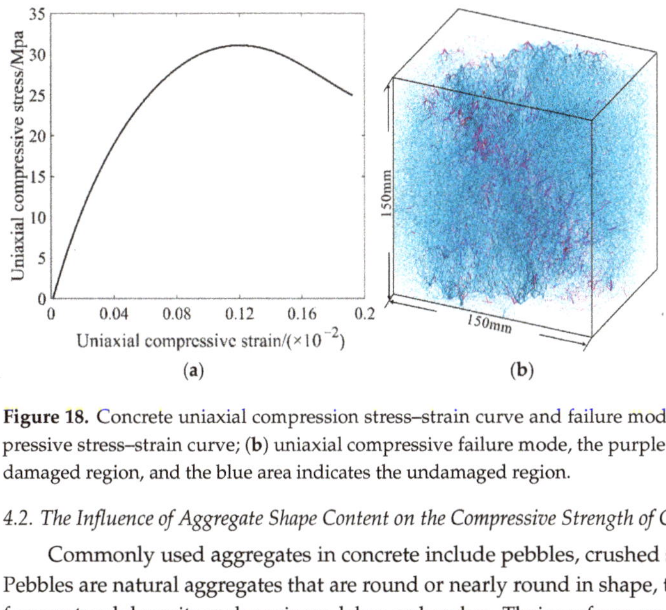

Figure 18. Concrete uniaxial compression stress–strain curve and failure mode: (**a**) uniaxial compressive stress–strain curve; (**b**) uniaxial compressive failure mode, the purple area represents the damaged region, and the blue area indicates the undamaged region.

4.2. The Influence of Aggregate Shape Content on the Compressive Strength of Concrete Specimens

Commonly used aggregates in concrete include pebbles, crushed stone, and gravel. Pebbles are natural aggregates that are round or nearly round in shape, typically obtained from natural deposits such as rivers, lakes, or beaches. Their surfaces are generally smooth, and their rounded shape contributes to better workability (flowability and ease of handling) of concrete. Crushed stone is an aggregate produced by mechanically crushing natural rocks. It is usually irregular and angular in shape, with rough surfaces. This type of aggregate bonds well with cement paste, enhancing the strength of the concrete. Gravel generally refers to irregularly shaped rock fragments formed through natural weathering or water flow. Compared to pebbles, gravel has a more irregular shape but smoother surfaces, falling between pebbles and crushed stone in terms of characteristics. In this study, spheres or ellipsoids are used to model pebbles, while irregular polyhedra are used to model crushed stone and gravel.

The coarse aggregates used in the experimental program were natural crushed stone, primarily composed of limestone. These aggregates exhibit irregular polyhedral shapes with pronounced angularity and complex geometrical features (Figure 14a). Therefore, it is necessary to investigate the influence of aggregate shape on the simulated compressive strength of concrete. To examine the effect of different aggregate geometries on the macroscopic mechanical behavior of concrete, three types of aggregate models were constructed in numerical simulations: spherical, ellipsoidal, and irregular polyhedral shapes. Among them, the irregular polyhedral aggregates were generated using a stochastic algorithm designed to closely approximate the geometrical characteristics of real crushed stone. The results indicate that aggregate shape has a noticeable influence on the simulated uniaxial compressive strength.

The 3D aggregate random generation method proposed in this study was used to create three models. Each model had dimensions of 150 mm × 150 mm × 150 mm and an aggregate volume fraction of 40%. The aggregate shapes included spherical, ellipsoidal, and irregular polyhedral forms (Figure 19). Figure 20 shows DEM three-dimensional numerical models of aggregates with different shapes. Figure 21 presents the compressive stress–strain curves of models with different aggregate shapes. The figure shows that for an aggregate volume fraction of 40%, specimens with polyhedral aggregates have the lowest compressive strength, while those with spherical and ellipsoidal aggregates exhibit similar compressive strength. This conclusion aligns well with previous findings [53–55].

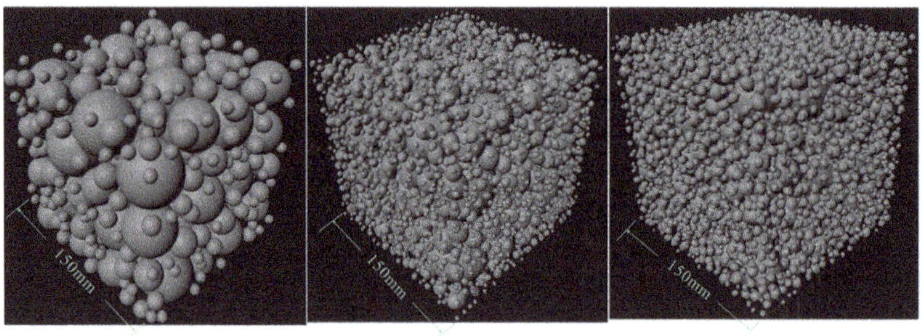

(a) (b) (c)

Figure 19. Cubic samples containing differently shaped aggregates: (**a**) spherical aggregates; (**b**) ellipsoidal aggregates; (**c**) polyhedral aggregates.

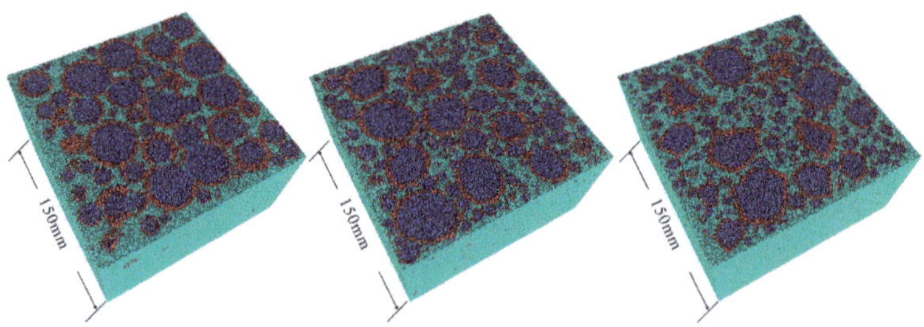

(a) (b) (c)

Figure 20. DEM three-dimensional numerical models of aggregates with different shapes: (**a**) spherical aggregate model; (**b**) ellipsoidal aggregate model; (**c**) polyhedral aggregate model. Blue, red, and green represent aggregate, ITZ, and mortar, respectively.

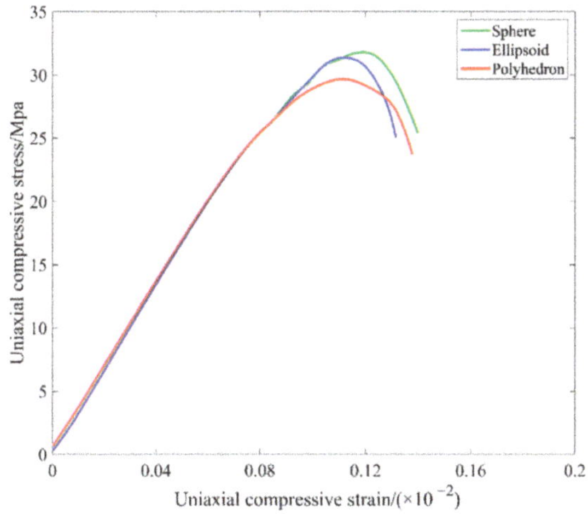

Figure 21. Compressive stress–strain curves obtained from simulations of concrete with different aggregate shapes.

4.3. The Influence of Aggregate Particle Size on the Compressive Strength of Concrete Specimens

A two-graded concrete was selected, with model dimensions of 150 mm × 150 mm × 150 mm. The aggregate particle sizes were randomly distributed within the ranges of 5–10 mm and 10–30 mm, with a total aggregate content of 40%. Aggregates in the 5–10 mm range were uniformly distributed, while the 10–30 mm aggregates were divided into three groups based on the maximum particle size: D_{max} = 20, 25, and 30 mm (Figure 22). This approach ensured that all other parameters remained constant, with only the maximum aggregate size varying. Figure 23 shows DEM three-dimensional numerical models with various maximum aggregate particle sizes. The stress–strain curves of specimens with different aggregate sizes are shown in Figure 24. It can be observed that the stress–strain curves in the elastic deformation region are quite similar. However, differences are evident in the peak stress and the softening stage, with some deviations near the yield point and peak stress. The compressive strength of the specimens decreases as the aggregate size increases, which is consistent with previous experimental results [52].

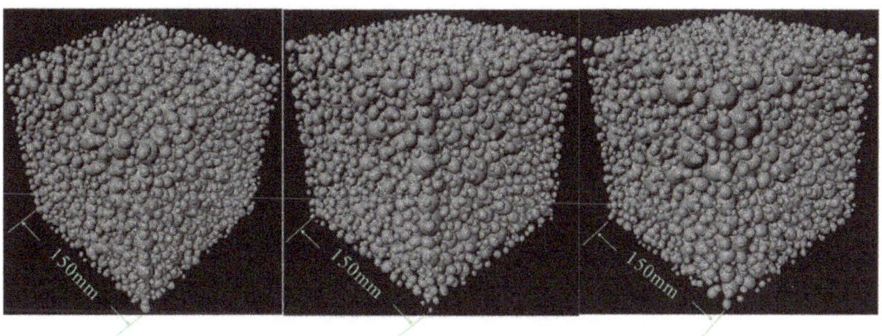

(a) (b) (c)

Figure 22. Cubic samples containing various maximum aggregate particle sizes: (**a**) D_{max} = 20 mm; (**b**) D_{max} = 25 mm; (**c**) D_{max} = 30 mm.

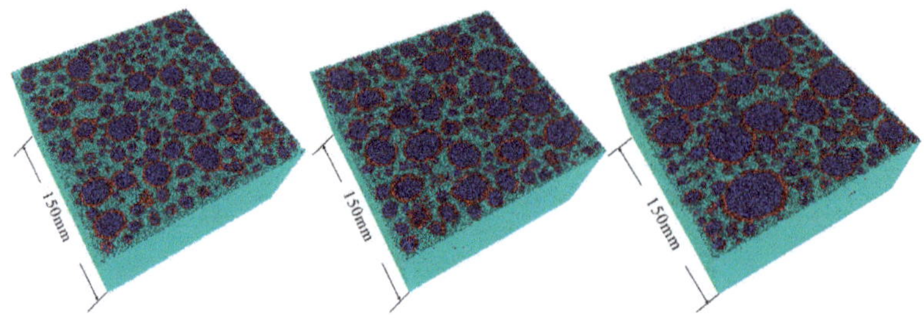

Figure 23. DEM three-dimensional numerical models with various maximum aggregate particle sizes: (**a**) D_{max} = 20 mm; (**b**) D_{max} = 25 mm; (**c**) D_{max} = 30 mm. Blue, red, and green represent aggregate, ITZ, and mortar, respectively.

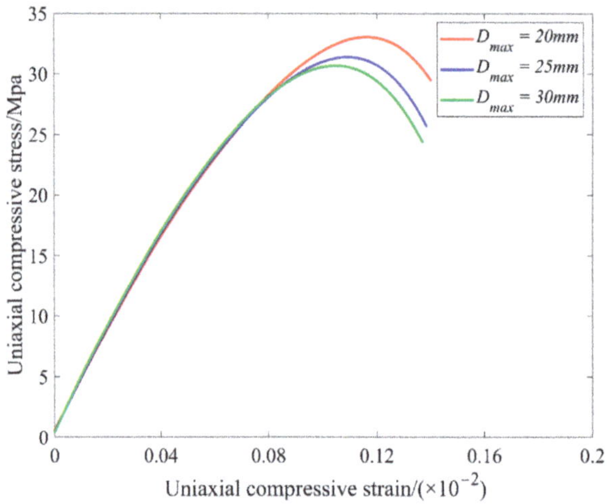

Figure 24. Compressive stress–strain curves for cubes with different maximum aggregate particle sizes.

When the load is below 80% of the peak stress, the concrete material remains in the linear elastic damage phase. At this stage, the interfacial transition zone (ITZ) develops some relatively stable microcracks that have not yet interconnected or penetrated one another. Once the load exceeds 80% of the peak stress, irreversible plastic damage occurs within the concrete, and cracks begin to propagate unstably. Microcracks extend and connect to form larger cracks. The increase in aggregate size weakens the bonding performance of the ITZ, leading to a decline in compressive strength. In the initial elastic stage of the stress–strain curves, the curves corresponding to different maximum aggregate sizes nearly overlap, indicating that aggregate size has a negligible effect on the overall elastic stiffness of concrete. This is primarily because the deformation in this stage is dominated by the synergistic action between the mortar matrix and the aggregate skeleton, while the interfacial transition zone (ITZ) has not yet developed significant microcracks—only isolated and stable microcracks may appear, which do not cause substantial stress disturbances. However, once the material enters the inelastic stage, the peak stress decreases significantly with increasing maximum aggregate size. This phenomenon can be attributed to the fact that, under the same volume fraction, larger aggregates possess a smaller specific surface area, resulting in more concentrated stress at the mortar–aggregate interfaces. This exacerbates stress concentration in the ITZ and weakens its mechanical performance, promoting earlier initiation and faster coalescence of interfacial microcracks. In contrast, smaller aggregates can form a more uniform and compact particle structure, which helps to restrain crack propagation. As the aggregate size increases, size effects become more pronounced, leading to more severe local stress concentration and crack instability. Consequently, concrete with larger aggregates tends to enter the softening stage more abruptly after reaching the peak stress, and the failure process exhibits more brittle characteristics.

5. Conclusions

This study systematically investigates the mechanism by which aggregate shape and size affect the compressive performance of concrete by constructing a 3D three-phase mesoscale model and employing the discrete element method. The main conclusions are as follows:

(1) To study the compressive performance of concrete, a three-dimensional three-phase mesoscale model composed of coarse aggregates, mortar, and the interfacial transition zone was constructed. The proposed aggregate generation method uses grid partitioning and conditional screening to achieve a random distribution of aggregates, avoiding aggregate overlap and significantly improving generation efficiency and accuracy. This model provides an effective tool for investigating the random distribution characteristics of aggregates in concrete.

(2) The 3D random aggregate generation method introduced in this study has significant advantages: the computational time required to generate a single aggregate decreases rapidly as the number of already generated aggregates increases, effectively avoiding computational redundancy.

(3) The discrete element method was used to simulate the effects of different aggregate shapes and maximum aggregate sizes on the compressive behavior of concrete. The simulation results reveal that increasing the maximum aggregate size from 20 mm to 30 mm leads to an 8% reduction in the peak compressive strength of concrete. Furthermore, under a constant aggregate volume fraction of 40%, the use of irregular polyhedral aggregates results in a decrease of approximately 7% in compressive strength compared to spherical and ellipsoidal aggregates. These findings indicate that larger aggregate sizes and more irregular shapes tend to exacerbate interfacial stress concentrations and promote unstable crack propagation, ultimately compromising the overall load-bearing capacity of concrete.

Additionally, this study considers only one type of coarse aggregate and assumes uniform properties for the interfacial transition zone (ITZ). The effects of aggregate type diversity and ITZ variability on crack development were not explored, limiting the model's ability to simulate interface damage. Certainly, the model still has some limitations when applied to more complex concrete systems, such as fiber-reinforced concrete or multiphase coarse-grained concrete. At present, the model is not yet capable of accurately capturing interfacial behavior and the interactions between fibers and the matrix. Future research should incorporate multi-scale modeling strategies and improve constitutive relationships for different material phases. Future research should address these factors to improve model adaptability and engineering relevance.

Author Contributions: Conceptualization, S.W. and H.Z.; methodology, D.W. and X.W.; software, D.W.; validation, H.Z.; formal analysis, S.W.; investigation, S.W. and M.C.; resources, H.Z.; data curation, H.Z.; writing—original draft preparation, S.W. and H.Z.; writing—review and editing, S.W.

and D.W.; visualization, H.Z.; supervision, S.W.; project administration, S.W.; funding acquisition, S.W. and X.W. All authors have read and agreed to the published version of the manuscript.

Funding: This research was funded by the National Key Research and Development Program of China, grant number 2023YFC2907600 and project ZR2022QD056 supported by Shandong Provincial Natural Science Foundation, and Qingdao Postdoctoral Science Foundation, grant number QDBSH20240101037.

Institutional Review Board Statement: Not applicable.

Informed Consent Statement: Not applicable.

Data Availability Statement: Data will be made available on request.

Conflicts of Interest: The authors declare no conflicts of interest.

References

1. Ren, Q.; Pacheco, J.; de Brito, J. Methods for the modelling of concrete mesostructures: A critical review. *Constr. Build. Mater.* **2023**, *408*, 133570. [CrossRef]
2. Xiong, Q.; Wang, X.; Jivkov, A.P. A 3D multi-phase meso-scale model for modelling coupling of damage and transport properties in concrete. *Cem. Concr. Compos.* **2020**, *109*, 103545. [CrossRef]
3. Bai, F.; Li, Y.; Liu, L.; Li, X.; Liu, W. An efficient and high-volume fraction 3D mesoscale modeling framework for concrete and cementitious composite materials. *Compos. Struct.* **2023**, *325*, 117576. [CrossRef]
4. Diehl, M.; Wang, D.; Liu, C.; Mianroodi, J.R.; Han, F.; Ma, D.; Kok, P.J.J.; Roters, F.; Shanthraj, P. Solving Material Mechanics and Multiphysics Problems of Metals with Complex Microstructures Using DAMASK—The Düsseldorf Advanced Material Simulation Kit. *Adv. Eng. Mater.* **2020**, *22*, 1901044. [CrossRef]
5. Wang, D.; Shanthraj, P.; Springer, H.; Raabe, D. Particle-induced damage in Fe–TiB2 high stiffness metal matrix composite steels. *Mater. Des.* **2018**, *160*, 557–571. [CrossRef]
6. Xiao, J.; Lv, Z.; Duan, Z.; Zhang, C. Pore structure characteristics, modulation and its effect on concrete properties: A review. *Constr. Build. Mater.* **2023**, *397*, 132430. [CrossRef]
7. Xia, Y.; Wu, W.; Yang, Y.; Fu, X. Mesoscopic study of concrete with random aggregate model using phase field method. *Constr. Build. Mater.* **2021**, *310*, 125199. [CrossRef]
8. Chen, T.; Xiao, S. Three-dimensional mesoscale modeling of concrete with convex aggregate based on motion simulation. *Constr. Build. Mater.* **2021**, *277*, 122257. [CrossRef]
9. Li, H.; Huang, Y.; Yang, Z.; Yu, K.; Li, Q.M. 3D meso-scale fracture modelling of concrete with random aggregates using a phase-field regularized cohesive zone model. *Int. J. Solids Struct.* **2022**, *256*, 111960. [CrossRef]
10. Wang, X.F.; Yang, Z.J.; Yates, J.R.; Jivkov, A.P.; Zhang, C. Monte Carlo simulations of mesoscale fracture modelling of concrete with random aggregates and pores. *Constr. Build. Mater.* **2015**, *75*, 35–45. [CrossRef]
11. Zhang, S.; Hamed, E.; Song, C. An image-based meso-scale model for the hygro-mechanical time-dependent analysis of concrete. *Comput. Mech.* **2023**, *72*, 1191–1214. [CrossRef]
12. Alwash, M.; Breysse, D.; Sbartaï, Z.M. Using Monte-Carlo simulations to evaluate the efficiency of different strategies for nondestructive assessment of concrete strength. *Mater. Struct.* **2017**, *50*, 90. [CrossRef]
13. Gong, F.; Yang, L.; Wang, Z.; Jia, J.; Ning, Y.; Ueda, T. Mesoscale discrete analysis of mechanical properties of recycled aggregate concrete based on Voronoi mesh. *Constr. Build. Mater.* **2023**, *370*, 130649. [CrossRef]
14. Wang, J.; Yu, X.; Fu, Y.; Zhou, G. A 3D Meso-Scale Model and Numerical Uniaxial Compression Tests on Concrete with the Consideration of the Friction Effect. *Materials* **2024**, *17*, 1204. [CrossRef] [PubMed]
15. Zhou, R.; Song, Z.; Lu, Y. 3D mesoscale finite element modelling of concrete. *Comput. Struct.* **2017**, *192*, 96–113. [CrossRef]
16. Ouyang, H.; Chen, X. 3D meso-scale modeling of concrete with a local background grid method. *Constr. Build. Mater.* **2020**, *257*, 119382. [CrossRef]
17. Yang, X.; Wang, M. Fractal dimension analysis of aggregate packing process: A numerical case study on concrete simulation. *Constr. Build. Mater.* **2021**, *270*, 121376. [CrossRef]
18. Xu, W.; Fu, J.; Hua, R.; Han, F. ITZ volume fraction and thermal conductivity of concrete: A unified random packing model for gravels and crushed rocks. *J. Build. Eng.* **2024**, *90*, 109457. [CrossRef]
19. Wang, J.; Jivkov, A.P.; Engelberg, D.L.; Li, Q. Image-Based vs. Parametric Modelling of Concrete Meso-Structures. *Materials* **2022**, *15*, 704. [CrossRef]

20. Song, Y.; Shen, C.; Damiani, R.M.; Lange, D.A. Image-based restoration of the concrete void system using 2D-to-3D unfolding technique. *Constr. Build. Mater.* **2021**, *270*, 121476. [CrossRef]
21. Pan, G.; Song, T.; Li, P.; Jia, W.; Deng, Y. Review on finite element analysis of meso-structure model of concrete. *J. Mater. Sci.* **2025**, *60*, 32–62. [CrossRef]
22. Xu, W.; Han, Z.; Tao, L.; Ding, Q.; Ma, H. Random non-convex particle model for the fraction of interfacial transition zones (ITZs) in fully-graded concrete. *Powder Technol.* **2018**, *323*, 301–309. [CrossRef]
23. Paruthi, S.; Husain, A.; Alam, P.; Khan, A.H.; Hasan, M.A.; Magbool, H.M. A review on material mix proportion and strength influence parameters of geopolymer concrete: Application of ANN model for GPC strength prediction. *Constr. Build. Mater.* **2022**, *356*, 129253. [CrossRef]
24. Ren, D.; Liu, X.; Cui, B.; Wang, E.; Ma, Q.; Yan, F.; Xie, W. Meso-crack evolution based constitutive model for concrete material under compression. *Compos. Part B Eng.* **2023**, *265*, 110956. [CrossRef]
25. Zhou, R.; Lu, Y.; Wang, L.-G.; Chen, H.-M. Mesoscale modelling of size effect on the evolution of fracture process zone in concrete. *Eng. Fract. Mech.* **2021**, *245*, 107559. [CrossRef]
26. Sun, Y.; Roubin, E.; Shao, J.; Colliat, J.-B. Meso-scale Finite Element modeling of the Fracture Process Zone evolution for concrete. *Theor. Appl. Fract. Mech.* **2023**, *125*, 103869. [CrossRef]
27. Baktheer, A.; Martínez-Pañeda, E.; Aldakheel, F. Phase field cohesive zone modeling for fatigue crack propagation in quasi-brittle materials. *Comput. Methods Appl. Mech. Eng.* **2024**, *422*, 116834. [CrossRef]
28. Schröder, J.; Pise, M.; Brands, D.; Gebuhr, G.; Anders, S. Phase-field modeling of fracture in high performance concrete during low-cycle fatigue: Numerical calibration and experimental validation. *Comput. Methods Appl. Mech. Eng.* **2022**, *398*, 115181. [CrossRef]
29. Wu, Z.; Lei, J.; Ye, C.; Yu, C.; Zhao, J. Coupled FDM-DEM simulations of axial compression tests on FRP-confined concrete specimens. *Constr. Build. Mater.* **2022**, *351*, 128885. [CrossRef]
30. Mansour, D.M.; Ebid, A.M. Predicting thermal behavior of mass concrete elements using 3D finite difference model. *Asian J. Civ. Eng.* **2024**, *25*, 1601–1611. [CrossRef]
31. Nitka, M. Static and dynamic concrete calculations: Breakable aggregates in DEM model. *J. Build. Eng.* **2024**, *89*, 109006. [CrossRef]
32. Nitka, M.; Tejchman, J. Modelling of concrete behaviour in uniaxial compression and tension with DEM. *Granul. Matter* **2015**, *17*, 145–164. [CrossRef]
33. Wang, P.; Gao, N.; Ji, K.; Stewart, L.; Arson, C. DEM analysis on the role of aggregates on concrete strength. *Comput. Geotech.* **2020**, *119*, 103290. [CrossRef]
34. Wang, Z.M.; Kwan, A.K.H.; Chan, H.C. Mesoscopic study of concrete I: Generation of random aggregate structure and finite element mesh. *Comput. Struct.* **1999**, *70*, 533–544. [CrossRef]
35. Chen, J.; Wang, H.; Dan, H.; Xie, Y. Random Modeling of Three-Dimensional Heterogeneous Microstructure of Asphalt Concrete for Mechanical Analysis. *J. Eng. Mech.* **2018**, *144*, 04018083. [CrossRef]
36. Ma, H.; Xu, W.; Li, Y. Random aggregate model for mesoscopic structures and mechanical analysis of fully-graded concrete. *Comput. Struct.* **2016**, *177*, 103–113. [CrossRef]
37. Sun, Y.; Wei, X.; Gong, H.; Du, C.; Wang, W.; Chen, J. A two-dimensional random aggregate structure generation method: Determining effective thermo-mechanical properties of asphalt concrete. *Mech. Mater.* **2020**, *148*, 103510. [CrossRef]
38. Ma, H.; Song, L.; Xu, W. A novel numerical scheme for random parameterized convex aggregate models with a high-volume fraction of aggregates in concrete-like granular materials. *Comput. Struct.* **2018**, *209*, 57–64. [CrossRef]
39. Wei, S.; Shen, J.; Yang, W.; Li, Z.; Di, S.; Ma, C. Application of the renormalization group approach for permeability estimation in digital rocks. *J. Pet. Sci. Eng.* **2019**, *179*, 631–644. [CrossRef]
40. Wei, S.; Wang, D.; Wang, X.; Zhang, H.; Cao, H.; Liu, Q. Research on the random generation model of 2D irregular aggregates in concrete based on fast grid region division method and its ground penetrating radar characteristics. *Case Stud. Constr. Mater.* **2024**, *21*, e03932. [CrossRef]
41. Marples, C.R.; Williams, P.M. The Golden Ratio in Nature: A Tour across Length Scales. *Symmetry* **2022**, *14*, 2059. [CrossRef]
42. Sen, S.K.; Agarwal, R.P. Golden ratio in science, as random sequence source, its computation and beyond. *Comput. Math. Appl.* **2008**, *56*, 469–498. [CrossRef]
43. Zhang, J.; Ma, R.; Pan, Z.; Zhou, H. Review of Mesoscale Geometric Models of Concrete Materials. *Buildings* **2023**, *13*, 2428. [CrossRef]
44. Wei, D.; Hurley, R.C.; Poh, L.H.; Dias-da-Costa, D.; Gan, Y. The role of particle morphology on concrete fracture behaviour: A meso-scale modelling approach. *Cem. Concr. Res.* **2020**, *134*, 106096. [CrossRef]
45. Masad, E.; Olcott, D.; White, T.; Tashman, L. Correlation of Fine Aggregate Imaging Shape Indices with Asphalt Mixture Performance. *Transp. Res. Rec. J. Transp. Res. Board* **2001**, *1757*, 148–156. [CrossRef]

46. Qiu, W.; Fu, S.; Ye, J. Auto-generation methodology of complex-shaped coarse aggregate set of 3D concrete numerical test specimen. *Constr. Build. Mater.* **2019**, *217*, 612–625. [CrossRef]
47. Wriggers, P.; Moftah, S.O. Mesoscale models for concrete: Homogenisation and damage behaviour. *Finite Elem. Anal. Des.* **2006**, *42*, 623–636. [CrossRef]
48. Jin, L.; Zhang, S.; Li, D.; Xu, H.; Du, X.; Li, Z. A combined experimental and numerical analysis on the seismic behavior of short reinforced concrete columns with different structural sizes and axial compression ratios. *Int. J. Damage Mech.* **2018**, *27*, 1416–1447. [CrossRef]
49. Liao, K.-Y.; Chang, P.-K.; Peng, Y.-N.; Yang, C.-C. A study on characteristics of interfacial transition zone in concrete. *Cem. Concr. Res.* **2004**, *34*, 977–989. [CrossRef]
50. Zhang, S.; Zhang, C.; Liao, L.; Wang, C. Numerical study of the effect of ITZ on the failure behaviour of concrete by using particle element modelling. *Constr. Build. Mater.* **2018**, *170*, 776–789. [CrossRef]
51. Li, J.; Zhang, J.; Qian, G.; Zheng, J.; Zhang, Y. Three-Dimensional Simulation of Aggregate and Asphalt Mixture Using Parameterized Shape and Size Gradation. *J. Mater. Civ. Eng.* **2019**, *31*, 04019004. [CrossRef]
52. Zheng, Y.; Zhang, Y.; Zhuo, J.; Zhang, P.; Hu, S. Mesoscale synergistic effect mechanism of aggregate grading and specimen size on compressive strength of concrete with large aggregate size. *Constr. Build. Mater.* **2023**, *367*, 130346. [CrossRef]
53. Zhou, X.; Xie, Y.; Long, G.; Li, J. Effect of surface characteristics of aggregates on the compressive damage of high-strength concrete based on 3D discrete element method. *Constr. Build. Mater.* **2021**, *301*, 124101. [CrossRef]
54. Ghosh, S.; Dhang, N.; Deb, A. Influence of aggregate geometry and material fabric on tensile cracking in concrete. *Eng. Fract. Mech.* **2020**, *239*, 107321. [CrossRef]
55. Wang, X.; Zhang, M.; Jivkov, A.P. Computational technology for analysis of 3D meso-structure effects on damage and failure of concrete. *Int. J. Solids Struct.* **2016**, *80*, 310–333. [CrossRef]

coatings

MDPI

Article

Effect of Macroscopic Composition on the Performance of Self-Compacting Concrete

He Liu [1,*], Wenxi Li [1], Haonan Zou [1], Wei Bian [2], Jingyi Zhang [3], Ji Zhang [4] and Peng Zhang [4]

1 School of Transportation and Geomatics Engineering, Shenyang Jianzhu University, Shenyang 110168, China; 18512453100@163.com (W.L.); 15114147960@163.com (H.Z.)
2 Key Laboratory of Highway Construction & Maintenance Technology in Loess Region, Shanxi Transportation Research Institute, Taiyuan 030006, China
3 School of Civil Engineering, Shenyang Urban Construction University, Shenyang 110167, China; dq_zjy@syucu.edu.cn
4 Liaoning New Development Highway Technology Maintenance Co., Ltd., Shenyang 110058, China; 13324131953@163.com (J.Z.); 19232408304@163.com (P.Z.)
* Correspondence: heliu@sjzu.edu.cn

Abstract: In recent years, there has been significant interest in the development of self-compacting concrete (SCC). This study views SCC as a two-phase composite material and introduces a new aggregate spacing coefficient model based on the concept of Fullman's mean free path and stereological theory. The validity of the aggregate spacing coefficient model was verified. The relationship between the fine and coarse aggregate coefficients and the properties of SCC are revealed. The results show that the slump and slump flow of SCC increase as the fine and coarse aggregate coefficients increase. The coarse aggregate spacing coefficient has a significant influence on the compressive strength and drying shrinkage of SCC. A significant linear relationship between the coarse aggregate spacing coefficient and SCC dry shrinkage properties is revealed. Compared to the conditional mixing proportion method, which considers the aggregate volume as a control factor, the aggregate spacing coefficient takes into account the aggregate volume and gradation, which can more accurately reflect the characteristics of the aggregate. Meanwhile, this new perspective on the macroscopic composition of SCC provides insights into the controlling factors of its performance.

Keywords: self-compacting concrete; Fullman's mean free path; stereological theory; aggregate volume fraction; aggregate spacing coefficient

1. Introduction

The concept of self-compacting concrete (SCC) was proposed at the end of the 20th century [1]. SCC is a kind of concrete that can be compacted under its own gravity. Even if there are dense steel bars, SCC can completely fill the formwork evenly without additional vibration [2]. As a kind of high-performance concrete, SCC has been developed for more than thirty years and has been widely studied by scholars. Many research results have been achieved in the areas of preparation technology [3], mix design methods [4], performance test evaluation [5], and engineering applications [6] of SCC. As the key properties of SCC, such as slump flow, compressive strength and dry shrinkage [7], all affect the quality of SCC, it is crucial to study the key influencing factors of SCC performance.

To investigate the correlation between the material composition and mix design of SCC and its properties, scholars have conducted numerous experiments and numerical simulations.

Liu and Chen [8] emphasized that SCC performance was significantly related to aggregate characteristics. Khaleel et al. [9,10] carried out test methods such as slump flow and L-box. The test results confirmed that as the maximum particle size of the coarse aggregate increases, the slump flow and passing ability of the SCC mixture descend and the compressive strength decreases. Krasna et al. [11] proposed factors influencing the compressive strength of SCC, including the particle size and coarse aggregate proportion. The results showed that with coarse aggregate proportion increases, the compressive strength of concrete increases. However, Liu [8,12] found that within a certain range, the maximum increase in particle size was beneficial to the compressive strength of SCC. Chen [13] established the relationship between slump flow and aggregate properties, which included the average particle size and volume of the aggregate. These results helped to develop the SCC design.

Zhao [14] defined a coarse aggregate gradation (A/B ratio, size 5–10 mm, coarse aggregate weight/size 10–20 mm for 4/6, 5/5, 6/4, 7/3). The results showed that with the change in the A/B ratio from 4/6 to 7/3, the initial slump flow, blocking, and segregation ratio of SCC decreased. Zhang [15] developed a model to predict the plastic viscosity threshold of the mixture. The model was forecasted according to the paste yield stress threshold, which was calculated by the coarse aggregate content of SCC. Results showed that a higher coarse aggregate content would lead to a lower yield stress and larger plastic viscosity. Santos [16] investigated the properties of SCC with different skeleton distribution forms. Compared with the discontinuous gradation skeleton SCC, the continuous gradation skeleton SCC showed better flowability and passing ability and higher early strength. However, both types of gradation demonstrated similar compressive strength at 365 days. Furthermore, the shape of the aggregate also affected SCC performance. As the content of the multi-angular elongated and flaky aggregate increases, the slump flow of the SCC mixture decreases [17]. Previous research focused on fresh and hardened properties of SCC using a traditional mix design and aggregate macroscopic morphology, which was tedious and time-consuming [18]. In addition, contradictions between the properties of high-performance concrete made SCC preparation more complicated.

Scholars have proposed another research idea regarding SCC as a two-phase composite material, such as aggregate and mortar. Su [19] presented a new method for SCC mix design. After determining the required aggregate, mortar was then filled into the aggregate voids. Basheerudeen [20] claimed that by maximizing the aggregate particles' packing density and cementitious materials, the concrete mixture performance was enhanced.

Researchers also simulated fluid–solid flow to study SCC, using the smoothed particle hydrodynamics (SPH) method [21]. Dhaheer [22] verified the ability of the SPH simulation method to predict the passing ability of SCC through steel bars gap. Alyhya [23] simulated the flow of SCC through a V-funnel by the three dimensional meshless smoothed particle hydrodynamics (SPH) method. The simulation results were consistent with the experimental results. Tran-Duc [24] revealed the relationship between concrete performance and coarse aggregate content by a simulation method. The results revealed that the yield stress and plastic viscosity of SCC were increased as the coarse aggregate content increased.

To summarize the above, the qualitative influence of aggregate properties on the performance of SCC has achieved many research results, such as the particle size, proportion, and gradation of aggregate. These research results have played a vital role in promoting the development of SCC. However, in the process of studying the performance of SCC, the traditional method mainly takes the aggregate volume as a parameter. Therefore, this paper establishes a model based on Fullman's mean free path and the theory of stereology. The concept of coarse and fine aggregate spacing coefficients is proposed, considering the aggregate volume and aggregate gradation at the same time. The validity of the coarse and fine aggregate spacing coefficients is verified. The quantitative relationship between the aggregate spacing coefficient and various properties of SCC is revealed. These findings will help us to understand the relationship between SCC performance and macroscopic composition parameters, measuring the key performance of SCC in a more refined way.

2. Material and Methods

SCC was prepared using raw materials including P.O42.5 cement, S95 slag, polycarboxylate superplasticizer, methyl cellulose ether, water, fine aggregates (graded river sand), and coarse aggregates (limestone aggregate) with an average size of 4.75 mm to 20 mm. The apparent density of river sand was 2.65 g/cm^3, while the coarse aggregate was 2.73 g/cm^3. The particle gradation results for river sand and gravel are shown in Table 1.

Table 1. Screening results of river sand and gravel particle composition.

	Passing Screening Rate/%								
	0.15 mm	**0.3 mm**	**0.6 mm**	**1.18 mm**	**2.36 mm**	**4.75 mm**	**9.5 mm**	**16 mm**	**19 mm**
river sand	1.2	12.4	31.5	20.5	15.0	19.4	0	0	0
gravel	0	0	0	0	0.6	3.6	41.3	/	54.5
	Rate of particles passing through each sieve/%								
	0.15 mm	0.3 mm	0.6 mm	1.18 mm	2.36 mm	4.75 mm	9.5 mm	16 mm	19 mm
river sand	1.2	13.6	45.1	65.6	80.6	100	0	0	0
gravel	0	0	0	0	0.6	4.2	45.5	/	100

2.1. Mix Proportion

In order to investigate the impact of the spacing coefficients of coarse and fine aggregates on the workability of fresh concrete, as well as the compressive strength and shrinkage of SCC, a series of experiments were conducted. Concretes with different aggregate volume fractions and spacing coefficients were designed. The mix proportions of each group of specimens are shown in Table 2.

Table 2. Materials and mix proportion of each specimen.

Item.	**Cementitious Materials (kg/m^3)**			**Water (kg/m^3)**	**SP (kg/m^3)**	**Aggregate**			
	C	**FA**	**GGBS**			**Δ_{ca}**	**Δ_{fa}**	**m_{ca} (kg/m^3)**	**m_{fa} (kg/m^3)**
1	338	78	104	156	6.24	0.9–1.8	1.09–1.41	1156–733	564–458
2	338	78	104	156	6.24	1.0–1.6	0.93–1.47	1086–798	638–443
4	338	78	104	156	6.24	1.0–1.6	0.83–1.66	1086–798	694–400
5	338	78	104	156	6.24	0.9–1.75	1.27	1156–748	499

2.2. Test Method

The test methods in this study were generally categorized into two types: the workability of fresh concrete, characterized by slump and slump flow, and the hardened proporties of SCC, characterized by the compressive strength and drying shrinkage performance of SCC.

2.2.1. Workability Test

According to Q/CR 596-2017 [25], the workability of concrete is measured by the slump and the slump flow test. After the collapse cylinder is lifted, the distance of the SCC mixture falling down is the slump. The slump flow is the average of the maximum diameter of the slump flow surface and the diameter in the vertical direction from the beginning of the slump to when the SCC mixture stops flowing, as shown in Figure 1.

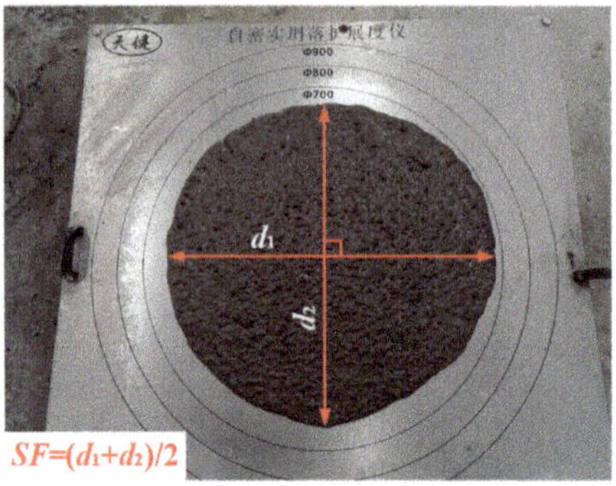

Figure 1. Test and calculation method of slump flow.

The slump flow is calculated according to Equation (1).

$$SF = \frac{d_1 + d_2}{2} \tag{1}$$

2.2.2. Compressive Strength Test

According to GB/T50081-2019 [26], the compressive strength of specimens with a size of $150 \times 150 \times 150$ mm^3 were tested by a hydraulic universal testing machine; the loading rate was 0.5 MPa/s. The compressive strength is calculated according to Equation (2).

$$f_c = F/A \tag{2}$$

where f_c is compressive strength, MPa; F is the failure load, N; and A is the bearing area, mm^2.

2.2.3. Drying Shrinkage Test

The drying shrinkage test of SCC was performed according to GB/T50082-2009 [27]. Prism specimens sized $100 \times 100 \times 515$ mm^3 were used, and two stainless steel sheets were pasted at both ends of each specimen. The shrinkage rate of concrete is calculated according to Formula (3).

$$\varepsilon_{st} = \frac{L_0 - L_t}{L_b} \tag{3}$$

where ε_{st} is the shrinkage rate; L_0 is the specimen initial length, mm; L_t is the specimen length at t area, mm; and L_b is the measuring scale distance, mm.

3. Establishment and Verification of Aggregate Spacing Model

3.1. Establishment of Aggregate Spacing Model

Concrete is a multi-component composite material consisting of coarse and fine aggregates, cement, water, and admixtures. This study considers SCC as a two-phase composite material consisting of coarse aggregate and mortar, with the coarse aggregate particles being uniformly dispersed in the continuous phase of mortar, as represented by Figure 2.

Furthermore, considering the mortar as a two-phase material consisting of fine aggregate and cement paste, the macroscopic structure of concrete and mortar can be described by aggregate spacing. As an important characteristic parameter, aggregate spacing can be characterized by the aggregate spacing coefficient, which has a significant influence on the workability, strength, and durability of SCC.

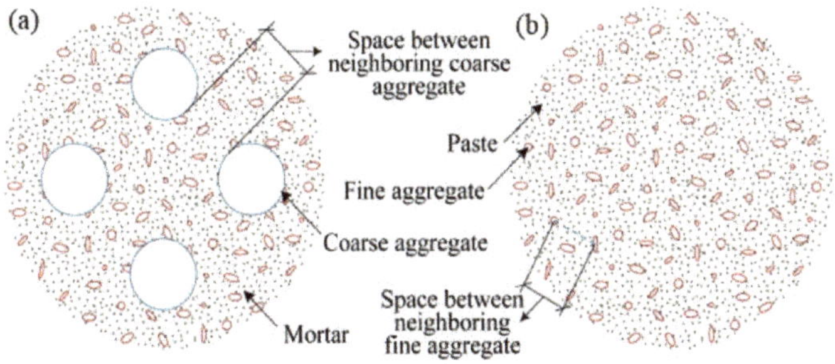

Figure 2. Schematics diagram of physical model for the compositions of cementitious material system. (**a**) concrete level, (**b**) mortar level.

When solving the distribution of solid particles and the distance between particles in a certain space system, the concept of Fullman's mean free path and the stereology theory are adopted. Therefore, we discussed the aggregate spacing coefficient based on the above theory [28,29]. For a binary composite system consisting of a dispersed phase of particles and a continuous phase, the mean free path between particles in the system can be calculated using the above theory. The calculation formula for the mean free path between particles in the system is shown in Equation (4).

$$\lambda = 4[1 - (V_V)_a]/(S_V)_a \tag{4}$$

where $(V_V)_a$ is the volume fraction of solid particle a in the unit volume binary system; $(S_V)_a$ is the surface area of solid particle a in the unit volume binary system, 1/m; and λ is the average free path between particles.

For the concrete system composed of mortar and coarse aggregate, if assuming that the maximum and minimum particle sizes of coarse aggregate are $D_{ca,max}$ and $D_{ca,min}$, respectively, the distribution of particle composition conforms to the Fuller distribution function, which is shown in Formula (5).

$$P_V(x) = (\sqrt{x} - \sqrt{D_{ca,\min}})/(\sqrt{D_{ca,\max}} - \sqrt{D_{ca,\min}}) \tag{5}$$

where P_V (x) is the cumulative volume (or mass) fraction of coarse aggregate particles and x is the coarse aggregate diameter of the corresponding grain size.

Assuming that the total volume of the coarse aggregate particles is denoted as V_{ca}, the total surface area S_{ca} of the aggregate particles in this system can be expressed using Formula (6).

$$S_{ca} = \frac{6V_{ca}}{\sqrt{D_{ca,\max} \times D_{ca,\min}}} \tag{6}$$

By combining Formulas (4) and (6), we can derive the average spacing expression between adjacent coarse aggregate particle surfaces, which occurs in the binary system of concrete composed of mortar and coarse aggregate, as shown in Formula (7).

$$\lambda_{ca} = \frac{2(1 - V_{ca})}{3V_{ca}} \times \sqrt{D_{ca,\max} \times D_{ca,\min}} \quad (7)$$

If using Formula (8) to express the average particle diameter $d_{ca,av}$ of the particles in the coarse aggregate system and defining the average coarse aggregate spacing coefficient Δ_{ca} as the ratio of the average spacing of adjacent coarse aggregates to the average diameter of the particles, then we can obtain the calculation formula of the coarse aggregate spacing coefficient in the concrete system, as shown in Formula (9):

$$d_{ca,av} = \frac{\sum_{i}^{n} d_{ca,i} m_{ca,i}}{\sum_{i}^{n} m_{ca,i}} \quad (8)$$

$$\Delta_{ca} = \frac{2(1 - V_{ca})}{3V_{ca}} \times \frac{\sqrt{D_{ca,\max} \times D_{ca,\min}}}{d_{ca,av}} \quad (9)$$

Similarly, we can obtain the spacing coefficient Δ_{fa} between sands in the mortar binary system, which is composed of sand fine aggregate and slurry, as shown in Formula (10).

$$\Delta_{fa} = \frac{2(1 - V_{fa})}{3V_{fa}} \times \frac{\sqrt{D_{fa,\max} \times D_{fa,\min}}}{d_{fa,av}} \quad (10)$$

where $d_{ca,I}$ and $m_{ca,i}$ represent the particle size and percentage of coarse aggregate for grade I, while $D_{ca,max}$, $D_{ca,min}$, and $D_{ca,av}$, represent the maximum, minimum, and average particle sizes of the coarse aggregate; $D_{fa,max}$, $D_{fa,min}$, and $D_{fa,av}$, represent the maximum, minimum, and average particle sizes of the fine aggregate; and V_{ca} and V_{fa} denote the volume fractions of coarse and fine aggregates, respectively.

Formulas (9) and (10) demonstrate that the coarse aggregate spacing coefficient in the concrete system is influenced by some factors, such as the volume fraction content and particle size distribution (maximum and minimum particle size, average particle size) of the coarse aggregate. Similarly, the fine aggregate spacing coefficient is influenced by the volume fraction content and particle size distribution of sand. Therefore, when the properties of cement paste and aggregate are certain, the aggregate spacing coefficient is closely related to the performance of concrete or mortar.

3.2. Validation of Aggregate Spacing Model

To verify the validity of the spacing coefficient aggregates, we tested the spacing coefficient of adjacent coarse aggregates. Firstly, we cut the SCC cube specimens with sizes of 100 × 100 × 100 mm^3 into two equal parts by a cutting machine and polished the cutting surface until smooth. After the surface dried, each side was divided into three equal parts using a colored pen, and these parts were connected by four straight lines, as shown in Figure 3. Then, we measured the length of each coarse aggregate passing through each straight line using a steel ruler, with a minimum scale of millimeters, and calculated the total number of coarse aggregates passing through each straight line. The total length of coarse aggregate passing through each straight line was crossed with the total number of coarse aggregates, and then we divided the total length of the spacing on each straight line by the total number of coarse aggregates passing through it and added 1; this way, the average spacing coefficient Δ_{ca} of adjacent aggregates on each straight line can be obtained. Through the average spacing Δ_{ca} and the average diameter $d_{ca,av}$ of the coarse aggregate, the spacing coefficient of the adjacent coarse aggregate on each straight line can be calculated. The aggregate spacing was the aggregate spacing coefficient multiplied by the average particle size of the coarse aggregate. Finally, the test results for the arithmetic mean of aggregate spacing calculated on the four straight lines are obtained. The average spacing test diagram of the adjacent aggregates is shown in Figure 3.

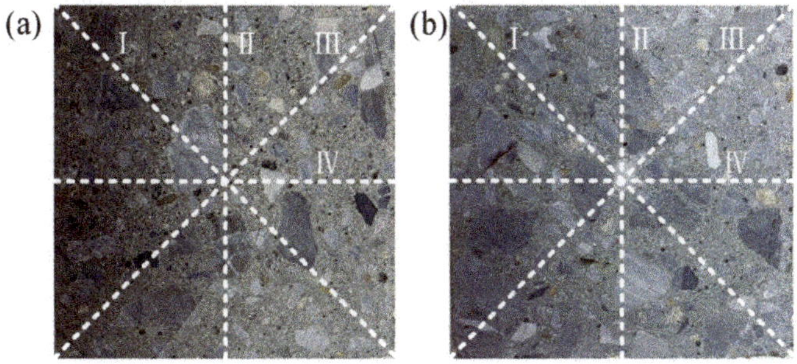

Figure 3. The average spacing test diagram of adjacent aggregates. (**a**) aggregate spacing coefficient 2.4, (**b**) aggregate spacing coefficient 1.8.

In Figure 3a, I is 8, II is 5, III is 8, IV is 6, $\lambda = (17.6 + 20 + 17.6 + 16.7)/4 = 18.6$ mm, the calculated value is $\Delta ca = 18.6/7.44 = 2.5$, and the error is 4.17%.

In Figure 3b, I is 10, II is 7, III is 10, IV is 8, $\lambda = (14.1 + 14.3 + 14.1 + 12.5)/4 = 13.8$ mm, the calculated value is $\Delta_{ca} = 13.8/7.44 = 1.85$, and the error is 2.78%.

The comparison between the test value and calculated value of the two specimens is shown in Figure 4.

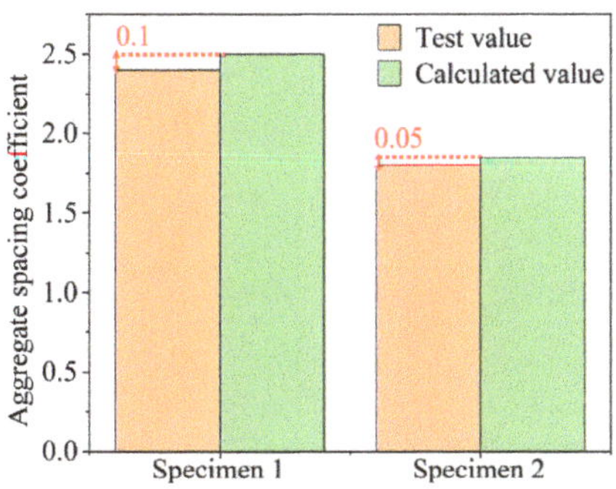

Figure 4. The comparison results of the calculated value and the test value of the aggregate spacing coefficient.

We calculated specimens with coarse or fine aggregate volume fractions of 0.1, 0.15, 0.2, 0.25, 0.3, 0.35, 0.4, 0.45, and 0.5. According to above method, the corresponding coarse or fine aggregate spacing coefficients were calculated. The relationship between the volume fraction and spacing coefficients is shown in Figure 5.

Figure 5 shows that for both coarse and fine aggregates, the corresponding spacing coefficient decreases as the aggregate volume fraction increases. In SCC, which is a two-phase composite system, more aggregate will lead to a smaller gap between aggregates.

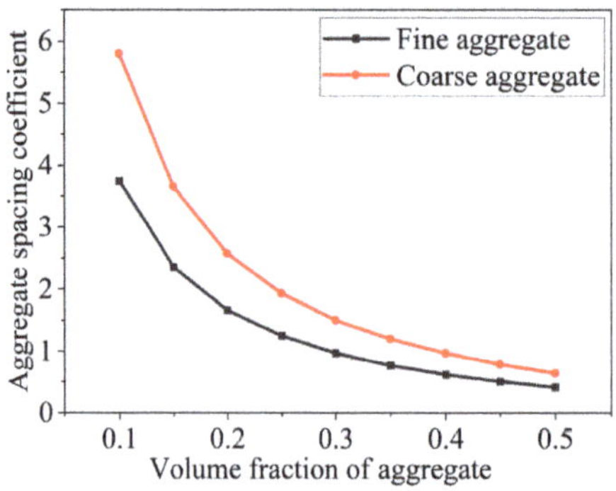

Figure 5. Relationship between the volume fraction of aggregate and the aggregate spacing coefficient in the system.

4. Result and Discussion

4.1. *Workability of Fresh Concrete*

SCC with fine aggregate spacing coefficients of 1.41, 1.27, and 1.09 when changing the coarse spacing coefficient from 0.9 to 1.8 was analyzed. The slump and slump flow of SCC were measured, as shown in Figure 6.

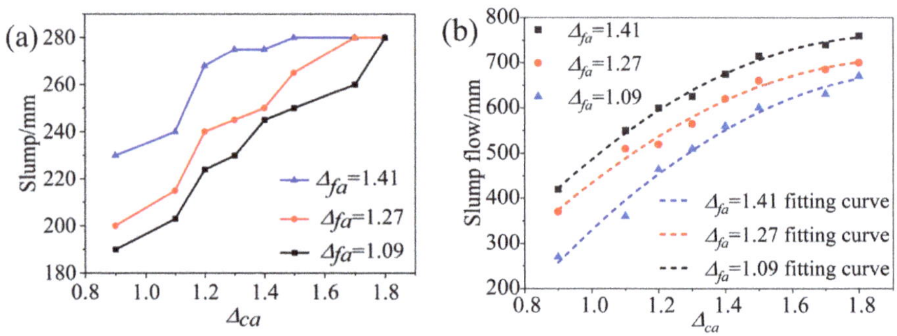

Figure 6. Relationship between coarse aggregate spacing coefficient and fluidity of fresh concrete. (**a**) Correlation between coarse aggregate spacing coefficient and slump, (**b**) correlation between coarse aggregate spacing coefficient and slump flow.

The results in Figure 6 demonstrate that when the fine aggregate spacing coefficient is 1.41, 1.27, and 1.09, the slump and slump flow of fresh SCC increase as the coarse aggregate spacing coefficient increases. The coarse aggregate particles disperse uniformly in the mortar continuous phase with a certain spacing. The larger the coarse aggregate spacing, the more mortar can flow over the coarse aggregate in the system and the better the flowability performance.

Figure 6a illustrates that for the mixture with an aggregate spacing coefficient of 1.41, when the coarse aggregate spacing coefficient increased to 1.3, the slump no longer increased and remained at 280 mm. Similarly, the slump of the mixture with a fine aggregate spacing coefficient of 1.27 also increased to 280 mm and remained unchanged. At this time, the coarse aggregate spacing coefficient was 1.5. This shows that when the slump increases to 280 mm, the workability of the SCC mixture cannot be accurately reflected. Figure 6b illustrates that the three types of fresh SCC all have the same trend; that is, they increase continuously with the coarse spacing coefficient. For the mixture with the fine aggregate spacing coefficient of 1.09, the slump flow increases most rapidly with the coarse aggregate spacing coefficient being between 1.1 and 1.2, and the growth rate is slow with the coarse aggregate spacing coefficient being between 0.8 and 1.1. For the mixture with the fine aggregate spacing coefficient of 1.27, the slump flow increases most rapidly with the coarse aggregate spacing coefficient being between 0.8 and 1.1, and the growth rate is slow between 1.1 and 1.2. The growth rate of the mixture slump flow is essentially unchanged, with a fine aggregate spacing coefficient of 1.09 when the coarse aggregate spacing coefficient is between 0.8 and 1.2. This is mainly because the fiction between aggregates decreases when the coarse aggregate spacing coefficient increases. This leads to the concrete dissipating less energy. This also increases the slump of SCC.

When the fine aggregate spacing coefficients are 1.41, 1.27, and 1.09, respectively, the fitting relationship between the coarse aggregate spacing coefficient and the fluidity of fresh SCC is shown in Table 3.

Table 3. Relationship between coarse aggregate spacing and SCC fluidity.

Δ_{fa}	Fitting Formula
1.41	$SF = -337.65\Delta_{ca}^2 + 1282.98\Delta_{ca} - 458.66$
1.27	$SF = -304.06\Delta_{ca}^2 + 1184.93\Delta_{ca} - 446.97$
1.09	$SF = -339.83\Delta_{ca}^2 + 1369.96\Delta_{ca} - 700.45$

Fresh SCC with fine aggregate spacing coefficients of 1.0, 1.3, and 1.6 was analyzed. The slump and slump flow of SCC were measured, as shown in Figure 7.

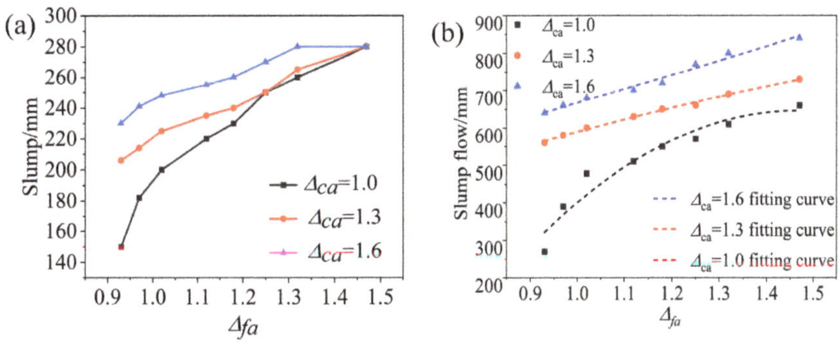

Figure 7. Relationship between fine aggregate spacing coefficient and fluidity of fresh concrete. (**a**) Correlation between fine aggregate spacing coefficient and slump, (**b**) correlation between fine aggregate spacing coefficient and slump.

The results in Figure 7 illustrate that the coarse aggregate spacing remains at 1.0, 1.3, and 1.6. The slump and slump flow of fresh SCC increase as the fine aggregate spacing coefficient increases. In the mortar system, fine aggregates are uniformly suspended in the cement paste. A higher spacing coefficient of the fine aggregates leads to a smaller fine aggregate volume fraction, an increase in the cement paste content, and a higher flowability of fresh SCC.

As shown in Figure 7a, the slump of SCC with a coarse aggregate spacing coefficient of 1.6 is most affected by the fine aggregate spacing coefficient. In addition, SCC with a coarse aggregate spacing coefficient of 1.0 is stable; when the slump increases to 280 mm, it no longer increases with the increase in the fine aggregate spacing coefficient. The slump flow of SCC increases as the fine aggregate spacing coefficient increases, as shown in

Figure 7b. The SCC with a coarse aggregate spacing coefficient of 1.3 grows the most uniformly. When the fine aggregate spacing is between 1.02 and 1.18, SCC with a coarse aggregate spacing coefficient of 1.0 slowly increases, and then it grows rapidly. When the fine aggregate spacing is between 0.93 and 1.02, the slump flow of SCC with an aggregate spacing coefficient of 1.6 increases rapidly and then slows down.

The coarse aggregate spacing coefficients are 1.6, 1.3, and 1.0. The fitting relationship between the fine aggregate spacing coefficient and the slump of fresh SCC is shown in Table 4.

Table 4. Relationship between fine aggregate spacing and SCC slump.

Δ_{fa}	Fitting Formula
1.6	$SF = -32.16\Delta_{fa}^2 + 301.48\Delta_{fa} - 333.23$
1.3	$SF = -93.50\Delta_{fa}^2 + 528.56\Delta_{fa} - 153.77$
1.0	$SF = -1132.61\Delta_{fa}^2 + 3319.33\Delta_{fa} - 1785.86$

The flow process of SCC could be regarded as the conversion of gravitational potential energy to friction energy and kinetic energy. With the increase in the coarse aggregate spacing coefficient, the friction between aggregates decreases; that is, the energy dissipated by the concrete phase decreases and more gravitational potential energy is converted into kinetic energy, which makes the mixture flow better. However, in the process of increasing the fine aggregate spacing, the fine aggregate volume and surface area decrease and the water adsorbed on the surface of aggregate decreases, which leads to a decrease in SCC viscosity and better flow.

4.2. Compressive Strength of Concrete

Figure 8 shows the compressive strength of SCC specimens with different coarse aggregate spacing coefficients and different fine aggregate spacing coefficients.

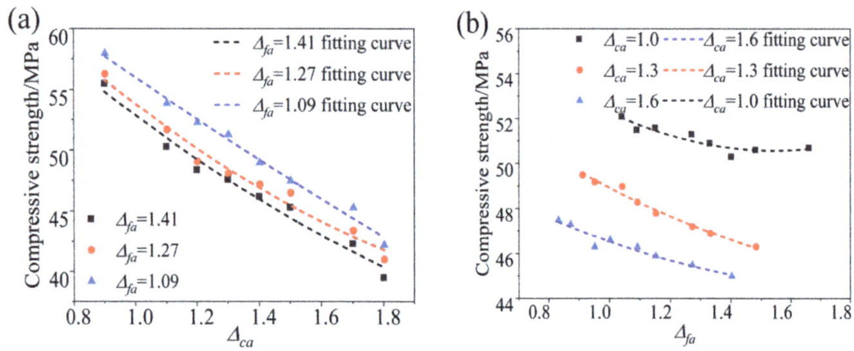

Figure 8. Relationship between aggregate spacing coefficients and compressive strength of SCC. (**a**) Correlation between coarse aggregate spacing coefficient and compressive strength, (**b**) correlation between fine aggregate spacing coefficient and compressive strength.

Figure 8a shows that when the fine aggregate spacing coefficients are 1.09, 1.27, and 1.41, the compressive strength of SCC specimens decreases with the increase in the coarse aggregate spacing coefficients, which decrease by 28.83%, 27.18%, and 27.24%, respectively. The fitting relationship between coarse aggregate spacing coefficients and the compressive strength of SCC is shown in Table 5.

Table 5. Relationship between coarse aggregate spacing and SCC compressive strength.

Δ_{fa}	Fitting Formula
1.41	$\sigma = 4.09\Delta_{ca}^2 - 27.11\Delta_{ca} + 75.92$
1.27	$\sigma = 5.44\Delta_{ca}^2 - 30.27\Delta_{ca} + 78.62$
1.09	$\sigma = 1.25\Delta_{ca}^2 - 19.96\Delta_{ca} + 74.70$

When the coarse aggregate spacing coefficients are 1.0, 1.3, and 1.6, the fitting relationship between the fine aggregate spacing coefficient and the compressive strength of SCC is as shown in Table 6. As shown in Figure 8b, the compressive strength of SCC with coarse aggregate spacing coefficients of 1.0, 1.3, and 1.6 decreases by 2.69%, 6.46%, and 5.26%, respectively.

Table 6. Relationship between fine aggregate spacing and SCC compressive strength.

Δ_{ca}	Fitting Formula
1.6	$\sigma = 2.52\Delta_{fa}^2 - 9.73\Delta_{fa} + 53.74$
1.3	$\sigma = 2.98\Delta_{fa}^2 - 12.97\Delta_{fa} + 58.91$
1.0	$\sigma = 5.86\Delta_{fa}^2 - 18.10\Delta_{fa} + 64.52$

Comparing Figure 8a,b, the variation in the coarse aggregate spacing coefficient has a more significant effect on SCC compressive strength. This is because the strength of SCC is influenced by the skeleton, which is composed of coarse aggregate. The fine aggregate is mainly used to fill the skeleton gap. As the coarse aggregate spacing coefficient increases, the volume fraction decreases and the skeleton cannot be formed, resulting in a significant decrease in strength.

When the aggregate spacing is small, the locked skeleton structure can be formed between the aggregates, which can provide greater strength when the SCC is under pressure. When the aggregate spacing increases, the locking force decreases, resulting in a decrease in SCC strength.

4.3. *Drying Shrinkage of SCC*

The shrinkage at different ages for different aggregate spacing coefficients is as shown in Figure 9.

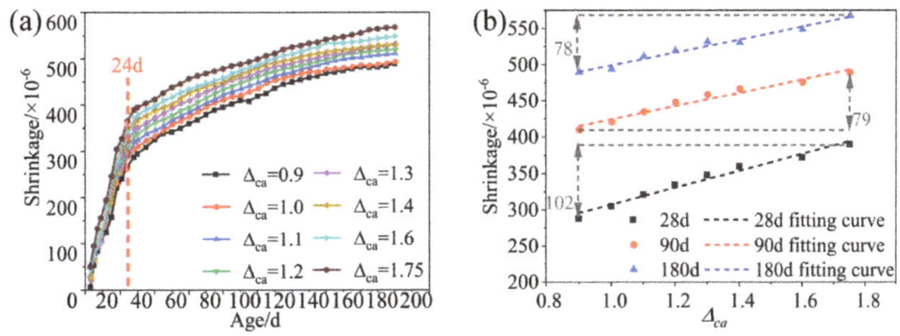

Figure 9. Relationship between aggregate spacing coefficient and drying shrinkage of concrete. (**a**) Correlation between curing age and drying shrinkage, (**b**) correlation between coarse aggregate spacing coefficient and drying shrinkage.

Figure 9a illustrates the drying shrinkage development of SCC with varying coarse aggregate spacing coefficients, from 0.9 to 1.75. As the coarse aggregate spacing coefficient increases, the shrinkage of SCC will increase gradually.

In SCC, the aggregates are inert materials and the main shrinkage is generated by the paste. As the aggregate spacing increases, the aggregate volume decreases, resulting in the inert phase volume decreasing, leading to an increase in the shrinkage of SCC.

Figure 9b illustrates the shrinkage of SCC under different coarse aggregate spacing coefficients, with ages of 28 days, 90 days, and 180 days. As the coarse aggregate spacing coefficient increases, the shrinkage of SCC at three different ages increases, where the shrinkage at 28 days is the largest. The shrinkage at 90 days increased by 79×10^{-6}, and the SCC shrinkage at 180 days increased the least, which was 78×10^{-6}.The changes in shrinkage values at 90 days and 180 days were very similar, which further shows that shrinkage develops rapidly in the early stage and slows down in the later stage.

The relationships between the coarse aggregate spacing coefficient and the drying shrinkage of SCC at different ages were fitted. The results are presented in Table 7.

Table 7. Relationship between coarse aggregate spacing and SCC drying shrinkage.

Age.	Fitting Formula
180 d	$y = 116.20\Delta_{ca} + 191.00$
90 d	$y = 91.79\Delta_{ca} + 333.02$
28 d	$y = 89.17\Delta_{ca} + 410.25$

In addition, according to Figure 9b, the drying shrinkage of SCC increases with the increase in the coarse aggregate spacing coefficient. As the coarse aggregate spacing coefficient increases, the volume fraction will decrease, leading to the need to fill the gap in the coarse aggregate. Therefore, the water content in the SCC suspension system increases, resulting in a large drying shrinkage of SCC.

5. Conclusions

In this paper, a model was established based on Fullman's mean free path and stereological theory, and the relationship between the aggregate volume and aggregate spacing coefficient was proposed. The quantitative relationship between aggregate spacing and various properties of SCC was obtained through experiments. The following conclusions can be drawn:

(1) A new aggregate spacing model was proposed. By comparing the results of the calculated value and the test value of the aggregate spacing coefficient, the validity of the model is verified.

(2) The slump and the slump flow of SCC increases as the aggregate spacing coefficient increases. The slump was maintained at 280 mm and then no longer decreased significantly, while the slump flow continued to increase.

(3) With the increase in the aggregate spacing coefficient, the compressive strength of SCC obviously decreased. And compared with the fine aggregate, the increase in the coarse aggregate spacing coefficient was more obvious in the decrease in compressive strength.

(4) An increase in the aggregate spacing coefficient will result in a significant increase in the drying shrinkage of SCC, which showed the fastest development in the early stage (3 d–24 d), slowing down in the later stage (24 d–180 d).

(5) In practical applications, for the case of different steel bar spacings in the project, choosing SCC with different aggregate spacing coefficients can make the design more convenient and accurate.

Author Contributions: Data curation, Funding acquisition, Methodology, Project administration, Supervision, Writing—original draft, H.L.; Investigation, Writing—original draft, Writing—review and editing, W.L.; Investigation, Original draft preparation, Data curation, H.Z.; Project ad-ministration, Formal analysis, Review and editing, W.B.; Writing—original draft, Writing—review and Editing, J.Z. (Jingyi Zhang); Supervision and Validation, J.Z. (Ji Zhang); Original draft preparation, Data curation, P.Z. All authors have read and agreed to the published version of the manuscript.

Funding: This work was supported by Key Laboratory of Highway Construction and Maintenance Technology in Loess Region, Ministry of Transport, PRC (KLTLR-Y23-10); Foundation of Liaoning Province Education Administration (grant numbers JYTMS20231556) and Liaoning transportation technology project (grant number: 202328).

Institutional Review Board Statement: Not applicable.

Informed Consent Statement: Not applicable.

Data Availability Statement: The data presented in this study are available on request from the corresponding author.

Conflicts of Interest: Ji Zhang and Peng Zhang were employed by the Liaoning New Development Highway Technology Maintenance Co., Ltd. The remaining authors declare that the research was conducted in the absence of any commercial or financial relationships that could be construed as a potential conflict of interest.

References

1. Okamura, H.; Ouchi, M. Self-compacting high performance concrete. *Prog. Struct. Eng. Mater.* **1998**, *1*, 378–383. [CrossRef]
2. Okamura, H.; Ozawa, K. Self-Compacting High Performance Concrete. *Struct. Eng. Int.* **1996**, *6*, 269–270. [CrossRef]
3. Liu, S.H. Key Techniques of Self-Compacting Concrete. *Adv. Mater. Res.* **2011**, *261–263*, 394–397. [CrossRef]
4. Shi, C.; Wu, Z.; Lv, K.; Wu, L. A review on mixture design methods for self-compacting concrete. *Constr. Build. Mater.* **2015**, *84*, 387–398. [CrossRef]
5. Goodier, C.I. Development of self-compacting concrete. *Proc. Inst. Civ. Eng.—Struct. Build.* **2003**, *156*, 405–414. [CrossRef]
6. Henault, J.W. Self-Consolidating Concrete: A Synthesis of Research Findings and Best Practices. *Lit. Rev.* **2014**, *8*, 11–20. [CrossRef]
7. Cu, Y.T.H.; Tran, M.V.; Ho, C.H.; Nguyen, P.H. Relationship between workability and rheological parameters of self-compacting concrete used for vertical pump up to supertall buildings. *J. Build. Eng.* **2020**, *32*, 101786. [CrossRef]
8. Liu, H.Q.; Chen, X. Influence of Aggregate on Workability of Self-Compacting Concrete. *Adv. Mater. Res.* **2012**, *482–484*, 1101–1104. [CrossRef]
9. Khaleel, O.; Al-Mishhadani, S.; Razak, H.A. The Effect of Coarse Aggregate on Fresh and Hardened Properties of Self-Compacting Concrete (SCC). *Procedia Eng.* **2011**, *14*, 805–813. [CrossRef]
10. Shobha, M.; Mohan, D.H.; Raju, P.S.N. Aggregate size and behaviour of self-compacting concrete. *Proc. Inst. Civ. Eng.—Constr. Mater.* **2006**, *159*, 147–152. [CrossRef]
11. Dehestani, M.; Nikbin, I.; Asadollahi, S. Effects of specimen shape and size on the compressive strength of self-consolidating concrete (SCC). *Constr. Build. Mater.* **2014**, *66*, 685–691. [CrossRef]
12. Nikbin, I.; Beygi, M.; Kazemi, M.; Amiri, J.V.; Rahmani, E.; Rabbanifar, S.; Eslami, M. A comprehensive investigation into the effect of aging and coarse aggregate size and volume on mechanical properties of self-compacting concrete. *Mater. Des.* **2014**, *59*, 199–210. [CrossRef]
13. Ma, K.; Feng, J.; Long, G.; Xie, Y.; Chen, X. Improved mix design method of self-compacting concrete based on coarse aggregate average diameter and slump flow. *Constr. Build. Mater.* **2017**, *143*, 566–573. [CrossRef]
14. Zhao, H.; Sun, W.; Wu, X.; Gao, B. The effect of coarse aggregate gradation on the properties of self-compacting concrete. *Mater. Des.* **2012**, *40*, 109–116. [CrossRef]
15. Zhang, J.; An, X.; Yu, Y.; Nie, D. Effects of coarse aggregate content on the paste rheological thresholds of fresh self-compacting concrete. *Constr. Build. Mater.* **2019**, *208*, 564–576. [CrossRef]

16. Santos, A.C.; Ortiz-Lozano, J.A.; Villegas, N.; Aguado, A. Experimental study about the effects of granular skeleton distribution on the mechanical properties of self-compacting concrete (SCC). *Constr. Build. Mater.* **2015**, *78*, 40–49. [CrossRef]
17. Wu, J.; Jia, Z.; Zhou, X. Discrete element analysis of the effect of aggregate morphology on the flowability of self-compacting concrete. *Case Stud. Constr. Mater.* **2023**, *18*. [CrossRef]
18. Chen, B.; Wang, L.; Feng, Z.; Liu, Y.; Wu, X.; Qin, Y.; Xia, L. Optimization of high-performance concrete mix ratio design using machine learning. *Eng. Appl. Artif. Intell.* **2023**, *122*. [CrossRef]
19. Su, N.; Hsu, K.-C.; Chai, H.-W. A simple mix design method for self-compacting concrete. *Cem. Concr. Res.* **2001**, *31*, 1799–1807. [CrossRef]
20. Basheerudeen, A.; Anandan, S. Particle Packing Approach for Designing the Mortar Phase of Self Compacting Concrete. *Eng. J.* **2014**, *18*, 127–140. [CrossRef]
21. Feiguo, C.; Wei, G. A review of smoothed particle hydrodynamics family methods for multiphase flow. *Chin. J. Theor. Appl. Mech.* **2021**, *53*, 2357–2373. [CrossRef]
22. Dhaheer, M.A.; Kulasegaram, S.; Karihaloo, B. Simulation of self-compacting concrete flow in the J-ring test using smoothed particle hydrodynamics (SPH). *Cem. Concr. Res.* **2016**, *89*, 27–34. [CrossRef]
23. Alyhya, W.; Kulasegaram, S.; Karihaloo, B. Simulation of the flow of self-compacting concrete in the V-funnel by SPH. *Cem. Concr. Res.* **2017**, *100*, 47–59. [CrossRef]
24. Tran-Duc, T.; Ho, T.; Thamwattana, N. A smoothed particle hydrodynamics study on effect of coarse aggregate on self-compacting concrete flows. *Int. J. Mech. Sci.* **2020**, *190*, 106046. [CrossRef]
25. *Q/CR 596-2017*; Specification of Self-Compacting Concrete for High-Speed Railway CRTSIII Slab Ballastless Track. China Railway Press: Beijing, China, 2019.
26. *GB/T 50081-2019*; Standard for Test Methods of Concrete Physical and Mechanical Properties. Ministry of Housing and Urban-Rural Development of the People's Republic of China: Beijing, China, 2019.
27. *GB/T 50082-2009*; Standard for Test Methods of Long-Term Performance and Durability of Ordinary Concrete. China Construction Industry Press: Beijing, China, 2009.
28. Chen, H.S.; Sun, W.; Stroeven, P. Regular dodecahedron model to calculate the average surface spacing between the nearest neighboring aggregate grains concrete. *J. Chin. Ceram. Soc.* **2003**, *31*, 1048–1052. [CrossRef]
29. Chen, H.S.; Sun, W.; Stroeven, P.; Stroeven, M. Stereological method of calculatmg the average value of surface spacing be-tween the neighboring aggregate grains in concrete. *J. Harbin Inst. Technol.* **2005**, *37*, 1511–1514.

coatings

MDPI

Article

Study on the Dynamic Properties of the Polyurethane Mixture with Open-Graded Gradation

Haisheng Zhao [1,2], Bin Wang [3], Peiyu Zhang [1], Yong Liu [3], Chunhua Su [1], Mingzhu Xu [3], Wensheng Zhang [4] and Shijie Ma [1,*]

1. Shandong Transportation Research Institute, Jinan 250102, China; zhaohaisheng@sdjtky.cn (H.Z.); zhangpeiyu@sdjtky.cn (P.Z.); suchunhua@sdjtky.cn (C.S.)
2. School of Highway, Chang'an University, Xi'an 710064, China
3. Dezhou Transportation Bureau, Dezhou 253057, China; bin13053498877@126.com (B.W.); jtj@dz.shandong.cn (Y.L.); 13053498877@126.com (M.X.)
4. Wanhua Chemical Group Co., Ltd., Yantai 265599, China; wszhangb@whchem.com

* Correspondence: mashijie@sdjtky.cn; Tel.: +86-186-6016-3082

Abstract

Polyurethane (PU) mixtures exhibit superior mechanical performance compared to traditional asphalt mixtures, owing to the excellent engineering properties of the PU binder. This study investigates the dynamic rheological properties of an open-graded polyurethane mixture (PUM–OGFC) in comparison with a dense-graded polyurethane mixture (PUM–AC). The time–temperature superposition principle and three rheological models (Standard Logistic Sigmoid (SLS), Generalized Logistic Sigmoid (GLS), and Havriliak–Negami (HN)) were employed to construct and analyze master curves. The results show that while PUM–AC possesses a higher dynamic modulus, PUM–OGFC exhibits a lower phase angle, indicating a more elastic response. Critically, PUM–OGFC demonstrated superior rutting resistance, as evidenced by its higher rutting parameter ($|E^*|/\sin\delta$). Aggregate gradation significantly influenced all rheological properties. The master curve analysis further revealed that PUM–OGFC exhibits greater temperature sensitivity than PUM–AC. The SLS and GLS models provided excellent fits for both dynamic modulus and phase angle data, whereas the HN model was suitable only for dynamic modulus. In summary, the open-graded structure, when combined with a PU binder, creates a high-performance composite with an exceptional balance of elasticity and rutting resistance, showcasing its potential for demanding pavement applications.

Keywords: PU mixture; aggregate gradation; dense-graded; open-graded; dynamic property

1. Introduction

Water or rain accumulates on the surface of traditional dense gradation asphalt pavement, creating several hazards to driving safety and structural integrity: (a) hydroplaning risk, water buildup can form a film that reduces tire–pavement surface contact, increasing skidding risk [1–14]; (b) water mist and glare, water spray from water buildup impairs driver visibility and increases the risk of glare [1,2,5,9,10,12,13,15–18]; (c) increased braking distance, the water film decreases the skid coefficient between the tire and pavement, leading to longer braking distance [19,20]; (d) reduced durability, water penetrating the pavement structure accelerates asphalt aging and can induce freeze–thaw damage, raveling, and other moisture-related distresses [1,3,5,10,12,14,21–24]; (e) freezing risk, water buildup freezes at low temperature, creasing ice layers that further increase skidding risk; and (f) reduced skid resistance, water erosion accelerates asphalt binder stripping and aging, which decreases skid resistance [25]. Porous asphalt mixture, commonly known as an open-graded friction course (OGFC), is designed to mitigate these water-related issues and enhance overall pavement performance. OGFC consists of a coarse aggregate skeleton with stone-on-stone contact, bonded by a high-viscosity asphalt film (≥8 μm), and is designed with a high air void content of 15%–25% [1,3–5,9,10,26]. OGFC offers several advantages: (a) rapid drainage, water on the pavement surface drains rapidly through the pores, minimizing water spray, splash, and glare [1,2,4–6,9–12,15,17,18,26], thereby significantly enhancing driver visibility; (b) anti-slip and wear-resistance, the exposed coarse aggregates create a significant macro-structure (mean profile depth, MPD > 1.5 mm) and yield a higher British Pendulum Number (BPN > 60), reducing accident risk in wet conditions [1,5,8,10,12,14,17,20,24,26–31]; (c) noise reduction, the high air void content absorbs tire–pavement interaction noise, typically reducing sound levels by 3–5 dB compared to dense-graded pavements, thereby improving driving comfort [1,2,4,5,10,15,16]; (d) rutting resistance, the stone skeleton enhances shear strength and provides greater high-temperature dynamic stability compared to dense-graded asphalt pavement [29,32–35]; (e) performance in special climates, OGFC is particularly suitable for rainy regions and areas prone to winter freezing [34,36,37]; and (f) environmental benefit, OGFC reduces runoff pollution, mitigates urban flooding, and aids in underwater recharge [1,12,24,27,32].

However, the widespread application of conventional OGFC is often limited by its mechanical performance, particularly at elevated temperatures. The dynamic modulus of OGFC is highly dependent on the coarse aggregate skeleton and the asphalt binder. As temperature rises, the asphalt binder softens, significantly weakening skeletal support and leading to a substantial reduction in dynamic modulus and rutting resistance [14,22,28,34,35,38]. This inherent temperature susceptibility underscores the need for advanced binder technologies to unlock the full, durable potential of porous pavement structures. For instance, the dynamic modulus at 60 °C is reduced by 40%–50% compared to that at 25 °C [39,40]. The dynamic modulus of epoxy asphalt OGFC at −10 °C is three to four times greater than that at 20 °C, but it is reduced by 75%–80% at 60 °C [39]. As the asphalt binder softens, the supporting function of coarse aggregate is weakened. The phase angle, which reflects the temperature sensitivity of OGFC, increases with rising temperature due to the enhanced viscous effect of the asphalt binder. The phase angle decreases at low temperature or high frequency, where the elastic properties of the asphalt binder dominate. Conversely, it increases at high temperature or low frequency, where the viscous property dominates [39,41]. The dynamic modulus values of OGFC also increase with loading frequency [8,12,38]. This occurs because the aggregate skeleton dominates the response at high frequency, while the viscosity of the asphalt binder dominates at low loading frequency, which is the characteristic of a viscoelastic material [39]. At high loading frequency, OGFC exhibits a high elastic modulus due to its frequency-dependent behavior, which helps reduce rutting damage [19,34,42]. For epoxy asphalt OGFC at 20 °C, the dynamic modulus increases from 2000 MPa at 0.1 Hz to 12,000 MPa at 25 Hz [39]. The dynamic modulus of OGFC can reduce tensile strain at the bottom of the asphalt layer by 12%–18% and pressure strain on the soil base by 10%–15%, thereby mitigating fatigue and rutting damage. For example, in the Mechanistic–Empirical Pavement Design Guide (MEPDG), the addition of an OGFC layer reduces bottom tensile strain by 15%–20%. The dynamic modulus is dominated by the coarse aggregate skeleton [43]. OGFC with a medium aggregate gradation (15%–25% passing the 4.75 mm sieve) composes more aggregate skeleton contact points; consequently, its dynamic modulus is greater than that of OGFC with a coarse aggregate gradation [44,45]. When the air void content in OGFC exceeds

 https://doi.org/10.3390/coatings16020153

22%, the aggregate skeleton and dynamic modulus decrease, which can induce raveling damage [5,46]. The dynamic modulus of OGFC decreases by 23.5% when the air void content increases from 18% to 24%, as the increased air void weakens the aggregate skeleton [47–49].

To address this limitation, the performance of OGFC can be significantly enhanced by using high-performance binders, such as crumb rubber-modified asphalt (CRMA) or polyurethane (PU). The performance of the OGFC mixture can be significantly enhanced by using high-viscosity asphalt binders (with a dynamic viscosity at 60 °C typically ≥20,000 Pa·s, according to ASTM D3381/D3381M), such as crumb rubber-modified asphalt (CRMA) and polyurethane (PU)-modified asphalt (PUMA), which improve durability and reduce temperature sensitivity [7,12,17,26,27]. CRMA improves the adhesive strength and fatigue resistance of OGFC mixtures. Similarly, PUMA enhances the durability and water damage resistance of porous polyurethane-modified asphalt mixture (PPAM) [29]. Blockage and permeability tests indicate that PPAM exhibits superior resistance to blockage compared to traditional OGFC [29]. Freeze–thaw tests demonstrate that PPAM possesses greater low-temperature stability than traditional OGFC; for instance, the dynamic modulus of OGFC decreased by 20%–35% after the test, whereas PPAM declined by less than 10% [34]. Furthermore, the Hamburg wheel-track test and dynamic modulus test reveal that PPAM offers greater resistance to deformation and fatigue [29,35]. The dynamic modulus values of PPAM exhibit a positive correlation with its rutting resistance and serve as an indicator of its viscoelastic property. Consequently, the dynamic modulus test can be utilized as a tool for selecting high-performance porous mixtures [34,35] and optimizing mixture design [50,51].

Notably, when PU replaced asphalt as the binder, the dynamic modulus of the PU mixture was greater than that of the corresponding asphalt mixture at elevated temperatures (>40 °C). For instance, the dynamic modulus of a stone matrix polyurethane mixture (SMPU–13) exceeded 8000 MPa at 60 °C, whereas the corresponding asphalt mixture dropped to 156 MPa at 50 °C [29]. The PU mixture also demonstrated a higher rutting factor, indicating superior resistance to permanent deformation at high temperatures [41]. Furthermore, the PU mixture exhibited a more stable phase angle and excellent viscoelastic properties [29]. The dynamic modulus of the asphalt mixture was significantly influenced by temperature and loading frequency, decreasing dramatically at high temperatures. In contrast, a Superpave polyurethane mixture (SUPU–20) maintained significantly higher dynamic modulus values under the same conditions [29]. Wang Huoming [52–54] found that porous PU mixture (PPM) with an open-graded structure exhibits superior performance, including deformation, fatigue, corrosion, and light/heat aging resistance, compared to both dense-graded and open-graded asphalt mixtures. Chen et al. [55] reported that PPM exhibits greater resistance to clogging from infiltrating soil suspended solids than OGFC. Cong et al. [56] found that PPM possesses a Marshall stability three times greater and a fatigue life one order of magnitude higher than those of OGFC. Sun [57] demonstrated that PPM exhibits superior resistance to high-temperature rutting, low-temperature cracking, and moisture damage. Similarly, Li [58] found that PPM offers improved resistance to high-temperature deformation, cracking, and moisture compared to traditional open-graded asphalt mixtures. While these studies confirm the outstanding macro-mechanical performance of PPM, a comprehensive understanding of their fundamental dynamic rheological behavior—the very basis for mechanistic design and performance prediction—remains less explored. Specifically, there is a lack of comparative studies that systematically evaluate how aggregate gradation (open-graded vs. dense-graded) influences the viscoelastic properties, rutting potential, and model-predicted responses of PU-based composites.

To bridge this knowledge gap, this study presents a comparative investigation into the fundamental dynamic rheological behavior of polyurethane mixtures with open-graded

https://doi.org/10.3390/coatings16020153

(PUM–OGFC, OGFC–13) and dense-graded (PUM–AC, AC–13) structures. Dynamic modulus and phase angle were characterized across a spectrum of temperatures and loading frequencies. The time–temperature superposition principle was applied to construct master curves, which were fitted using three rheological models (SLS, GLS, HN). Furthermore, the Kramers–Kronig (K–K) relations were employed to derive and evaluate corresponding phase angle master curves. This work aims to elucidate how aggregate gradation governs the viscoelastic response and performance potential of PU-based composites, providing critical insights for the mechanistic design of high-performance, sustainable porous pavements.

2. Materials and Methods

Two aggregate gradations were selected for mixture preparation. The first was a conventional dense-graded asphalt concrete (AC–13) gradation, representing a standard, continuously graded structure. The second was an open-graded friction course (OGFC–13) gradation, characterized by a gap-graded design with a high proportion of coarse aggregate and limited fine content to create an interconnected air void network. These corresponding aggregate gradations were plotted in Figure 1. The binder, which was provided by Wanhua Chemical Group Co., Ltd. (Yantai, China), was one kind of wet-setting PU binder; the optimum PU binder contents were determined by the Marshal method for asphalt mixture. The optimum PU binder contents were 4.9% and 4.7% for PUM–OGFC and PUM–AC, respectively.

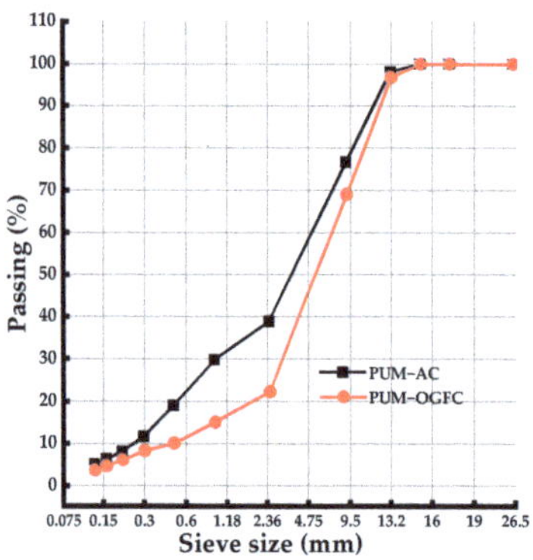

Figure 1. Aggregate gradation for PUM–OGFC and PUM–AC.

After the aggregate gradation and optimum PU binder content were determined, specimens would be compacted by the Superpave gyratory compactor (SGC) to a size of 175 mm in height and 150 mm in diameter. The specimens would be kept in a certain environment (35 °C and 70% RH) for 7 days. After the setting process, the specimens were cut into special sizes for the dynamic modulus test, which was 150 mm in height and 100 mm in diameter. A dynamic modulus test was performed on selected conditions (5 °C, 15 °C, 25 °C, 35 °C, 45 °C, 55 °C, and 25, 20, 10, 5, 2, 1, 0.5, 0.2, 0.1 Hz for each test temperature) by the asphalt mixture performance tester (UFA102QK00, Controls S.p.A., Milan, Italy). All specimens of PUM–OGFC and PUM–AC were tested at the same test temperature and loading frequency; average values for three duplicates for each PU mixture were adopted for the following analysis.

Scholars developed various master curve models for analyzing dynamic modulus test results: the SLS model [59], the GLS model [60], and the HN model [61], selected in

https://doi.org/10.3390/coatings16020153

this paper for analyzing dynamic modulus test results. These three models were listed in Equations (1)–(3).

$$log(\left|E^{*}\right|) = \delta + \frac{\alpha}{1 + e^{\beta + \gamma \cdot log(f_r)}} \tag{1}$$

where $|E^*|$—dynamic modulus (MPa); f_r—load frequency at the reference temperature (Hz); δ, α, β, and γ—fitting parameters.

$$log(|E^*|) = \delta\prime + \frac{\alpha\prime}{\left[1 + \lambda \cdot e^{\beta' + \gamma \cdot log(f_r)}\right]^{\frac{1}{\lambda}}} \tag{2}$$

where δ', α', β', and γ'–fitting parameters; λ–added parameter to account for the asymmetric shape of the function.

$$E^{*}(\omega) = |E_0| + \frac{|E_\infty| - |E_0|}{\left[1 + (\frac{\omega_0}{i \cdot \omega})^{\alpha''}\right]^{\beta''}} \tag{3}$$

where $i^2 = -1$; $\omega_0 = 1/\tau_0$; ω—angular frequency; τ_0—determines the horizontal position of the real or imaginary part of the dynamic modulus along the frequency axis; $|E_\infty|$—dynamic modulus as ω approaches ∞; $|E_0|$—dynamic modulus as ω approaches 0; and α'' and β''—fitting parameters.

There were no determined models developed for fitting phase angle data, but the Hilbert integral transforms [62] can be introduced to transform dynamic modulus master curve models to fit corresponding phase angle data; this process is known as the K–K relation, which is shown in Equation (4) [63].

$$\varphi(\omega) = \frac{\pi}{2} \cdot \frac{dlog(\left|E^{*}(\omega)\right|)}{dlog(\omega)} \tag{4}$$

Using Equation (4), the SLS, GLS, and HN models in Equations (1)–(3) can be transformed into corresponding mathematical models for the phase angle, which are shown in Equations (5)–(7).

$$\varphi(\omega) = -\frac{\pi}{2} \cdot \frac{\alpha \cdot \gamma \cdot e^{\beta + \gamma \cdot log(\omega)}}{\left(1 + e^{\beta + \gamma \cdot log(\omega)}\right)^2} \tag{5}$$

$$\varphi(\omega) = -\frac{\pi}{2} \cdot \frac{\alpha' \cdot \gamma' \cdot e^{\beta' + \gamma' \cdot log(\omega)}}{\left(1 + \lambda \cdot e^{\beta' + \gamma' \cdot log(\omega)}\right)^{\frac{1}{\lambda} + 1}} \tag{6}$$

$$\tan \varphi(\omega) = \frac{\sin\left(\frac{\alpha'' \cdot \pi}{2}\right)}{\left(\frac{\omega}{\omega_0}\right)^{\alpha''} + \cos\left(\frac{\alpha'' \cdot \pi}{2}\right)} \tag{7}$$

Based on the time–temperature superposition principle (TTSP), measured data points at different test temperatures can be shifted to the selected temperature to build a smooth master curve, which can be used to characterize the viscoelastic property of different asphalt mixtures. The shift factor is used as the shifted distance between the measured data point at different temperatures and master curves at the reference temperature, as shown in Equation (8).

$$log(f_r) = log(f) + log(\alpha_T) \tag{8}$$

where $log(f)$—frequency in experiment temperature; $log(f_r)$—reduced frequency in reference temperature; and α_T—shift factor.

https://doi.org/10.3390/coatings16020153

The temperature shift factor α_T at temperature T can be calculated using the Williams–Landel–Ferry (WLF) [64–67] empirical equations. The equation is described in Equation (9):

$$log(\alpha_T) = \frac{-C_1 \cdot (T - T_0)}{C_2 + (T - T_0)} \tag{9}$$

where C_1 and C_2—constants; T—test temperature; and T_0—reference temperature (20 °C in this paper).

The master curve fitting process is to minimize the error between experimental data points and predicted data points using different models. The error-minimizing process is performed by the nonlinear regression process, as shown in Equations (10) and (11).

$$E^*_{Error} = \sum_{i}^{i=n} \left(logE^*_{Exp}(\omega_i) - logE^*_{Prd}(\omega_i) \right)^2 \tag{10}$$

$$\delta_{Error} = \sum_{i}^{i=n} (\delta_{Exp}(\omega_i) - \delta_{Pre}(\omega_i))^2 \tag{11}$$

where E^*_{Exp}—experimental dynamic modulus; E^*_{Pre}—predicted dynamic modulus using different mathematical models; δ_{Exp}—experimental phase angle; and δ_{Pre}—predicted phase angle using different mathematical models.

Indexes, including S_e/S_y (the ratio of the standard error (S_e) to the standard deviation (S_y)) minimization, R^2, SSE (Sum of Squared Error), and $Error^2$, were introduced to estimate the regression accuracy, shown in Equations (12)–(16).

$$S_e = \sqrt{\frac{1}{n-p-1} \cdot \sum_{i}^{n} |\hat{x}_i - x_i|} \tag{12}$$

$$S_y = \sqrt{\frac{1}{n-1} \cdot \sum_{i}^{n} |\hat{x}_i - \overline{x}_i|} \tag{13}$$

where x_i—measured dynamic modulus; $\hat{x}_i$—predicted dynamic modulus; and $\overline{x}_i$—mean value of the measured dynamic modulus.

$$R^2 = 1 - \frac{(n-p-1) \cdot S_e^2}{(n-1) \cdot S_y^2} \tag{14}$$

where n—sample size, and p—number of parameters to be estimated.

$$SSE = \sum_{i}^{i=n} \frac{\left(E^*_{Exp}(\omega_i) - E^*_{Prd}(\omega_i) \right)^2}{\left(E^*_{Exp}(\omega_i) \right)^2} \tag{15}$$

$$Error^2 = \sum_{i}^{i=n} \left(E^*_{Exp}(\omega_i) - E^*_{Prd}(\omega_i) \right)^2 \tag{16}$$

3. Results

3.1. Dynamic Modulus and Phase Angle Results of Two PU Mixtures

3.1.1. Dynamic Modulus Results

The dynamic modulus results of PUM–OGFC and PUM–AC under different test temperatures and loading frequencies were plotted separately in Figure 2a,b, and the

 https://doi.org/10.3390/coatings16020153

dynamic modulus results at the same test temperatures for PUM–OGFC and PUM–AC were shown together in Figure 2c–e.

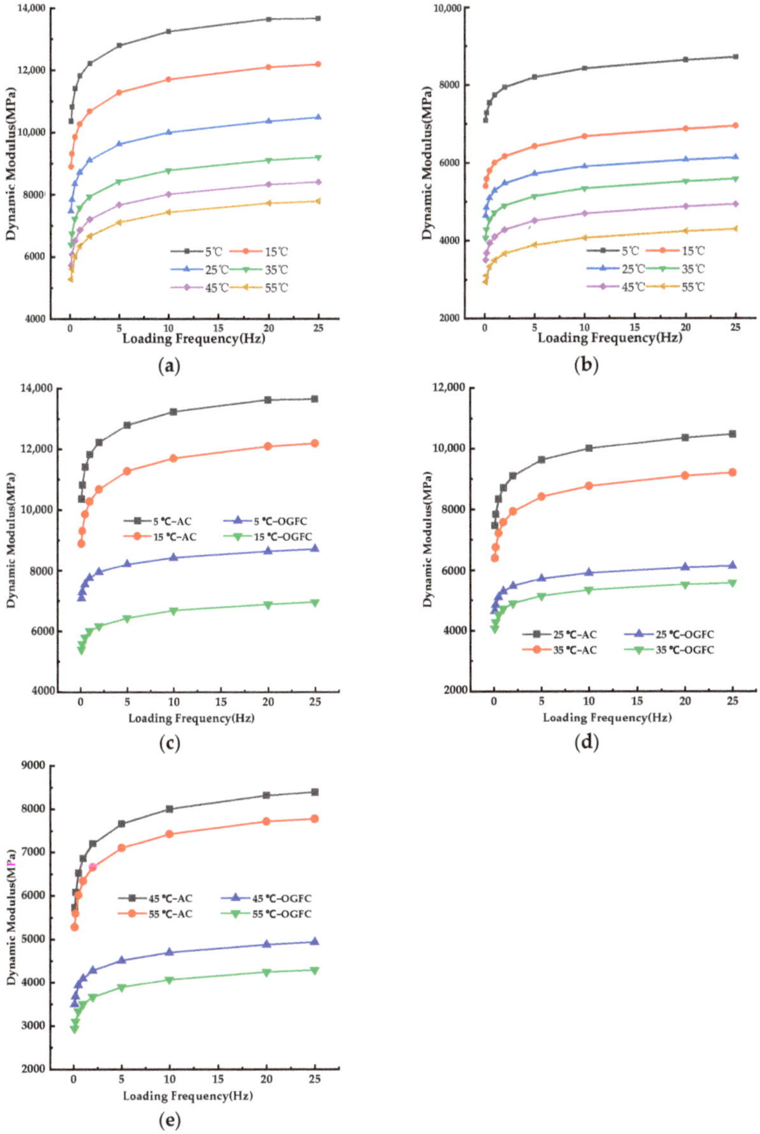

Figure 2. Dynamic modulus $|E^*|$ values for PUM–OGFC and PUM–AC. (**a**) $|E^*|$ of PUM–AC across various test temperatures; (**b**) $|E^*|$ of PUM–OGFC across various test temperatures; (**c**) comparison $|E^*|$ of PUM–OGFC and PUM–AC at 5 °C and 15 °C; (**d**) comparison $|E^*|$ of PUM–OGFC and PUM–AC at 25 °C and 35 °C; (**e**) comparison $|E^*|$ of PUM–OGFC and PUM–AC at 45 °C and 55 °C.

3.1.2. Phase Angle Results

The phase angle results of PUM–OGFC and PUM–AC under different test temperatures and loading frequencies were plotted separately in Figure 3a,b, and the phase angle

 https://doi.org/10.3390/coatings16020153

results at the same test temperatures for PUM–OGFC and PUM–AC were shown together in Figure 3c–e.

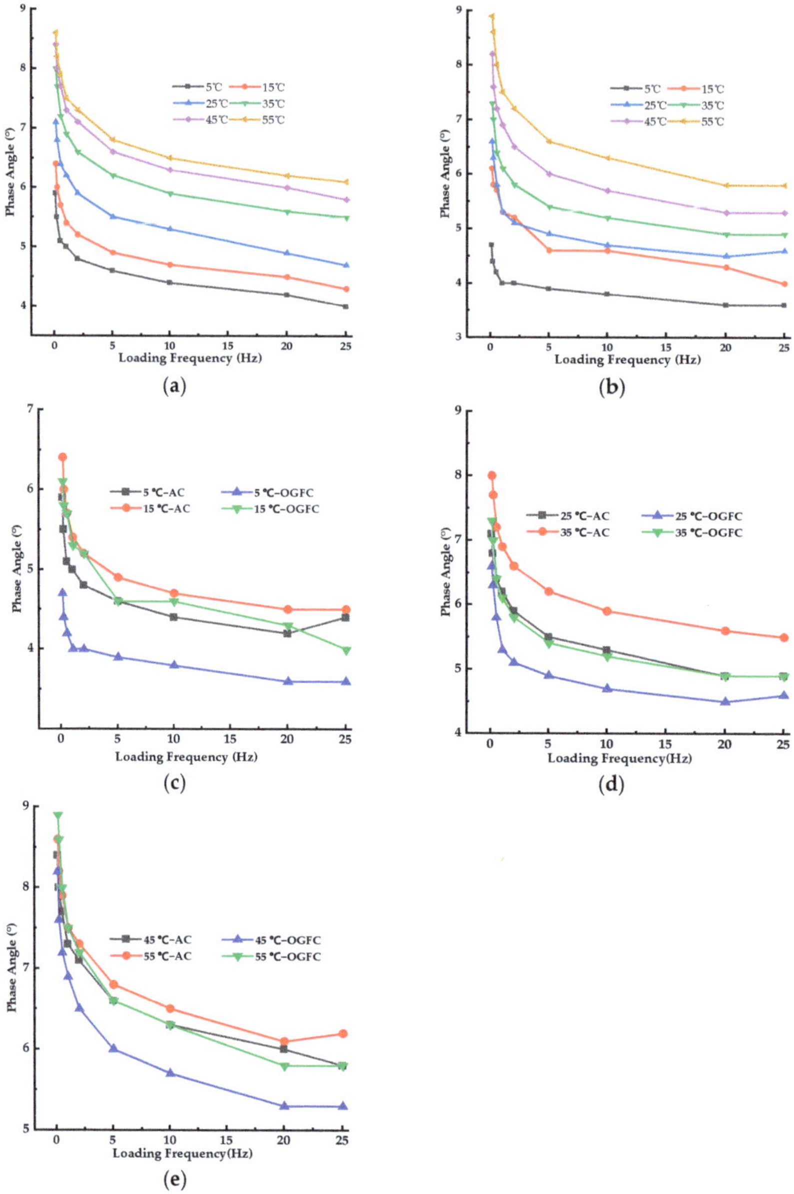

Figure 3. Phase angle (δ) values for PUM–OGFC and PUM–AC. (**a**) δ of PUM–AC across various test temperatures; (**b**) δ of PUM–OGFC across various test temperatures; (**c**) comparison of δ of PUM–OGFC and PUM–AC at 5 °C and 15 °C; (**d**) comparison of δ of PUM–OGFC and PUM–AC at 25 °C and 35 °C; (**e**) comparison of δ of PUM–OGFC and PUM–AC at 45 °C and 55 °C.

https://doi.org/10.3390/coatings16020153

3.2. Rutting Parameter (|E*|/sin(δ)) Results

The parameter $|E^*|/\sin(\delta)$ represents the rutting resistance of asphalt mixtures, respectively. Values of rutting parameters for the PUM–OGFC and PUM–AC, obtained across various test temperatures and loading frequencies, are presented separately in Figure 4.

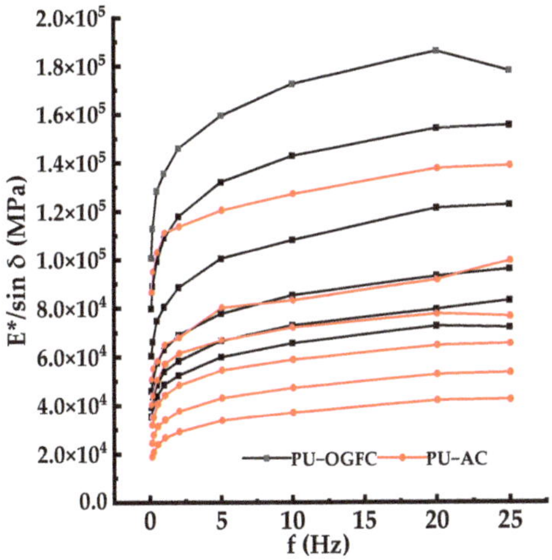

Figure 4. Rutting parameters results for the PUM–OGFC and PUM–AC.

3.3. Black Space Diagram Results

The black space diagrams for the PUM–OGFC and PUM–AC were plotted in Figure 5.

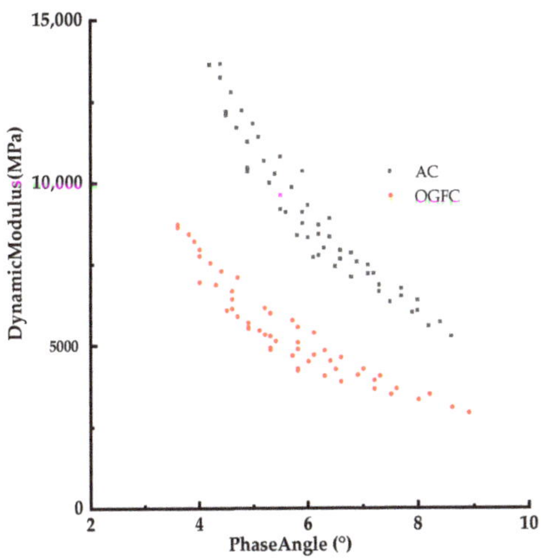

Figure 5. The black space diagrams for the PUM–OGFC and PUM–AC.

3.4. Master Curves Model Results

3.4.1. Dynamic Modulus Master Curve

The master curves of the dynamic modulus $|E^*|$ for PUM–OGFC and PUM–AC, constructed using SLS, GLS, and HN models with shifted measured dynamic modulus

https://doi.org/10.3390/coatings16020153

data, are presented in Figure 6a–f, respectively. A comparison of the three master curves for each PU mixture is shown in Figure 6g,h.

The shift factors used to generate the master curves for PUM–OGFC and PUM–AC are presented in Figure 7a,b, respectively.

A comparison of the predicted and measured dynamic modulus $|E^*|$ data for PUM–OGFC and PUM–AC is presented in Figure 8a,b, respectively. The corresponding linear regression parameters are summarized in Tables 1 and 2.

Table 1. Linear regression parameters between predicted and measured $|E^*|$ for PUM–AC.

Model Type	Equation	R^2
SLS model	Y = 0.9962 × X + 32.58	0.9959
GLS model	Y = 1.0046 × X − 25.85	0.9839
HN model	Y = 1.0008 × X + 5.31	0.9843

Table 2. Linear regression parameters between predicted and measured $|E^*|$ for PUM–OGFC.

Model Type	Equation	R^2
SLS model	Y = 0.9868 × X + 69.51	0.9902
GLS model	Y = 0.9806 × X + 89.51	0.9893
HN model	Y = 0.9776 × X + 135.01	0.9581

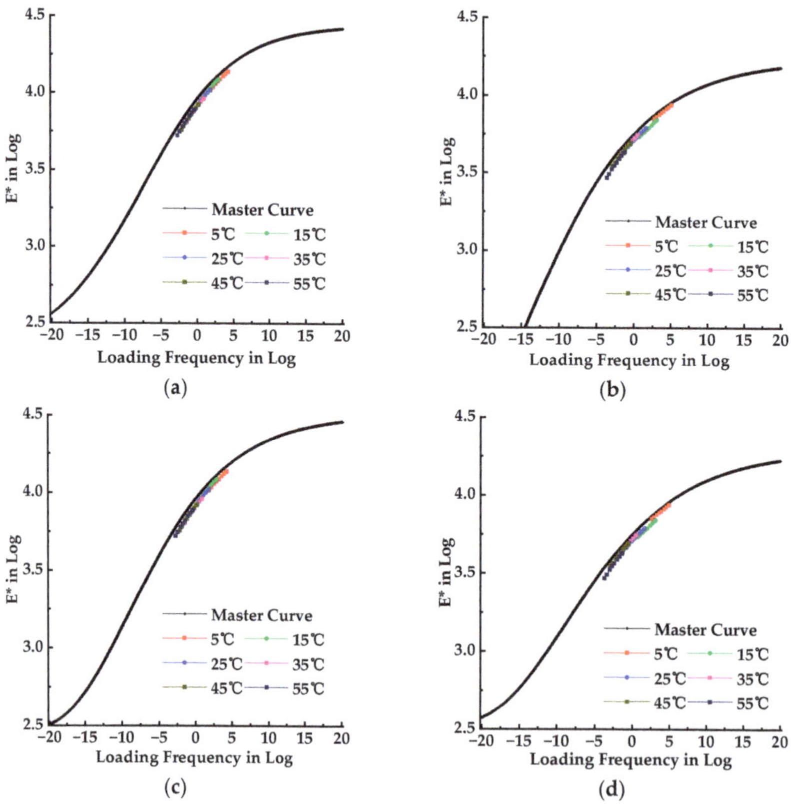

Figure 6. *Cont.*

https://doi.org/10.3390/coatings16020153

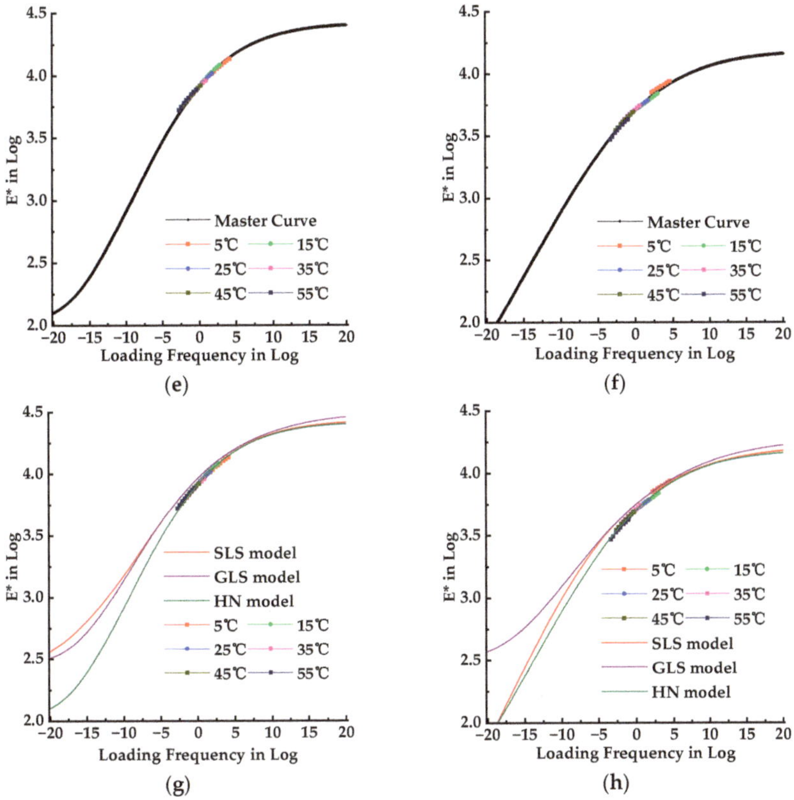

Figure 6. Master curves of dynamic modulus for PUM–OGFC and PUM–AC constructed using different models. (**a**) PUM–AC with SLS model; (**b**) PUM–OGFC with SLS model; (**c**) PUM–AC with GLS model; (**d**) PUM–OGFC with GLS model; (**e**) PUM–AC with HN model; (**f**) PUM–OGFC with HN model; (**g**) model comparison for PUM–AC; (**h**) model comparison for PUM–OGFC.

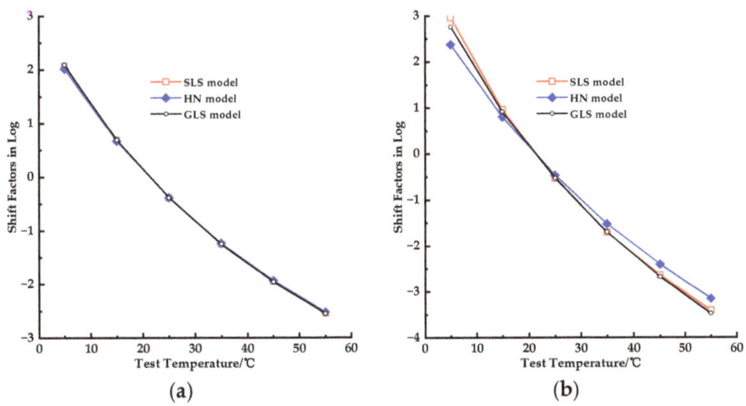

Figure 7. Shift factors for the dynamic modulus master curves of PUM–OGFC and PUM–AC. (**a**) PUM–AC; (**b**) PUM–OGFC.

https://doi.org/10.3390/coatings16020153

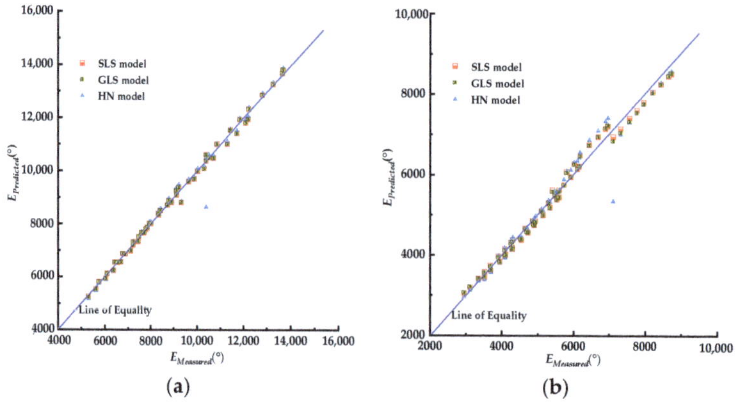

Figure 8. Comparison of predicted and measured dynamic modulus |E^*| for PUM–OGFC and PUM–AC. (**a**) PUM–AC; (**b**) PUM–OGFC.

3.4.2. Phase Angle Master Curve

The phase angle (δ) master curves for PUM–OGFC and PUM–AC, constructed using SLS, GLS, and HN models, are presented in Figure 9a–f. A comparison of the three models for each PU mixture is shown in Figure 9g,h.

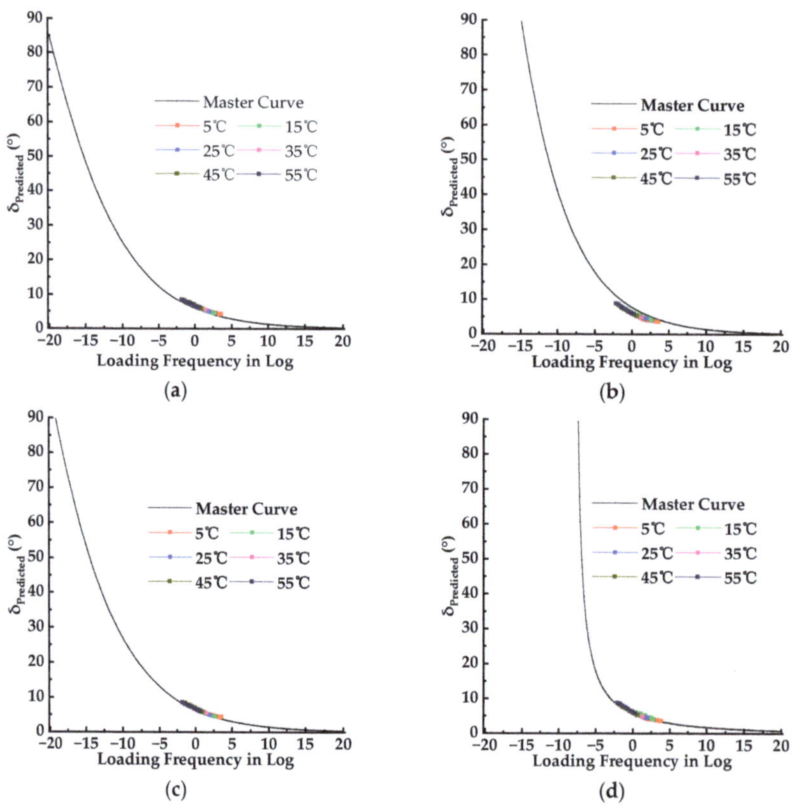

Figure 9. *Cont.*

https://doi.org/10.3390/coatings16020153

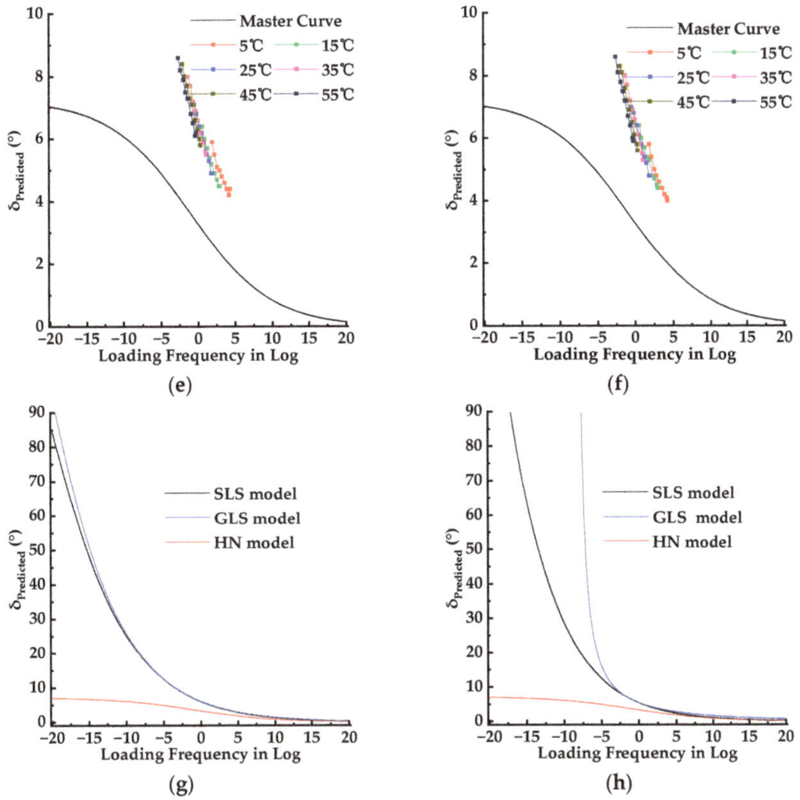

Figure 9. Phase angle (δ) master curves for PUM–OGFC and PUM–AC, constructed using different models. (**a**) PUM–AC with the SLS model; (**b**) PUM–OGFC with the SLS model; (**c**) PUM–AC with the GLS model; (**d**) PUM–OGFC with the GLS model; (**e**) PUM–AC with the HN model; (**f**) PUM–OGFC with the HN model; (**g**) model comparison for PUM–AC; (**h**) model comparison for PUM–OGFC.

The shift factors used to construct the phase angle master curve are presented in Figure 10a,b for PUM–OGFC and PUM–AC, respectively.

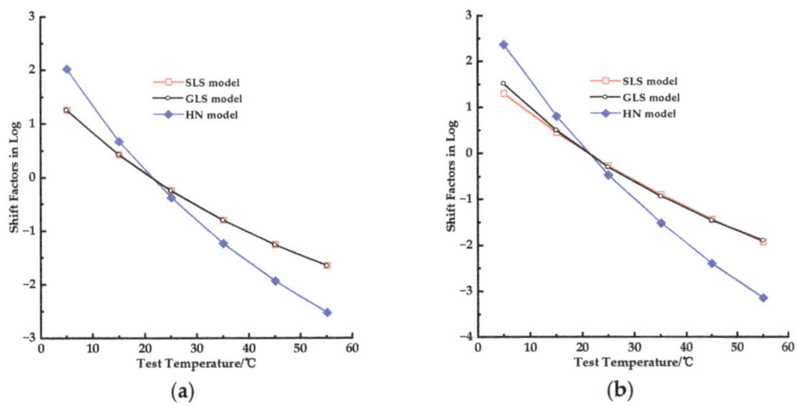

Figure 10. Shift factors for the phase angle (δ) master curves of PUM–OGFC and PUM–AC. (**a**) PUM–AC; (**b**) PUM–OGFC.

https://doi.org/10.3390/coatings16020153

A comparison of the predicted and measured phase angle (δ) data for PUM–OGFC and PUM–AC is presented in Figure 11a,b, respectively. The corresponding linear regression parameters are summarized in Tables 3 and 4.

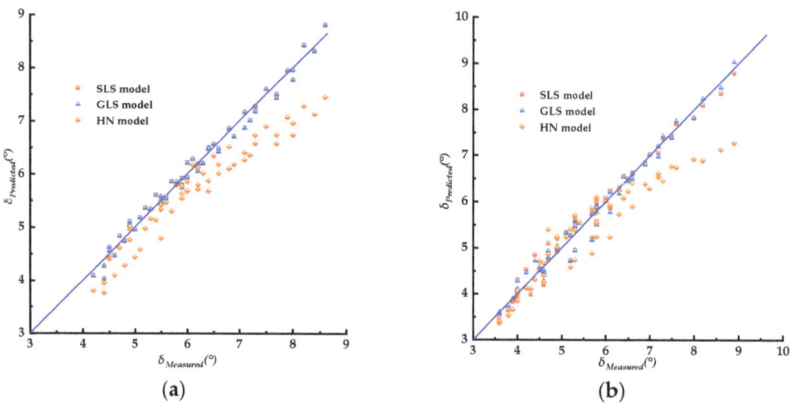

Figure 11. Comparison of predicted and measured phase angle (δ) for the PUM–OGFC and PUM–AC. (**a**) PUM–AC; (**b**) PUM–OGFC.

Table 3. Linear regression parameters between predicted and measured phase angle data of PUM–AC.

Model Type	Equation	R^2
SLS model	Y = 0.9865 × X + 0.08	0.9850
GLS model	Y = 0.9863 × X + 0.08	0.9850
HN model	Y = 0.7821 × X + 0.88	0.9319

Table 4. Linear regression parameters between predicted and measured phase angle data of PUM–OGFC.

Model Type	Equation	R^2
SLS model	Y = 0.9831 × X + 0.09	0.9758
GLS model	Y = 0.9808 × X + 0.11	0.9799
HN model	Y = 0.7386 × X + 1.22	0.8840

4. Discussion

4.1. Comparing Dynamic Modulus and Phase Angle of Two PU Mixtures

4.1.1. Dynamic Modulus Comparison

As shown in Figure 2a,b, the dynamic modulus ($|E^*|$) of both PUM mixtures exhibited characteristic viscoelastic behavior: it increased with loading frequency and decreased with rising temperature. The sensitivity to these factors was quantified. The increasing rate from 0.1 Hz to 25 Hz was more pronounced at higher temperatures, ranging from 31.8% to 47.2% for PUM–AC and from 22.9% to 46.1% for PUM–OGFC. Furthermore, $|E^*|$ decreased with increasing test temperature across all loading frequencies. The magnitude of this decrease was greater at lower frequencies, rising steadily from 43% to 49% as loading frequencies decreased from 25 Hz to 0.1 Hz for PUM–AC and from 50.7% to 58.6% for PUM–OGFC. Therefore, the dynamic modulus of both PUM mixtures is influenced by both test temperature and loading frequency. Temperatures had a more significant influence on $|E^*|$ than loading frequency. Specifically, high temperature and low loading frequencies (corresponding to slow vehicle speed) reduce the dynamic modulus of both PUM mixtures.

https://doi.org/10.3390/coatings16020153

As shown in Figure 2c–e, the dynamic modulus of PUM–OGFC was consistently lower than that of PUM–AC at all test temperatures. For instance, the $|E^*|$ value for PUM–AC at 55 °C was greater than that for PUM–OGFC at 15 °C. Therefore, aggregate gradation profoundly influences the dynamic modulus of the PU mixture. This result indicates that the aggregate skeleton, in combination with the PU binder, contributes significantly to load bearing. The denser AC gradation, with its lower air void content and more continuous aggregate skeleton, provides a higher resistance to elastic deformation under load compared to the open-graded structure.

However, it is critical to distinguish between resistance to elastic deformation and resistance to permanent deformation (rutting). While the higher $|E^*|$ suggests PUM–AC may offer advantages in terms of instantaneous stiffness, the overall rutting performance is governed by both the stiffness and the viscous characteristics of the material, as evaluated by the phase angle and rutting parameter ($|E^*|/\sin(\delta)$) in the subsequent sections. The following discussion will demonstrate that the PUM–OGFC mixture compensates for its lower stiffness with a markedly more elastic response, leading to superior long-term rutting resistance.

4.1.2. Phase Angle Comparison

Figure 3a shows that the phase angle (δ) of PUM–AC decreased with increasing loading frequency across all test temperatures, with a reduction rate ranging from 29.1% to 33.8%. This rate of decrease did not vary consistently with temperature. Conversely, the phase angle increased with temperature, with an increase rate of 45.8% to 54.9% across all loading frequencies. This rate of increase varied steadily with loading frequency. The phase angle demonstrated a higher percentage change in response to variations in temperature and loading frequency compared to the dynamic modulus, highlighting its strong dependence on test conditions. As shown in Figure 3b, the phase angle of PUM–OGFC exhibits a trend similar to that of PUM–AC. Specifically, the phase angle decreased with loading frequency (by 25.5% to 37.8%) and increased with temperature (by 60% to 95.5%). These results indicate that the phase angle of PUM–OGFC is even more sensitive to changes in loading frequency and temperature than that of PUM–AC.

Figure 3c–e shows that at a given temperature, the phase angle of PUM–AC was slightly larger than that of PUM–OGFC. However, the absolute phase angle values for both PU mixtures were less than 10°, regardless of gradation. The phase angle represents the viscoelastic behavior of a material: lower values indicate more elastic solid-like behavior, while higher values indicate more viscous fluid-like behavior. The consistently low phase angle (<10°) indicated that both PUM–OGFC and PUM–AC exhibit a predominantly elastic and highly recoverable response, even at 55 °C. This pronounced elastic character is a key indicator of the material's potential to resist the accumulation of permanent deformation.

4.2. *Comparing Rutting Parameter ($|E^*|/sin(\delta)$)*

The NCHRP report 513 established a strong correlation between the dynamic modulus of asphalt mixtures and field permanent deformation behavior, a finding subsequently incorporated into the Superpave performance evaluation system. Research indicates that the rutting resistance of a hot mixed asphalt mixture correlates well with the rutting parameter ($|E^*|/\sin(\delta)$). Consequently, this parameter is a reliable indicator for evaluating the high-temperature rutting resistance of hot mixed asphalt mixture (HMA). Thus, higher $|E^*|/\sin(\delta)$ values correspond to greater rutting resistance.

As shown in Figure 4, PUM–OGFC and PUM–AC exhibited similar trends: the rutting parameters ($|E^*|/\sin(\delta)$) increased with loading frequency and decreased with temper-

 https://doi.org/10.3390/coatings16020153

ature. Consequently, both PU mixtures exhibit maximum rutting resistance at low temperatures and high loading frequencies. For PUM–AC, increasing the loading frequency from 0.1 Hz to 25 Hz raised $|E^*|/\sin(\delta)$ by approximately 94.2% to 111.8% across the tested temperatures. Conversely, increasing the temperature from 5 °C to 55 °C reduced it by approximately 61.6% to 65.9% across all loading frequencies. Thus, for PUM–AC, the loading frequency had a greater influence on $|E^*|/\sin(\delta)$ than temperature. Similarly, for PUM–OGFC, $|E^*|/\sin(\delta)$ increased by approximately 60.3% to 123.7% as loading frequency increased from 0.1 Hz to 25 Hz, and decreased by 69.4% to 78.2% as temperature increased from 5 °C to 55 °C. Therefore, the loading frequency also had a greater effect on PUM–OGFC than temperature.

More critically, under identical temperature and frequency conditions, PUM–OGFC consistently exhibited higher $|E^*|/\sin(\delta)$ values than PUM–AC. For example, the value for PUM–OGFC at 55 °C exceeded that of PUM–AC at 35 °C. This direct comparison conclusively demonstrates the superior rutting resistance of the PUM–OGFC mixture.

Furthermore, this finding underscores the unique advantage of the polyurethane binder system. Despite the PUM–OGFC mixture's higher air void content (which typically compromises stiffness), the synergistic effect of the strong, elastic polyurethane binder and the interlocking open-graded skeleton leads to a composite material with an exceptionally favorable balance of stiffness and elasticity, as quantified by the rutting parameter.

4.3. Black Space Diagram Comparing

The black space diagram [68] is a plot of dynamic modulus versus phase angle, used to analyze the rheological properties of asphalt and compare the viscoelastic performance of HMA [37,69]. It also assesses the stiffness and relaxation capability of HMA [30]. In this diagram, a curve shifted to the right or a right-side inflection indicates viscous status in the asphalt binder or HMA. A typical HMA exhibits a peak phase angle value in its black space plot, reflecting the interaction between asphalt binder and aggregate [70]. Consequently, a well-defined curve shape and a distinct inflection point in the black space diagram are characteristic of typical rheological properties in the asphalt binders or HMA.

Figure 5 shows that the black space diagrams for PUM–OGFC and PUM–AC lack the well-defined curve shape and distinct inflection points characteristic of typical HMA. This distinct, near-linear clustering of data points indicates that the relationship between $|E^*|$ and δ for these polyurethane mixtures is remarkably consistent across the wide range of temperatures and loading frequencies tested. Unlike HMA, the viscoelastic response of the PUM composites does not traverse a broad, curved path in the black space, suggesting a fundamentally different and simpler relaxation mechanism. Furthermore, the entire dataset for PUM–OGFC is positioned at lower phase angles compared to PUM–AC at similar stiffness levels. This clear separation underscores the significant influence of aggregate gradation on the viscoelastic character of the PU composite, with the OGFC structure consistently promoting a more elastic-dominant response.

4.4. Master Curves Comparing

4.4.1. Dynamic Modulus Master Curve Analyzing

This section details the use of three master curve models—SLS, GLS, and HN models—to fit equation parameters and construct the corresponding master curves. The measured data points were shifted to a selected reference temperature using the WLF shift factor equation. During the fitting process, the sum of squared error (SSE, denoted as $error^2$) between the predicted and measured $|E^*|$ values was minimized. The coefficient of R^2 and Se/Sy were used as additional good-of-fit criteria. The quality of the model fits

https://doi.org/10.3390/coatings16020153

was evaluated using the coefficient of determination (R^2) and additional statistical metrics (*error*^{2}, Se/Sy, and SSE), as defined in Supplementary Material Table S1.

Figure 6a,b show that the $|E^*|$ master curves for PUM–OGFC and PUM–AC have similar shapes but different peak values. This finding confirms the discussion in Section 4.1.1 that PUM–AC exhibits higher stiffness than PUM–OGFC. The measured data points are clustered on the right side of the master curves, slightly offset from them. The R^2 values for both PU mixtures approach 1, indicating that the SLS model provides a good fit to the measured $|E^*|$ data. As summarized in Supplementary Table S1, the *error*^{2}, Se/Sy, and *SSE* values were lower for the PUM–AC fits than for the PUM–OGFC fits across all models, indicating a marginally more precise prediction for the dense-graded mixture's dynamic modulus data.

All three models successfully generated continuous master curves for both PUM–OGFC and PUM–AC, as shown in Figure 6. A consistent, key observation is that the master curves for PUM–AC are positioned at significantly higher $|E^*|$ values than those for PUM–OGFC across all models (Figure 6a,c,e), providing model-based confirmation of the higher stiffness of the dense-graded mixture. Furthermore, the statistical metrics (R^2, *Error*^{2}, SSE) uniformly indicate that all models fit the experimental data for PUM–AC more precisely than for PUM–OGFC (see Supplementary Table S1). This suggests the dynamic modulus of the dense-graded mixture is more predictable within these viscoelastic frameworks. The SLS model (Figure 6a,b) provides a very good fit for both mixtures ($R^2 \approx 1$). The error metrics (*Error*^{2}, SSE) are notably lower for PUM–AC than for PUM–OGFC. The GLS model (Figure 6c,d) yields a similar quality of fit to the SLS model, with comparably high R^2 values. Consistent with the SLS results, the GLS model also shows lower fitting errors for PUM–AC. The HN Model (Figure 6e,f) provides an excellent fit, with R^2 values of 0.9855 for PUM–AC and 0.9675 for PUM–OGFC. While its R^2 for PUM–OGFC is relatively lower, the HN model produces the lowest absolute SSE values for both materials, indicating it minimizes the absolute prediction error most effectively. A direct comparison of the master curves from the three models for each mixture is shown in Figure 6g,h. The curves exhibit similar shapes, with minor differences in predicted values across the frequency range.

For PU–OGFC mixtures, the shift factor fitted with the SLS and GLS models shows minimal difference, but their absolute values are consistently larger than those fitted with the HN model (Figure 7b). While for PU–AC in Figure 7a, all shift factors for the three models exhibited nearly the same values. The shift factor, a function of temperature, facilitates the translation of data from various test temperatures to a referenced temperature to form a continuous, smooth curve. It also indicates the quality of this translation, reflecting the time–temperature superposition principle [71]. The absolute shift factor value at a given test temperature reflects the mixture's temperature dependency, with higher values indicating greater temperature dependency [72]. The activation energy, which is related to the shift factor [73], represents the energy barrier overcome when data measured at different temperatures are shifted to form the master curve. A higher activation energy indicates that more energy is required to shift the data, implying greater temperature sensitivity. A critical finding is that the absolute values of a_T are larger for PUM–OGFC than for PUM–AC across all three models. This indicates the dynamic modulus of the open-graded mixture exhibits greater temperature dependency, implying a higher activation energy for its viscoelastic relaxation.

It was generally observed from Figure 8a that for PUM–AC, all the predicted $|E^*|$ values for the three models are closely clustered around the line of equality, with only an outlier from the HN model. Linear fitting was used to evaluate the correlation between predicted and measured $|E^*|$ values. According to Table 1, all three models predict PUM–

 https://doi.org/10.3390/coatings16020153

AC's $|E^*|$ values with high accuracy (high R^2 values). The SLS model has higher R^2 values than the HN model, while the GLS model showed slightly lower R^2 values than the HN model. The HN model has the smallest intercept values and slope values closest to 1, indicating it provides the most accurate prediction for PUM–AC. Figure 8b shows that for PUM–OGFC, the predicted values also cluster around the line of equality but exhibit greater dispersion than those for PUM–AC. The linear regression results in Table 2 show that all three models also predict the $|E^*|$ of PUM–OGFC with high precision, as indicated by high R^2 values. The SLS and GLS models have higher R^2 values and smaller, nearly identical slope and intercept values compared to the HN model. The R^2 values are higher for PUM–AC than for PUM–OGFC. The SLS model, with its smaller slope and intercept values, predicts the $|E^*|$ of PUM–AC more accurately than that of PUM–OGFC. The R^2 values from this linear regression are consistently higher for PUM–AC, reaffirming that aggregate gradation influences the predictability of the dynamic modulus, with the behavior of the dense-graded mixture being more readily captured by these models.

4.4.2. Phase Angle Master Curve Analyzing

Numerous models, such as the SLS, GLS, and HN models used in this paper, have been developed to construct dynamic modulus master curves. However, few models existed to construct a master curve for the simultaneously measured phase angle data. Consequently, the K–K relations [74,75] were employed to derive a phase angle master curve from the corresponding dynamic modulus data. The fitting procedure mirrored that used for the dynamic modulus; the WLF equation was also used to fit the shift factors.

The phase angle master curves constructed using the SLS, GLS, and HN models are presented in Figure 9. Both the SLS and GLS models successfully generated curves within the theoretical range of 0–90°, capturing the characteristic decrease in phase angle with increasing loading frequency (Figure 9a–d). In contrast, the HN model produced master curves with very low ultimate values (below 8°), which substantially underestimated the measured phase angles and showed poor agreement with the experimental data, indicating its inadequacy for this purpose (Figure 9e,f).

For both the SLS and GLS models, the master curves of PUM–AC were consistently positioned below those of PUM–OGFC across the frequency range (Figure 9a–d), confirming the more elastic-dominant behavior of the dense-graded mixture. Statistically, the fits for PUM–AC also demonstrated slightly higher R^2 values and lower $Error^2$, Se/Sy, and SSE values compared to PUM–OGFC, indicating a marginally better prediction precision for the dense-graded mixture.

The fitted shift factors are plotted in Figure 10a,b. The values for the SLS and GLS models are nearly identical and are smaller in absolute value than the HN model. Furthermore, the absolute shift factor values for PUM–AC are smaller than those for PUM–OGFC. Consequently, PUM–OGFC exhibits greater temperature sensitivity than PUM–AC and is more prone to high-temperature deformation. Therefore, aggregate gradation type affects PU mixture performance, with the AC gradation exhibiting superior performance.

The predicted versus measured phase angle data for all three models are shown in Figure 11a,b. Figure 11a shows that for PUM–AC, the data points predicted by the SLS and GLS models are distributed around the line of equality, whereas those predicted by the HN model show significantly greater dispersion. This result is consistent with the previous discussion on the phase angle master curves. Similarly, Figure 11b shows that for PUM–OGFC, the data points predicted by the SLS and GLS models are distributed around the line of equality but exhibited greater dispersion than those for PUM–AC. The data points from the HN model are also located far from the line of equality and exhibit

https://doi.org/10.3390/coatings16020153

the highest dispersion. The linear regression process was used to analyze the correlation between predicted and measured phase angles; the results are listed in Tables 3 and 4. For PUM–AC, the HN model has the lowest prediction precision, while the GLS and SLS models have identical prediction precision, evidenced by their slope and intercept values. For PUM–OGFC, the HN and SLS models also show the lowest and highest precision, respectively. Furthermore, the R^2 values are higher for PUM–AC than for PUM–OGFC, while the slope and intercept values are close to 1 and 0, respectively. Therefore, aggregate gradation type significantly influences the prediction precision of the PU mixture's phase angle. Overall, the GLS model provides the most precise phase angle predictions for both PUM–OGFC and PUM–AC.

5. Conclusions

This paper utilized open-graded (OGFC–13) and dense-graded (AC–13) aggregate gradation to prepare PU mixtures and investigate the dynamic properties of the open-graded PU mixture (PUM–OGFC). Based on complex dynamic modulus test results, the dynamic modulus, phase angle, rutting resistance, and rheological properties of PUM–OGFC and PUM–AC were compared. The SLS, GLS, and HN models, in conjunction with the WLF equation, were used to construct dynamic modulus master curves for both PU mixtures. The K–K relations were employed to derive phase angle master curve equations from the dynamic modulus models, which were then used to construct the phase angle master curves. The main conclusions are as follows:

(1) The dynamic modulus of PUM–AC at 55 °C was greater than that of PUM–OGFC at 15 °C. The phase angle of PUM–OGFC was slightly lower than that of PUM–AC across all temperatures and frequencies. All phase angles were below 10 °, indicating predominantly elastic behavior at 55 °C. The phase angle of PUM–OGFC was more sensitive to variations in temperatures and frequencies than that of PUM–AC.
(2) PUM–OGFC exhibited higher rutting resistance. The rutting parameter ($|E^*|/\sin(\delta)$) for PUM–OGFC at 55 °C exceeded that for PUM–AC at 35 °C. The parameter was more sensitive to loading frequency than to temperature.
(3) Unlike typical HMA, PUM–OGFC and PUM–AC did not exhibit defined black space diagram shapes. Aggregate gradation significantly influenced the rheological property of the PU mixture.
(4) The SLS, GLS, and HN models accurately fitted the dynamic modulus data of both PU mixtures. PUM–OGFC exhibited greater temperature sensitivity than PUM–AC. Aggregate gradation influenced the fitting result and prediction accuracy.
(5) The SLS and GLS models provided a more precise fit to the phase angle data than the HN model. Consistent with the dynamic modulus results, PUM–OGFC showed greater temperature sensitivity, and aggregate gradation affected the prediction accuracy.

This study reveals that the introduction of a PU binder fundamentally alters the conventional framework for understanding the gradation–performance relationship in paving mixtures. For PU-based composites, the exceptional performance originates from the binder's intrinsic mechanical properties and its unique interaction with the aggregate structure, rather than strictly following the traditional skeleton or densification principles of asphalt mixtures.

In conventional asphalt mixtures, open-graded (OGFC) designs rely primarily on point contacts between coarse aggregates to form a load-bearing skeleton. However, the thermoplastic and viscoelastic nature of asphalt binder can lead to skeleton instability under prolonged loading at high temperatures. In contrast, for PU mixtures, the situation is transformed. The reactive PU binder cures to form a tough, highly elastic three-dimensional

 https://doi.org/10.3390/coatings16020153

network. Within PUM–OGFC, this elastic network not only provides strong adhesion at aggregate contact points but, more critically, it fills and reinforces the large void spaces, creating a continuous elastic matrix that supports the aggregate skeleton. This means load transmission is no longer reliant solely on limited aggregate point contacts but is achieved through a synergistic mechanism combining the "aggregate skeleton" and the "PU elastic matrix." This synergy results in a composite material that achieves an exceptional balance of high stiffness, high elasticity, and superior resistance to permanent deformation, surpassing the performance of traditional materials.

Conversely, in the dense-graded PUM–AC, although the PU binder strengthens the mastic, the compact structure limits the formation of an extensive, effective spatial network of the elastic binder. The performance improvement here stems more from the increased modulus of the mastic itself. The rheological data from this study, specifically the higher rutting parameter ($|E^*|/\sin(\delta)$) and lower phase angle for PUM–OGFC, confirm that the "spatial synergy" between the PU binder and the open-graded skeleton translates the excellent material properties of PU into macroscopic engineering performance advantages more effectively than the "filling and reinforcement" mechanism in the dense-graded mixture.

Therefore, a key insight from this research is that for high-performance reactive binders like polyurethane, an open-graded structure is not a mechanical disadvantage but an ideal framework to fully utilize its high elasticity and strong adhesive properties. This provides a new theoretical basis and a viable technical pathway for designing sustainable pavement materials that combine excellent drainage function with high resistance to rutting. Beyond the laboratory findings, this study underscores the need to correlate lab performance with long-term field behavior through future pilot sections. While the established rheological parameters provide a sound basis for mix design, their validation ultimately requires real-world performance data under complex traffic and environmental conditions. Future work should therefore focus on bridging this gap by developing reliable transfer functions between lab tests and field monitoring data. Addressing this challenge is crucial to translate the present mechanistic insights into robust, practical criteria for sustainable pavement design [76].

Supplementary Materials: The following supporting information can be downloaded at: https://www.mdpi.com/article/10.3390/coatings16020153/s1, Table S1: The dynamic modulus master curve fitting results for PUM–AC and PUM–OGFC. Table S2: The phase angle master curve fitting results for PUM–AC and PUM–OGFC.

Author Contributions: Conceptualization, H.Z.; methodology, S.M.; software, Y.L. and M.X.; validation, H.Z., Y.L. and M.X.; formal analysis, H.Z.; investigation, H.Z. and C.S.; resources, B.W., W.Z. and P.Z.; data curation, H.Z., C.S. and P.Z.; writing—original draft preparation, H.Z.; writing—review and editing, S.M.; visualization, H.Z., C.S., M.X. and B.W.; supervision, S.M.; project administration, W.Z. and S.M.; funding acquisition, W.Z.; All authors have read and agreed to the published version of the manuscript.

Funding: This research received no external funding.

Institutional Review Board Statement: Not applicable.

Informed Consent Statement: Not applicable.

Data Availability Statement: The data presented in this study are available on request from the corresponding author.

Acknowledgments: We thank Guang Li and Gen Li for their assistance with experiments and valuable discussion.

https://doi.org/10.3390/coatings16020153

Conflicts of Interest: Authors Bin Wang, Yong Liu, Mingzhu Xu was employed by the company of Dezhou Transportation Bureau. Author Wensheng Zhang was employed by the company of Wanhua Chemical Group Co. Ltd. The remaining authors declare that the research was onducted in the absence of any commercial or financial relationships that could be construed as apotential conflict of interest.

Abbreviations

The following abbreviations are used in this manuscript:

AC	Dense-Graded Asphalt Concrete
BPN	British Pendulum Number
CRMA	Crumb Rubber-Modified Asphalt
DSR	Dynamic Shear Rheometer
FTIR	Fourier Transform Infrared Spectroscopy
GLS	Generalized Logistic Sigmoid (model)
HMA	Hot Mixed Asphalt (mixture)
HN	Havriliak–Negami (model)
K–K	Kramers–Kronig (relations)
MPD	Mean Profile Depth
OGFC	Open-Graded Friction Course
PPAM	Porous Polyurethane-Modified Asphalt Mixture
PU	Polyurethane
PUM–AC	Polyurethane Mixture with Dense-Graded (Asphalt Concrete)
PUM–OGFC	Polyurethane Mixture with Open-Graded Friction Course
PUMA	Polyurethane-Modified Asphalt
R^2	Coefficient of Determination
SMPU–13	Stone Matrix Polyurethane Mixture (13 mm nominal size)
SLS	Standard Logistic Sigmoid (model)
SSE	Sum of Squared Errors
SUPU–20	Superpave Polyurethane Mixture (20 mm nominal size)
TTSP	time–temperature superposition principle

References

1. Hou, X.; Lv, S.; Chen, Z.; Xiao, F. Applications of Fourier transform infrared spectroscopy technologies on asphalt materials. *Measurement* **2018**, *121*, 304–316. [CrossRef]
2. Kumar, A.; Choudhary, R.; Kumar, A. A review of the application of crosslinking additives in asphalt binder modification for rheological changes and storage stability. *Eur. J. Environ. Civ. Eng.* **2024**, *28*, 2821–2861. [CrossRef]
3. Zhou, J.; Chen, X.; Xu, G.; Zhang, H. Evaluation on low temperature characteristics of SBS/CR modified asphalt binder under different aging conditions. *IOP Conf. Ser. Mater. Sci. Eng.* **2019**, *652*, 12019. [CrossRef]
4. Fu, Q.; Sun, F.; Xu, G.; Chen, X. Performance Evaluation of SBS+CR Composite Modified Asphalt Binder and Mixture for Railway Substructure. In Proceedings of the 19th COTA International Conference of Transportation Professionals (CICTP 2019), Nanjing, China, 6–8 July 2019; pp. 928–940.
5. Qian, C.; Fan, W.; Yang, G.; Han, L.; Xing, B.; Lv, X. Influence of crumb rubber particle size and SBS structure on properties of CR/SBS composite modified asphalt. *Constr. Build. Mater.* **2020**, *235*, 117517. [CrossRef]
6. Wu, S.; He, R.; Chen, H.; Li, W.; Li, G. Rheological Properties of SBS/CRP Composite Modified Asphalt under Different Aging Treatments. *Materials* **2020**, *13*, 4921. [CrossRef]
7. Li, L.; Wu, C.; Cheng, Y.; Wang, H.; Liang, J.; Zhao, W. Synergistic effect of waste rubber powder on low-temperature toughness and high-temperature rheological properties of SBS modified asphalt. *Constr. Build. Mater.* **2023**, *365*, 130112. [CrossRef]
8. Wang, W.; Wang, Z.; Guo, L. High-Temperature Property Evaluation and Index Research of Modified Asphalt before and After Aging. *J. Phys. Conf. Ser.* **2022**, *2152*, 12032. [CrossRef]
9. Šernas, O.; Vaitkus, A.; Škulteckė, J. Performance of crumb rubber bitumen and asphalt modified in the wet process alone and in combination with SBS polymer. *Road Mater. Pavement Des.* **2023**, *24*, 107–123. [CrossRef]

 https://doi.org/10.3390/coatings16020153

10. Huang, J.; Yan, K.; Wang, M.; Shi, K.; Li, Y.; Zhang, Y. Performance evaluation of SBS-modified asphalt mixtures incorporating waste tire rubber and HDPE. *Constr. Build. Mater.* **2024**, *430*, 136423. [CrossRef]
11. Wang, X.; Hui, Y.; Xu, X.; Xu, T.; Zhao, T.; Li, B. Multi-scale study on the whole process adhesion behaviour of steel slag and SBS/CR composite modified asphalt. *Int. J. Pavement Eng.* **2024**, *25*, 2419937. [CrossRef]
12. Lin, P.; Huang, W.; Tang, N.; Xiao, F. Performance characteristics of Terminal Blend rubberized asphalt with SBS and polyphosphoric acid. *Constr. Build. Mater.* **2017**, *141*, 171–182. [CrossRef]
13. Qian, C.; Fan, W.; Liang, M.; He, Y.; Ren, S.; Lv, X.; Nan, G.; Luo, H. Rheological properties, storage stability and morphology of CR/SBS composite modified asphalt by high-cured method. *Constr. Build. Mater.* **2018**, *193*, 312–322. [CrossRef]
14. Ren, S.; Liu, X.; Xu, J.; Lin, P. Investigating the role of swelling-degradation degree of crumb rubber on CR/SBS modified porous asphalt binder and mixture. *Constr. Build. Mater.* **2021**, *300*, 124048. [CrossRef]
15. Jiang, Z.; Easa, S.M.; Hu, C.; Zheng, X. Understanding damping performance and mechanism of crumb rubber and styrene-butadiene-styrene compound modified asphalts. *Constr. Build. Mater.* **2019**, *206*, 151–159. [CrossRef]
16. Li, H.; Cui, C.; Temitope, A.A.; Feng, Z.; Zhao, G.; Guo, P. Effect of SBS and crumb rubber on asphalt modification: A review of the properties and practical application. *J. Traffic Transp. Eng.* **2022**, *9*, 836–863. [CrossRef]
17. Wang, H.; Huang, Y.; Jin, K.; Zhou, Z. Properties and mechanism of SBS/crumb rubber composite high viscosity modified asphalt. *J. Clean. Prod.* **2022**, *378*, 134534. [CrossRef]
18. Zhao, Z.; Wu, S.; Xie, J.; Yang, C.; Wang, F.; Li, N.; Liu, Q.; Amirkhanian, S. Effect of direct addition of asphalt rubber pellets on mixing, performance and VOCs of asphalt mixtures. *Constr. Build. Mater.* **2024**, *411*, 134494. [CrossRef]
19. Loderer, C.; Partl, M.N.; Poulikakos, L.D. Effect of crumb rubber production technology on performance of modified bitumen. *Constr. Build. Mater.* **2018**, *191*, 1159–1171. [CrossRef]
20. Brondani, C.; Menezes Vestena, P.; Faccin, C.; Lisboa Schuster, S.; Pivoto Specht, L.; da Silva Pereira, D. Moisture susceptibility of asphalt mixtures: 2S2P1D rheological model approach and new index based on dynamic modulus master curve changes. *Constr. Build. Mater.* **2022**, *331*, 127316. [CrossRef]
21. Lagos-Varas, M.; Movilla-Quesada, D.; Raposeiras, A.C.; Monsalve-Cárcamo, P.; Castro-Fresno, D. Rheological analyses of binders modified with triple combinations of Crumb-Rubber, Sasobit and Styrene-Butadiene-Styrene. *Case Stud. Constr. Mater.* **2023**, *19*, e02235. [CrossRef]
22. Liu, X.; Bi, Y.; An, Y.; Zhou, S.; Li, F. Secondary-Reaction Method to Improve Performance of a Compound Modified Rubber Asphalt after Short-Term Aging. *J. Transp. Eng. Part B Pavements* **2021**, *147*, 04021038. [CrossRef]
23. Zhang, Y.; Deng, X.; Xiao, P.; Qian, P.; Zhang, Y.; Kang, A. Properties and interaction evolution mechanism of CR modified asphalt. *Fuel* **2024**, *371*, 131886. [CrossRef]
24. Li, H.; Jia, M.; Zhang, X.; Wang, Z.; Liu, Y.; Yang, J.; Yang, B.; Sun, Y.; Wang, H.; Ma, H. Laboratory Investigation on Fumes Generated by Different Modified Asphalt Binders. *Transp. Res. Part D Transp. Environ.* **2023**, *121*, 103828. [CrossRef]
25. Azimi Alamdary, Y.; Singh, S.; Baaj, H. Effect of aggregates containing iron sulphide on asphalt ageing. *Road Mater. Pavement Des.* **2019**, *22*, 623–638. [CrossRef]
26. Pei, Y.; Jiang, S.; Ding, Z.; Cheng, L.; Li, P.; Jiang, X. Preparation and performance analysis of high-viscosity asphalt containing high-content SBS and crumb rubber by oxygen-free high-temperature treatment. *Constr. Build. Mater.* **2023**, *402*, 132763. [CrossRef]
27. Vural Kok, B.; Furtana Yalcin, B.; Yilmaz, M.; Yalcin, E. Performance evaluation of bitumen modified with styrene–isoprene-styrene and crumb rubber compound. *Constr. Build. Mater.* **2022**, *344*, 128304. [CrossRef]
28. Yang, X.; You, Z.; Hasan, M.R.M.; Diab, A.; Shao, H.; Chen, S.; Ge, D. Environmental and mechanical performance of crumb rubber modified warm mix asphalt using Evotherm. *J. Clean. Prod.* **2017**, *159*, 346–358. [CrossRef]
29. Zhang, Q.; Wang, S.; Lv, R.; Wu, J.; Qi, H. Viscoelastic mechanical performance of dense polyurethane mixtures based on dynamic and static modulus testing and creep testing. *Constr. Build. Mater.* **2022**, *320*, 126207. [CrossRef]
30. Oshone, M.; Dave, E.; Daniel, J.S.; Rowe, G.M. Prediction of phase angles from dynamic modulus data and implications for cracking performance evaluation. *Road Mater. Pavement Des.* **2017**, *18*, 491–513. [CrossRef]
31. Zhang, F.; Wang, L.; Li, C.; Xing, Y. Predict the Phase Angle Master Curve and Study the Viscoelastic Properties of Warm Mix Crumb Rubber-Modified Asphalt Mixture. *Materials* **2020**, *13*, 5051. [CrossRef]
32. Yan, K.; Shi, K.; Wang, M.; Huang, J.; Li, Y.; Ouyang, Z. Effect of high-density-polyethylene (HDPE) and waste tire rubber (WTR) on laboratory performance of styrene butadiene styrene (SBS) modified asphalt. *Road Mater. Pavement Des.* **2024**, *26*, 1360–1375. [CrossRef]
33. Bi, Y.; Guo, F.; Zhang, J.; Pei, J.; Li, R. Correlation analysis between asphalt binder/asphalt mastic properties and dynamic modulus of asphalt mixture. *Constr. Build. Mater.* **2021**, *276*, 122256. [CrossRef]
34. Nguyen, T.H.; Ahn, J.; Lee, J.; Kim, J.H. Dynamic Modulus of Porous Asphalt and the Effect of Moisture Conditioning. *Materials* **2019**, *12*, 1230. [CrossRef] [PubMed]

 https://doi.org/10.3390/coatings16020153

35. Hajibandeh, E.; Shalaby, A. Dynamic Modulus and Rutting Performance of As-Built Hot-Mix Asphalt Using Small-Scale Specimens. *J. Mater. Civ. Eng.* **2020**, *32*, 04020089. [CrossRef]
36. Lachance-Tremblay, É.; Perraton, D.; Vaillancourt, M.; Di Benedetto, H. Degradation of asphalt mixtures with glass aggregates subjected to freeze-thaw cycles. *Cold Reg. Sci. Technol.* **2017**, *141*, 8–15. [CrossRef]
37. Mensching, D.J.; Rowe, G.M.; Sias Daniel, J. A mixture-based Black Space parameter for low-temperature performance of hot mix asphalt. *Road Mater. Pavement Des.* **2016**, *18*, 404–425. [CrossRef]
38. Zhang, J.; Cao, W.; Li, W.; Lou, S.; Zhang, W.; Hou, Z. Performance assessment of composite modified bitumen integrated with desulfurized crumb rubber and SBS modifier as waterproofing adhesive layer. *Int. J. Adhes. Adhes.* **2024**, *134*, 103802. [CrossRef]
39. Wang, S.; Liu, C.; Zhu, X.; Peng, X.; Pan, Q.; Zhao, Z.; Zhang, Z. Influence of gradation on volume characteristics and engineering performance of asphalt mixture. *Mater. Today Commun.* **2024**, *38*, 107615. [CrossRef]
40. Du, J.; Ren, D.; Ai, C.; Zhang, J.; Qiu, Y. Effect of aggregate gradation on crack propagation in asphalt mixtures at low temperatures based on the Eshelby equivalent inclusion theory. *Constr. Build. Mater.* **2021**, *290*, 123181. [CrossRef]
41. Wang, H.; Zhan, S.; Liu, G. The Effects of Asphalt Migration on the Dynamic Modulus of Asphalt Mixture. *Appl. Sci.* **2019**, *9*, 2747. [CrossRef]
42. Shi, C.; Zhou, W.; Wang, T.; Cai, X.; Wang, H.; Yang, J.; Hong, J.; Gong, M. Fatigue performance characterization of styrene-butadiene-styrene and crumb rubber composite modified asphalt binders with high crumb rubber contents. *J. Clean. Prod.* **2022**, *331*, 129979. [CrossRef]
43. Chen, J.; Wang, J.; Li, M.; Zhao, Z.; Ren, J. Mesoscopic Mechanical Properties of Aggregate Structure in Asphalt Mixtures and Gradation Optimization. *Materials* **2023**, *16*, 4709. [CrossRef] [PubMed]
44. Cheng, Z.; Li, X.; Yang, Q.; Liang, N.; Chen, L.; Zheng, S.; Wang, D. Study on Compaction Properties and Skeleton Structural Characteristics of Porous Asphalt Mixture. *Sustainability* **2023**, *15*, 13911. [CrossRef]
45. Lv, Q.; Huang, W.; Zheng, M.; Sadek, H.; Zhang, Y.; Yan, C. Influence of gradation on asphalt mix rutting resistance measured by Hamburg Wheel Tracking test. *Constr. Build. Mater.* **2020**, *238*, 117674. [CrossRef]
46. Qian, C.; Fan, W. Evaluation and characterization of properties of crumb rubber/SBS modified asphalt. *Mater. Chem. Phys.* **2020**, *253*, 123319. [CrossRef]
47. Ren, J.; Xu, Y.; Huang, J.; Wang, Y.; Jia, Z. Gradation optimization and strength mechanism of aggregate structure considering macroscopic and mesoscopic aggregate mechanical behaviour in porous asphalt mixture. *Constr. Build. Mater.* **2021**, *300*, 124262. [CrossRef]
48. Kusumawardani, D.M.; Wong, Y.D. Evaluation of aggregate gradation on aggregate packing in porous asphalt mixture (PAM) by 3D numerical modelling and laboratory measurements. *Constr. Build. Mater.* **2020**, *246*, 118414. [CrossRef]
49. Taghipoor, M.; Hassani, A.; Karimi, M.M. Investigation of material composition, design, and performance of open-graded asphalt mixtures for semi-flexible pavement: A comprehensive experimental study. *J. Traffic Transp. Eng.* **2024**, *11*, 92–116. [CrossRef]
50. Zhang, D.; Birgisson, B.; Luo, X. A new dynamic modulus predictive model for asphalt mixtures based on the law of mixtures. *Constr. Build. Mater.* **2020**, *255*, 119348. [CrossRef]
51. Zhang, J.; Bao, L. Determination of asphalt mixture's viscoelastic constitutive parameters for pavement response analysis using dynamic modulus transformation. *Constr. Build. Mater.* **2022**, *315*, 125729. [CrossRef]
52. Wang, H.; Li, R.; Wang, X.; Ling, T.; Zhou, G. Strength and road performance for porous polyurethane mixture. *China J. Highw. Transp.* **2014**, *27*, 27–35.
53. Wang, H.; Hu, Q.; Li, R. Study of the water-heat stability for porous polyurethane mixture. *Highw. Eng.* **2014**, *39*, 246–250.
54. Lu, G.; Renken, L.; Li, T.; Wang, D.; Li, H.; Oeser, M. Experimental study on the polyurethane-bound pervious mixtures in the application of permeable pavements. *Constr. Build. Mater.* **2019**, *202*, 838–850. [CrossRef]
55. Chen, J.; Yin, X.; Wang, H.; Ding, Y. Evaluation of durability and functional performance of porous polyurethane mixture in porous pavement. *J. Clean. Prod.* **2018**, *188*, 12–19. [CrossRef]
56. Cong, L.; Wang, T.; Tan, L.; Yuan, J.; Shi, J. Laboratory evaluation on performance of porous polyurethane mixtures and OGFC. *Constr. Build. Mater.* **2018**, *169*, 436–442. [CrossRef]
57. Sun, M. Research on Performance of Polyurethane Porous Elastic Pavement Mixture. Master's Thesis, Southeast University, Dhaka, Bangladesh, 2016.
58. Li, T.; Lu, G.; Wang, D.; Hong, B.; Tan, Y.; OESER, M. Key properties of high-performance polyurethane bounded pervious mixture. *China J. Highw. Transp.* **2019**, *32*, 158–169.
59. Zhang, J.; Zhang, X.; Liang, M.; Jiang, H.; Wei, J.; Yao, Z. Influence of different rejuvenating agents on rheological behavior and dynamic response of recycled asphalt mixtures incorporating 60% RAP dosage. *Constr. Build. Mater.* **2020**, *238*, 117778. [CrossRef]
60. Zhang, Y.; Luo, W.; Liu, X. Experimental studies on the dynamic viscoelastic properties of basalt fiber-reinforced asphalt mixtures. *Sci. Eng. Compos. Mater.* **2021**, *28*, 489–498. [CrossRef]

 https://doi.org/10.3390/coatings16020153

61. Havriliak, S.; Negami, S. A Complex Plane Analysis of α-dispersions in Some Polymer Systems. In *Journal of Polymer Science Part C Polymer Symposia*; Wiley Subscription Services, Inc.: New York, NY, USA, 1966; pp. 99–117.
62. Tschoegl, N.W. *The Phenomenological Theory of Linear Viscoelastic Behavior*; Springer Science & Business Media: Berlin/Heidelberg, Germany, 1989.
63. Jongepier, R.; Kuilman, B.; Schmidt, R.; Puzinauskas, V.; Rostler, F. Characteristics of the rheology of bitumens. In *Association of Asphalt Paving Technologists Proc*; Michigan University: Ann Arbor, MI, USA, 1969.
64. Rowe, G.; Baumgardner, G.; Sharrock, M. Functional forms for master curve analysis of bituminous materials. In *Advanced Testing and Characterization of Bituminous Materials, Two Volume Set*; CRC Press: Boca Raton, FL, USA, 2009; pp. 97–108.
65. Olard, F.; Di Benedetto, H.; Eckmann, B.; Triquigneaux, J.-P. Linear Viscoelastic Properties of Bituminous Binders and Mixtures at Low and Intermediate Temperatures. *Road Mater. Pavement Des.* **2011**, *4*, 77–107. [CrossRef]
66. Dobson, G.; Monismith, C.; Puzinauskas, V.; Busching, H. The dynamic mechanical properties of bitumen. In *Association of Asphalt Paving Technologists Proc*; Michigan University: Ann Arbor, MI, USA, 1969; pp. 123–139.
67. Su, N.; Xiao, F.; Wang, J.; Amirkhanian, S. Precision Analysis of Sigmoidal Master Curve Model for Dynamic Modulus of Asphalt Mixtures. *J. Mater. Civ. Eng.* **2018**, *30*, 04018290. [CrossRef]
68. Airey, G.D. Use of Black Diagrams to Identify Inconsistencies in Rheological Data. *Road Mater. Pavement Des.* **2011**, *3*, 403–424. [CrossRef]
69. Christensen, D. Analysis of creep data from indirect tension test on asphalt concrete. *J. Assoc. Asph. Paving Technol.* **1998**, *67*, 458–492.
70. Mensching, D.J.; Rowe, G.M.; Daniel, J.S.; Bennert, T. Exploring low-temperature performance in Black Space. *Road Mater. Pavement Des.* **2015**, *16*, 230–253. [CrossRef]
71. Qin, X.; Ma, L.; Wang, H. Comparison analysis of dynamic modulus of asphalt mixture: Indirect tension and uniaxial compression test. *Transp. A Transp. Sci.* **2018**, *15*, 165–178. [CrossRef]
72. Arefin, M.S.; Quasem, T.; Nazzal, M.; Dessouky, S.; Abbas, A.R. Accuracy of MEPDG Dynamic Modulus Predictions for Short-Term and Long-Term Aged Asphalt Mixtures. *J. Transp. Eng. Part B Pavements* **2019**, *145*, 04019025. [CrossRef]
73. Ding, Z.; Su, J.; Li, P.; Bing, H. Mechanism of HCB-Modified Asphalt and Dynamic Properties of Mixtures. *Appl. Sci.* **2020**, *10*, 4971. [CrossRef]
74. Chailleux, E.; Ramond, G.; Such, C.; de La Roche, C. A mathematical-based master-curve construction method applied to complex modulus of bituminous materials. *Road Mater. Pavement Des.* **2011**, *7*, 75–92. [CrossRef]
75. Rouleau, L.; Deü, J.F.; Legay, A.; Le Lay, F. Application of Kramers–Kronig relations to time–temperature superposition for viscoelastic materials. *Mech. Mater.* **2013**, *65*, 66–75. [CrossRef]
76. Loizos, A.; Gkyrtis, K.; Plati, C. Modelling Asphalt Pavement Responses Based on Field and Laboratory Data. In *Accelerated Pavement Testing to Transport Infrastructure Innovation*; Springer International Publishing: Cham, Switzerland, 2020; pp. 438–447.

 https://doi.org/10.3390/coatings16020153

coatings

MDPI

Article

Settlement Prediction for Cast-in-Place Tubular Piles with Large Diameters Based on the Load Transfer Approach

Jiujiang Wu [1,2], Lin Xiao [1], Jifeng Lian [3] and Lijuan Wang [4,*]

1 Shock and Vibration of Engineering Materials and Structures Key Laboratory of Sichuan Province, Southwest University of Science and Technology, Mianyang 621010, China; wujiujiang1988@swust.edu.cn (J.W.); xiaolin_1999@mails.swust.edu.cn (L.X.)
2 Department of Civil and Environmental Engineering, Western University, London ON N6A 5B9, Canada
3 School of Emergency Science, Xihua University, Chengdu 610039, China; lianjf@mail.xhu.edu.cn
4 School of Physics and Engineering Technology, Chengdu Normal University, Chengdu 611130, China
* Correspondence: 071074@cdnu.edu.cn

Abstract: Large-diameter cast-in-place tubular piles offer high efficiency and adaptability for various engineering applications. Despite their widespread use, the bearing behavior of these piles remains complex due to the interactions with the internal soil core, and the related theoretical framework is not yet fully developed. In this study, a simplified load transfer model is proposed based on the pile–soil interaction mechanism of large-diameter tubular piles. Comprehensive load transfer models for the skin friction and end resistance of both the pile body and the soil core are established, supported by a detailed theoretical analysis. A novel three-criteria approach is introduced for the first time to enhance settlement predictions for large-diameter tubular piles by considering the displacement coordination mechanism of the internal soil core, addressing the limitations of traditional two-criteria methods. The proposed methods are validated through two engineering case studies, demonstrating their effectiveness and confirming their rationality and applicability in practical scenarios.

Keywords: layered soils; cast-in-place tubular pile with large diameter; load transfer function; modified interactive procedure; settlement prediction

1. Introduction

Large-diameter cast-in-place tubular piles are recognized as an efficient and versatile foundation solution for complex geological conditions, including saturated soft soil, clay, silt, and medium-dense sand [1]. Their applications span highways, railways, hydraulic dams, foundation pit support, and civil building foundations, offering substantial advantages in soil improvement and foundation reinforcement. For instance, Liu et al. [2] introduced the use of large-diameter cast-in-situ concrete pipe (PCC) piles for embankments over soft clay, demonstrating superior efficiency and cost-effectiveness compared with traditional methods. Mahfouz [3] emphasized their economic and structural advantages in highway foundation projects. Zhou et al. [4] documented the successful implementation of PCC piles in high-speed railway foundations, meeting strict settlement requirements with consistent performance. Chen et al. [5] investigated the seismic behavior of coral sand and found that drainage PCC piles significantly reduced liquefaction effects such as pore pressure, bending moments, and deformation when compared with conventional pile foundations. In the context of transportation infrastructure, particularly in pavement engineering, large-diameter cast-in-place tubular piles have proven to be a reliable and adaptable solution, ensuring long-term stability and safety under diverse soil conditions.

Constructed with a twin-tube mold, cast-in-place tubular piles are formed by pouring concrete into the annular space during casing extraction, with controlled vibration ensuring high-quality in situ reinforced concrete. These piles are designed to transfer substantial superstructure loads to deeper soil layers, effectively controlling settlement and differential settlement within permissible limits [3]. The pile body consists of cast-in-place reinforced concrete formed in a single process, with the in situ soil core retained within the interior, as illustrated in Figure 1. Compared with conventional large-diameter bored piles, cast-in-place tubular piles offer notable advantages. Their larger diameter and thin-walled design reduce concrete consumption while maintaining load-bearing capacity, lowering material costs and carbon emissions. The construction process retains the soil core within the tubular structure, minimizing ground disturbance and reducing risks to nearby structures. Additionally, vibration-assisted installation ensures better concrete consolidation, enhancing pile integrity and durability. These features make cast-in-place tubular piles a more efficient, sustainable, and technically superior alternative to traditional bored piles.

Figure 1. Construction details of cast-in-place tubular piles. (**a**) Mold used for cast-in-place tubular piles. (**b**) Completed cast-in-place tubular pile in ground. (**c**) On-site construction.

The load transfer mechanism of large-diameter thin-walled cast-in-place tubular piles becomes complex due to the influence of the internal soil core, and the load-bearing theory for cylindrical piles remains incomplete. For pile foundations, settlement calculation has always been one of the key challenges in design theory. Accurately predicting pile foundation settlement is crucial for evaluating the bearing capacity of foundations. Currently, the load transfer method is one of the most widely used theoretical approaches [6–10]. By selecting different load transfer models and parameters, the load transfer method can adapt to the settlement analysis of single piles under various soil conditions. This method is highly regarded in engineering due to its clear concept and simple calculations. For instance, Dias et al. [11] improved the traditional load transfer formula by defining different unloading paths and adjusting displacement variables. Liu et al. [12] developed a three-step calculation program to simulate the load-displacement response of a single pile in unsaturated expansive soil. Li et al. [13] developed an analytical method to predict the nonlinear load-displacement behavior of pile groups in clays. Singh et al. [14] proposed a nonlinear load transfer method to evaluate the load-displacement response of conical piles under static axial loads. Dong et al. [15] introduced a hyperbolic model to improve the transfer function method, providing a solution for the load-settlement relationship of large-diameter composite piles. Zhou et al. [16] and Liu et al. [17] introduced models for specific conditions, further refining the load transfer approach.

It should be noted that the solution of load transfer functions for tubular piles currently relies mostly on analytical methods, which are generally based on homogeneous soils. However, when solving segmented load transfer functions, analytical methods often require iterative solutions of differential equations for each segment, leading to substantial computational effort. Additionally, analytical methods become challenging to apply when dealing with more complex transfer functions, such as hyperbolic models. In contrast, the displacement coordination method, as a common approach for solving load transfer function [18], can effectively describe the complex properties of pile periphery medium, such as stratification, nonlinear and other characteristics, with clear concepts, simple calculation, ease of use, and reliable performance. Traditional displacement coordination methods are primarily designed for settlement analysis of solid pile foundations and are not fully applicable to tubular piles that contain an internal soil core. These methods typically neglect the influence of the soil core, which is a critical factor in the load transfer and settlement behavior of such piles. As of now, no studies have explored the application of displacement coordination methods to iteratively solve the settlement of tubular piles while accounting for the effect of the internal soil core. The omission of this factor can lead to significant errors in settlement predictions, as conventional methods fail to capture the complex interactions between the pile and the soil core.

Based on an analysis of the pile–soil interaction mechanism of large-diameter cast-in-place tubular piles, this paper proposes a simplified load transfer mechanics model. Corresponding load transfer functions for lateral friction resistance and bottom resistance are developed, along with detailed theoretical derivations. Furthermore, this study introduces a modified iterative procedure with three criteria for the first time, which improves upon the traditional procedure with two criteria. The proposed modification specifically addresses the displacement coordination of the internal soil core in tubular piles, enabling the displacement coordination method to be successfully applied to the settlement calculation of such piles. This advancement resolves the current lack of methods capable of iteratively solving the settlement of tubular piles while considering the influence of the soil core. Finally, the proposed approaches are validated through two practical engineering cases.

2. Materials and Methods

2.1. Simplified Model of Load Transfer Mechanics

Based on the load transfer mechanism of cast-in-place thin-walled concrete caisson foundations [19–21], their load transfer mechanical model can be simplified, as shown in Figure 2. When the upper load P is applied to the cap of the tubular pile, it is divided into two components, with the force N_0 transferred to the top concrete ring of the pile and the force N_{c0} acting on the top of the soil core, as shown in Figure 2a. For the pile part, N_0 is balanced by the pile external friction τ, pile internal friction τ', and pile bottom resistance N_n. For the soil core, N_{c0} is balanced by the lateral friction resistance τ_c' and the end resistance of the soil core N_{cn}.

The basic concept of the load transfer approach is to divide the pile into many elastic units, each of which is connected with the soil by nonlinear springs, to simulate the load transfer relationship between the pile and soil [6,22,23]. According to this principle, the tubular pile and soil core are divided into n units along the depth direction for analysis, as shown in Figure 2b. Pile element i bears axial force N_{i-1}, external friction resistance τ, internal friction resistance τ' from upper unit $i-1$ and reaction force N_i from bottom unit $i+1$. The soil core unit i bears the axial force N_{Ci-1}, internal friction resistance τ'_{ci} transmitted by the upper unit $i-1$ and the reaction force N_{ci} provided by the bottom unit $i+1$.

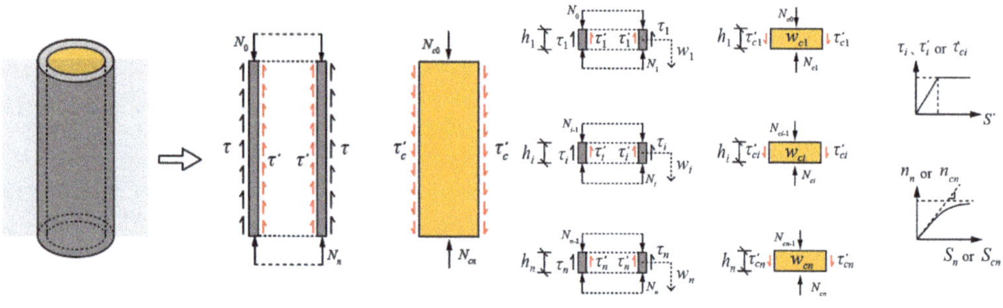

Figure 2. Load transfer analysis model of tubular piles with large diameters. (**a**) Stress analysis of tubular pile. (**b**) Load transfer model.

2.2. Load Transfer Function and Parameters

2.2.1. External Friction Resistance

Assuming that the load transfer function of the side friction of the foundation follows an elastoplastic bilinear model [24], as shown in Figure 2b, the equation for its load transfer function can be expressed as:

$$\tau(z) = \begin{cases} K_s(z)S(z), & \tau(z) < \tau_u(z) \\ \tau_u(z), & \tau(z) \geq \tau_u(z) \end{cases}, \tag{1}$$

where $\tau(z)$ is the external skin friction at depth z; $K_s(z)$ is the stiffness coefficient of the external skin friction at depth z, representing the slope of the curve between skin friction and relative displacement before the ultimate skin friction is reached, as shown in Figure 2b; $S(z)$ is the relative displacement of pile soil at depth z; and $\tau_u(z)$ is the ultimate skin friction at depth z.

As the relative displacement between the pile body and the surrounding soil continues to develop, once it exceeds the soil's ultimate deformation value, relative plastic slip occurs between the pile and the soil. The stiffness coefficient of the skin friction depends on various factors, including the soil conditions, the pile installation method, and the geometry of the pile, and it can be related to the in situ lateral earth pressure coefficient K_0. At this point, the ultimate side friction resistance τ_u can be determined using Coulomb's friction law [25,26], as shown in Equation (2).

$$\tau_u = K_0(\frac{K}{K_0})\tan[\varphi'(\frac{\delta}{\varphi'})]\sigma' + c'(\frac{c}{c'}), \tag{2}$$

where K_0 is the coefficient of static lateral pressure of soil, $K_0 = 1 - \sin\varphi'$; K is the lateral earth pressure coefficient. For a foundation with a small displacement, K/K_0 should be 0.7~1.2, while for a foundation with a large displacement, K/K_0 should be 1.0~2.0 [26]. φ' and δ are the effective internal friction angles of the soil and pile–soil interface friction angles, respectively; δ/φ' can be 0.75; σ' is the effective overburden pressure at the depth under consideration; c' and c are the effective, cohesive forces of the soil and pile–soil interface cohesive forces, respectively; and c/c' can be 0.8~1.0 [25].

Finally, τ_u can be expressed as:

$$\tau_u = K\sigma' \tan\delta + c, \tag{3}$$

where K is the lateral earth pressure coefficient, σ' is the effective overburden pressure at the depth under consideration, and δ is the friction angle of the pile–soil interface.

For the elastic state, Randolph and Worth [27] suggested the transfer stiffness coefficient K_s of pile external friction resistance in a homogeneous isotropic elastic soil medium:

$$K_s = \frac{2\pi G_s}{\ln(r_m/r_0)}, \tag{4}$$

where K_s is the transfer stiffness coefficient of external friction resistance; G_s is the shear modulus of the soil around the pile; r_0 is the outer radius of the pile; and r_m is the shear influence radius, which is related to the buried depth of the pile and the soil distribution. For homogeneous soil, $r_m = 2.5l(1-v)$, where l is the pile length and v is Poisson's ratio of the soil.

When the pile is located in stratified soil, r_m can be expressed as [28]:

$$r_m = 2.5\frac{\sum\limits_{i=1}^{n_s} G_{si}h_i}{G_{sm}}\left(1 - \frac{\sum\limits_{i=1}^{n_s} v_{si}h_i}{l}\right), \tag{5}$$

where r_m is the shear influence radius; G_{si} is the shear modulus of the i-th soil layer; G_{sm} is the maximum shear modulus in the soil layers; h_i is the thickness of soil layer i; n_s is the total number of soil layers; v_{si} is the Poisson's ratio of the i-th soil layer; and l is the pile length.

2.2.2. Internal Frictional Resistance

Assuming that the load transfer function of the internal frictional resistance also follows the elastic-plastic bilinear model, as shown in Figure 2b, the load transfer function equation can be expressed as:

$$\tau'(z) = \begin{cases} K'_s(z)S'(z), & \tau'(z) < \tau_u(z) \\ \tau_u(z), & \tau'(z) \geq \tau_u(z) \end{cases}, \tag{6}$$

where $\tau'(z)$ is the internal friction resistance at pile depth z; $K'_s(z)$ is the internal friction transfer stiffness coefficient at depth z; $S'(z)$ is the relative displacement of pile soil at depth z; $\tau_u(z)$ is the ultimate value of wall-soil friction resistance at depth z, and its value can be calculated from the following equation:

$$S'(z) = S(z) - S_c(z), \tag{7}$$

where $S'(z)$ is the relative displacement of pile soil at depth z; $S(z)$ is the relative displacement of pile soil at depth z; $S_c(z)$ is the displacement of the soil core at depth z, and its value is the sum of the compression of the soil core at depths z and below.

For the value of the stiffness coefficient K'_s of the internal frictional resistance of a tubular pile, Wang et al. [29] assumed the stiffness coefficient of the internal frictional resistance to be equal to the stiffness coefficient of the external frictional resistance K_s of a tubular pile at the same depth in the calculation. This paper also referred to the above views in the calculation, as shown in Equation (8).

$$K_s' = K_s = \frac{2\pi G_s}{\ln(r_m/r_0)}, \tag{8}$$

2.2.3. Pile Bottom Resistance

Assuming that the load transfer function of the bottom resistance of a tubular pile follows the hyperbolic model [30–33], as shown in Figure 2b, its load transfer function equation can be expressed as:

$$N_n = \frac{A_p S_n}{f_b + g_b S_n}, \tag{9}$$

where N_n is the pile bottom resistance value; A_p is the horizontal cross-sectional area of the pile; S_n is the pile bottom displacement; and f_b and g_b are hyperbolic parameters.

According to the research results of Randolph et al., f_b can be expressed as Equation (10), and K_b is the pile bottom resistance stiffness coefficient:

$$f_b = \frac{1}{K_b}, \tag{10}$$

where f_b is the parameter of the hyperbola, and K_b is the stiffness coefficient of pile tip resistance.

For solid confined circular piles, the pile bottom can usually be assumed to be a circular rigid block acting on an infinite elastic half-space body, and the bottom resistance stiffness coefficient can be expressed as Equation (11) [27,34]:

$$K_b = \frac{4G_b}{\pi r_0(1 - v_b)}, \tag{11}$$

where K_b is the stiffness coefficient of pile tip resistance; r_0 is the outer radius of the pile; G_b and v_b are divided into the shear modulus and Poisson's ratio of the pile bottom soil.

The pile bottom of the tubular pile acts on the infinite elastic half-space body under pressure as a rigid ring. The outer radius of the ring is the outer radius of the tubular pile r_0, the inner diameter $r_0 - d$, and d is the pile wall thickness. The deformation of a semi-infinite elastic body under a uniform circular load requires a series of integral operations. In order to simplify the calculation, the deformation of the semi-infinite elastic body under ring loading can be simplified as the difference in deformation under two circular loads (radius r_0 and $r_0 - d$), and the relationship between displacement and load can be expressed as Equation (12) [35]:

$$S_n = \frac{N_n(1 - v_b)}{4G_b(r_0 + r_0 - d)} = \frac{N_n(1 - v_b)}{4G_b(2r_0 - d)}, \tag{12}$$

The pile bottom stiffness coefficient of the tubular pile is:

$$K_b = \frac{N_n}{A_p S_n} = \frac{N_n}{A_p} \times \frac{4G_b(2r_0 - d)}{N_n(1 - v_b)} = \frac{4G_b(2r_0 - d)}{A_p(1 - v_b)}, \tag{13}$$

By substituting Equation (13) into Equation (10), we can conclude that the first parameter f_b in the hyperbolic function of tubular pile bottom resistance is:

$$f_b = \frac{A_p(1 - v_b)}{4G_b(2r_0 - d)}, \tag{14}$$

The reciprocal of the second parameter g_b in the hyperbolic function of the end resistance of the tubular pile is the asymptote of the hyperbolic function, and g_b can be expressed as:

$$g_b = \frac{R_b}{n_{bu}}, \tag{15}$$

where R_b is the bottom resistance failure ratio, 0.75~0.80 [32], and n_{bu} is the unit ultimate bearing capacity of the pile bottom soil.

According to the cavity expansion theory proposed by Vesic [36], the per unit ultimate bearing capacity n_{bu} of pile bottom soil can be estimated, as shown in Equation (16):

$$n_{bu} = c' N_c + \sigma'_{nb} N_q, \tag{16}$$

where c' is the effective cohesion of the soil; N_c and N_q are parameters related to the friction angle of the soil, and their values can be calculated from Equations (17) and (18), respectively. σ'_{nb} is the average effective pressure at the side of the pile bottom, and its value can be calculated by Equation (19):

$$N_c = (N_q - 1)\cot\varphi', \tag{17}$$

$$N_q = \left(\tan\varphi' + \sqrt{1 + \tan^2\varphi'}\right)^2 e^{2\psi\tan\varphi'}, \tag{18}$$

$$\sigma'_{nb} = \frac{1 + 2K_0}{3}\sigma'_{vb}, \tag{19}$$

where σ'_{vb} is the vertical effective stress at the pile tip; ψ is the included angle between the compaction core boundary of the pile bottom and the horizontal plane in the Janbu failure mode [37] for the pile end soil, and its value decreases with increasing pile end soil compressibility.

2.2.4. Soil Core Lateral Friction Resistance and Bottom Resistance

As mentioned above, the pile internal friction resistance τ' and the soil core lateral friction resistance τ'_c are a pair of equal and opposite forces. Therefore, the soil core lateral friction resistance τ'_c can be solved by referring to the pile internal friction resistance τ' in Section 2.2.2.

For the tip resistance of the soil core, it is assumed to follow a hyperbolic distribution [31], as shown in Figure 2b, with its load transfer function equation expressed as:

$$N_{cn} = \frac{A_{cp} S_{cn}}{f_{cb} + g_{cb} S_{cn}}, \tag{20}$$

where, N_{cn} is the resistance value of the soil core bottom, and A_{cp} is the horizontal cross-section area of the soil core. S_{cn} is the displacement of the soil core bottom, f_{cb} and g_{cb} are hyperbolic parameters, and $g_{cb} = g_b$.

As shown in Figure 3, the soil core in the pile can be regarded as a series of overlapping disk soil units. Taking a certain soil unit as the object of analysis, the upper part of the disk unit of the soil core bears uniform load σ_z, the lower part bears uniform load $\sigma_z + d\sigma_z$, and the lateral friction resistance τ'_c of the soil core acts on all sides. The vertical equilibrium equation of the soil core disk unit is shown in Equation (21):

$$\pi r_c^2 \sigma_z + 2\pi r_c \tau'_z dz = \pi r_c^2 (\sigma_z + d\sigma_z), \tag{21}$$

where r_c is the radius of the soil core.

The above equation can be further simplified as:

$$\frac{d\sigma_z}{dz} = \frac{2\tau_z'}{r_c}, \tag{22}$$

According to the research results of Randolph et al. [38] and Lehane et al. [39] on tubular piles, there is a certain proportional relationship between the lateral friction resistance τ_c' of the soil core unit and the axial stress σ_z. If $\tau_z' = \beta\sigma_z$ is set, the following equation can be obtained:

$$\frac{d\sigma_z}{dz} = \frac{2\beta\sigma_z}{r_c}, \tag{23}$$

where β is the undetermined coefficient, and its specific value can be obtained from the following formula [39]:

$$\beta = \frac{\sin\varphi' \sin(\Delta - \delta)}{1 + \sin\varphi' \cos(\Delta - \delta)}, \tag{24}$$

where Δ is the conversion angle, and $\sin\Delta = \sin\delta / \sin\varphi$.

Integrating Equation (22) and assuming uniform load p_c at $z = 0$ (top of soil core), Equation (25) can be obtained:

$$\sigma_z = p_c e^{\eta z}, \tag{25}$$

where p_c is the uniformly distributed load borne at the top of the soil core (at $z = 0$), and $\eta = 2\beta / r_c$.

For the soil core, the displacement S_{cn} at the bottom of the soil core can be obtained by integrating the compression displacement of all soil core units, as shown in Equation (26):

$$S_{cn} = \int_0^h \frac{\sigma_z}{E_0} dz = \left.\frac{p_c e^{\eta z}}{\eta E_0}\right|_{z=0}^{z=h} = \frac{p_c e^{\eta h}}{\eta E_0} - \frac{p_c}{\eta E_0} \approx \frac{\sigma_z(h)}{\eta E_0}, \tag{26}$$

where E_0 is the compression modulus of the soil.

Thus, the bottom resistance stiffness coefficient of the tubular pile–soil core K_b' is:

$$K_b' = \frac{\sigma_z(h)}{\Delta s} = \eta E_0 = \frac{2\beta E_0}{r_c}, \tag{27}$$

Finally, f_{cb} can be expressed as:

$$f_{cb} = \frac{1}{K_b'} = \frac{r_c}{2\beta E_0}, \tag{28}$$

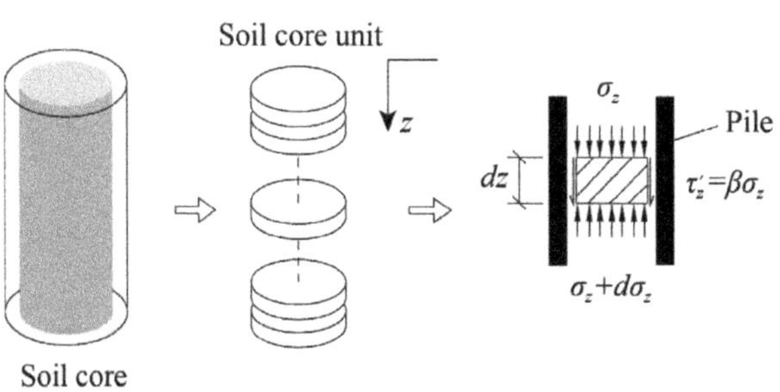

Figure 3. Analysis model of the soil core.

2.3. Modified Displacement Coordination Iterative Approach

The displacement coordinated iterative approach can effectively describe the complex properties of the medium around the pile, such as stratification and nonlinear characteristics, and does not need to solve the complex stiffness matrix, saving calculation time and resources, which has been accepted by a vast number of scholars and engineers [26,34]. In view of the structural characteristics of tubular piles, based on the traditional two-criterion displacement coordination iterative approach, this paper proposes a three-criterion displacement coordination iterative approach suitable for the load transfer characteristics of tubular piles, and its specific steps are described as follows:

(a) Along the depth direction, the tubular pile and soil core are divided into n units, as shown in Figure 2. The more units there are, the more accurate the calculation result will be, but the more units there are, the more time and resources will be needed to perform the calculation. In addition, when dividing units, the dividing line of units should coincide with the dividing line of the soil layer. For pile unit i, the following equilibrium equation can be established:

$$N_{i-1} = N_i + 2\pi r_0 h_i \tau_i + 2\pi r_c h_i \tau_i', \tag{29}$$

$$S_{i-1} = S_i + w_i, \tag{30}$$

where S_i is the top displacement of pile unit I, h_i is the length of unit I, and w_i is the compression displacement of pile unit i, which can be calculated by the following formula:

$$w_i = \frac{h_i(N_i + N_{i-1})}{2EA_p}, \tag{31}$$

where E is the elastic modulus of the pile.

For soil core unit i, the following equilibrium equation can be established:

$$N_{ci-1} = N_{ci} - 2\pi r_c h_i \tau_{ci}', \tag{32}$$

$$S_{ci-1} = S_{ci} + w_{ci}, \tag{33}$$

where S_{ci} is the top displacement of soil core unit I, and w_{ci} is the compression displacement of soil core unit i, which can be calculated by the following formula:

$$w_{ci} = \frac{h_i(N_{ci} + N_{ci-1})}{2E_0 A_{cp}}, \tag{34}$$

(b) We assume that under the upper load P, the pile tip displacement S_n and the soil core bottom displacement S_{cn} are small and that S_{cn} is slightly less than or equal to S_n.

(c) By substituting S_n and S_{cn} into Equations (9) and (20), the pile bottom resistance N_n and soil core bottom resistance N_{cn} can be obtained. In addition, we assume that the displacement in the middle of pile unit n is a small value, which may be slightly greater than S_n. The external friction resistance τ_n of unit n can be calculated according to Equation (1), as shown in Equation (35):

$$\tau_n = K_s S_n, \tag{35}$$

For soil core unit n:

$$m_{cn} = \frac{N_{cn-1} + 3N_{cn}}{8E_0 A_{cp}} h_n + S_{cn}, \tag{36}$$

$$\tau'_{cn} = K'_s(m'_n - m_{cn}), \tag{37}$$

$$N_{cn-1} = N_{cn} - 2\pi r_c h_n \tau'_{cn}, \tag{38}$$

where m_{cn} is the displacement in the middle of soil core unit n.

Substituting Equation (37) into Equation (38), the following formula can be obtained:

$$N_{cn-1} = N_{cn} - 2\pi r_c h_n K'_s(m'_n - m_{cn}), \tag{39}$$

Set $\zeta_1 = 2\pi r_c h_n K'_s$, $\zeta_2 = h_n/8E_0A_{cp}$. Substituting Equation (36) into Equation (39), the following formula can be obtained:

$$N_{cn-1} = N_{cn} - \zeta_1\left[m'_n - \zeta_2(N_{cn-1} + 3N_{cn}) - S_{cn}\right], \tag{40}$$

Equation (41) can be obtained by solving the above formula.

$$N_{cn-1} = \frac{N_{cn} - \zeta_1(m'_n - S_{cn} - 3N_{cn}\zeta_2)}{1 - \zeta_1\zeta_2}, \tag{41}$$

Substituting the transformation result of N_{cn-1} into Equation (36), Equation (42) can be obtained:

$$m_{cn} = \frac{N_{cn}\zeta_2 - \zeta_1\zeta_2(m'_n - S_{cn} - 3N_{cn}\zeta_2)}{1 - \zeta_1\zeta_2} + 3N_{cn}\zeta_2 + S_{cn}, \tag{42}$$

Therefore, according to Equations (6) and (7), the internal friction resistance of pile unit n and the friction resistance of soil core unit n can be solved as:

$$\tau'_n = -\tau'_{cn} = K'_s(S_n - m_{cn}), \tag{43}$$

N_{n-1} and N_{cn-1} can be solved by substituting τ_n and τ'_{cn} into Equations (29) and (32).

(d) The compression Δw_n in the middle of pile unit n is calculated by Equation (44):

$$\Delta w_n = [\frac{(N_{n-1} + N_n)/2 + N_n}{2}]\frac{h_n/2}{EA_p} = \frac{h_n(N_{n-1} + 3N_n)}{8EA_p}, \tag{44}$$

Thus, the displacement of the middle position of pile unit n can be updated as m_n:

$$m_n = S_n + \Delta w_n, \tag{45}$$

(e) Compare the m_n obtained in step (d) with the assumed value m'_n in step (c). If $|m_n - m'_n| \leq 1 \times 10^{-6}$, m_n is selected as the displacement of the final middle position of pile unit n; otherwise, reset the values in step (c) and repeat steps (c) through (d) until the inequality $|m_n - m'_n| \leq 1 \times 10^{-6}$ is satisfied.

(f) S_{n-1} and S_{cn-1} are calculated according to Equations (30) and (33); thus far, all variables of pile unit n and soil core unit n have been obtained. For pile unit $n-1$ and soil core unit $n-1$, the end displacement is S_{n-1} and S_{cn-1}, so all variables can be calculated by repeating steps (c) to (e). Similarly, all variable values of pile and soil core unit $n-2$ to top unit 1 can be calculated.

(g) For the tubular pile under vertical stress, its boundary condition requires that the displacement of the top of the soil core S_{c0} be equal to the displacement of the top of pile S_0, which has been calculated in the above steps. If $|s_0 - s_{c0}| \leq 1 \times 10^{-6}$, then the assumed soil core bottom displacement S_{cn} in step (b) is selected as the final value under the action of load P; otherwise, the S_{cn} value is reset, and steps (b) to (f) are repeated until $|s_0 - s_{c0}| \leq 1 \times 10^{-6}$. The adjustment strategy is if $S_0 - S_{c0} > 0$, then S_{cn} is increased, and vice versa.

(h) After determining S_{cn}, the applied load P can be compared with the calculated value $N_0 + N_{c0}$. If $|[(N_0 + N_{c0}) - P]/P| > 1 \times 10^{-8}$, the pile end displacement S_n in step (b) needs to be readjusted, and steps (b) to (g) are repeated until $|[(N_0 + N_{c0}) - P]/P| \leq 1 \times 10^{-8}$ is satisfied. The adjustment strategy is as follows: if $N_0 + N_{c0} > P$, increase it; otherwise, decrease it.

(i) For the next load applied on the pile top, repeat steps (b) to (h) until the final load is loaded.

Through the above iterative calculation steps, the distribution and changes in displacement, internal and external friction resistance, and soil core bottom resistance of tubular piles in layered strata under various loads can be calculated. The process can be simplified, as shown in Figure 4. Compared with the traditional two-criteria displacement iteration approach of pile foundations, the modified interactive procedure proposed in this paper introduces three criteria:

① $|m_n - m'_n| \leq 1 \times 10^{-6}$ m,
② $|s_0 - s_{c0}| \leq 1 \times 10^{-6}$ m,
③ $|[(N_0 + N_{c0}) - P]/P| \leq 1 \times 10^{-8}$.

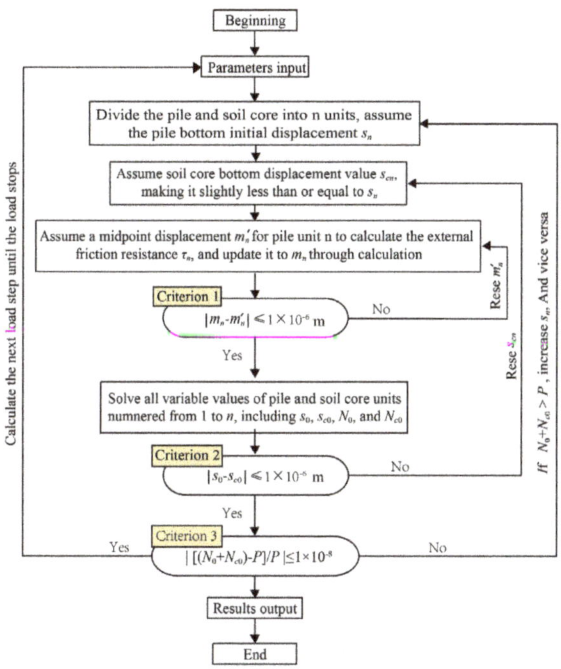

Figure 4. Flowchart of the interactive process.

2.4. Practical Engineering Case Verification

2.4.1. Engineering Case 1

This paper verifies the above approach combined with the static load test of the cast-in-place concrete tubular pile soft foundation reinforcement project of Jingyi Road in Nanjing Dachang reported in the literature [40]. The length of the test pile is 7.8 m, the outer diameter is 1.00 m, and the wall thickness is 0.12 m. The site soil is simplified into two layers, namely, loam layer 1A and loam layer 1B. The loam 1A layer near the surface is 4.5 m thick, and the elastic modulus of the pile body of the 1B layer is 20 GPa.

The depth and distribution of each soil layer at the site are shown in Figure 5. The soil layer near the surface is loam 1A, with a thickness of 4.50 m, and the soil layer below is loam 1B, with a thickness of 3.30 m. All the soil layers are distributed horizontally. According to the principle of the load transfer approach mentioned above, the pile and soil core are divided into units according to the soil layer distribution. The pile and soil core located in soil layer 1 are divided into 4 units with lengths of 1.00, 1.00, 1.25, and 1.25 m from top to bottom, as shown in Figure 5. Meanwhile, the pile and soil core located in soil layer 2 are divided into 3 units with a length of 1.10 m.

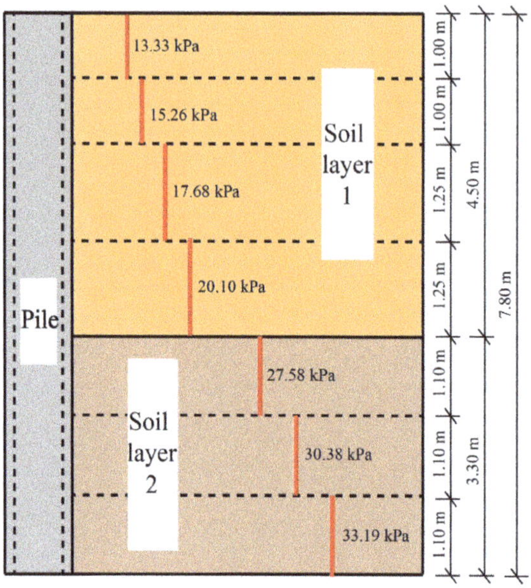

Figure 5. Layout of soil layers and interactive elements.

According to the soil parameters in Table 1, the average limit lateral friction resistance of each soil layer in each pile and soil core unit along the depth can be calculated by Equation (2) above, where K/K_0 is 1.2, δ/φ' is 0.75 and c/c' is 0.95. The final values of ultimate lateral friction resistance are shown in Figure 5 for the calculation of foundation load transfer.

Table 1. Parameters of soil in situ.

Soil	Compression Modulus E_0 (MPa)	c (kPa)	φ' (°)	ρ (kg/m^3)	v
Loam 1A	11.5	28.5	30.6	980	0.4
Loam 1B	14.2	16.6	32.8	980	0.4

According to the load transfer model described in this paper and the modified displacement coordination approach, the load settlement (Q-s) curve of the foundation under various loads can be obtained by iteration, and the theoretical calculation results are compared with the test results, as shown in Figure 6. Through comparison, the theoretical analysis results based on the proposed approach can be seen to be very close to the test results, which explains the rationality of the load transfer model and the modified displacement coordination approach proposed in this paper for cast-in-place tubular piles with large diameters in layered sites.

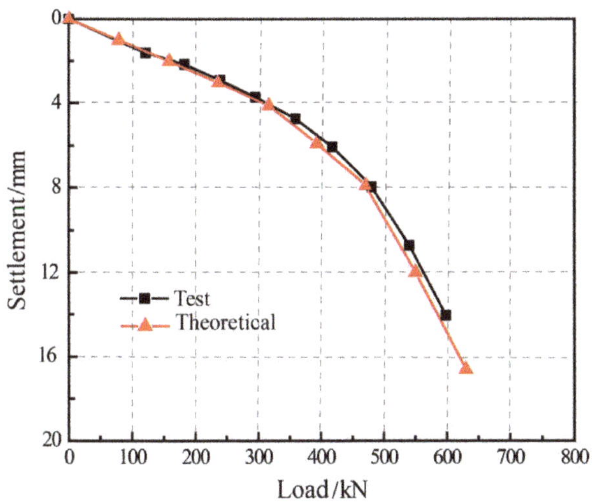

Figure 6. Results comparison.

2.4.2. Engineering Case 2

To further verify this approach, engineering case 2 is the field static load test of a tubular pile at the south end of Baxi Harbor Bridge in the second phase of the Hangzhou-Nanjing Expressway (Changxing Section) reported in the literature [1]. The length of the test pile is 18 m, the outer diameter is 1.00 m, and the wall thickness is 0.12 m. The soil layer of the site is clay, silt, mucky silty clay, and silty clay from top to bottom. The pile concrete grade is C25, the elastic modulus is E = 28 GPa, and the specific parameters of each soil in the soil layer are shown in Table 2.

Table 2. Parameters of soil in situ.

Soil	Compression Modulus E_0 (MPa)	c (kPa)	φ' (°)	ρ (kg/m^3)	v
clay	3.26	-	-	1800	0.4
silt	7.81	21.6	6.0	1860	0.35
mucky silty clay	1.94	16.8	4.38	1750	0.5
silty clay	5.85	42.0	16.7	2040	0.4

The depth and distribution of each soil layer at the site are shown in Figure 7. All soil layers are distributed horizontally, and the thicknesses are 1.30, 2.70, 13.5 and 0.5 m from top to bottom along the pile. According to the principle of the load transfer approach mentioned above, the pile and soil core are divided into units according to the soil layer distribution. The pile and soil core located in soil layer 1 are divided into 1 unit with a length of 1.30 m. The piles and soil cores in soil layer 2 are divided into two units, each of which is 1.35 m in length. The piles and soil cores in soil layer 3 are divided into 10 units, each of which is 1.35 m in length. The pile and soil core in soil layer 4 are divided into one unit with a length of 0.50 m. According to the soil parameters in Table 2, the average ultimate lateral friction resistance of each soil layer in each pile and soil core unit along the depth can be calculated by Equation (2) above, where K/K_0 is 1.2, δ/φ' is 0.75 and c/c' is 0.9. The final values of ultimate lateral friction resistance are shown in Figure 7 for the calculation of foundation load transfer.

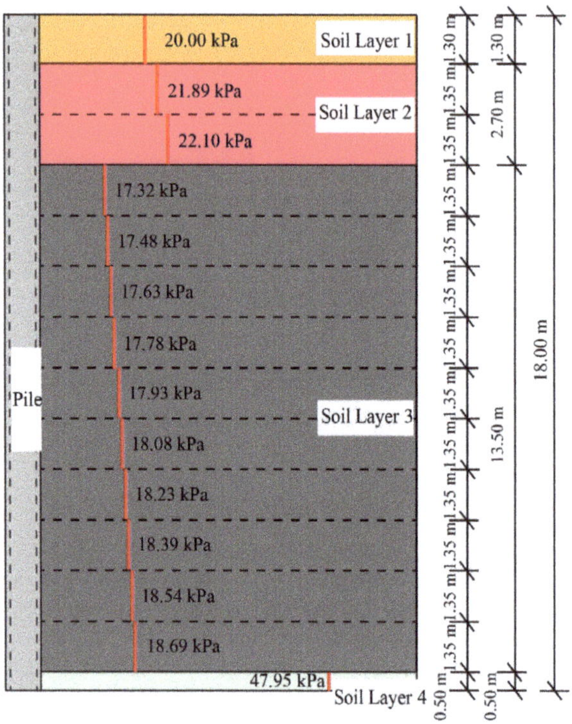

Figure 7. Layout of soil layers and interactive elements.

According to the load transfer model described in this paper and the modified displacement coordination approach, the load settlement (Q-s) curve of the foundation under various loads can be obtained by iteration, and the theoretical calculation results are compared with the test results, as shown in Figure 8. The comparison shows that the theoretical analysis results based on the proposed approach are very close to the test results, especially in the initial stage of loading. Therefore, the rationality and applicability of the load transfer model and the modified displacement coordination approach of cast-in-place tubular piles in layered sites are illustrated, and the relevant load transfer and settlement calculation results are of great practical value for analyzing the load transfer mechanism of tubular piles and guiding the engineering design and testing of piles.

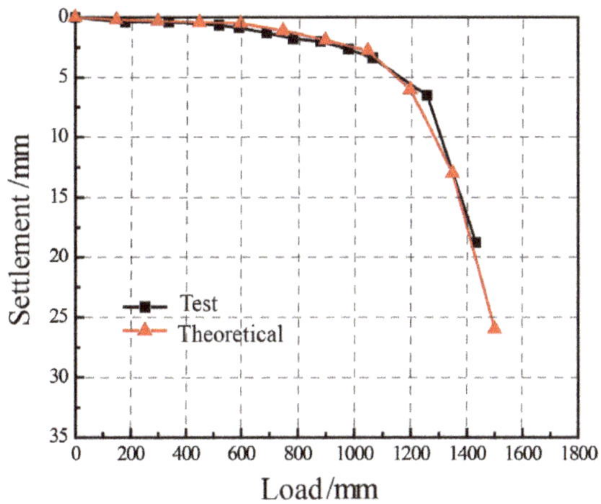

Figure 8. Results comparison.

3. Conclusions

This study develops a simplified load transfer mechanics model for large-diameter cast-in-place tubular piles, based on an in-depth analysis of their pile–soil interaction mechanisms. The model employs bilinear load transfer functions to characterize the internal and external friction resistance and the friction resistance of the soil core, while the end resistance is described using a hyperbolic function. These formulations effectively capture the nonlinear behavior of the pile–soil system and address the complexities arising from stratified soil conditions.

To overcome the limitations of traditional two-criteria iterative methods, a three-criteria (the criteria are as follows: $|m_n - m_n'| \leq 1 \times 10^{-6}$ m, $|s_0 - s_{c0}| \leq 1 \times 10^{-6}$ m, $|[(N_0 + N_{c0}) - P]/P| \leq 1 \times 10^{-8}$) displacement coordination iterative approach is introduced. This approach incorporates the displacement compatibility of the soil core and considers the layered soil structure, enabling efficient and accurate predictions of load transfer and settlement behavior. The proposed method avoids complex stiffness matrix calculations, offering a computationally efficient framework for practical engineering applications.

The validity of the proposed method is demonstrated through two engineering case studies. Field validation serves as a critical bridge between theory and practice, enabling methods to be tested under real and complex working conditions. This allows for a comprehensive assessment of their actual performance and potential issues. The results reveal a strong correlation between theoretical predictions and experimental measurements, particularly in capturing load settlement behavior under varying load conditions. This reinforces the rationality and reliability of the approach, highlighting its practical value for analyzing load transfer mechanisms and guiding the design, construction, and performance evaluation of tubular piles in layered soil environments.

Unlike traditional methods, which are often conservative and lead to significant deviations from actual conditions, the proposed model offers higher accuracy in settlement prediction. Traditional two-criteria methods, which overlook the influence of the soil core, tend to overestimate settlement, causing overdesign and unnecessary material usage. In contrast, the three-criteria approach proposed in this paper enables the use of smaller diameter piles that provide the same load-bearing capacity as larger ones. This results in reduced material consumption, lower construction costs, and decreased excavation volumes, ultimately promoting sustainability and reducing carbon emissions in engineering projects. In addition, this model demonstrates strong applicability to ordinary soils and a variety of soil scenarios that conform to the Mohr–Coulomb criterion. However, it is important to note that soils with special properties, such as expansive soils and loess, are not currently addressed in this study. These types of soils, which may exhibit behaviors such as negative skin friction, represent potential areas for future research to further enhance the model's applicability and accuracy.

Future work could explore the application of the model to dynamic loading conditions or more complex geological settings. These advancements would further enhance the applicability and versatility of the proposed approach in addressing diverse engineering challenges.

Author Contributions: Conceptualization, J.W.; Methodology, J.W. and L.W.; Software, L.W.; Validation, J.W. and L.X.; Investigation, J.W. and L.X.; Data curation, L.X.; Writing—original draft, J.W.; Writing—review & editing, L.X. and J.L.; Supervision, J.W.; Project administration, L.W.; Funding acquisition, J.W. All authors have read and agreed to the published version of the manuscript.

Funding: This research was funded by the National Foreign Expert Project (Grant No. DL2023036001L), Sichuan Science and Technology Program (Grant No. 2024JDHJ0019), and Nature Science Foundation of Sichuan Province (Grant No. 2024NSFSC0164).

Institutional Review Board Statement: Not applicable.

Informed Consent Statement: Not applicable.

Data Availability Statement: The original contributions presented in this study are included in the article. Further inquiries may be directed to the respective author.

Conflicts of Interest: The authors declare that the research was conducted in the absence of any commercial or financial relationships that could be construed as a potential conflict of interest.

References

1. Wang, Z.; Gong, X.N.; Ding, Z.X.; Zhou, J. Research on influence of soil core on load bearing character of large-diameter tubular pile using cat-in-situ concrete. *Chin. J. Rock Mech. Eng.* **2005**, *24*, 916–3921. (In Chinese)
2. Liu, H.L.; Chu, J.; Deng, A. Use of large-diameter, cast-in situ concrete pipe piles for embankment over soft clay. *Can. Geotech. J.* **2009**, *46*, 915–927. [CrossRef]
3. Mahfouz, A.H. New technique of ground improvement: Driven cast-in-situ thin wall concrete pipe piles (PCC). *Geomech. Geoengin.* **2024**, *6*, 119–129. [CrossRef]
4. Zhou, M.; Liu, H.L.; Chen, Y.M.; Hu, Y.X.; Hassan, R. First application of cast-in-place concrete large-diameter pipe (PCC) pile-reinforced railway foundation: A field study. *Can. Geotech. J.* **2015**, *53*, 708–716. [CrossRef]
5. Chen, Z.X.; Wang, B.X.; Wang, C.L.; Wang, Y.H.; Xiao, P.; Li, K.Y. Performance of a subgrade-embankment-seawall system reinforced by drainage PCC piles and ordinary piles subjected to lateral spreading. *Geofluids* **2023**, *8*, 4489478. [CrossRef]
6. Feng, S.J.; Fang, J.C.; Zhao, Y.; Zhang, Z.J.; Wang, Y. Thermomechanical analysis of energy piles using a load-transfer approach considering soil coupling effects. *Comput. Geotech.* **2024**, *168*, 106147. [CrossRef]
7. Liang, X.; Cheng, Q.G.; Wu, J.J.; Chen, J.M. Model test of the group piles foundation of a high-speed railway bridge in mined-out area. *Front. Struct. Civ. Eng.* **2016**, *10*, 488–498. [CrossRef]
8. Wu, J.J.; Cheng, Q.G.; Wen, H.; Wang, L.J.; Li, Y.; Zhang, J.L. A Load Transfer Approach to Rectangular Closed Diaphragm Wall. *Proc. Inst. Civ. Eng.-Geotech. Eng.* **2016**, *169*, 509–526. [CrossRef]
9. Wu, J.J.; El Naggar, M.H.; Cheng, Q.G.; Wen, H.; Li, Y.; Zhang, J.L. Iterative Load Transfer Procedure for Settlement Evaluation of Lattice-Shaped Diaphragm Walls in Multilayered Soil. *Comput. Geotech.* **2020**, *120*, 103409. [CrossRef]
10. Wu, J.J.; Zhang, Y.; Li, Y.; Wen, H.; Wang, L.J. Uplift behavior of nodular diaphragm wall: Experiment and theory. *Int. J. Numer. Anal. Met.* **2024**, *48*, 1197–1226. [CrossRef]
11. Dias, T.G.S.; Bezuijen, A. Load-Transfer Method for Piles Under Axial Loading and Unloading. *J. Geotech. Geoenviron.* **2018**, *144*, 04017096. [CrossRef]
12. Liu, Y.L.; Vanapalli, S.K. Load displacement analysis of a single pile in an unsaturated expansive soil. *Comput. Geotech.* **2019**, *106*, 83–98. [CrossRef]

13. Li, L.; Li, J.P.; Wang, Y.; Gong, W.B. Analysis of nonlinear load-displacement behaviour of pile groups in clay considering installation effects. *Soils Found.* **2023**, *60*, 752–766. [CrossRef]
14. Singh, S.; Patra, N.R. Axial behavior of tapered piles using cavity expansion theory. *Acta Geotech.* **2020**, *15*, 1619–1636. [CrossRef]
15. Dong, Y.X.; Feng, Z.J.; He, J.B.; Yang, Y.D. An analytical solution for the load-settlement relation of hollow composite piles with ultra-large diameters. *Phys. Chem. Earth* **2023**, *132*, 103466. [CrossRef]
16. Zhou, Z.J.; Zhang, Z.P.; Chen, C.R.; Xu, F.; Xu, T.Y.; Zhu, L.X.; Liu, T. Application of load transfer method for bored pile in loess area. *Eur. J. Environ. Civ. Eng.* **2022**, *26*, 4457–4475. [CrossRef]
17. Liu, J.; He, J.; Wu, Y.P.; Yang, Q.G. Load transfer behaviour of a tapered rigid pile. *Geotechnique* **2012**, *62*, 649–652. [CrossRef]
18. Coyle, H.M.; Reese, L.C. Load Transfer for Axially Loaded Piles in Clay. *J. Soil Mech. Found. Eng. Div.* **1966**, *92*, 1–26. [CrossRef]
19. Chen, Z.; Zhang, Q.; Xing, Y.; Wang, S.; Zhao, Y.; Cui, W. Analysis of the response of a single pile using the disturbance state concept theory. *Int. J. Geomech.* **2022**, *22*, 04022180. [CrossRef]
20. Sengupta, I.; Patra, N.R.; Rajesh, S. Response of piled raft foundations using the disturbed state concept theory: Analysis and interpretation. *Ocean Eng.* **2024**, *309*, 118566. [CrossRef]
21. Wang, X.; Wang, X.Z.; Yang, G.Q.; Yang, X.; Zhang, D. Research on the Load Transfer Law of Cross-Sections of Pile-Supported Reinforced Embankments Based on the Finite Element Method. *Sustainability* **2022**, *14*, 7831. [CrossRef]
22. Bhartiya, P.; Basu, D.; Chakraborty, T. Nonlinear settlement of piled rafts in sandy soil. *Int. J. Geomech.* **2021**, *21*, 04021214. [CrossRef]
23. Huang, M.; Jiang, S.; Xu, C.S.; Xu, D. A new theoretical settlement model for large step-tapered hollow piles based on disturbed state concept theory. *Comput. Geotech.* **2020**, *124*, 103626. [CrossRef]
24. Zhang, J.; Shang, Y.Q.; Ye, B. Analytic Calculation of P-S curve of bolts based on bilinear load-transfer function. *Chin. J. Rock Mech. Eng.* **2005**, *24*, 1072–1076. (In Chinese)
25. Zhang, L.M.; Xu, Y.; Tang, W.H. Calibration of models for pile settlement analysis using 64 field load tests. *Can. Geotech. J.* **2008**, *45*, 59–73. [CrossRef]
26. Zhang, Q.Q.; Zhang, Z.M. A simplified nonlinear approach for single pile settlement analysis. *Can. Geotech. J.* **2012**, *49*, 1256–1266. [CrossRef]
27. Randolph, M.F.; Wroth, C.P. Analysis of deformation of vertically loaded piles. *J. Geotech. Eng. Div.* **1978**, *104*, 1465–1488. [CrossRef]
28. LEE, C.Y. Pile group settlement analysis by hybrid layer approach. *J. Geotech. Eng.* **1993**, *119*, 984–997. [CrossRef]
29. Wang, Z.; Zhou, J.; Gong, X.N. Analysis of axial load–transfer of large-diameter tubular pile sing cast-in-situ concrete. *Chin. J. Geotech. Eng.* **2005**, *27*, 1185–1189. (In Chinese)
30. Kondner, R.L. Hyperbolic stress-strain response: Cohesive soils. *J. Soil Mech. Found. Div.* **1963**, *89*, 115–143. [CrossRef]
31. Castelli, F.; Maugeri, M. Simplified nonlinear analysis for settlement prediction of pile groups. *J. Geotech. Geoenviron. Eng.* **2002**, *128*, 76–84. [CrossRef]
32. Zhang, Q.Q.; Li, L.P.; Chen, Y.J. Analysis of compression pile response using a softening model, a hyperbolic model of skin friction, and a bilinear model of end resistance. *J. Eng. Mech.* **2014**, *140*, 102–111. [CrossRef]
33. Liu, P.; Jiang, C.; Sun, M.L. Analytical model for lateral response of piles considering inhomogeneous disturbance characteristics of homogeneous sand. *Comput. Geotech.* **2023**, *161*, 105593. [CrossRef]
34. Timoshenko, S.P.; Goodier, J.N. Theory of Elasticity, 3rd ed. *J. Appl. Mech. Sep.* **1970**, *37*, 888. [CrossRef]
35. Sun, X.L.; Yang, M.; Mo, H.O. Displacement of base-enlarged tension piles caculated by load transfer method. *Chin. J. Geotech. Eng.* **2008**, *30*, 1815–1820. (In Chinese)
36. Vesic, A.S. Analysis of ultimate loads of soil mechanics and foundations. *J. Soil Mech. Found. Eng.* **1973**, *99*, 45–73. [CrossRef]
37. Janbu, N. Static bearing capacity of friction piles. In Proceedings of the European Conference on Soil Mechanics and Foundation Engineering, Vienna, Austria, 22–24 March 1976; pp. 479–488.
38. Randolph, M.F.; Leong, E.C.; Houlsby, G.T. One dimensional analysis of soil plugs in pipe piles. *Geotechnique* **1991**, *41*, 587–598. [CrossRef]
39. Lehane, B.M.; Gavin, K.G. Base resistance of jacked pipe piles in sand. *J. Geotech. Geoenviron.* **2001**, *127*, 473–780. [CrossRef]
40. Fei, K.; Liu, H.L.; Gao, Y.F.; Feng, T.G. Load transfer mechanism for field pour concrete thin wall cased pile (PCC). *Rock Soil Mech.* **2004**, *25*, 764–768. (In Chinese)

coatings

MDPI

Article

A New Method for Evaluating Liquefaction by Energy-Based Pore Water Pressure Models

Jianlei Zhang [1,2,*], Qiangong Cheng [3,4,*], Haozhen Fan [1], Mengjie Dai [1], Yan Li [5], Jiujiang Wu [6] and Yufeng Wang [3]

[1] School of Engineering, China University of Petroleum (Beijing) at Karamay, Karamay 834000, China
[2] Xinjiang Key Laboratory of Multi-Medium Pipeline Safety Transportation, Urumqi 830011, China
[3] Department of Geological Engineering, Southwest Jiaotong University, Chengdu 611756, China
[4] Key Laboratory of High-Speed Railway Engineering, Ministry of Education, Chengdu 611756, China
[5] Chengdu Surveying Geotechnical Research Institute Co., Ltd. of MCC, Chengdu 610023, China
[6] Shock and Vibration of Engineering Materials and Structures Key Laboratory of Sichuan Province, Southwest University of Science and Technology, Mianyang 621010, China
* Correspondence: zhangjianlei@cupk.edu.cn (J.Z.); chengqiangong@home.swjtu.edu.cn (Q.C.)

Abstract: Liquefaction-induced damage can be mitigated through remediation methods, contingent upon a thorough evaluation of liquefaction, which necessitates comprehensive investigation. This paper presents a novel energy-based pore pressure model for the assessment of liquefaction potential, utilizing cyclic triaxial numerical tests. In this model, the energy of the earthquake is quantified using the Arias intensity. The validity of the energy-based pore pressure model was corroborated by the results of cyclic triaxial tests. Based on the validated model, a new methodology that incorporates permeability and the shear stress reduction coefficient was proposed for the evaluation of liquefaction potential. This new approach was further validated through centrifuge tests and numerical simulations. The findings indicate that the proposed method can accurately predict the generation and accumulation of excess pore pressure, thereby demonstrating its efficacy in evaluating ground liquefaction potential.

Keywords: liquefaction potential; pore water pressure model; Arias intensity; numerical simulation

1. Introduction

Soil liquefaction is a significant contributor to the destruction of superstructures, primarily due to the lateral spreading and softening of the ground that occurs as a result of liquefaction during earthquake events [1]. Consequently, it is imperative to assess the liquefaction potential of the ground prior to the construction of foundations, which can effectively mitigate the damage associated with soil liquefaction. However, the assessment of liquefaction potential presents a complex engineering challenge, influenced by various factors, including the characteristics of saturated sand, the initial stress state, and the magnitude, duration, and frequency of earthquake ground motion [2]. Currently, the conventional methods for assessing the liquefaction potential of saturated sand are categorized into three primary approaches: the first is the stress method, which is based on the shear stress that can be generated in the ground [3]; the second is the strain method, which focuses on the shear strain induced by earthquake events in the ground [4]; and the third is the energy method, which is predicated on the energy dissipated at the site [5]. Among these methodologies, the simplified evaluation method developed by Seed and Idriss [3], which is grounded in the stress method, is the most widely utilized.

The energy method employs simplified procedures that are predicated on standard penetration values and ground motion, offering several advantages [6,7]. Experimental results have demonstrated a strong correlation between the cumulative dissipated energy per unit volume of soil (referred to as unit dissipated energy) and the excess pore water pressure generated under undrained cyclic loading [6]. This relationship emerges from the fact that the energy dissipated per unit is dependent on the area of the shear stress-strain hysteresis loop, whereas the excess pore water pressure is closely related to the shear strain induced by cyclic shear [6]. Consequently, it is justifiable to utilize unit dissipated energy as a means to evaluate the excess pore water pressure that contributes to soil liquefaction. As the earthquake ground motion propagates toward the surface, both amplification and attenuation of amplitude occur, which are contingent upon the frequency range. The stress method assesses liquefaction potential based on the maximum horizontal acceleration at the surface; however, it fails to account for the acceleration amplification and attenuation resulting from the propagation of ground motion. Furthermore, this method selectively interprets certain aspects of the amplitude while disregarding others. In contrast, the energy method employs dissipated energy to evaluate soil liquefaction potential, defined as the difference between total energy and emitted energy. Both total energy and emitted energy are influenced by the amplitude, frequency, and duration of earthquake ground motion. Thus, the energy method effectively incorporates considerations of amplitude amplification and attenuation. In summary, the energy method provides a more effective and accurate assessment of soil liquefaction potential induced by earthquakes.

Based on the energy method, researchers have developed a range of pore water pressure models over the past few decades [5–9]. Initially, several pore water pressure models were established through cyclic undrained tests conducted under regular loading conditions. However, these models differ significantly from the irregular periods and asymmetric amplitudes characteristic of earthquake ground motions. Consequently, the pore water pressure models formulated under regular loading are not appropriate for calculating the generation and accumulation of pore water pressure during earthquake ground motions [10,11]. Seed and Idriss [3] proposed a method for assessing liquefaction potential subjected to earthquake ground motion, utilizing results from cyclic undrained tests performed under constant stress dynamic loading. In this approach, the constant stress is set at 0.65 times the peak value of the dynamic stress, and the number of cycles in the test is determined by the magnitude of the earthquake. This equivalent method demonstrates a degree of rationality in strength estimation and is user-friendly, leading to its continued application. However, Xie and Wu [10] identified that the ground motion pattern and the stress history resulting from irregular and asymmetric dynamic loading significantly affect the liquefaction potential of saturated sand. The equivalent method proposed by Seed and Ishihara [3] does not adequately consider the influence of ground motion patterns and neglects the impact of stress history on pore water pressure. Therefore, it is deemed unsuitable for evaluating soil liquefaction potential [10]. In response to these limitations, some researchers have proposed pore water pressure models specifically designed for irregular loading conditions to enhance the evaluation of liquefaction potential. Meng et al. [11] developed a pore water pressure model for irregular loading conditions based on the existing model designed for constant amplitude loading, which necessitated a limited number of parameters. Fu et al. [12] introduced an incremental model of pore water pressure under irregular loading derived from in-situ liquefaction tests that utilized acceleration, burial depth, and sand type as fundamental indices. Wang et al. [13] proposed a revised incremental pore water pressure model that effectively predicts pore water pressure generation under dynamic loading, as corroborated by the results of shaking tests. The aforementioned pore water pressure models facilitate the prediction of pore water pressure generation under irregular loading conditions. However, these models do not adequately account for the effects of ground motion type and the influence of stress history on pore water pressure.

In this study, a novel energy pore water pressure model is developed based on the relationship between pore water pressure and the Arias intensity (I_a) [14] of ground motion using cyclic triaxial numerical tests. A total of 1000 numerical tests were performed utilizing 100 different ground waves, each with 10 varying peak ground accelerations. Furthermore, the mean period was incorporated into the model to account for the frequency effects on pore water pressure. Consequently, the model is capable of capturing the influence of ground motion patterns as well as the effects of stress history. Based on this model, a liquefaction evaluation method incorporating the influence of permeability and stress reduction coefficients has been proposed specifically for Fujian standard sand by theoretical analysis. Finally, centrifuge shaking table tests and numerical simulations are conducted to validate the reliability and precision of the proposed evaluation method.

2. Pore Water Pressure Modelling

2.1. Proposal of Pore Water Pressure Model

During the course of the study, the authors observed that the shape of the pore water pressure curve obtained from both physical and numerical cyclic triaxial tests under dynamic loading closely resembles the shape of the I_a curve of ground motion, as illustrated in Figure 1. In the cyclic triaxial tests, sand samples with varying relative densities were consolidated using different levels of consolidation pressure (σ_0) and consolidation pressure ratios (k_0). For specific sand parameters and test results, please refer to reference [15].

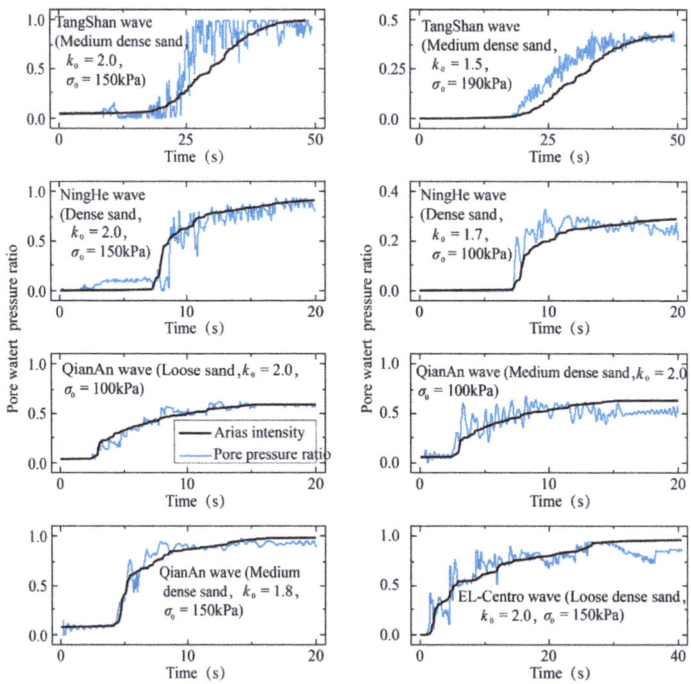

Figure 1. Relationship between excess pore pressure ratio and Arias intensity (I_a).

Consequently, this paper proposes a pore water pressure model that is based on the I_a curve of ground motion. In this model, the shape of pore water pressure time history is determined by the shape of the I_a curve. The peak value of pore water pressure is determined by the I_a input into the soil unit. The model incorporates the stress history induced by dynamic loading as well as the effects of the ground motion pattern. A product form is utilized to represent the primary parameters that affect the generation of pore water pressure in saturated sand, including earthquake energy, mean period, and the material properties of saturated sand. It is important to note that the interactions between these parameters are not considered in this model. The specific formulation of the model is presented in Equation (1):

$$Ud(t) = \alpha F(Ia, \sigma_0, k_0) Ia(t)/Ia, 1.0g \tag{1}$$

$$Ia = \frac{\pi}{2g}\int_0^{Td} [a(t)]^2 dt \tag{2}$$

where α is a parameter related to sand, $I_{a,1.0g}$ is the Arias intensity of ground motion when the amplitude is 1.0 g, g is the acceleration of gravity, T_d is the duration of the ground motion, and a is the acceleration of the ground motion.

2.2. Numerical Model of Cyclic Triaxial Test

The stress state of the soil unit prior to and during the earthquake is illustrated in Figure 2. Before the earthquake, the soil unit exists in a state of consolidation, influenced by both vertical and horizontal stresses, which represent the initial stress conditions. During the earthquake, the soil unit experiences additional horizontal shear stresses alongside the existing consolidation pressures.

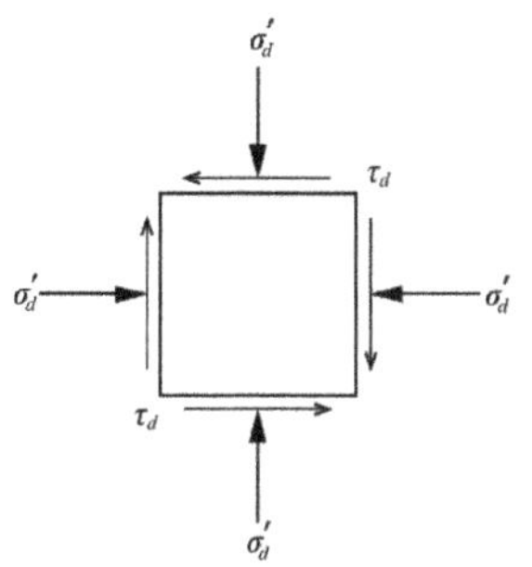

Figure 2. Stress conditions of soil element under earthquake.

To date, the predominant method for investigating the stress state of soil units subjected to dynamic loading is the soil unit test, for example, the unidirectional and bidirectional undrained cyclic triaxial tests. These tests utilize the shear stress on the 45° surface of the soil unit to replicate the dynamic shear stress experienced during an earthquake (Figure 3). The shear stress states for both the unidirectional and bidirectional dynamic triaxial tests are illustrated in Figure 3. Following the completion of soil unit consolidation, a dynamic load of $\sigma_d/2$ is applied in both the x and y directions, resulting in the generation of shear stress of $\sigma_d/2$ on the 45° surface of the soil unit (Figure 3). In the unidirectional cyclic triaxial test, the dynamic load $\sigma_d/2$ is applied solely in the y direction, leading to the generation of both shear stress and spherical stress of $\sigma_d/2$ on the 45° surface of the soil unit, as depicted in Figure 3. It is important to note that the pore water pressure within the soil unit is influenced only by the consolidation pressure and dynamic shear stress; the spherical stress does not contribute to the generation of pore water pressure. However, it does impact the fluctuations of pore water pressure under dynamic loading conditions. Consequently, the bidirectional dynamic triaxial test is more effective in capturing the stress state of the soil unit under dynamic loading (Figure 4). This paper applied the bidirectional cyclic triaxial numerical test to investigate the stress state of the soil unit and established a pore water pressure model based on numerical simulations.

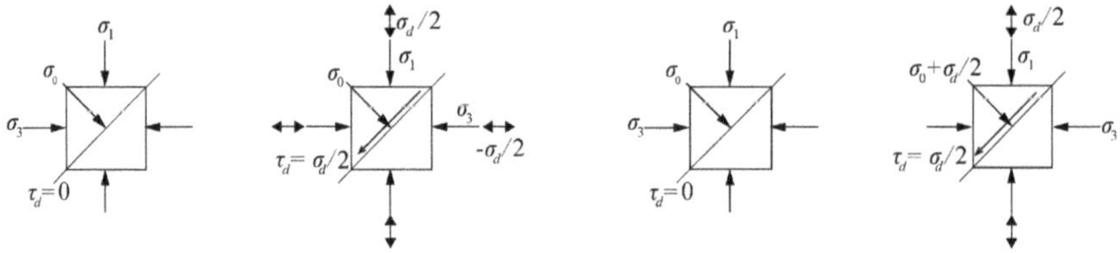

Figure 3. Stress conditions of soil element in two-directional cyclic triaxial test and unidirectional cyclic triaxial test.

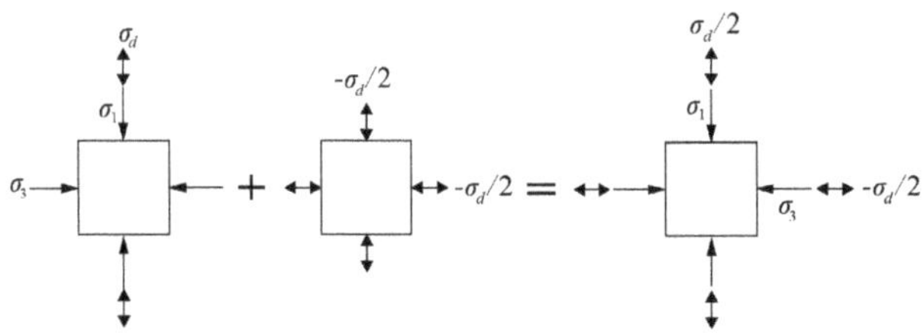

Figure 4. Relationship of soil element stress conditions in uni-directional and two-directional cyclic triaxial test.

In this paper, the saturated Fujian standard sand with relative densities (D_r) of 60% (medium density) and 90% (dense) was utilized to create soil samples for bidirectional cyclic triaxial testing. The soil unit tests were simulated using the open-source software OpenSees 2.5.0, which was developed by UC Berkeley. The software is a state-of-the-art open-source finite element software specifically designed for modeling geotechnical systems and computational simulation in earthquake engineering [16]. In the numerical model, the SSPquadUP element [17] was employed to represent the soil unit. This element is based on the porous media theory, fluid–solid coupling principles, and Darcy's law. The soil element functions as a fully coupled fluid–solid element. The fluid–solid coupling methodology for the solid and liquid phases is based on the u-p formulation established by Zienkiewicz and Shiomi [18], which is derived from Biot's porous media theory [19], as illustrated in Equations (3) and (4). This soil element effectively simulates the dynamic interactions between the liquid and solid phases under dynamic loading conditions.

$$\int_{\Omega} B^T \sigma' d\Omega + M\ddot{u} + Qp = f^s \tag{3}$$

$$Q^T \dot{u} + S\dot{p} + H\dot{p} = f^p \tag{4}$$

where B is the strain-displacement matrix, σ' is the effective stress tensor, M is the mass matrix, u is the displacement vector, Q is the discrete gradient operator of soil–water coupling, p is the pore water pressure vector, S is the compression matrix, H is the permeability coefficient matrix, and f^s and f^p represent the given boundary of volume force in the water–soil mixture and liquid phase, respectively.

The multi-yield surface plastic constitutive model, referred to as the PDMY02 material, was employed to simulate the soil material. This model was originally proposed by Elgamal et al. [20,21] for liquefaction-induced large deformation and cyclic flow. The yield function associated with the soil constitutive model is presented in Equation (5). Distinct material parameters have been selected for Fujian standard sand, corresponding to relative densities of 60% and 90%, in accordance with the references provided [22,23], as presented in Table 1.

$$f = \frac{3}{2}[s - (p' + p'_0)\alpha] : [s - (p' + p'_0)\alpha] - M^2(p' + p'_0)^2 = 0 \tag{5}$$

where $s = \sigma' - p'\delta$ is the deviatoric stress tensor, δ is the second-order identity tensor, p' is the mean effective shear stress, $p_{0'}$ is a small normal number, α is the second-order deviatoric tensor, which can define the center of the yield surface in the deviatoric stress space. M is the size of the yield surface, and ":" is the product of double compression tensors.

Table 1. Parameters of Fujian standard sand.

D_r	Density (kg·m^{-3})	Reference Pressure (kPa)	Shear Modulus at Pr (MPa)	Bulk Modulus at Pr (MPa)	Friction Angle (°)	Phase Transformation Angle (°)	c_1	c_3	d_1	d_3
60%	19.38	80	6.5×10^4	1.6×10^5	31	31	0.087	0.18	0.0	0.0
90%	19.98	800.087	1.1×10^5	2.4×10^5	33	26	0.28	0.05	0.1	0.05

The bidirectional cyclic triaxial numerical test model is illustrated in Figure 5. The numerical testing procedure was conducted in three steps:

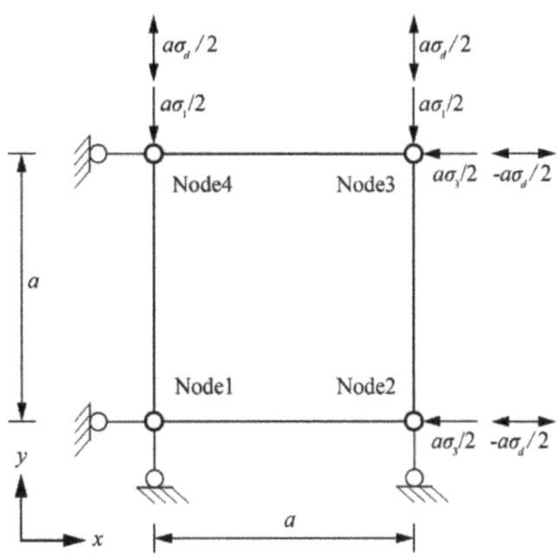

Figure 5. Numerical model of two-directional cyclic triaxial test.

Step 1: Soil element consolidation. The boundary conditions for the numerical model were established in accordance with the physical test. The degrees of freedom for nodes 1 and 4 in the x-direction, as well as for nodes 1 and 2 in the y-direction, are constrained. Additionally, the degrees of freedom for nodes 2 and 3 in the x-direction have been coupled using a penalty function, ensuring that the displacements of nodes 2 and 3 in the x-direction are equivalent. A similar approach was applied to the degrees of freedom for nodes 3 and 4 in the y-direction. In this way, the concentrated load exerted on the soil nodes will be transmitted to the soil element as a uniform load. The degree of freedom concerning pore water pressure at all nodes remains unconstrained, and consolidation stress is applied to the model to simulate the drained consolidation of the soil unit. The consolidation pressure, denoted as $-a\sigma_3/2$, is applied to nodes 2 and 3 in the x-direction. Concurrently, the consolidation pressure, $-a\sigma_1/2$, is applied to nodes 3 and 4 in the y-direction. A static analysis is conducted to complete the consolidation process.

Step 2: Apply dynamic pressure. Prior to the application of the pressure, the degrees of freedom for pore water pressure at all fixed nodes are constrained to simulate an undrained condition. A dynamic pressure with an amplitude of $-a\sigma_d/2$ is applied to nodes 2 and 3 in the x-direction. Concurrently, a dynamic pressure with an amplitude of $a\sigma_d/2$ is applied to nodes 3 and 4 in the y-direction. A dynamic analysis is subsequently conducted. During this analysis, the amplitude of the dynamic pressure is incrementally increased until the saturated sand approaches a state of soil liquefaction.

Step 3: Input of different dynamic pressures. The seismic waves undergo an initial baseline correction utilizing the EQSignal v1.2.0 software, followed by a filtering process aimed at enhancing both processing efficiency and accuracy. Nodal pressure was applied in the form of ground waves with varying amplitudes to node 2 and node 3 in the x-direction, as well as to node 3 and node 4 in the y-direction, in order to simulate the application of pressure. The subsequent step was reiterated until the cyclic triaxial test was completed for all 100 ground waves.

2.3. The Relationship Between Pore Water Pressure and Arias Intensity

This study involved the selection of 100 distinct ground motion records derived from 10 seismic events, which include the Tabas Earthquake in 1978, the San Fernando Earthquake in 1971, the Coalinga Earthquake in 1983, Northridge Earthquake in 1994, Loma Prieta Earthquake in 1989, Kobe Earthquake in 1995, Kern County Earthquake in 1952, Imperial Valley Earthquake in 1979, El Centro Earthquake in 1940, and the Chi-Chi Earthquake in 1999. The selected 100 ground motions were utilized to conduct a bidirectional cyclic triaxial numerical test aimed at elucidating the relationship between pore water pressure and Arias intensity. The consolidation pressure was set at 100 kPa, and the consolidation ratio was maintained at 1.0 during the numerical testing. The duration of the ground motions ranged from 18 s to 100 s. The predominant frequencies varied from 0.06 Hz to 1.24 Hz, while the mean period ranged from 0.2 s to 1.3 s. The Arias intensity of the ground motions ranged from 4 m/s to 117 m/s, with an amplitude of 1.0 g.

Figure 6 shows the relationship between the cyclic stress ratio (*CSR*) and pore water pressure in saturated Fujian standard soil subjected to earthquake ground motions. The relationship between the cyclic stress ratio (*CSR*) and pore water pressure was analyzed using data fitting software called CurveFitter, which possesses the capability to determine the optimal curve for a specified function along with its associated parameters. In this study, five potential functions—linear, quadratic, cubic, parabolic, and power functions—were evaluated for curve fitting. The function exhibiting the highest correlation coefficient (R^2) was chosen to accurately represent the relationship between the *CSR* and pore water pressure. The curves presented in the subsequent text have all been generated utilizing this methodology; therefore, further elaboration is unnecessary. A power function relationship exists between the cyclic stress ratio and pore water pressure in saturated sand. The R^2 for medium-dense sand (D_r = 60%) is 0.627, accompanied by a root mean square error of 16.7. In contrast, the R^2 for medium-dense sand (D_r = 90%) is 0.641, with a root mean square error of 20.4. While a power function relationship between cyclic stress ratio and pore water pressure in saturated sand is observed, it is important to note that this relationship is not particularly strong.

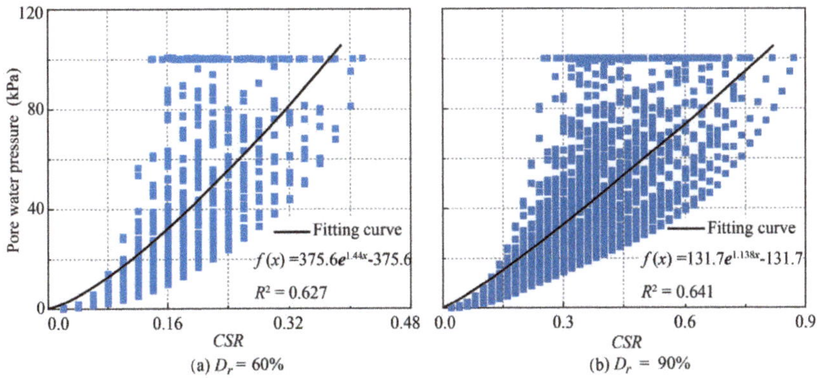

Figure 6. Relationship between pore water pressure and cyclic stress ratio (*CSR*) in the two-directional cyclic triaxial numerical test for (**a**) $D_r = 60\%$ and (**b**) $D_r = 90\%$.

The Arias intensity and mean period of ground motions significantly influence the generation and accumulation of pore water pressure in saturated sand during soil unit tests. An increase in Arias intensity correlates with a higher energy level of ground motion, which subsequently reduces the cyclic stress ratio necessary for soil liquefaction. Conversely, a shorter mean period results in a greater number of shear cycles during the shaking duration, thereby also decreasing the cyclic stress ratio required for soil liquefaction. Consequently, it is essential to consider both the Arias intensity and mean period of ground motions when modifying the relationship between the cyclic stress ratio (*CSR*) and pore water pressure in saturated sand. This paper introduces a new parameter, termed the modified cyclic stress ratio (M_{csr}), to account for the effects of Arias intensity ($I_{a,1.0g}$) and mean period (T_m) on the cyclic stress ratio (*CSR*). The proposed relationship is expressed as (M_{csr}). $M_{csr} = (CSR)^a \times (I_{a,1.0g})^b/(T_m)^c$, where *a*, *b*, and c are parameters that relate to the material properties of the saturated sand.

A power function relationship has been established between the M_{csr} and the pore water pressure of saturated Fujian standard sand with different relative densities. For medium-dense Fujian standard sand ($D_r = 60\%$), when $a = 1.866$, $b = 0.78$, and $c = 0.63$, yield a significant power function relationship, with the highest correlation coefficient of 0.990. Similarly, for medium-dense Fujian standard sand, a strong power function relationship is observed between the M_{csr} and the pore water pressure, characterized by $a = 1.440$, $b = 0.64$, and $c = 0.51$, resulting in a correlation coefficient of 0.986, which indicates a very strong correlation, as illustrated in Figure 7. A comparable power function relationship has also been identified in dense Fujian standard sand ($D_r = 90\%$). The equations representing these relationships are as follows:

(1) For the $D_r = 60\%$:

$$T_m = \int_0^{Td} [af(t)]^2 \times 1/f(t)\mathrm{d}t \tag{6}$$

$$M_{csr} = \frac{CSR^{1.866} \times I_{a,1.0g}^{0.78}}{T_m^{0.63}} \tag{7}$$

$$U_d(t) = 31.06e^{1.44\times \mathrm{Mcsr}} - 31 \tag{8}$$

(2) For the D_r = 90%:

$$M_{csr} = \frac{CSR^{1.440} \times I_{a,1.0g}^{0.64}}{T_m^{0.51}} \quad (9)$$

$$U_d(t) = 31.93e^{0.46 \times \text{Mcsr}} - 32 \quad (10)$$

where a_f is the amplitude with a frequency of f, f is the frequency corresponding to the acceleration a_f, and e is the Euler number.

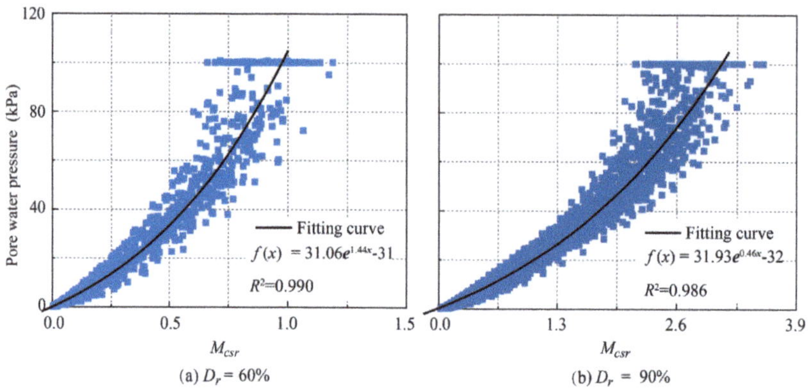

Figure 7. Relationship between pore water pressure and the modified cyclic stress ratio (M_{csr}) in the two-directional cyclic triaxial numerical test for (**a**) D_r = 60% and (**b**) D_r = 90%.

2.4. Criteria for Liquefaction

Figure 7 shows the relationship between the M_{csr} and pore water pressure when the pore water pressure ratio ranges from 0.8 to 1.0. This finding suggests that Equations (8) and (10) are inadequate for accurately calculating pore water pressure as the sand approaches a liquefaction state. Consequently, it is essential to develop a new equation to ascertain whether the saturated sand has reached a liquefaction state.

The relationship between the *CSR* required for liquefaction and the Arias intensity (with an amplitude of 1.0 g) is illustrated in Figure 8 for saturated Fujian standard sand with different relative densities. It is observed that the *CSR* required for the liquefaction of saturated Fujian standard sand decreases as the Arias intensity increases, indicating an inverse relationship between these two parameters. Although a correlation exists between the *CSR* and the Arias intensity—evidenced by R^2 values of 0.622 for a D_r of 60% and 0.632 for a D_r of 90%—there remains a significant margin of error, suggesting considerable variability in the data. Furthermore, the cyclic shear stress induces cyclic shear strain in the saturated sand, which in turn generates pore water pressure. The number of cyclic shears per time unit is influenced by the mean period, which also significantly affects the liquefaction behavior of saturated sand. Consequently, the mean period should be considered when studying the liquefaction criteria for saturated sands.

To establish more accurate liquefaction criteria, the mean period of ground motion is used to modify the Arias intensity. The modified Arias intensity, denoted as MI_a, is defined by the equation $MI_a = (I_a,1.0\text{g})^d/(T_m)^e$, where d and e are parameters associated with the material properties of saturated sand. The relationships between MI_a and the *CSR* required for liquefaction are shown in Figure 9 for saturated Fujian standard sand with different relative densities. The relationship between *CSR* and MI_a was analyzed and fitted using MATLAB as the saturated sand approached the liquefaction threshold. For medium-dense Fujian standard sand (D_r = 60%), the correlation coefficient is the largest (R^2 = 0.937), with d = 0.90 and e = 0.70. For dense Fujian standard sand (D_r = 90%), the correlation coefficient between the two variables is maximized at R^2 = 0.935, with d = 0.91 and e = 0.71, as illustrated in Figure 9. The relationships between the MI_a and the CSR required for liquefaction are represented by the following equations:

(1) For the D_r = 60%:

$$MI_a = \frac{I_{a,1.0g}^{0.9}}{T_m^{0.7}} \tag{11}$$

$$y = MI_a = 1.208x^{-1.922} = 1.208 \times CSR^{-1.922} \tag{12}$$

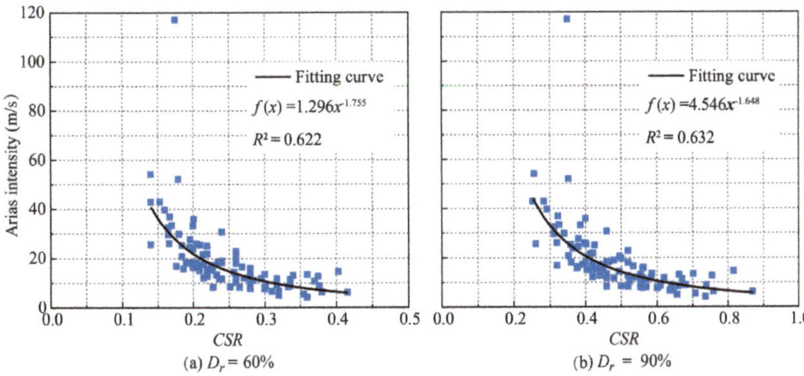

Figure 8. Relationship between I_a and CSR in soil element when initial liquefaction triggered in saturated soil for (**a**) D_r = 60% and (**b**) D_r = 90%.

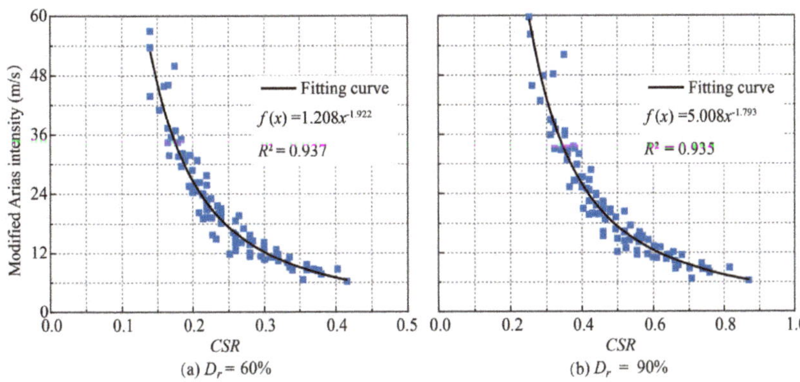

Figure 9. Relationship between MI_a and CSR in soil element when initial liquefaction triggered in saturated soil for (**a**) D_r = 60% and (**b**) D_r = 90%.

According to Equation (12), the criterion of soil liquefaction is L_{iqu}:

$$L_{iqu} = 1/1.208 \times (CSR^{1.922})MI_a = 1 \tag{13}$$

(2) For the D_r = 90%:

$$MI_a = \frac{I_{a,1.0g}^{0.91}}{T_m^{0.71}} \tag{14}$$

$$y = MI_a = 5.008x^{-1.793} = 5.008 \times CSR^{-1.793} \tag{15}$$

According to Equation (15), the criterion of soil liquefaction is L_{iqu}:

$$L_{iqu} = 1/5.008 \times (CSR^{1.793})MI_a = 1 \tag{16}$$

In Equations (13) and (16), the expressions $(I_{a,1.0g})^{0.9} \times (CSR)^{1.922}$ and $(I_{a,1.0g})^{0.91} \times (CSR)^{1.793}$ represent the Arias intensity of the ground motion when the amplitude of the input ground motion corresponds to the value of the *CSR*. The theoretical value is expected to be $(I_{a,1.0g})^{0.9} \times (CSR)^{2.0}$. A discrepancy exists between the numerical results and the theoretical value, with the error being less than 10.04%. Consequently, the parameter L_{iqu} in Equations (13) and (16) aligns with the principles of Arias intensity theory, suggesting that the application of these equations for evaluating soil liquefaction is both reasonable and accurate. When the values of MI_a and *CSR* conform to Equations (13) and (16), it indicates that the saturated Fujian standard sand has attained a liquefaction state. Therefore, Equations (13) and (16) serve as the liquefaction criteria for saturated Fujian standard sand with relative densities of 60% and 90%, respectively.

The pore water pressure of saturated sand, along with the Arias intensity, mean period, and the *CSR* of the ground motion, adheres to Equations (8) and (10) when the pore water pressure ratio is less than 0.8, as shown in Figure 9. Conversely, when the pore water pressure ratio exceeds 0.8, employing Equations (8) and (10) to calculate the pore water pressure value results in significant errors. Consequently, the pore water pressure model has been divided into two equations, utilizing a pore water pressure ratio of 0.8 as the demarcation point.

(1) For the D_r = 60%:

$$U_d(t) = \begin{cases} (31e^{1.44\times Mcsr} - 31) \times I_a(t)/I_{a,\max}, & (U_d \le 0.8) \\ L_{iqu} \times I_a(t)/I_{a,\max}, & (U_d > 0.8) \end{cases} \tag{17}$$

(2) For the D_r = 90%:

$$U_d(t) = \begin{cases} (32e^{0.46\times Mcsr} - 32) \times I_a(t)/I_{a,\max}, & (U_d \le 0.8) \\ L_{iqu} \times I_a(t)/I_{a,\max}, & (U_d > 0.8) \end{cases} \tag{18}$$

2.5. *Effect of Consolidation Pressure and Consolidation Ratio*

The consolidation pressure and the consolidation ratio significantly affect the pore water pressure during the cyclic triaxial test of saturated sand. Consequently, it is essential to account for both the consolidation pressure and the consolidation ratio when developing a model for pore water pressure in the context of the cyclic triaxial test.

This study conducted two series of cyclic triaxial tests to investigate the effects of consolidation pressure and consolidation pressure ratio. In the first series of tests, the consolidation ratio (k_0) was maintained at 1.0 while the consolidation pressure (σ_0) was incrementally increased from 50 kPa to 200 kPa, with an interval of 10 kPa. In the second series, the horizontal consolidation pressure (σ_3) was fixed at 100 kPa, and the consolidation ratio was varied from 1.0 to 2.0, with an increment of 0.1. Consequently, the vertical consolidation pressure (σ_1) was raised from 100 kPa to 200 kPa, also with an interval of 10 kPa.

2.5.1. Effect of Consolidation Pressure on the Pore Water Pressure Model

The relationship between the minimum modified CRS (M_{csr})and the consolidation pressure necessary for sand liquefaction under various ground motions are presented in Figure 10 for saturated sand with differing densities. The modified CRS denoted as M_{csr}, is defined by the equation M_{csr} = $CRS \times (I_a,1.0g)^f/(T_m)^g$, where f and g are parameters associated with the material properties of saturated sand. It is observed that the M_{csr} required for the liquefaction of saturated sand increases linearly with the consolidation pressure, indicating a strong linear correlation between these two variables. The correlation coefficients are 0.956 and 0.962 for sand with relative densities of 60% and 90%, respectively. The relationships are expressed by the following equations:

(1) For the D_r = 60%:

$$CSR \times I_{a,1.0g}^{0.41}/T_m^{0.35} = 0.831 \times \sigma_0/100 + 0.107 \quad (19)$$

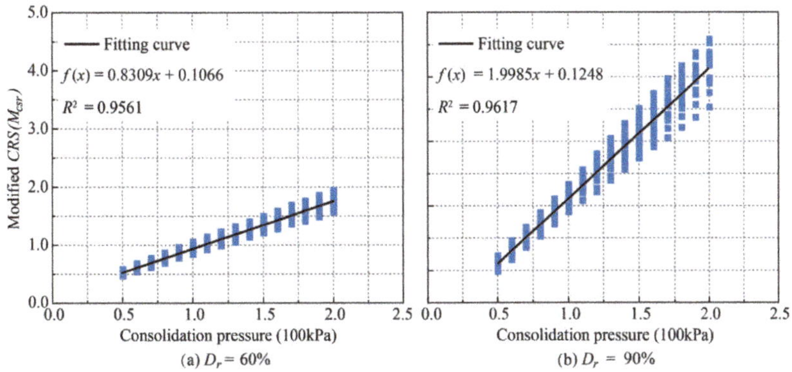

Figure 10. Relationship between the modified CRS (M_{csr}) and consolidation pressure for (**a**) D_r = 60% and (**b**) D_r = 90%.

The ratio of CSR at 100 kPa to the initial effective stress (σ_0) serves as a parameter representing the influence of consolidation pressure on the pore water pressure model, denoted as $k_{\sigma o}$.

The value of $k_{\sigma o}$ can be derived from Equation (19).

$$k_{\sigma_0} = CSR_{100}/CSR_{\sigma_0} = \frac{0.831 \times \sigma_{100}/100 + 0.107}{0.831 \times \sigma_0/100 + 0.107} \quad (20)$$

(2) For the D_r = 90%:

$$CSR \times I_{a,1.0g}^{0.47}/T_m^{0.32} = 1.999 \times \sigma_0/100 + 1.248 \quad (21)$$

The $k_{\sigma o}$ can be derived from Equation (21):

$$k_{\sigma_0} = CSR_{100}/CSR_{\sigma_0} = \frac{1.999 \times \sigma_{100}/100 + 0.125}{1.999 \times \sigma_0/100 + 0.125} \quad (22)$$

2.5.2. Effect of Consolidation Ratio on Pore Water Pressure Model

The relationship between the minimum M_{csr} and the consolidation pressure ratio under different ground motions is presented in Figure 11 for saturated sand with differing densities. It is observed that the M_{csr} required for the liquefaction of saturated sand increases linearly with the consolidation pressure ratio, indicating a strong linear correlation between these two variables, as depicted in Figure 11. In comparison to sand with a relative density of 60%, the M_{csr} exhibited greater variability with increasing consolidation pressure ratio for sand with a relative density of 90%. The correlation coefficients for the sands with relative densities of 60% and 90% are 0.908 and 0.885, respectively. The relationships are expressed by the following equations:

(1) For the D_r = 60%:

$$CSR \times I_{a,1.0g}^{0.37} / T_m^{0.15} = 2.053 \times k_0 - 1.357 \quad (23)$$

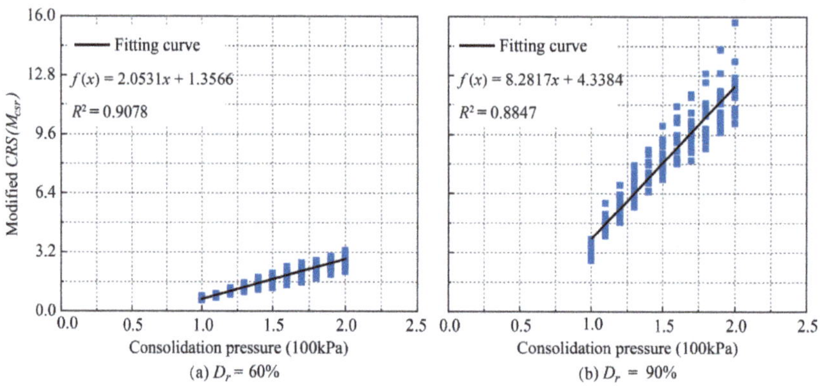

Figure 11. Relationship between the modified *CRS* (M_{csr}) and consolidation pressure ratio for (**a**) D_r = 60% and (**b**) D_r = 90%.

The ratio of *CSR* at 1.0 to k_0 serves as a parameter for assessing the impact of consolidation pressure on the pore water pressure model (k_{k0}). The value of k_{k0} can be derived from Equation (23).

$$k_{\sigma_0} = CSR_{100}/CSR_{k_0} = \frac{8.282 \times k_{1.0} - 4.338}{8.282 \times k_0 - 4.338} \quad (24)$$

(2) For the D_r = 90%:

$$CSR \times I_{a,1.0g}^{0.63} / T_m^{0.32} = 8.282 \times k_0 - 4.338 \quad (25)$$

The k_{k0} can be derived from Equation (25):

$$k_{k_0} = CSR_{1.0}/CSR_{k_0} = \frac{8.282 \times k_{1.0} - 4.338}{8.282 \times k_0 - 4.338} \quad (26)$$

The *CSR* can be determined using Equation (27), which is derived from Equations (17), (19), (21), and (23) at different consolidation pressures and consolidation ratios for medium-dense sand. Similarly, the *CSR* can be calculated using the same methodology.

$$CSR_{100,1.0} = CSR \times k_{\sigma_0} \times k_{k_0} \quad (27)$$

Substituting Equation (27) into Equations (7), (9), (13), and (16) yields Equation (28), which incorporates the effects of consolidation pressure and consolidation ratio on the pore water pressure model, as well as the liquefaction criterion under different ground motions. Similarly, the pore water pressure model can be calculated using the same methodology.

(1) For the D_r = 60%:

$$U_d(t) = \begin{cases} \left[31.93e^{0.46\times\frac{CSR^{1.793}\times I_{a,1.0g}^{0.64}}{T_m^{0.51}}} - 31.93\right]\frac{I_a(t)}{I_{a,\max}}(L_{iqu} \leq 0.8) \\ \frac{L_{iqu}I_a(t)}{I_{a,\max}}(0.8 < L_{iqu} \leq 1.0) \\ \frac{I_a(t)}{I_{a,\max}}(L_{iqu} > 1.0) \end{cases} \quad (28)$$

(2) For the D_r = 90%:

$$U_d(t) = \begin{cases} \left[31e^{1.44\times\frac{CSR^{1.866}\times I_{a,1.0g}^{0.78}}{T_m^{0.63}}} - 31\right]\frac{I_a(t)}{I_{a,\max}}(L_{iqu} \leq 0.8) \\ \frac{L_{iqu}I_a(t)}{I_{a,\max}}(0.8 < L_{iqu} \leq 1.0) \\ \frac{I_a(t)}{I_{a,\max}}(L_{iqu} > 1.0) \end{cases} \quad (29)$$

2.5.3. Validation of the Pore Water Pressure Model

Figure 12 presents a comparison between the values obtained from the pore water pressure model introduced in this paper and those derived from the pore water pressure model proposed by Sun et al. [15], which is based on cyclic triaxial tests conducted on saturated Fujian standard sand. Additionally, it includes values calculated using the pore water pressure model developed by Sun et al. [15].

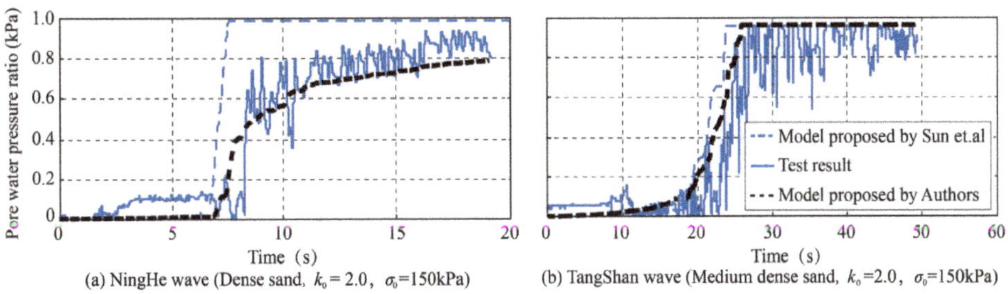

Figure 12. Comparison of computed and experimental pore pressure time histories for (**a**) NingHe wave and (**b**) TangShan wave [15].

The pore water pressure model proposed in this study demonstrates a high degree of accuracy in predicting the pore water pressure of saturated dense and medium-dense Fujian standard sand. For saturated dense Fujian standard sand, the values generated by the proposed model are slightly lower than the experimental results; however, there is a strong overall correlation between the model's predictions and the experimental data. Additionally, the model's predictions regarding the initial liquefaction time align closely with the results obtained from physical testing. In the case of medium-dense Fujian standard sand, the values produced by the proposed model exhibit a commendable agreement with the experimental findings reported by Sun Rui, as well as with the model developed by Sun Rui [23]. Consequently, the pore water pressure model proposed in this paper is capable of accurately calculating the pore water pressure of saturated Fujian standard sand and is applicable for predicting pore water pressure under earthquake ground motions.

3. Liquefaction Potential Assessment Methodology

The soil units subjected to earthquake ground motions in both the free field and the undrained cyclic triaxial test are in a state of stress that is essentially identical. It is important to acknowledge that the soil in the free field is also influenced by its permeability, and the *CSR* remains undetermined at this stage. Consequently, it is essential to evaluate the impact of soil permeability on pore water pressure to accurately assess the potential for soil liquefaction and to derive the *CSR*.

3.1. Cyclic Stress Ratio

The cyclic stress ratio (*CSR*) is defined as $CSR = \tau/\sigma_{v0}$, where τ is the dynamic shear stress and σ_{v0} is the vertical effective stress in the soil. The vertical effective stress at varying depths within the free field can be determined by utilizing the effective bulk weight of the soil in conjunction with the depth. Consequently, it is only necessary to ascertain the dynamic shear stress within the soil to calculate the *CSR*.

In this paper, the dynamic shear stress [24] at varying depths is calculated utilizing the shear stress reduction function in conjunction with the peak ground acceleration of the site (Equations (28) and (29)). To date, numerous models of the shear stress reduction function [25–28] have been proposed for different soil types. However, a limited number of shear stress reduction functions specifically designed for liquefaction sites exist. Based on the shear stress reduction function proposed by Golesorkhi et al. [24], Idriss et al. [25] developed a shear stress reduction function that is tailored to the unique characteristics of liquefied soil (Figure 13). In their study, Idriss et al. [25] emphasized that the shear stress reduction function is also influenced by both the depth of the soil and the magnitude of the earthquake. The relationship between the earthquake magnitude, depth, and the shear stress reduction function is articulated through Equations (30) and (31). The shear stress reduction function effectively accounts for the impacts of soil liquefaction, burial depth, and earthquake magnitude, thereby yielding more accurate predictive results.

$$\tau = \tau_{\mathrm{max,r}} \times r_d \tag{30}$$

$$\tau_{\mathrm{max,r}} = \left(\frac{\gamma_0 z}{g}\right) \times a_{\mathrm{max}} = \left(\frac{a_{\mathrm{max}}}{g}\right) \times \sigma_{v0} \tag{31}$$

where $\tau_{max,r}$ is the maximum acceleration at the surface of the rigid body, r_d is the shear stress reduction function, z is the depth of the soil layer, γ_0 is the effective bulk weight of the soil, and a_{max} is the maximum acceleration at the surface.

$$r_d = e^{\alpha(z)+\beta(z)\times M} \tag{32}$$

$$\alpha(z) = -1.012 - 1.126\sin\left(\frac{z}{11.73} + 5.133\right) \tag{33}$$

$$\beta(z) = 0.106 + 0.118\sin\left(\frac{z}{11.28} + 5.142\right) \tag{34}$$

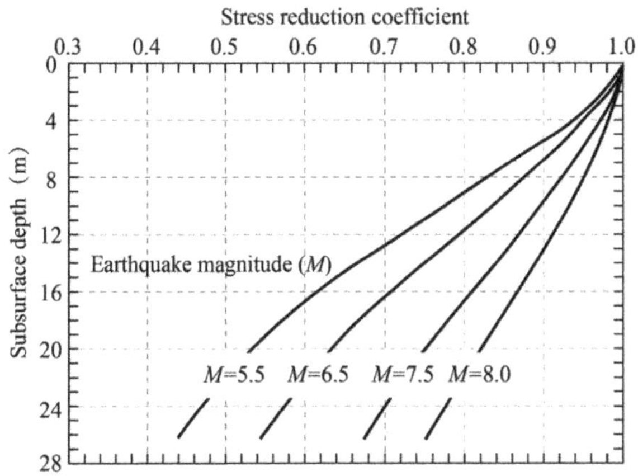

Figure 13. Curve of shear stress reduction function [23].

In the equation, M is the magnitude of the earthquake, and the parameter within the sine term is expressed in radians. The maximum depth of the soil layer to which Equation (32) is applicable is 34 m. It is important to note that the uncertainty associated with the shear stress reduction function increases with depth; therefore, the maximum recommended depth for the shear stress reduction function proposed by Idriss et al. [25] is 20 m. The assessment of liquefaction in deep soils entails specific conditions, necessitating further detailed studies to appropriately adapt the shear stress reduction function [26].

The dynamic shear stress at the site is determined based on the maximum acceleration amplitude observed at the surface (Equation (32)). However, it is important to note that, due to the phenomenon of soil liquefaction, the peak ground acceleration at the surface of a liquefiable site is significantly lower than that at the surface of a non-liquefiable site when subjected to dynamic loading. Consequently, the peak ground acceleration at the surface is not an appropriate metric for evaluating the liquefaction potential of a liquefiable site. If the site conditions are known, this paper selects the total stress method (without considering the effect of pore water pressure [2]) for site response analysis to determine the peak ground acceleration at the surface of the liquefied site. If the site conditions are unknown, this paper selects the empirical relationship proposed by Idriss [25], which is applicable to certain soft soils, to calculate the peak ground acceleration at the surface of the liquefied site. The amplitude of the ground motion input in the model is adjusted by applying the stress reduction coefficient to account for its effects.

3.2. *Effect of Free-Field Permeability Coefficients*

In the undrained triaxial test, the soil unit was in an undrained state; therefore, the influence of the permeability coefficient on the pore water pressure did not need to be considered. At the actual site, however, the soil unit is in a drained state, and the influence of the permeability coefficient must be considered when calculating the pore water pressure. Many researchers have found that the variable permeability coefficient can effectively improve the accuracy of numerical simulation compared to the fixed permeability coefficient [29–31]. For example, Shahir et al. [29] established the relationship between the permeability coefficient (k) and the excess pore water pressure ratio (r_u) as shown in (Equation (35)). The accuracy of the numerical simulation for soil liquefaction was effectively improved by considering Equation (33). Therefore, the relationship between k and r_u is widely used [32,33] in numerical simulations of soil liquefaction.

Therefore, in this study, the variable permeability coefficient model proposed by Shahir et al. [29] is employed to consider the influence of the permeability coefficient on the pore water pressure model (Figure 14). Furthermore, the dissipation stage of pore water pressure is not included in the assessment of liquefaction potential. This omission is justified by the observation that the liquefaction of soil or the pore water pressure reaches its maximum value prior to the dissipation stage, which does not impact the peak value of the pore water pressure. In the cumulation stage, the pore water pressure model is integrated into the permeability coefficient equation to formulate a revised pore water pressure model that incorporates the effects of the permeability coefficient. Subsequently, during the liquefaction stage, the reduction in pore water pressure resulting from the permeability coefficient is directly deducted from the established pore water pressure model. Consequently, this paper establishes a method for assessing liquefaction potential based on the energy pore water pressure model, focusing solely on the cumulative stage of pore water pressure and the liquefaction stage. The model parameters were selected in accordance with reference [31], with values of $\alpha = 20$ and $\beta_1 = 2$. Additionally, the initial permeability coefficients for saturated Fujian standard sand were set at 0.001 and 0.0001 for relative densities of 60% and 90%, respectively.

$$\frac{k}{k_0} = \begin{cases} 1 + (\alpha - 1)r_u^{\beta_1}, & r_u < 1.0, \text{ cumulation phase} \\ \alpha, & r_u < 1.0, \text{ liquefaction phase} \\ 1 + (\alpha - 1)r_u^{\beta_2}, & r_u < 1.0, \text{ dissipation phase} \end{cases} \tag{35}$$

where k_0 is the initial permeability coefficient, and α, β_1, β_2 are the model parameters.

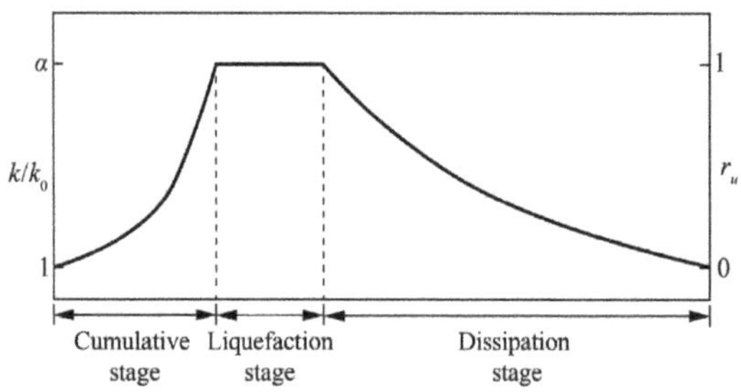

Figure 14. Schematic of permeability coefficient function.

3.3. Validation and Application of Liquefaction Risk Assessment Methodology

The centrifuge shaking table test, along with numerical simulations, are employed to validate the liquefaction potential assessment method developed based on the energy pore water pressure model. The far-field distance from structures was applied to validate the method in the centrifuge shaking test [34]. In the centrifuge shaking table test, the upper soil layer consists of saturated Fujian standard sand with a relative density of 60% and a thickness of 13 m, while the lower soil layer comprises saturated Fujian standard sand with a relative density of 90% and a thickness of 7 m. For more detailed information, please refer to reference [34]. The OpenSees software was utilized to construct a free-field model of the soil layer corresponding to the centrifuge shaking table test prototype, as illustrated in Figure 15. The soil unit and parameters used in the numerical simulation are consistent with those employed in the two-way dynamic triaxial numerical tests.

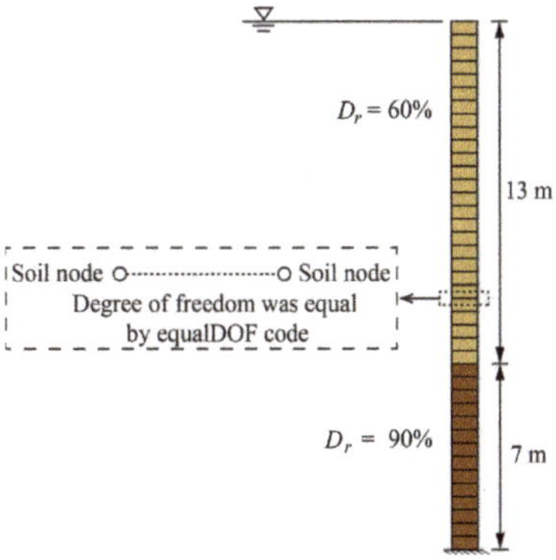

Figure 15. Numerical model of free field.

The bottom of the free-field model is fixed to facilitate the application of dynamic loads. Nodes of soil units that share the same horizontal elevation are constrained using a penalty function to establish periodic boundaries, thereby simulating an effectively infinite free field [21,22]. The top surface of the model is designed to be permeable, while the remaining boundaries are characterized as impermeable.

Figure 16 presents a comparison between the calculated values derived from the energy pore water pressure model and the results obtained from the centrifuge shaking table test as well as numerical simulations. The energy pore water pressure model demonstrates the capability to effectively calculate the excess pore water pressure ratio of saturated Fujian standard sand at varying depths subjected to Taft wave with acceleration of 0.13 g. The model exhibits a higher accuracy in predicting the variation of the excess pore water pressure ratio when the burial depth of the soil is relatively shallow. However, as the burial depth increases, the model tends to predict the initial liquefaction time to occur later than that observed in the centrifuge tests and numerical simulations.

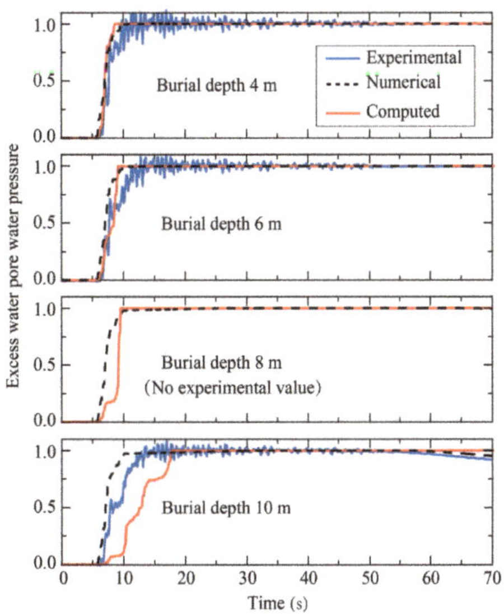

Figure 16. Comparison of experimental, numerical, and computed pore pressure.

Figure 17 shows the comparison between the calculated and numerically simulated maximum excess pore water pressure ratios at burial depths of 4 m and 10 m, subjected to varying amplitudes of the Ninghe and Qian'an ground motions.

At a burial depth of 4 m, the calculated values derived from the pore water pressure model developed in this study align closely with the numerical simulation results. This congruence suggests that the model is capable of accurately predicting excess pore water pressure in soil at shallower depths, thereby facilitating the assessment of liquefaction potential in such contexts. Conversely, at a burial depth of 10 m, the calculated values from the energy pore water pressure model, influenced by ground motions, exceed the numerical simulation values. This discrepancy indicates that the energy pore water pressure model proposed in this research adopts a more conservative approach in predicting excess pore water pressure at greater depths, which aligns with the requirements for engineering assessments.

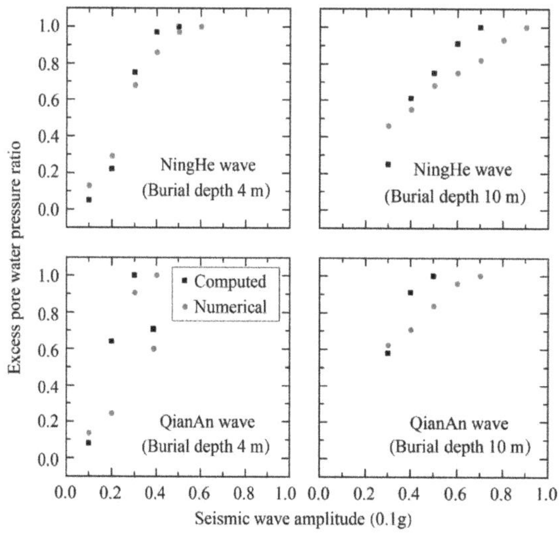

Figure 17. Comparison of numerical and computed maximum excess pore water pressure ratio.

4. Conclusions

In this paper, OpenSees software is employed to simulate bi-directional dynamic triaxial tests aimed at investigating the relationship between the pore water pressure of saturated Fujian standard sand and the Arias intensity of ground motion. Additionally, a novel energy-based pore water pressure model is developed, incorporating the influences of the permeability coefficient, reduction coefficient, waveform, and various other factors to improve and enhance its precision. A methodology for assessing the liquefaction potential of saturated Fujian standard sand is proposed based on the model. The principal conclusions drawn from this research are as follows:

(1) The results of bi-directional dynamic triaxial tests showed that there was a significant correlation between the accumulation of pore water pressure and the Arias intensity of ground motion in the saturated sand. Additionally, the mean period of the ground motion and the cyclic stress ratio also influence this relationship.

(2) The consolidation pressure and the consolidation ratio of the test significantly influence the modeling of pore water pressure. As both the consolidation pressure and

the consolidation ratio increase, the liquefaction resistance of the soil mass exhibits a linear increase.

(3) A novel energy pore water pressure model has been developed based on the results of bi-directional dynamic triaxial tests. The model is capable of accurately predicting the accumulation characteristics of pore water pressure observed during cyclic triaxial tests conducted on dense and medium-dense saturated Fujian standard sand. As the relative density increases, the sand liquefaction potential decreases, making the sand less likely to approach liquefaction.

(4) A methodology for assessing the liquefaction potential has been developed based on the energy pore water pressure model. This method demonstrates enhanced accuracy in predicting the maximum excess pore water pressure ratio of the soil with a shallow burial depth of 4 m. Conversely, the assessment method tends to be conservative when applied to soil bodies with a greater burial depth of 10 m.

To gain a comprehensive understanding of the sand liquefaction potential, upcoming scientific studies will concentrate on clarifying how particle gradation and different types of sand influence sand liquefaction potential. The discrete element method could be an appropriate approach for future research topics.

Author Contributions: Conceptualization, Q.C.; methodology, J.Z.; software, J.Z.; validation, J.Z. and Y.L.; data curation, H.F. and M.D.; writing—original draft preparation, J.Z.; writing—review and editing, J.Z., J.W., Y.W. and Q.C.; supervision, Q.C. All authors have read and agreed to the published version of the manuscript.

Funding: This research was funded by several funding sources, including the Natural Science Foundation of Xinjiang Uygur Autonomous Region (grant number 2022D01F38), the Research Foundation of China University of Petroleum-Beijing at Karamay (grant number XQZX20220007), the "Tianchi Talent" Introduction Plan (grant number 2021592120), the Xinjiang Tianshan Innovation Team for Research and Application of High-Efficiency Oil and Gas Pipeline Transportation Technology (grant number 2022TSYCTD0002), and the National Natural Science Foundation of China (grant numbers 41530639, 42007247, and 41761144080).

Institutional Review Board Statement: Not applicable.

Informed Consent Statement: Not applicable.

Data Availability Statement: Data is available upon reasonable request.

Acknowledgments: The authors extend their sincere appreciation to Elgamal from the Department of Structural Engineering, University of California, San Diego, for his valuable guidance in developing numerical models.

Conflicts of Interest: Author Yan Li was employed by Chengdu Surveying Geotechnical Research Institute Co., Ltd. of MCC. The remaining authors declare that the research was conducted in the absence of any commercial or financial relationships that could be construed as a potential conflict of interest.

References

1. Hamada, M. Large Ground Deformations and Their Effects on Lifelines: 1964 Niigata Earthquake. In *Case Studies of Liquefaction and Lifelines Performance During Past Earthquake*; Technical Report NCEER-92-0001; National Centre for Earthquake Engineering Research: Buffalo, NY, USA, 1992.
2. Youd, T.L.; Driss, I.M. Liquefaction Resistance of Soils: Summary Report from the 1996 NCEER and 1998 NCEER/NSF Workshops on Evaluation of Liquefaction Resistance of Soils. *J. Geotech. Geoenviron. Eng.* **2001**, *127*, 297–313. [CrossRef]
3. Seed, H.B.; Idriss, I.M. Simplified procedure for evaluating soil liquefaction potential. *J. Soil Mech. Found.* **1971**, *97*, 1249–1274. [CrossRef]

4. Dobry, R.; Ladd, R.S.; Yokel, F.Y.; Chung, R.M.; Powell, D. *Prediction of Pore Water Pressure Buildup and Liquefaction of Sands During Earthquakes by the Cyclic Strain Method*; National Bureau of Standards Building Science Series; US Department of Commerce: Washington, DC, USA, 1982; Volume 138.
5. Davis, R.O.; Berrill, J.B. Pore pressure and dissipated energy in earthquakes-field verification. *J. Geotech. Geoenviron. Eng.* **2001**, *127*, 269–274. [CrossRef]
6. Fardad, A.P.; Noorzad, R. Energy-based evaluation of liquefaction of fiber-reinforced sand using cyclic triaxial testing. *Soil Dyn. Earthq. Eng.* **2018**, *104*, 45–53. [CrossRef]
7. Nemat-Nasser, S.; Shokooh, A. A unified approach to densification and liquefaction of cohesionless sand in cyclic shearing. *Can. Geotech. J.* **1979**, *16*, 659–678. [CrossRef]
8. Towhata, I.; Ishihara, K. Shear work and pore water pressure in untrained shear. *Soils Found.* **1985**, *25*, 73–84. [CrossRef] [PubMed]
9. Law, K.T.; Cao, Y.L.; He, G.N. An energy approach for assessing seismic liquefaction potential. *Can. Geotech. J.* **1990**, *27*, 320–329. [CrossRef]
10. Xie, D.Y.; Wu, Z.H. Effect of irregular dynamic impulse history on liquefaction characteristics of saturated sand. *Chin. J. Rock Mech. Eng.* **1987**, *9*, 1–12.
11. Meng, S.J.; Liu, T.H. Study of pore-pressure model of saturated sands under irregular load. *Chin. J. Rock Mech. Eng.* **2014**, *33*, 3050–3055.
12. Wang, T.S.; Lin, Y.; Tong, J.; Qin, X. Generation and dissipation of pore pressure in saturated sand layer due to earthquake loading. *Chin. J. Geotech. Eng.* **1983**, *5*, 87–102.
13. Fu, H.Q. An Approach of Dynamic In-Situ Liquefaction Test and Characteristics of Liquefaction Soil. Ph.D. Thesis, Institute of Engineering Mechanics, Harbin, China, 2016.
14. Sun, R. Study on Seismic Ground Motion on Liquefiable Soil Layer and Site Liquefaction Detection. Ph.D. Thesis, Institute of Engineering Mechanics, Harbin, China, 2006.
15. Arias, A. A measure of earthquake intensity. In *Seismic Design for Nuclear Power Plants*; Hansen, R.J., Ed.; MIT Press: Cambridge, MA, USA, 1970.
16. Mazzoni, S.; McKenne, F.; Scott, M.; Fenves, G. *Open System for Earthquake Engineering Simulation User Manual Version 2.1.0*; University of California, Pacific Earthquake Engineering Center: Berkeley, CA, USA, 2009.
17. McGann, C.R.; Arduino, P.; Mackenzie-Helnwein, P. Stabilized single-point 4-node quadrilateral element for dynamic analysis of fluid saturated porous media. *Acta Geotech.* **2012**, *7*, 297–311. [CrossRef]
18. Zienkiewicz, O.C.; Shiomi, T. Dynamic behavior of saturated porous media: The generalized Biot formulation and its numerical solution. *Int. J. Numer. Anal. Met.* **1984**, *8*, 71–96. [CrossRef]
19. Biot, M.A. Theory of elasticity and consolidation for a porous anisotropic solid. *J. Appl. Phys.* **1955**, *26*, 182–185. [CrossRef]
20. Elgamal, A.; Yang, Z.H.; Parra, E. Computational modeling of cyclic mobility and post-liquefaction site response. *Soil Dyn. Earthq. Eng.* **2002**, *22*, 259–271. [CrossRef]
21. Yang, Z.H.; Elgamal, A.; Parra, E. Computational model for cyclic mobility and associated shear deformation. *J. Geotech. Geoenviron. Eng.* **2003**, *129*, 1119–1127. [CrossRef]
22. Zhang, J.L.; Cheng, Q.G.; Li, Y.; Qiu, Y.; Wang, Y.; Wu, J. Performance of rectangular closed diaphragm walls in gently sloping liquefiable deposits subjected to different earthquake ground motions. *Earthq. Eng. Eng. Vib.* **2021**, *20*, 905–923. [CrossRef]
23. Zhang, J.L.; Cheng, Q.G.; Li, Y.; Zhang, E.M.; Wang, Y.F.; Wu, J.J.; Xie, S.Y. Mechanism of liquefaction mitigation by rectangular closed diaphragm walls in sloping liquefiable deposits. *Soil Dyn. Earthq. Eng.* **2021**, *142*, 106582. [CrossRef]
24. Golesorkhi, R. Factors Influencing the Computational Determination of Earthquake-Induced Shear Stresses in Sandy Soils. Ph.D. Thesis, University of California, Berkeley, CA, USA, 1989.
25. Idriss, I.M.; Boulanger, R.W. Semi-empirical procedures for evaluating liquefaction potential during earthquakes. *Soil Dyn. Earthq. Eng.* **2006**, *26*, 115–130. [CrossRef]
26. Wang, K.; Sun, R.; Yuan, X.M. Influence factors and formula for dynamic stress reduction coefficient. *Chin. J. Rock Mech. Eng.* **2018**, *37*, 177–189.
27. Idriss, I.M. Earthquake ground motions at soft soil sites. In Proceedings of the Second International Conference on Recent Advances in Geotechnical Earthquake Engineering & Soil Dynamics, St. Louis, MO, USA, 11–15 March 1991; Volume 3, pp. 2265–2271.
28. Seed, R.B.; Dickenson, S.E.; Mok, C. Site effects on strong shaking and seismic risk: Recent developments and their impact on seismic design codes and practice. In *Structures Congress XII 1994*; ASCE: New York, NY, USA, 1994; Volume 1, pp. 573–578.
29. Shahir, H.; Mohammadi-Haji, B.; Ghassemi, A. Employing a variable permeability model in numerical simulation of saturated sand behavior under earthquake loading. *Comput. Geotech.* **2014**, *55*, 211–223. [CrossRef]

30. Arulanandan, K.; Sybico, J.; Schofield, A.N. Post-liquefaction settlement of sands. In Proceedings of the Wroth Memorial Symposium, Oxford, UK, 27–29 July 1992.
31. Taiebat, M.; Shahir, H.; Pak, A. Study of pore pressure variation during liquefaction using two constitutive models for sand. *Soil Dyn. Earthq. Eng.* **2007**, *27*, 60–72. [CrossRef]
32. Wang, Y.; Gao, G.Y.; Gu, X.Q.; Song, J. A numerical study of the influence of permeability coefficient on the liquefaction-induced settlement of sands. *Rock Soil Mech.* **2017**, *38*, 1813–1818.
33. Rahmani, A.; Ghasemi, F.O.; Pak, A. Investigation of the influence of permeability coefficient on the numerical modeling of the liquefaction phenomenon. *Sci. Iran.* **2012**, *19*, 179–187. [CrossRef]
34. Li, Y.; Cheng, Q.G.; Zhang, J.L.; Lyu, B.; Wang, Y.F.; Wu, J.J. Seismic behavior of rectangular closed diaphragm walls serving as bridge foundations in gently sloping liquefiable deposit: Dynamic centrifuge tests. *J. Geotech. Geoenviron. Eng.* **2019**, *145*, 04019105. [CrossRef]

coatings

MDPI

Article

Experimental Study on Pipeline–Soil Interaction in Translational Landslide

Tianjun Xue [1], Lingxin Liu [1], Jianlei Zhang [1,2,*], Mengjie Dai [1], Gengyuan Shi [1] and Xinze Li [1]

1 China University of Petroleum (Beijing) at Karamay, Karamay 834000, China; 2023015296@st.cupk.edu.cn (T.X.); 2022016291@st.cupk.edu.cn (L.L.); 2023216912@st.cupk.edu.cn (M.D.); 2022216845@st.cupk.edu.cn (G.S.); lixinze@cupk.edu.cn (X.L.)

2 Xinjiang Key Laboratory of Multi-Medium Pipeline Safety Transportation, Urumqi 830011, China

* Correspondence: zhangjianlei@cupk.edu.cn

Abstract: Pipelines in landslide-prone areas are highly susceptible to damage or rupture under soil movement, posing severe threats to social stability and national security. However, research on pipeline failure mechanisms across different landslide types remains insufficient. Therefore, this study employs large-scale indoor model tests to investigate the interaction mechanisms between pipelines and soil (pipeline–soil interaction) in translational landslide zones through comparative experiments. The results indicate that: (1) The failure process of translational landslides is characterized by initial sliding at the slope crest under loading, which progressively drives the lower soil mass, ultimately resulting in global slope instability. The sliding mass displacement exhibits a top-to-bottom reduction pattern. (2) Pipelines traversing slopes laterally significantly enhance slope stability by providing measurable anti-sliding resistance. (3) Pipeline displacement under sliding mass action occurs in the downslope direction, yet its trajectory deviates from the sliding mass movement. (4) Strain analysis reveals that the pipeline experiences peak strain in the middle region of the sliding mass and at the sliding-non-sliding interface, with the middle region being the primary location for initial yielding and fracture. This study advances the understanding of pipeline-sliding mass interaction mechanisms in translational landslides and offers critical insights for improving pipeline safety and reliability.

Keywords: translational landslide; pipeline; pipeline–soil interaction; model test

1. Introduction

Pipelines serve as a crucial means of transporting oil and gas, playing an irreplaceable strategic role in ensuring social stability and safeguarding national security [1–4]. As a result, pipeline engineering is widely regarded as the lifeline of national energy supply. With the rapid development of pipeline engineering towards large-diameter and long-distance systems, pipelines inevitably traverse complex and varied terrains [5,6]. During the construction of long-distance buried pipelines, geological instability along the routes has become increasingly prominent, potentially triggering multiple geohazards [7,8]. Among these, landslides are one of the most common types of geohazards that pose severe threats to pipeline safety [3,9–11]. Notably, over 96% of pipeline accidents caused by ground movement are attributed to landslide-related disasters [9]. Under the influence of landslides, pipelines may experience local buckling, tensile failure, or rupture due to excessive loads, leading to leaks and subsequent catastrophic incidents such as fires and explosions, which endanger both safety and the environment.

Currently, both domestic and international scholars have conducted experimental research on the interaction between pipelines and soil in landslide-prone areas. These experiments primarily include small-to-medium-scale indoor model tests under 1 g of gravitational acceleration, large-scale outdoor model tests, and centrifuge testing.

Lin et al. [12] conducted indoor experiments to explore the mechanical and deformation characteristics of pipelines subjected to landslides. Their results revealed that the axial stress distribution along the pipeline approximates a saddle-shaped pattern, with stress peaks observed at the center and lateral boundaries of the landslide width, while maximum deformation occurs at the center of the landslide width. Feng et al. [13] performed a large-scale outdoor test to study the deformation and failure mechanisms of slopes, specifically examining the transition from continuous sliding to shallow drift sliding occurring above the pipeline, as well as the deep compressive deformation occurring beneath it. The findings indicate that the presence of the pipeline contributes to improved slope stability by redistributing deformation towards the peripheries and mitigating the maximum displacement in the direction of the landslide.

To investigate the degradation of soil strength induced by water infiltration, several researchers have incorporated permeation or simulated rainfall systems into their experiments. For instance, Calvetti et al. [14] conducted small-scale 1 g model tests by decomposing relative displacement and tensile forces into horizontal and vertical components to analyze plane strain and loading states during pipeline–soil interaction. By horizontally pulling the pipeline under seepage conditions and varying pipeline diameters and burial depths, they identified critical characteristics of pipe–soil interaction and displacement patterns. Lian et al. [15] developed an indoor slope instability model to study rainfall-induced interactions between buried gas pipelines and soil, building on flume experiments. By monitoring sectional strain in pipelines during rainfall-induced slope failure, they quantified the effects of soil on pipeline behavior. Guan et al. [16] investigated the interaction mechanisms and failure evolution of micropile-reinforced landslide-pipeline systems through large-scale indoor physical model tests. Mohammad et al. [17] conducted 10 displacement-controlled pulling tests on a 324 mm-diameter, 6.35 mm-wall-thickness steel pipe segment (2.4 m long) embedded in dense sandy slopes at varying burial depths. They systematically analyzed the effects of embedment ratio and slope inclination on the relationship between lateral soil forces and relative pipeline displacement responses, with comparative evaluations against horizontal ground conditions.

To enhance the accuracy of mechanical analyses for pipelines in landslide zones, centrifuge testing has been adopted by researchers. This method generates body stress gradients that are equivalent to prototype conditions, ensuring similarity in effective stress and pore pressure at corresponding depths. Nasser et al. [18] conducted a series of centrifuge tests to study pipeline–soil interactions under axial, oblique, and lateral loading in dense sand. Zhang et al. [19] induced slope instability through crest loading in 30 g centrifuge tests. They performed experiments on six coarse-grained dense sand slopes with pipelines positioned at various locations relative to the slope crest, as well as on two medium-dense fine sand slopes, to investigate the influence of particle size distribution on pipeline–soil interactions. A novel methodology was proposed for calculating the ultimate external forces acting on buried pipelines at different slope positions during landslide failure.

However, current experimental studies on pipeline–soil interactions in slopes primarily emphasize factors such as pipeline burial depth, crossing configurations, sliding mass volume, and slope inclination. However, there is a notable lack of systematic research regarding the impact of different landslide types. Different landslide classifications demonstrate different sliding mechanisms that exert diverse effects on pipelines. For example, in the case of translational landslides, the upper sliding mass drives the overall movement of the slope, while the lower slope body provides partial resistance, thereby slowing the progression of the landslide. In contrast, retrogressive landslides originate from the lower sliding mass, which destabilizes the upper slope due to the loss of basal support, consequently accelerating the sliding process. Therefore, this study conducted a series of large-scale indoor tests based on the Xiaozhai Village landslide case. It conducts comparative experiments aimed at elucidating the mechanisms of pipeline–soil interaction during translational landslides. Two distinct experimental setups were implemented: (1) slope instability tests conducted in the absence of pipelines, and (2) slope instability tests that included a PVC pipeline positioned laterally across the slope. The results of this investigation contribute to a deeper understanding of pipeline–soil interactions in the context of translational landslides, thereby offering a theoretical framework for mitigating pipeline hazards in areas susceptible to landslides.

2. Experimental Section

2.1. Experimental Scheme

This study conducted two comparative experimental groups based on the Xiaozhai Village landslide prototype: (1) a slope instability test conducted without the inclusion of pipelines (Experiment-L) and (2) a slope instability test incorporating a PVC pipeline that traverses the slope laterally (Experiment-LP). A comparative analysis of these two experimental groups was performed to systematically examine the impact of pipelines on slope-instability patterns, thereby clarifying the mechanisms of pipeline–soil interaction in the context of translational landslides.

The Xiaozhai Village landslide is located on the right side of the Shazi–Xingren Highway in Qinglong County, Guizhou Province, near the China–Myanmar gas pipeline project. During the construction of the pipeline, a significant amount of excavated soil was deposited alongside the pipeline route, which considerably altered the original slope angle and increased the surcharge load at the slope crest. This disturbance disrupted the natural equilibrium of the slope. Ultimately, the failure of the slope was triggered by external factors, including rainfall, resulting in a landslide event. The landslide caused the gas pipeline to fracture, which subsequently led to leakage and combustion [19].

As illustrated in Figure 1, the landslide spans an elevation range from 1380 m at the toe to 1400 m at the crown, resulting in a relative height differential of 20 m. The predominant direction of sliding is oriented at 80°, with slope angles varying between 30° and 40° (averaging 35°), indicating a profile that is gentle in the upper section and steep in the lower section. The dimensions of the landslide are 40 m in length, 40 m in average width, and a thickness ranging from 4 to 8 m, which corresponds to a total volume of 9.6×10^3 m^3, thereby categorizing it as a small-scale landslide. The sliding mass exhibits an irregular semi-elliptical shape with distinct chair-like contours [20]. The slope is composed of near-horizontal, thin-to-medium-bedded Triassic fine sandstone, which is covered by a thin layer of gravelly soil. A cross-sectional view of the landslide is illustrated in Figure 2.

Figure 1. Pipeline–soil interaction in the Xiaozhai Village landslide: (**a**) Pipeline route; (**b**) Pipeline failure caused by landslide.

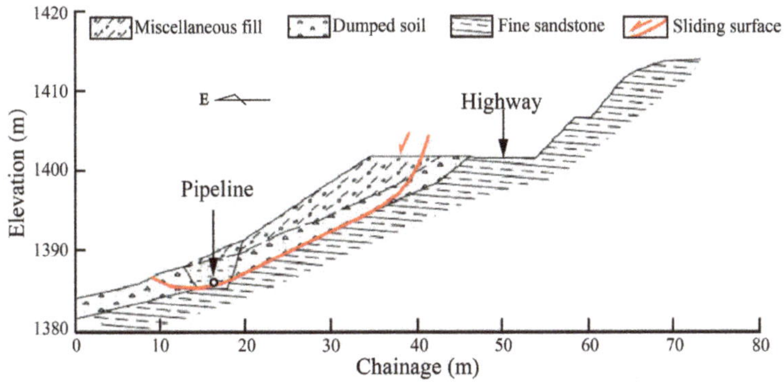

Figure 2. Cross-sectional view of Xiaozhai Village landslide.

2.2. Experimental Material

The soil samples utilized in this study were obtained from the Karamay Campus of China University of Petroleum (Beijing). The soil density was measured at 2142 kg/m^3 using the ring knife method. Shear strength parameters, specifically cohesion (11.59 kPa) and the internal friction angle (25.71°), were measured using an electric direct shear apparatus, as shown in Figure 3. The soil of the Xiaozhai Village landslide has a density of 2100 kg/m^3, cohesion of 10 kPa, and an internal friction angle of 24°. These parameters demonstrate a strong correlation with the in situ shear strength values of the soil from the Xiaozhai Village landslide, thereby confirming the representativeness of the selected soil samples.

The polyvinyl chloride (PVC) pipeline utilized in the experiments had a nominal outer diameter of 30 mm and a wall thickness of 1 mm. The mechanical properties of the pipeline material were characterized as follows: Young's modulus of 2.4 GPa, Poisson's ratio of 0.38, yield strength of 40 MPa, and tensile strength of 50 Mpa. Although there are differences in mechanical properties between PVC pipelines and steel pipelines in the plastic stage, their mechanical properties in the elastic stage are generally consistent. This study mainly focuses on the deformation characteristics and strain distribution of pipelines in the elastic stage. Therefore, although the model has an idealized representation of the landslide and pipeline, this comparative experiment can reveal the mechanism of pipeline–soil interaction.

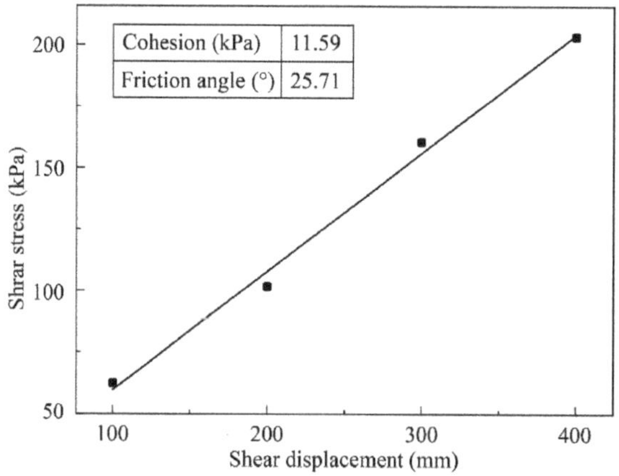

Figure 3. The relationship between the shear stress and shear displacement in the direct shear test.

2.3. Model Construction

2.3.1. Landslide Model Construction

The experimental model is constructed within a model box measuring 2.4 m in length, 1.8 m in width, and 1.5 m in height. The sliding mass in Xiaozhai Village measures 40 m in length, with an average width of approximately 40 m. A 12 m-wide non-sliding region has been established on both sides of the landslide to reduce the impact of boundary constraints on the pipeline. To ensure the accuracy of the experimental results, the geometric similarity ratio was set at 1:30. The other similarity ratios are summarized in Table 1. According to the principle of similarity, the reduced-scale soil box model established for the experiment may exhibit size effects. The geometric scaling of the soil model can affect the sliding displacement rate and crack formation, potentially leading to deviations from the prototype landslide. However, the effect of geometric scaling on the mechanism of pipeline–soil interaction is relatively minor. The model box was constructed with steel plates and tempered glass panels to enable the observation of landslide dynamics. The indoor landslide model comprises four key components: the bedrock, the slip zone, the sliding mass, and the non-sliding regions. The parameters for each component are determined based on the landslide in Xiaozhai Village and the experimental scale, as detailed in Figure 4.

Table 1. Similar parameters.

Physical Parameters	Units	Similitude Relationships	Scaling Ratios
Geometric dimensions	m	C_l	1:30
Gravity	m/s^2	C_g	1:1
Density	Kg/m^3	C_ρ	1:1
Stress	N/m^2	$C_\sigma = C_l C_g C_\rho$	1:30
Strain	/	C_ε	1:1

During the sliding process, the bedrock and non-sliding regions remain in a state of rest. To model this phenomenon, three support walls were constructed using bricks and concrete according to prototype dimensions. Wooden plates with a thickness of 2 cm were affixed to these walls to replicate the characteristics of bedrock (Figure 5). The sliding zone refers to the shear zone that exists between the sliding mass and the stable bedrock beneath it. Consequently, in the modeling process, the slip zone is established at the interface between the bedrock layer and the sliding mass layer. Due to its inherently low shear strength, characterized by minimal cohesion and a low internal friction angle in relative to the sliding mass [21–23], a plastic film was positioned across the width of the landslide. This film was subsequently covered with a 2 cm-thick layer of sand to create a weak slip zone (Figure 5).

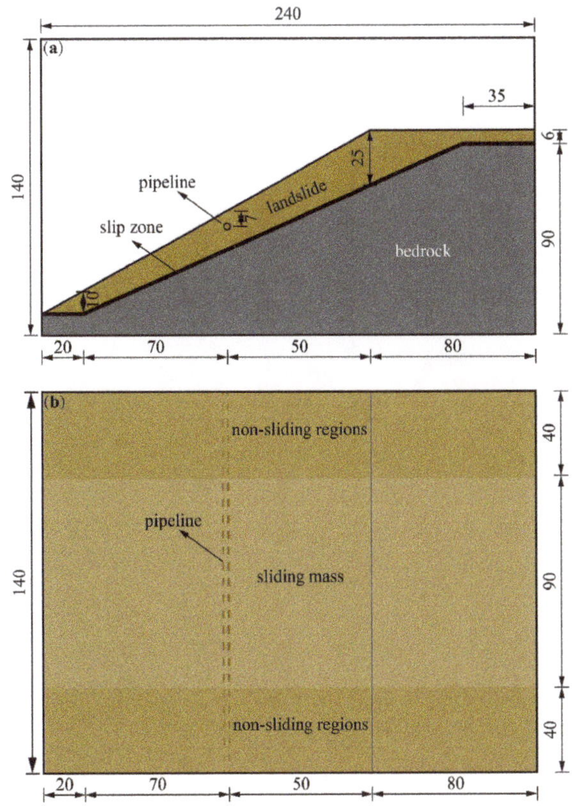

Figure 4. The indoor model test: (**a**) Side view; (**b**) Top view.

The configuration of the sliding mass has been simplified based on the experimental prototype, as illustrated in Figure 4. The sliding mass has a width of 1.2 m and an average thickness of 20 cm, which is segmented into four layers for installation, each with a thickness of 5 cm in the Experiment-L. In accordance with the principles governing the relationship between density and volume, the soil required for each layer is deposited and compacted using compaction tools to replicate the density characteristics of an actual sliding mass. The lengths of the non-sliding regions flanking the model are 0.4 m, with an average thickness of 21 cm, and the installation procedure employed is consistent with that of the sliding mass.

The methodology employed for the Experiment-LP closely resembles that of Experiment-L. The pipeline is installed at a burial depth of 1.8 m within both the sliding mass and the non-sliding regions, according to the experimental prototype. Field investigations indicate that when the length of the pipeline in the non-sliding regions exceeds 9 m, the pipeline remains essentially stationary. Consequently, bricks are positioned approximately 0.3 cm from the sliding mass to establish the boundary of the pipeline in the Experiment-LP. After the installation of the pipeline, the upper layer of the sliding mass is covered to a predetermined thickness. The configuration of the pipeline is illustrated in Figure 5.

Figure 5. The construction process of the slip zone.

2.3.2. Measurement of Experimental Data

During the experiments, displacements at both the surface and internal positions of the sliding mass, the pipeline displacements, and strain distributions along the pipeline were systematically observed. All electrical signals were recorded using a DH5922D dynamic signal-acquisition system. The DH5922D dynamic signal-acquisition system was employed for parameter monitoring with a measurement accuracy of ±0.1% full-scale range and a sampling rate of 204.8 kHz per channel. This system is capable to monitor the strain by the strain gauges (resistance: 120 ± 0.1 Ω, gauge factor: 2.0 ± 0.1%, strain limit: 2%) and the displacement using the laser-displacement sensors (measuring range: 400 ± 200 mm, repeatability: 400/800 μm, linearity: ±0.2% full scale range).

The surface displacements of the sliding mass were qualitatively recorded using three sand marker strips (Figure 6): one located directly above the pipeline, one positioned diagonally above the pipeline, and one situated diagonally below the pipeline. These strips enabled the visual monitoring of displacement patterns at the landslide surface.

Internal displacements within the sliding mass and the displacements of the pipeline were measured using paired monitoring blocks connected by wires. One block was embedded in the sliding mass, while the corresponding block was mounted on a platform atop the test box. A laser-displacement sensor was utilized to monitor the movements of the block mounted on the platform, thereby facilitating the indirect quantification of internal displacements within the sliding mass and the displacements of the pipeline. The specific configurations of this setup are illustrated in Figure 6.

There are 16 channels in the DH5922D dynamic signal-acquisition system. The laser-displacement sensors require four channels. Four strain gauges are required at each cross-section of the pipeline in the experiment. Every strain gauge needs one channel. Therefore, three cross-sections of the pipeline can be monitored. Three cross-sections exhibiting typical mechanical responses were selected: (a) the mid-span of the pipeline, which tends to exhibit significant mechanical responses; (b) the interface between the sliding mass and the non-sliding regions, which serves as a dynamic-static boundary, also influences the pipeline strain distribution; (c) the cross-section at the midpoint between two aforementioned cross-sections was selected to reveal the strain distribution patterns along the pipeline. Strain gauges were mounted on an identical pipeline prior to experimental testing to calibrate the strain gauges and the signal-acquisition system. In the pipeline analysis, strain criteria are commonly adopted as failure indicators. This study defines pipeline strain-failure indicators as tensile strain exceeding 2% or compressive strain exceeding 1.4%, in accordance with the ALA–ASCE guidelines.

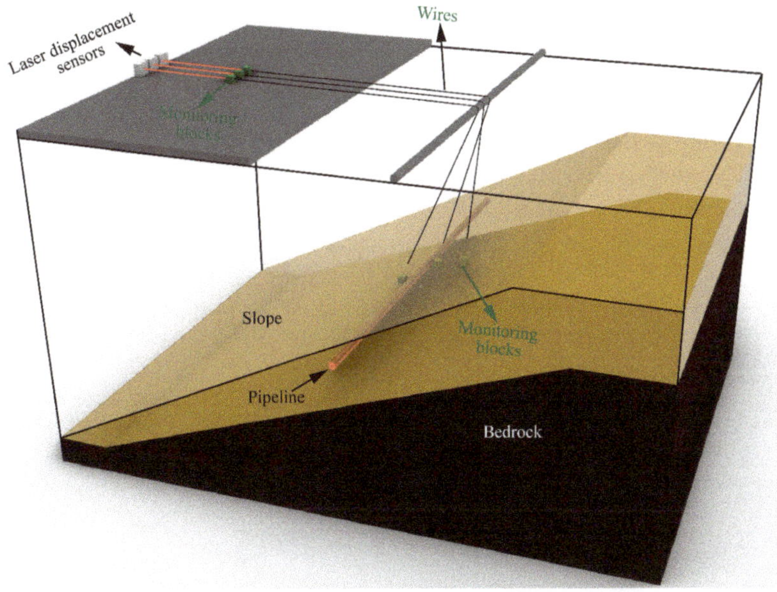

Figure 6. Schematic diagram of landslide and pipeline-displacement monitoring.

The initiation of translational landslides was simulated by applying surcharge loads at the crest of the slope, in accordance with the landslide case. The entire failure mechanism was documented using high-speed cameras. A schematic representation of the experimental setup is presented in Figure 7.

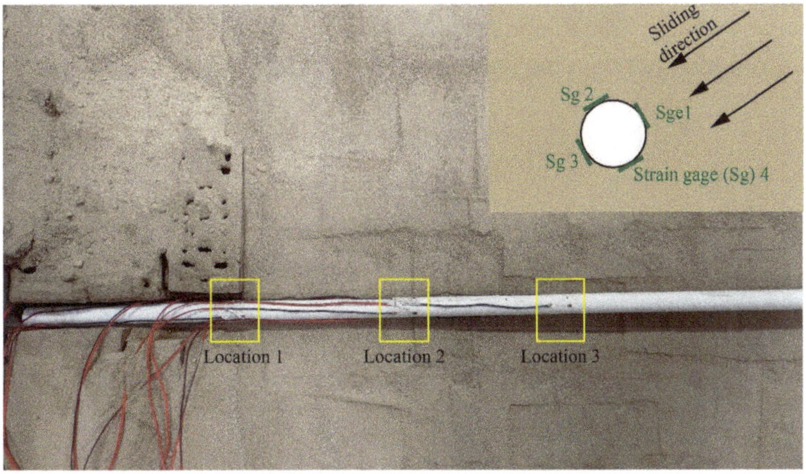

Figure 7. Setting of pipeline strain monitoring.

2.3.3. Load Application

Translational landslides were initiated by applying surcharge loading at the crest of the slope. Each increment of loading involved the placement of a 30 kg mass bag. After placing each bag, a stabilization period was observed to allow for the stabilization of landslide displacements, pipeline movements, and strains. Additional bags were then sequentially added until slope instability was observed. To maintain load stability, the mass bags were positioned on an inclined wooden board rather than being vertically stacked, thereby preventing any unintended redistribution of the load during the testing process (Figure 7).

3. Experimental Results and Discussion

3.1. Soil Response of Landslide Mass

During the experimental process, the evolution of the sliding mass is primarily characterized by two key indicators: the development of cracks and the variations in displacement. Accordingly, this section will focus on: (1) the progressive crack propagation patterns within both the sliding mass and adjacent stable regions, and (2) the displacement response characteristics of the sliding mass subjected to incremental loading conditions.

3.1.1. Development of Landslide Cracks

Figure 8 illustrates the patterns of crack propagation observed during the loading process in Experiment-L at a surcharge of 30 kg (the first loading stage) and 60 kg (the second loading stage).

Figure 8. Patterns of crack propagation during the loading process in the Experiment-L: (**a**) at a surcharge of 30 kg after 1 min and 50 s; (**b**) at a surcharge of 60 kg after 2 min and 25 s; and (**c**) at a surcharge of 60 kg after 3 min and 50 s.

The slope exhibited stable behavior throughout the first loading stage, with monitoring data confirming no signs of distress, such as surface cracking, internal displacement, or pipeline strain (Figure 8a). However, when the load increased to 60 kg, tension cracks developed at the interface between the sliding and non-sliding regions near the crest of the slope. These cracks propagated downward, resulting in localized sliding (Figure 8b). This phenomenon aligns with the failure mechanism associated with translational landslides, where instability originates at the crest and progressively destabilizes the mid-slope and toe areas. The sliding mass experienced rapid stabilization due to the redistribution of stress, affecting only a minor volume of soil.

Figure 9 presents patterns of crack propagation observed during the loading process in Experiment-L at a surcharge of 90 kg (the third loading stage). During the third loading stage, global slope instability phenomena were observed. The upper sliding mass exerted pressure on the lower section, resulting in cracks that extended entirely through the slope. Compared to earlier stages, this phase exhibited significantly larger volumes of sliding material and an increased thickness of the sliding mass. The accumulation of displaced soil at the base of the slope ultimately led to the cessation of landslide movement.

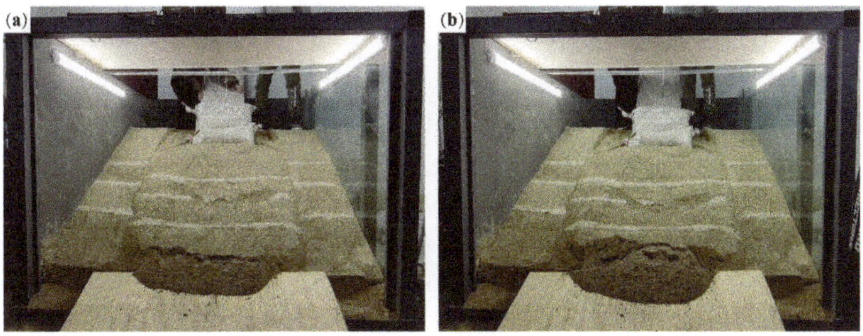

Figure 9. Patterns of crack propagation during the loading process in the Experiment-L: (**a**) at a surcharge of 90 kg after 4 min and 5 s; (**b**) at a surcharge of 90 kg after 5 min and 5 s.

Figures 10 and 11 depict the progression of cracks within the pipeline–slope system in the Experiment-LP at surcharges of 30 kg and 60 kg. Both the slope and the pipeline exhibited stability at a surcharge of 30 kg, as shown in Figure 10a. However, at a surcharge of 60 kg, initial cracks were observed at the interface between the sliding and non-sliding regions (Figure 10b), with minimal propagation toward the slope face, as depicted in Figure 11. Although transient sliding was observed initially, the system rapidly stabilized, leading to significantly shorter crack lengths compared to scenarios without the pipeline. This result confirms that the pipeline can improve the stability of the landslide.

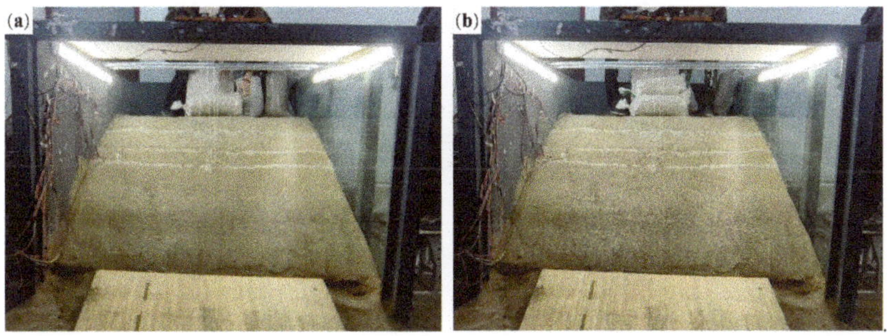

Figure 10. Patterns of crack propagation during the loading process in the Experiment-LP: (**a**) at a surcharge of 30 kg after 1 min and 25 s; (**b**) at a surcharge of 60 kg after 3 min and 55 s.

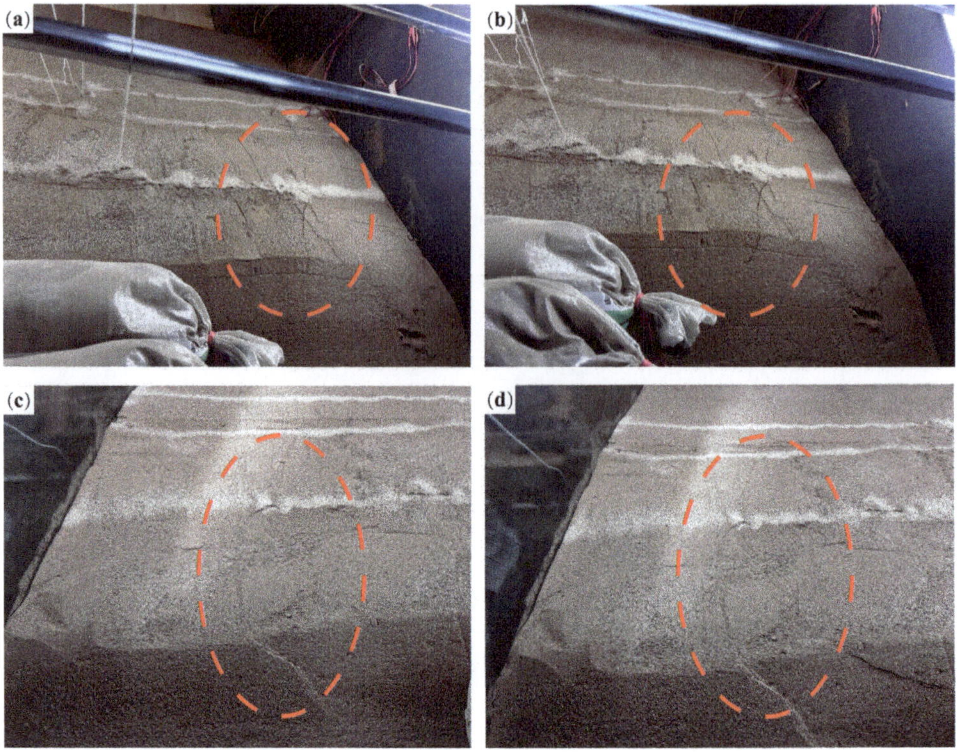

Figure 11. The crack of landslide in the Experiment-LP at a surcharge of 60 kg: (**a**) the right side after 1 min and 25 s; (**b**) the right side after 3 min and 55 s; (**c**) the left side after 1 min and 25 s; (**d**) the left side after 3 min and 55 s.

Figure 12 illustrates the development of deformation and cracks in the landslide in the Experiment-LP at a surcharge of 90 kg. During this loading stage, significant deformation was observed, characterized by notable cracks and shear displacements at the crest. The downward compression of the overlying soil imposed critical compressive and bending stresses on the pipeline, exceeding the ultimate bearing capacity of the pipeline, and leading to its fracture. Additionally, soil accumulated above the pipeline and at the landslide toe, effectively preventing any further sliding.

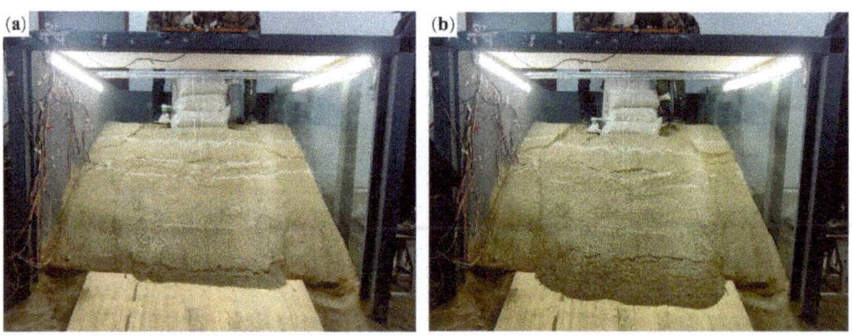

Figure 12. Patterns of crack propagation during the loading process in the Experiment-LP: (**a**) at a surcharge of 90 kg after 4 min and 5 s; (**b**) at a surcharge of 90 kg after 5 min and 25 s.

In the early phase of the third loading stage, significant sliding occurred along pre-existing fissures. Compared to the landslide that occurred in the absence of a pipeline, the observed sliding displacements were notably smaller, while soil accumulation above the pipeline became more pronounced (Figure 12a,b). The pipeline initially provided effective resistance to slope movement. However, upon fracture, this restraining capacity was lost, triggering large-scale sliding across the entire landslide area. It is noteworthy that the pre-fracture soil accumulation above the pipeline, combined with subsequent compaction during sliding, progressively decelerated the landslide movement until complete stabilization was achieved. This phenomenon suggests that the pipeline's structural integrity enables improvements in slope stability; its presence may also influence the failure mechanism and post-failure behavior of landslides.

3.1.2. Mechanisms of Landslide Movement

The surface displacement of slopes without pipelines and those with embedded pipelines are compared in Figure 13 at a surcharge of 60 kg. In the Experiment-L, the surface displacements of the upper (Line U), middle (Line M), and lower parts (Line L) of the sliding mass are 22, 15, and 16 mm, respectively. In the Experiment-LP, the surface displacements for the same sliding mass are 7, 5, and 5 mm, respectively. The upper part of the sliding mass exhibited larger displacements compared to the middle and lower parts of the sliding mass. In addition, slopes without pipelines developed significantly larger displacements, characterized by overall sliding from the crest to the toe. In contrast, slopes with embedded pipelines maintained a relatively stable condition, thereby confirming the sliding resistance of the pipeline.

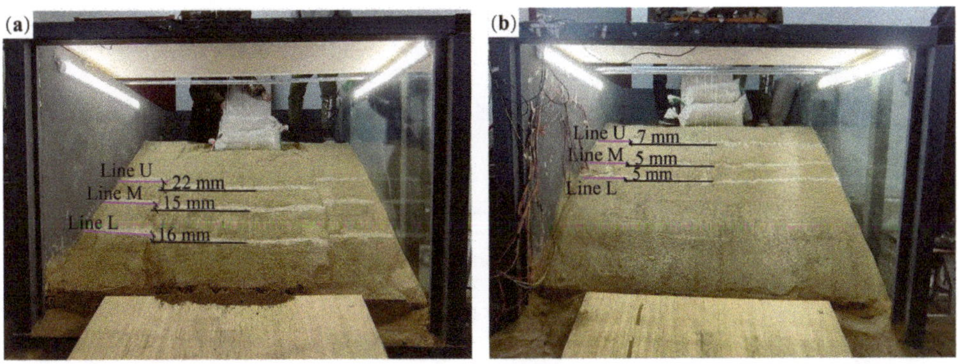

Figure 13. Displacement of the slope surface at the surcharge of 60 kg: (**a**) in the Experiment-L; (**b**) in the Experiment-LP.

The surface displacements of a slope without pipelines at a surcharge of 90 kg are presented in Figure 14. The surface displacements of the upper (Line U), middle (Line M), and lower parts (Line L) of the sliding mass are 324, 267, and 210 mm at the end of the experiment, respectively. The anti-sliding effect of the pipeline causes soil to accumulate above the pipeline. The accumulation of soil reached its maximum near the center of the landslide, with a noticeable decrease in thickness as one moves toward the lateral regions, as shown in Figure 14. The displacement time history of the slope without pipelines (Figure 15) indicates that the upper part of the landslide experienced greater displacements compared to the lower part. This observed accumulation can be attributed to the lateral movement of soil into non-sliding regions during the sliding event, coupled with a greater volume of mobilized soil in the area subjected to central loading, which ultimately resulted in an arc-shaped deposition morphology.

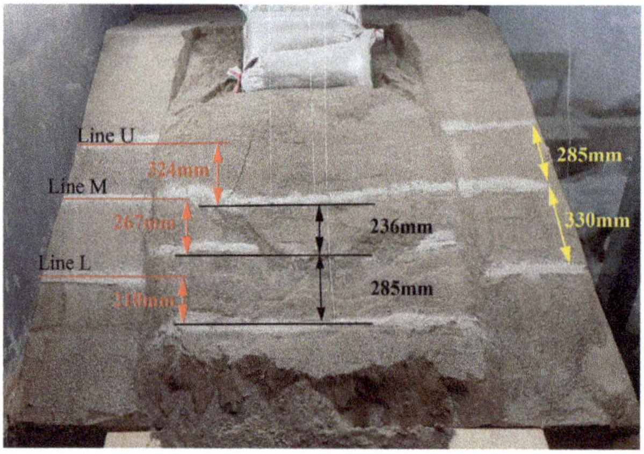

Figure 14. Displacement of the slope surface at the surcharge of 90 kg in the Experiment-L.

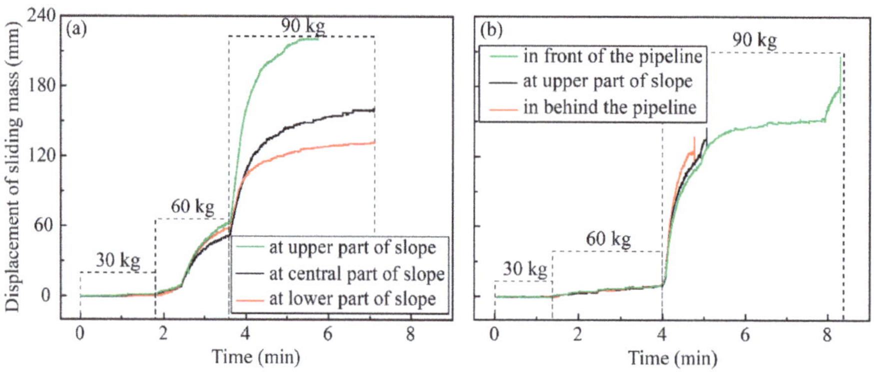

Figure 15. Displacement time histories of slope mass: (**a**) in the Experiment-L; (**b**) in the Experiment-LP.

In the slope test without pipelines (refer to Figure 14), at a surcharge of 60 kg (5–6.6 min), resulted in greater displacements in the upper slope compared to the mid and toe regions. At a surcharge of 90 kg, displacements in the upper slope increased significantly and occurred prior to movements in the mid and toe areas. This observation is consistent with the translational landslide mechanism, which is influenced by thrust originating from the crest.

The displacement time history of the slope in Experiment-L and Experiment-LP are compared in Figure 15. In the slope test conducted within the pipeline, negligible displacements were recorded during both the initial and secondary loading phases (lasting less than 4 min), indicating that the slope maintained a state of stability at a surcharge of 60 kg. Upon reaching the third loading stage, a significant landslide event was initiated, resulting in a sliding phase. The landslide persisted for a duration of approximately 4 to 5 min, during which soil displacement at three designated locations measured around 130 mm. Furthermore, it was observed that the displacement at the lower section of the pipeline exceeded that of both the upper section of the slope and the upper portion of the pipeline. This observation confirms that the presence of the pipeline obstructed the downward progression of the landslide. Following the pipeline rupture, the overlying soil mass, which had previously stabilized above the pipeline, became destabilized and initiated gradual downslope movement. This process ultimately led to a measurable increase in soil displacement after an initial period of stabilization, demonstrating the sliding resistance effect of the pipeline prior to failure.

3.2. Pipeline Response

3.2.1. The Deformation of Pipeline

Figure 16 illustrates the pipeline fracture condition revealed during post-experimental excavation following stabilization of the landslide. While surface observations initially indicated successful sliding resistance through substantial soil deposition above the pipeline, subsequent excavation exposed critical mid-slide fractures. These structural failures resulted from translational landslide forces acting along the failure plane, demonstrating that while the pipeline exhibited sufficient shear resistance to prevent complete displacement, it ultimately succumbed to bending stresses induced by differential ground movement.

Figure 16. Fracture of the pipeline in the Experiment-LP.

3.2.2. The Strain of Pipeline

Figure 17 illustrates the strain time history of the pipeline at three different locations: the midpoint of the sliding mass, the interface between the sliding and non-sliding regions, and an intermediate position between these two points. In the first stage of the loading process (30 kg), minimal strain was recorded across all measurement positions. This result indicated that the slope maintains stability with creep displacement below instrument detection thresholds. In the subsequent stage (60 kg), a distinct differential strain distribution pattern was observed. The maximum strain was recorded at the interface between the sliding and non-sliding regions, exceeding the values noted at other locations. The magnitude of strain diminished with increasing distance from this interface. The elliptical deformation was predominant at the mid-sliding mass, and the bending deformation was small. The bending-induced tensile and compressive strains were more pronounced at the interface between the sliding and non-sliding regions. Notably, compressive strains were detected at positions 1 and 2, while symmetrical tensile strains were present at positions 3 and 4 during the first two loading stages, suggesting that bending occurred primarily in the direction of sliding. At a surcharge of 90 kg, there was a significant increase in strain of the pipeline at the midpoint of the sliding mass, exceeding the ultimate strain capacity of the pipeline and leading to fracture, as illustrated in Figure 17c. The reason for the pipeline rupture occurring at the midpoint of the sliding mass is that the pipeline in the non-sliding area is constrained, which leads to a certain degree of resistance near the non-sliding area, hindering the sliding of the slope. At the midpoint of the sliding mass, the influence of the constraint on the pipeline is minimal. Therefore, the pipeline undergoes significant deformation subjected to a larger load, which ultimately leads to the rupture of the pipe. In addition, following the fracture, the redistribution of strain revealed modified deformation patterns.

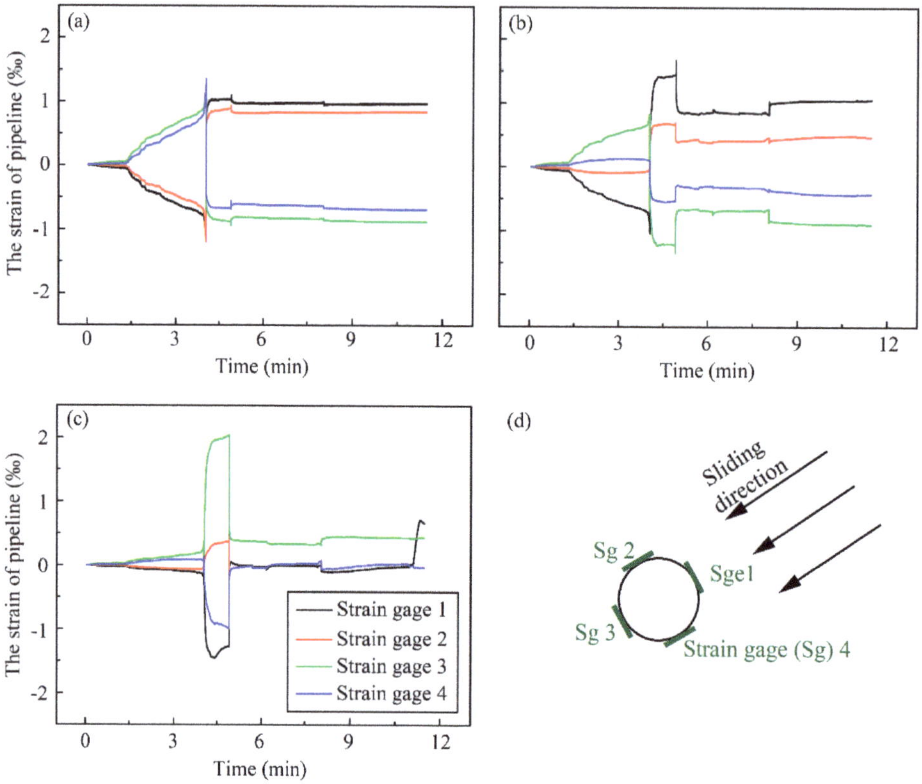

Figure 17. Strain time histories of pipeline in the Experiment-LP at: (**a**) the interface between sliding and non-sliding regions; (**b**) the quarter position of the slide mass; (**c**) the midpoint of the sliding mass; (**d**) the correlation between the positioning of the strain gauge and the sliding direction.

3.3. *Discussion of Pipeline–Soil Interaction*

The interaction between pipelines and sliding masses involves two principal dimensions. Firstly, pipelines have a significant influence on slope stability and the evolution of landslide movement patterns. A pipeline that traverses a slope laterally served as a flexible beam, constrained by non-sliding regions, which effectively mitigates slope creep and increases the load threshold necessary for triggering instability, thereby contributing to enhanced slope stability (Figure 18a). During an instability event, the pipeline changes the evolution of landslide movement patterns by: (1) decreasing overall sliding displacements across the upper, middle, and lower sections of the sliding mass; (2) altering patterns of soil accumulation, leading to a concentration of deposition in proximity to the pipeline; and (3) redirecting the movement of internal soil particles. Under the influence of gravitational and thrust forces, soil particles navigate around the pipeline in a circumferential manner, resulting in their accumulation beneath the depth at which the pipeline is buried.

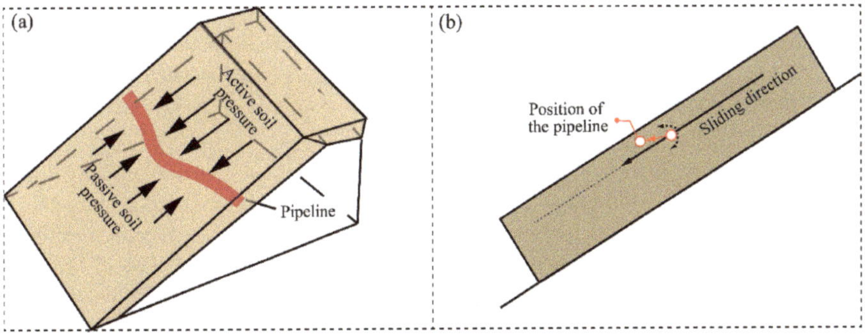

Figure 18. Pipeline–soil interaction in the landslide: (**a**) pipeline–soil interaction force; (**b**) pipeline movement direction.

In contrast, the sliding masses have significant impacts on pipelines. During a sliding event, the pipeline situated within the sliding mass is subjected to thrust forces, which result in considerable deformation. The peak strain of pipelines is observed at the midpoint of the sliding mass due to the thrust generated by the landslide, while increased strain at the interface between the sliding and non-sliding regions is attributed to shear stresses. The strains of the pipeline increase with increased sliding displacement of the sliding mass, ultimately leading to fracture when the strain of the pipeline reaches the material's strength limits. Furthermore, the accumulation of soil beneath the pipeline creates upward vertical forces that are perpendicular to the sliding plane, thereby decreasing the burial depth of pipelines (Figure 18b).

4. Conclusions

This research studied the interaction mechanisms between laterally traversing pipelines and sliding masses in translational landslides by conducting large-scale indoor model experiments. The comparative analysis provided valuable insights into the interactions between pipelines and sliding masses, focusing on the displacement and crack development in both pipeline-free and pipeline-embedded landslides, as well as the displacement and strain characteristics of the pipelines. The principal findings are summarized as follows:

(1) The crack in the slope initially develops at the crest when subjected to a top load, which is subsequently followed by downward sliding. The upper sliding mass exerts a thrust on the lower stable soil, leading to the propagation of cracks toward the toe of the slope. Once a critical loading threshold is reached, the lower slope becomes destabilized due to the thrust from the upper mass, resulting in global sliding and eventual slope failure.

(2) Pipelines that traverse the sliding mass are subjected to thrust loads in the direction of sliding, which results in peak strain of pipelines occurring at the midpoint of the sliding mass. This location is identified as the most susceptible to fracture of pipelines. Furthermore, shear forces at the interface between the sliding and non-sliding regions contribute to significant strain in these areas.

(3) When constrained by non-sliding regions, laterally embedded pipelines behave similarly to beam structures, providing resistance against thrust, shear, and bending

moments. This structural behavior increases the load threshold for slope instability, thereby enhancing the overall stability of the slope.

(4) The presence of pipelines alters the evolution of landslide movement patterns. The pipeline decreases the overall sliding displacements of the sliding mass and changes the sliding direction of the sliding mass. The movements of the sliding mass divide into upward and downward movements around the pipeline. The downward movements around the pipeline exert upward pressure on the pipeline, effectively reducing its burial depth of pipeline.

The results indicate that it is essential to avoid routing pipelines through landslide-prone areas. It is safer to lay the pipeline at the top of the slope rather than at the foot of the slope. Additionally, soil-displacement monitoring should be established on the slope to provide early warnings based on the displacement of the sliding mass. To gain a comprehensive understanding of the pipeline–soil interaction, upcoming scientific studies will concentrate on the research of microstructural parameters. The discrete-element method could be an appropriate approach for the research topics in these future studies.

Author Contributions: Conceptualization, J.Z.; methodology, J.Z.; investigation, T.X., L.L., G.S. and J.Z.; validation, J.Z.; data curation, T.X., L.L., and M.D.; writing—original draft preparation, T.X., L.L., and J.Z.; writing—review and editing, J.Z., X.L.; supervision, J.Z. All authors have read and agreed to the published version of the manuscript.

Funding: This research was funded by several funding sources, including the Research Foundation of China University of Petroleum-Beijing at Karamay (grant number XQZX20220007), the Natural Science Foundation of Xinjiang Uygur Autonomous Region (grant number 2022D01F38), the "Tianchi Talent" Introduction Plan (grant number 2021592120), the Xinjiang Tianshan Innovation Team for Research and Application of High-Efficiency Oil and Gas Pipeline Transportation Technology (grant number 2022TSYCTD0002).

Institutional Review Board Statement: Not applicable.

Informed Consent Statement: Not applicable.

Data Availability Statement: Data are available upon reasonable request.

Conflicts of Interest: The authors declare no conflicts of interest.

References

1. Huang, W.H. The development of oil and gas storage and transportation technology in China. *Oil Gas Storage Transp.* **2012**, *31*, 411–415.
2. Peng, S.B.; Liao, W.; Liu, E.B. Pipe–soil interaction under the rainfall-induced instability of slope based on soil strength reduction method. *Energy Rep.* **2020**, *6*, 1865–1875. [CrossRef]
3. Yan, Y.; Xiong, G.L.; Zhou, J.J.; Wang, R.; Huang, W.; Yang, M.; Wang, R.C.; Geng, D. A Whole Process Risk Management System for the Monitoring and Early Warning of Slope Hazards Affecting Gas and Oil Pipelines. *Front. Earth Sci.* **2022**, *9*, 812527. [CrossRef]
4. Fa-You, A.; Chen, T.H.; Yang, C.; Wu, Y.F.; Yan, S.Q. Study on Disaster Mechanism of Oil and Gas Pipeline Oblique Crossing Landslide. *Sustainability* **2023**, *15*, 3012. [CrossRef]
5. Cruz, A.M.; Krausmann, E. Vulnerability of the oil and gas sector to climate change and extreme weather events. *Clim. Chang.* **2013**, *121*, 41–53. [CrossRef]
6. Xiao, J.Z.; Kong, W.C.; Wang, X.L.; Li, M. Numerical modeling and assessment of natural gas pipeline separation in China: The data from Henan Province. *Pet. Sci.* **2020**, *17*, 268–278. [CrossRef]
7. Huang, W.H.; Chen, J.D.; Fu, C.; Huang, Y. Approach for natural gas to be a primary energy source in China. *Front. Eng. Manag.* **2019**, *6*, 467–476. [CrossRef]
8. Niu, H.T. Smart safety early warning model of landslide geological hazard based on BP neural network. *Saf. Sci.* **2020**, *123*, 104572.

9. Zahid, U.; Godio, A.; Mauro, S. An analytical procedure for modelling pipeline landslide interaction in gas pipelines. *J. Nat. Gas Sci. Eng.* **2020**, *81*, 103474. [CrossRef]
10. Zhang, J.; Liang, B.F.; Pan, B.; Shen, K.R.; Chen, L.Y. Mechanical Response Analysis of Gas Pipeline under Traction Landslide. *Press. Vessel Technol.* **2020**, *37*, 19–25+38+58.
11. Zhang, H.; Liu, X.B. Design strain calculation model for oil and gas pipelines subject to geological hazards. *Oil Gas Storage Transp.* **2017**, *36*, 91–97.
12. Lin, D.; Lei, Y.; Xu, K.; Huang, R. An experiment on the effect of a transverse landslide on pipelines. *Acta Pet. Sin.* **2011**, *32*, 728–732.
13. Feng, W.K.; Huang, R.Q.; Liu, J.T.; Xu, X.; Luo, M. Large-scale field trial to explore landslide and pipeline interaction. *Soils Found.* **2015**, *55*, 1466–1473. [CrossRef]
14. Calvetti, F.; Prisco, C.; Nova, R. Experimental and Numerical Analysis of Soil–Pipe Interaction. *J. Geotech. Geoenviron. Eng.* **2004**, *130*, 1292–1299. [CrossRef]
15. Lian, J.Q.; Shi, R.; Fan, R.D.; Li, X. Study on the Pipe–soil Interaction under Instability of Slope Based on Flume Experiment. *J. Phys. Conf. Ser.* **2022**, *2381*, 012060. [CrossRef]
16. Guan, W.; Wu, H.G.; Wu, D.Y.; Tang, L.; Wei, H. Study on Interaction Mechanism of Natural Gas Pipe-Landslide System Reinforced by Micropile Groups Based on Model Test. *Geofluids* **2022**, *2022*, 8436297. [CrossRef]
17. Katebi, M.; Wijewickreme, D.; Maghoul, P.; Roy, K. Lateral force-displacement response of buried pipes in slopes. *Géotechnique* **2023**, *73*, 375–387. [CrossRef]
18. Daiyan, N.; Kenny, S.; Phillips, R.; Popescu, R. Investigating pipeline–soil interaction under axial–lateral relative movements in sand. *Can. Geotech. J.* **2011**, *48*, 1683–1695. [CrossRef]
19. Zhang, W.Y.; Askarinejad, A. Behaviour of buried pipes in unstable sandy slopes. *Landslides* **2019**, *16*, 283–293. [CrossRef]
20. He, W.G.; Li, H.Z.; Jia, L.L.; Yao, G.Z. Deformation monitoring and mechanism analysis on landslide along China-Myanmar Natural Gas Pipeline: Case of Shazi Town Landslide in west of Guizhou Province. *Yangtze River* **2020**, *51*, 138–143.
21. Jahromi, F.H.; Jafarzadeh, F.; Zakaria, S.M. Experimental study of burial depth effect on embedded pipe deformations in sandy slopes under dynamic landsliding. *Soil Dyn. Earthq. Eng.* **2018**, *114*, 281–297. [CrossRef]
22. Wang, D.Y. Study on Improvement of Pipe-Soil Spring Model in Mountain Slope Area. *J. Phys. Conf. Ser.* **2022**, *2381*, 012090. [CrossRef]
23. Jiang, K.; Wang, D.; Yu, Z.F.; Huang, D. Analysis of Pipeline Damage Caused by Lateral Landslides: Taking Two Explosion Accidentsin the Qinglong Section of China-Myanmar Pipeline in Guizhou as an Example. *Sci. Technol. Eng.* **2023**, *23*, 8988–8995.

MDPI AG
Grosspeteranlage 5
4052 Basel
Switzerland
Tel.: +41 61 683 77 34

Coatings Editorial Office
E-mail: coatings@mdpi.com
www.mdpi.com/journal/coatings